The
Undersea

N. C. Flemming, General Editor

Dr. Edward J. Wenk, jr.
Editorial Consultant

The Undersea

Macmillan Publishing Co., Inc.
New York

Macmillan Publishing Co., Inc.
866 Third Avenue, New York, N.Y. 10022

First American Edition 1977

This book was designed and produced by
George Rainbird Limited,
36 Park Street,
London, W1Y 4DE

House Editor: Donald M. Berwick
Picture Research: Jonathan Moore, Elizabeth Fulton
Designer: Ruth Prentice

Library of Congress Cataloging in Publication Data
The Undersea
    Bibliography: p.
    Includes index
    1. Oceanography 2. Marine resources
    I. Flemming, Nicholas Coit.
GC16.U53   1977   551.4'6   76–56827
ISBN 0-02-538740-5

Printed and bound in Great Britain

Introduction   6

1   **The Ocean Floor**   22
    David G. Roberts
    Institute of Oceanographic Studies, Wormley,
    England

2   **The Water Itself**   46
    Dr. Jens Meincke
    Institute of Oceanography, University of Kiel,
    Germany

3   **Plant Life**   66
    Professor Charles A. Yentsch
    Bigelow Marine Science Laboratory,
    Boothbay Harbor, Maine

4   **Salt-Water Animals**   90
    John Gage
    Scottish Marine Biological Association,
    Dunstaffnage Marine Laboratory, Oban, Scotland

5   **The Ocean's Resources**   126
    Robert Barton
    Editor of *Offshore Services* and author
    of *Oceanology Today*

6   **Using Ocean Space**   166
    Robin Clarke
    Author of *The Science of War and Peace*

7   **Underwater
    Archaeologists**   202
    Colin Martin
    Director, Institute of Maritime Archaeology,
    St. Andrew's University, Scotland
    Dr. N. C. Flemming
    A Principal Scientific Officer, Institute of
    Oceanographic Studies, Wormley, England

8   **Diving and Divers**   230
    H. G. Delauze, X. Fructus, C. Lemaire, A. Tocco
    Comex Diving Organization, Marseille, France

9   **Submarine Craft**   262
    Dr. John P. Craven
    Dean of Marine Studies, University of Hawaii

10 **Marine Law and Politics**   282
    Robin Churchill
    British Institute of International and Comparative
    Law, London

    Bibliography   303
    Acknowledgments   308
    Index   311

Contents

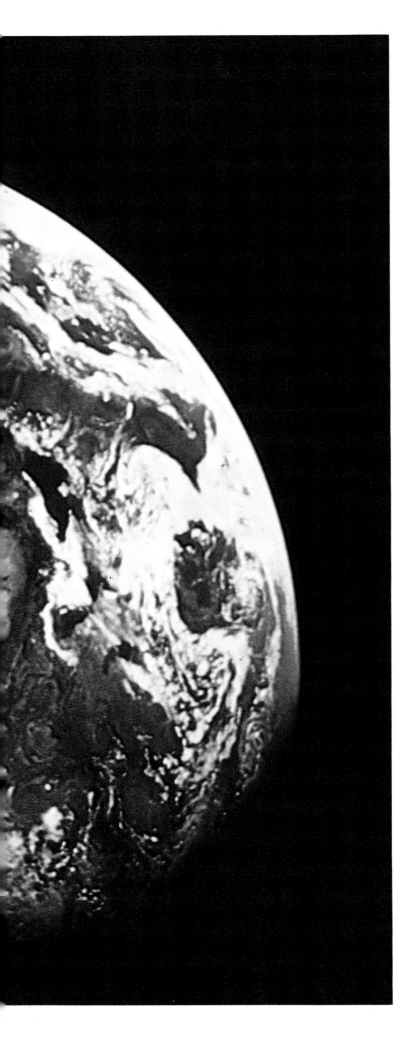

This book is about the relationship between people and the sea, about how they can enjoy, use, and abuse the sea itself, the life in it, and the sea floor. *The Undersea* is not in any way meant to be an introductory textbook of oceanography, and indeed the word "oceanography" seldom occurs in its pages. Oceanography is a scientific discipline that has grown up around the physical study of the sea. It encompasses mathematical and physical theories of wave motions, surface and deep-ocean currents, internal waves, the tides, and the circulation of water throughout entire ocean basins. Such studies require the comprehension of a rare blend of advanced mathematics, electronic instrumentation, and seamanship. But in most oceanographic laboratories physical oceanography in the strict sense is combined with marine biology, marine chemistry, sea-floor geology, and the study of the interaction between the sea and the atmosphere. This assemblage of marine sciences includes not one but many fields of interest.

Nor is science the end of the story. It is more the beginning, for the pursuit of marine science has to be paid for, and the results are used by industries, governments, planning authorities, military strategists, and ordinary people like you and me. Every taxpayer pays something toward oceanographic research, and everybody is affected by the results of the research. This book, then, has been written for everybody. It is about all the marine sciences and their applications. And, in particular, it is about their importance to us in the last quarter of the twentieth century.

Let me tell you a little about the authors of the ten chapters of *The Undersea*. Each of them is an authority in one or another field of marine science, technology, or law. They all work directly at, on, under, or with the sea, and one problem for the editor of the book has simply been to keep track of the contributors, who have often gone off to sea and vanished for weeks or months at a time. But at least they know what they are talking about. In spite of the necessity for detailed editing in order to avoid too much unintentional duplication, and to ensure continuity from chapter to chapter, their strong opinions as individuals come across clearly – sometimes, in fact, to the point of idiosyncrasy. It would have been easier to produce a book written in a more uniform style, by less knowledgeable people, thus avoiding many complicated issues. I have rejected that approach. The ideas in this book are the authors', not necessarily mine. But all of us have tried hard to express those ideas clearly – not always an easy task in view of the technical subject matter.

The "blue" planet: a view of the earth from space

The plan of the book is as simple and logical as the subject matter permits. We start with natural phenomena – the form of the ocean basins, the properties of sea water, and the behavior of plants and animals that live in the sea. Then we examine the modern industries that extract such material resources as oil and fish from the ocean and the ocean floor, and we review the other major activities that compete for use of the sea – for instance, shipping, recreation, and waste disposal. Special chapters are then devoted to undersea archaeology, commercial diving, and submarine craft. Finally, we have a necessarily brief discussion of the international laws of the sea and their relationship to political and military pressures: a subject that's relevant to all the previous sections of the book. Events in 1976 have forced us to realize that the worldwide Law of the Sea Conference may fail to reach agreement on important issues, or that agreement may come too late to prevent irrevocable damage to the environment or dangerous international conflict.

Since the Law of the Sea Conference is still in session as I write these words, there can be no conclusive summing up of its results. We are left, as the author of chapter 10 makes very clear, facing an uncertain future. I suppose, in a way, that does sum it up.

The aim in each chapter is to begin the story right at what is happening *now*, to introduce the reader as quickly as possible to the area of action with which today's scientists, engineers, and politicians are concerned. No chapter starts with a historical review of the subject, and this is a

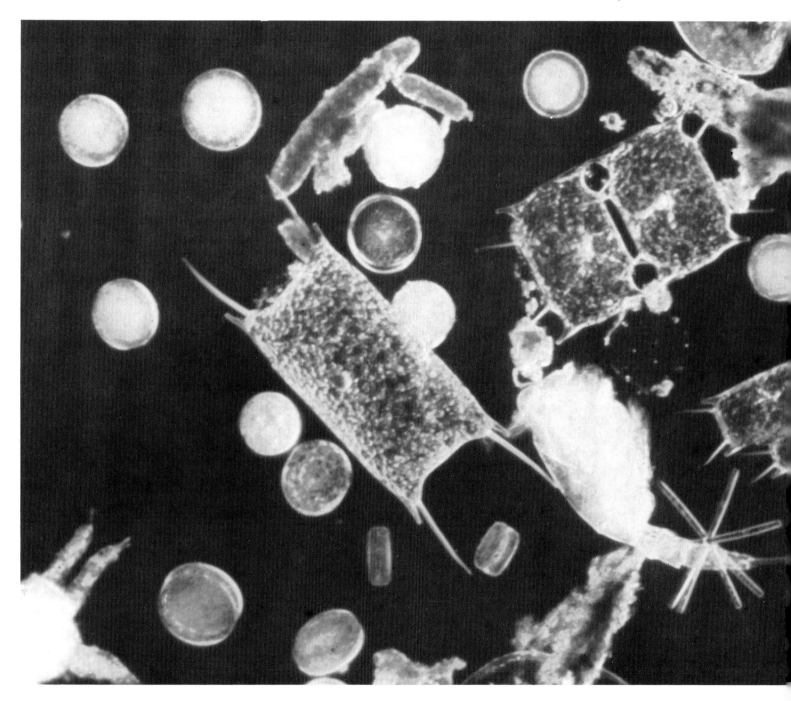

matter of deliberate policy. Many books about the sea take the more leisurely historical approach, but in *The Undersea* we emphasize the present and the future. Naturally, the writers do refer to the origins of techniques and ideas where necessary, but discussion of what happened in the past is subsidiary to the main argument: the state of the marine world in the closing years of the twentieth century. We have also kept references to scientific papers or the names of scientists down to a minimum. In a book for the general reader, the academic procedure of formal text references is neither necessary nor advisable. The entertaining technique of introducing scientists and politicians personally as "dramatis personae" is more justifiable but very space-consuming. There are hundreds of expert scientists, engineers, lawyers, industrialists, and government officials who deserve credit for their oceanography and allied sciences and technologies. But where does one stop? How many lines does Dr. X get if Professor Y got six lines and a mention of his pretty wife? The range of factual material to which we have tried to do justice *readably* in this book is enormous, and so there remains little room for references to individual achievements. Since many of those who might have received ample mention are my friends, I wish to acknowledge them directly here, to thank them, and to apologize. Most of them will find their names in the bibliography, however. I have made the bibliography as up-to-date and thorough as space and time permit.

Plants of the plankton (enlarged 250 times)

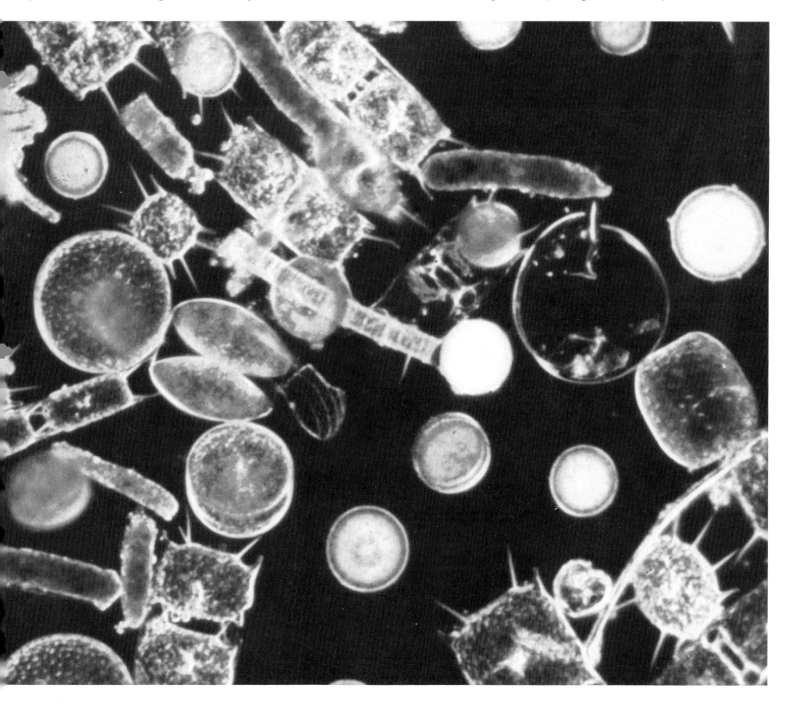

Another aim of *The Undersea* is to be truly international. Any contributor to a book of this type will tend to view the world from the stance of his national tradition; but as editor I have tried to minimize the effects of such "provincialism." The authors of these ten chapters, who, incidentally, come from four different countries, are also obviously aware of the virtues of universality. They seem to have made a special effort to select examples of ocean currents or fishing fleets or oil fields or whatever from all over the world, so that all oceans and countries get fair treatment. We use the metrical system of measurements, by the way, except when discussing territorial waters, which are conventionally expressed in nautical miles.

One glaring imbalance remains: the authors and editor do come from rich industrial nations, and none of them can express the viewpoint or attitudes of the people of developing nations from, as it were, the inside. Perhaps the role of ocean exploitation in helping the economies of developing countries warrants greater stress than it gets here. The subject is referred to several times, particularly in the chapter on law, but it is not treated at length. The problem, in a nutshell, is that advanced ocean exploitation involves massive investment in technology, and the developing countries inevitably depend on aid from the industrial countries to achieve this. Even the development of coastal fisheries or coastal tourism must often depend on foreign or United Nations aid. In consequence, the task of ensuring that the developing countries have a share in the wealth of the oceans tends to be seen as only a subsidiary part of the general problems of international aid and the law of the sea.

The prosperity and happiness of the entire world may depend on what happens under the sea in the next decade. From the depths of the sea we may wrest enough food to make the difference between getting by and enduring mass starvation; we may extract enough oil and minerals to keep industrial societies going and permit industrial benefits to come to the developing countries, or we may quarrel so bitterly over these resources as to trigger off another world war; we may promote the purity and productivity of the oceans, or we may pollute and sterilize them to a point where they become useless for food production and recreation; we may regulate the undersea arms race, or we may pour an ever-increasing proportion of our wealth into subsurface preparations for war.

There are two aphorisms that are bitingly relevant to the problem of managing and exploiting marine resources. The first is this: "For evil to triumph it requires only that men of goodwill should do nothing." And the second is this: "Good-

Underwater welder at work

will is not enough." The implications of the first saying are obvious. If we leave the investment of resources, political negotiations, and management of projects to those who stand to gain most, men of goodwill have only themselves to blame if they do not like the results. That is not meant to be a criticism of any particular political system. Everybody has an interest in what happens under the sea, and we owe it to ourselves to give the many enlightened people who are active in politics and industry the support they need in working for sensible policies. This means *we* must work, too. It means we must keep up to date with current affairs and, at the very least, protest when something seems to go wrong.

A submersible and its attendant divers

But "goodwill is not enough." This statement, in the context of the undersea, implies that we have a responsibility to try to understand the ocean from a scientific and technical point of view, to understand the political and financial pressures affecting the protagonists in the drama of military and resource management, and to express our goodwill not alone in protest, but in constructive ideas or even action. Those words may sound impracticably idealistic, but it is a fact that many people live by them and see nothing odd in being lifeguards, or nature-reserve guides, or scuba-safety instructors on a part-time basis. Moreover, students everywhere are already participating in scientific studies of the sea and in pollution monitoring. All over the world, local administration and government increasingly involve volunteer

workers. It will take a combination of the right spirit of willingness with a slight increase in the awareness of the importance of the ocean on the part of people of goodwill to guarantee the future of the undersea world – and to guarantee that future at every level of business and government.

The Undersea, I hope, will promote understanding and awareness. It is not a sensational book. It does not focus on easy issues that make the headlines for a few days and then fade away. (Thus, for instance, this sentence is our sole mention of the so-called Bermuda Triangle.) But we do, of course, discuss much-publicized matters when they are valid subjects for discussion: e.g., the fate of the great whales, oil spills, the risk of undersea

12 nuclear warfare. As such issues arise, they are set clearly in the context of broader facts about the nature of ocean basins, water, living plants and animals, and various industrial and military techniques and processes.

If our book had been written ten years ago, it would, I imagine, have been permeated by a double sense of wonder – wonder at the diversity, richness, and beauty of nature, and wonder at the marvels of modern technology and the vision of a magic future. In the last decade, however, there has been a revolution not just among scientists, but in everybody's attitude to science. This revolution in our ideas is becoming a political fact. We have discovered two awkward truths. First, the wonders of nature, we have learned, are linked in a complex and balanced network which, if not exactly fragile, has a limited capacity to absorb and adapt to the assaults of industrialization, pollution, tourist development, agriculture, and the sheer impact of people. Secondly, it has become evident that technology and computers generally produce socially or ecologically undesirable side effects, and that to expect a technological fire brigade to rush forward and correct the trouble is almost a contradiction in terms. At the worst, pessimists today predict that the world is likely to progress through a period of crisis and cures, with each successive cure worse than the crisis it displaces.

In short, we have learned that we have the power to destroy the globe in many different ways, and we cannot take for granted that fish and forests will always be there for us to enjoy. Whatever is not taken for granted, whatever is known to be able to die, should be nurtured and valued with a conscious intensity that must be quite different, but should be no less satisfying, than the breathless wonder of the old careless days. But we are not sure that we want such responsibility. There are people who simply want to turn back and recapture the original innocence – as if it were possible.

The peculiar situation of the ocean is that, unlike much of the land, we have not yet transformed it totally by industrialization or contaminated it with pollution and overcrowding. Still, it is an essential part of earth's climatic and biological stability, and we have begun to exploit it systematically. Because of our experience with the land and the mistakes we have made, we cannot fail to know that if we repeat the same mistakes we can spoil this last great reservoir of wealth and beauty. Will we act upon that knowledge? Is it possible that marine technology can be applied so wisely that unpleasant side effects are foreseen and prevented? Or will there be a series of bigger and bigger mistakes, each followed by more desperate countermeasures, as has so often happened on land?

The editor and writers of this book are not cynics. It is our view that an understanding of the diversity of nature and an intelligent anticipation of the consequences of technological action can help mankind to benefit immensely from the material wealth of the oceans even while preserving their beauty and variety.

The Undersea, then, is an optimistic book. Yet it is not naïve or innocent. It is less directly about the ocean than about what people do in and to the ocean. The ocean may be innocent, but thinking people are not. Science may be innocent, but real-life scientists are not. Accordingly, this book does not have the innocence of those internationalists who profess to believe that the risk of pollution is sure to be reduced if only the profits from ocean exploitation are distributed to the poor countries; it does not have the innocence of the nationalists who insist that *they* can fully exploit the ocean without harming it provided they own it, but that foreign exploitation is synonymous with pollution; it does not have the innocence of some scientists who claim to be able to study the ocean with total objectivity, devoid of national or personal bias in observation or interpretation; it does not have the innocence of militarists who argue that peace on earth is best maintained by nuclear submarines' having the right to cruise wherever they wish; it does not have the innocence of the Marxists who seem to believe that men become wise and kind if they are paid by the state, and that a state-controlled tanker produces less pollution than a capitalist one; it does not have the innocence of capitalists who believe the reverse; finally, it does not have the innocence of the environmentalists who are convinced that if your heart is in the right place, you will be granted the ability to make instantly correct decisions about the most complex problems that have ever confronted the human community.

How can any organization, let alone an individual, comprehend the incredible variety of the sea itself and of all the technological advances that are appearing almost daily? Even as I write, the cover of an oil-trade magazine is ablaze with this headline: "Under Construction or Planned: 43 Semi-Submersible Drilling Rigs, 25 Drill Ships, 43 Jack-Up Rigs, 170 Fixed Oil or Gas Production Platforms, 39 Deep-Sea Mooring Terminals, 45 Subsea Production Systems." Taking only the first three figures, which apply to new exploratory offshore drilling rigs, and ignoring the rest, the worldwide total of rigs to be built within the next couple of years adds up to 111. Since the present world total is 306, the number in operation by 1980

Archaeologists mapping a wreck on the sea floor

will undoubtedly have increased by more than a third (allowing for a few dropouts) to over 400. And the same swift pace of growth affects every field of activity. Fishing vessels are getting bigger and more powerful, dredgers roam farther and farther offshore in the search for sand, gravel, and minerals, and engineers are constantly seeking to improve the economics of extracting fresh water and minerals from sea water. Meanwhile, paralleling the feverish rate of development on the technical front, politicians, lawyers, diplomats, and military advisers sit through month after month of committee work in connection with the Law of the Sea Conference, the largest international conference ever held, where 150 nations have been doggedly trying to negotiate a peaceful allocation of the space and resources of nearly three-quarters of the earth's surface.

Many banks, trading organizations, oil companies, and engineering contractors have been planning their operations on a global basis for years. But it is not literally true that they are globally oriented. In actual fact, they are in competition with one another, as are the nation-states of the world, and this makes it difficult to arrive at international agreement on the means to regulate exploitation of the undersea. Nearly everyone gives lip-service to the truism that we live in a closed and limited planet, but this realization is generally trivialized as an impractical basis for political action. Such myopia is hardly surprising. The idea of a closed world whose frontiers are limited, not limitless, does not appeal to the human spirit. But the proof that the world is finite, with only one world ocean, was provided long ago – four and a half centuries ago, in fact – when Ferdinand Magellan sailed his doomed fleet around the globe.

Just imagine that an all-wise king in about 1523 could have foreseen the bloodshed and suffering that would result from the competitive nationalist manner in which the many genuine benefits of European culture were to be transmitted around the world after Magellan's voyage. Imagine that he called together the heads of the most advanced states – Russia, Sweden, France, Holland, Spain, Portugal, England, parts of Italy – and they held a conference. The gist of the king's opening speech might well have been something like this: "If we compete with one another to acquire gold, trading monopolies, spices, silks, slaves, and territory in the lands that have just been discovered, we will all surely lose. The fighting among ourselves will create antagonism among the natives, will kill our own citizens, will lead to the loss of valuable cargoes at sea, and will lay waste to rich parts of the world, wasting resources that we can ill afford to throw away. So let us agree on a common policy of peaceful trade with the new lands, so that everybody in the world can benefit. We are, all of us, human beings. Magellan has proved that the world is a globe of limited area, and that every part of it can be reached easily by sea. This is therefore the only sensible policy."

As far as we know, nobody had such a thought. Anyhow, any such proposal made before our own time would have been rejected with cynical derision by rulers and citizens alike. But in the oceans we are attempting to conduct the imaginary king's exercise today. And we of today cannot, as *The Undersea* emphasizes, fail to know what is at stake.

In order to make a clear and readable book, I have tried, with the authors' cooperation, to separate the interwoven strands of the undersea world so that each subject can be treated in as uncomplicated a manner as possible. Nevertheless, as I pointed out earlier, you will find in every chapter allusions, direct or indirect, to subjects treated elsewhere; in fact, one of our themes might have been the interrelatedness of phenomena in the sea. Partly because the world ocean is one connected fluid medium, and because water and living species can travel from region to region, everything that happens in the sea seems to interact sooner or later with something unexpected. "Interrelatedness" is, of course, the watchword of ecology, and it applies to all natural phenomena on the land, in the sea, or in the atmosphere. Scientific understanding has often progressed by means of a deliberate simplification of natural situations, an elimination of unwanted effects until the problem to be solved has shrunk to manageable proportions. Ecology seeks to reverse this approach by admitting the endless skein of interacting cause and effect between natural processes, and setting out to see and study the whole system.

It is obvious that the ecological approach, especially in studying the sea, can lead to staggering complexities, and ecological solutions to problems are bound to involve scientific and mathematical understanding of a high order. In the face of this complexity, many people, not surprisingly, have taken emotional escape routes. Some of the escapists argue that since modern industrial society isn't worth saving, there's no point in trying to do something about it; the correct course for society is back to the paleolithic cave. Others, more actively inclined, suggest that we should lower our standard of living, redistribute the available food, and proceed to live as farmers. Such attitudes may be understandable, but they are not likely to

An oil rig being towed into position

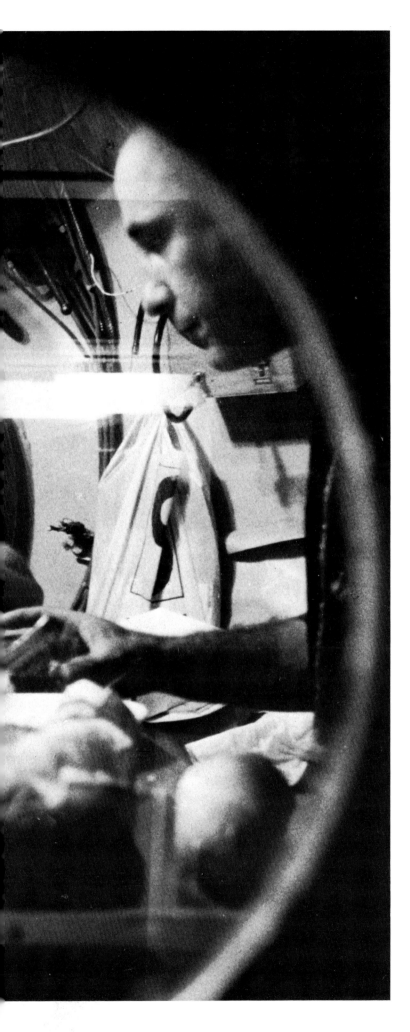

receive general approval. An awareness of inter-relationships in the modern industrial world is here to stay, and what most of us want is not to escape from thinking about them, but to improve our understanding of them.

Whether in the ocean or on land, interrelated-ness can be expressed symbolically in simple diagrams or models. It is possible to use these to develop a habit of mind for tracking chain re-actions, at least in general terms, across the con-ventional boundaries of academic disciplines and government departments. Once the branching and twisting links have been identified in prin-ciple, we can start to build rigorous mathematical models to see how they work, how changes in one part or another of a given system might affect the working of the whole. To a limited extent, this predictive technique has already proved its va-lidity in a number of studies made by scientists, sociologists, and economists. The most famous such study, the so-called Club of Rome Report (1972) on the results of an economic-social model of the whole world, appears to predict inevitable doom by the year 2000. But later models have produced widely varying results – which demon-strates the primitive stage of a technique aimed at forecasting specific trends on the basis of in-numerable wildly fluctuating complexities.

In a more modest corner of the field, however, models of interactions can be highly instructive. Marine scientists, for example, have been building increasingly elaborate models of food chains in different environments – estuaries, coastal seas, coral reefs, the open ocean, etc. Other groups are constructing models of population changes and employment trends in special coastal localities, such as around estuaries. Still others have been analyzing economic and environmental models of different programs for the development of marshlands, or the effect upon the marine environ-ment of boosting tourism on a deserted coast or on tropical islands. These models may be fairly ele-mentary and inadequate, but with luck they will evolve into more general marine models, enab-ling us to predict the long-term effects of large-scale industrial or social decisions.

The accuracy of such predictions, and the value of the decisions based on them, will depend en-tirely on the accuracy of the facts fed into the com-puter. To get these facts right is, in broad terms, the social justification for all scientific research at sea. Oceanographers and other scientists have been studying the oceans for centuries for all sorts of reasons – improvements to navigation, fisheries, or pure curiosity – but only in the last decade has it

Divers at ease in a pressure chamber

become possible to contemplate building the kind of predictive model I am discussing. As soon, however, as one realizes the possible implications of such a predictive technique for the orderly planning of marine exploitation, it becomes apparent that, in spite of the vast accumulation of data already available, we still do not have enough. We do not understand the exact speed and nature of all the thousands of processes that link physical events to one another within the water, nor do we understand the links between the ocean and the atmosphere. Similarly, we have much to learn about the precise chemical and physical balance between marine life and seawater, and between the animals and plants themselves.

Until now, societies based on intensive agriculture and technology have tended to behave as if there were always another field, country, or continent to go to after the ruination of this one. For the last thirty years or so, though knowing the worst, we have been behaving as if there were perhaps another *planet* to go to. That is an improbable dream. This planet and the oceans on it are a single closed system. Earth's only energy input is sunlight, plus a certain quantity of cosmic particles and radiation; its only output is heat, plus a few water molecules, a little gas, and some bits of rocketry destined to wander around the solar system. Apart from that we are in a closed box, where nothing comes in or goes out. We have to live with what we have. Everything we degrade is destroyed forever; everything one person throws out of his personal patch lands in some other person's. There is no ultimate dumping ground.

The knowledge that interconnectedness exists within our closed system is certainly a necessary step toward preserving that system. But can predictive models of the marine environment really assist the preservation by suggesting ways to avoid the great mistake of using the wrong technology in the wrong way in the wrong place? The answer, I believe, is yes. There are those who maintain that tomorrow's technology can correct the errors of today's, but this seems to me to underestimate the irreversible damage that may be done in some cases, and to disregard such economic problems as investment in plant (which cannot be written off) and in human skills (which take time to reestablish). What we need – and do not yet have – is enough comprehension of how marine technology and production interact with natural marine phenomena to be able to foresee the consequences of alternative courses of action. The closer we come to that goal, the better our chances of forestalling tragic mistakes.

Life in a coral reef

And so *The Undersea* ranges widely over the fields of geophysics, oceanography, marine technology, marine biology, archaeology, law, and a little politics – a seemingly strange mixture, perhaps, but an essential one. Boundaries between fields of knowledge are arbitrary conveniences; it is an enlightening experiment to slice the cake in new directions and produce unfamiliar patterns and designs. As I now review the patterns that emerge in this book – the shifting design of the earth beneath the sea and the kaleidoscope of moving continents, the patterns of microscopic plants woven into the migration routes of the great whales, the indescribable floating architecture of oil rigs and the intricate workings of oceanographic instruments – I am more than ever convinced that the living oceans will endure, that man will not fail to solve the problems he himself is creating. Nor need I apologize for my optimism.

Optimism that sees problems and tries to solve them is surely preferable to a pessimism that depresses the spirit and precludes action. If *The Undersea* encourages even a few readers to give more thought to the present crisis of the oceans, and to support efforts to keep the crisis under control, I shall feel well rewarded.

A U.S. attack submarine shows her flag

1
The Ocean Floor

David G. Roberts

Of the nine planets, only one has an ocean. The division of the earth's surface into land (roughly 30%) and ocean (70%) is the result of three phenomena: the emission of large quantities of vapor from molten rocks during millions of years of prehistoric volcanic activity; the fact that the earth's temperature is such that most of the vapor has liquefied; and the segregation of the rocks of the earth's crust into uplifted continents and depressed ocean basins, with a total vertical range of 20,000 meters.

No other planet is like the earth. Mercury has lost its water, is covered with craters, and has no surface erosion or volcanic activity. Venus, with a surface temperature of 477°C, can have no liquid water, and since Mars has a general temperature of about −43°C, it can have very little unfrozen water. As for the outer planets, they are so different in composition and temperature because of their distance from the sun that their physical features can in no way be comparable with ours.

So the ocean is unique, at least within our planetary system. At first sight, its shape and position seem self-explanatory: if you have water and depressions on a planet, the water flows into the depressions, leaving the rest of the surface dry. Until about 20 years ago, most geologists believed, indeed, that the form of the ocean basins is fixed, and that geological processes consist of more or less local forces that have acted upon the earth's crust, uplifting a mountain range in one place, depressing a valley in another. Almost all geological changes were thought of as vertical movements, with just a certain amount of shearing, sliding, or thrusting. There were many geological

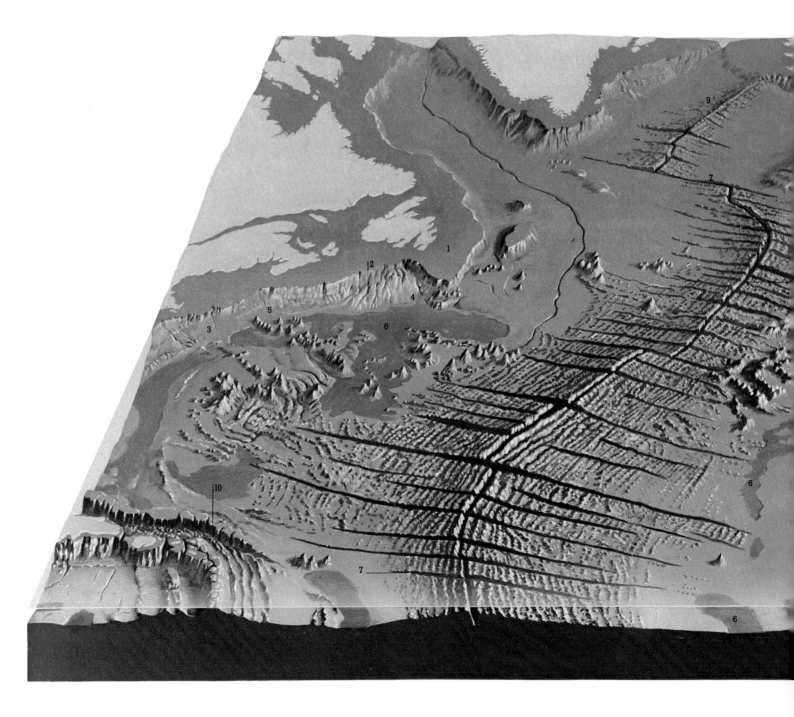

puzzles, such as strangely complementary configurations in Scotland and Canada, and in central Africa and Brazil; similar fossil bones found in such widely separated places as South America, South Africa, and Australia; and so on. But the suggestion that continents could move about horizontally – what we now call "continental drift" – was not taken seriously.

Today we know better. We believe that the static view of the earth's crust is no longer tenable. The crust, in fact, is in a perpetual state of change, for it is a thin layer of rocky material forming the top of about a dozen rigid plates – which are like 100-kilometer-thick pieces of a huge jigsaw puzzle – that never stop moving in relation to one another. The stresses between them as they draw together or apart at relative speeds of 2 to 10 centimeters a year are

responsible for many of the geological features of earth's surface, including the ocean basins. This is the generally accepted theory of "plate tectonics." It is accepted because it fits the facts that have been discovered in recent years about the tectonics (i.e., the geological forces) that have shaped – and are still shaping – both the land and the ocean floor.

Before proceeding further, one point of definition must be stressed: in speaking of the *ocean floor*, geologists are not usually referring to the spot where you step off your favorite beach into the sea, or even to the bottom several kilometers off the coast. The continents have what we call "margins" of varying widths, as we shall see in the next section. The true ocean floor consists of the basin that lies beyond these continental margins.

A Tour of the Bottom

23

The ocean floor and its margins are geologically in motion and constantly changing, but we cannot discuss the dynamic aspects of the floor unless we know what it looks like today. In brief, then, the actual material at the bottom may be either solid rock or – more widely – unconsolidated sediments of clay, silt, sand, or gravel that have been washed off the continents by rivers, or of mud formed from countless billions of skeletons of planktonic animals. The solid rock itself may be sedimentary – i.e., composed of sediments compacted under great pressure thousands of meters below the present sea floor. Or the rock may be igneous – formed of molten magma welling up from within the earth. Most of the igneous rock beneath the sea floor is of a kind called basalt.

Now imagine yourself on a flight westward from Europe across a completely drained Atlantic Ocean. First you would fly over the almost flat area of the continental shelf, viewing exposed rocks in many places (because of strong tidal currents) and beds of gravel and sand in others. At the edge of the continental shelf begins the continental slope, which descends fairly sharply from a depth of 200 to as much as 3,500 meters. Here and there, cut into the slope, are a number of very deep canyons. The foot of the slope blends into a broad, very gently sloping sedimentary area that we call the continental rise. This is the outer part of the continental margin. It merges with the true ocean floor, usually beginning with hundreds of kilometers of flatland known as the "abyssal plain." Flying on, you will eventually see that the ocean bed below you is rising, until it becomes a huge mountain range, the Mid-Atlantic Ridge, cut by many transverse fractures and with a broad valley – the median rift valley – running through its crest. On the western side of the ridge, the abyssal plain is succeeded by the continental rise, and so on – a mirror image of the eastern Atlantic.

The Mid-Atlantic Ridge, with its median valley, is only part of a vast mid-ocean chain of mountains that extends all around the globe, through every ocean, for over 60,000 kilometers. It generally rises about 3,000 meters above the abyssal plain, and even breaks surface occasionally, as at Iceland, the Azores, and the Galapagos Islands.

The other great oceans have features comparable to those of the Atlantic, but with spectacular variations caused by some of the tectonic forces that we shall be examining in this chapter. In the western Pacific, for example, the abyssal plains do not blend smoothly into the continental margin (which includes the rise, slope, and shelf) but are bordered by deep trenches. This is where we find the greatest depths at one point,

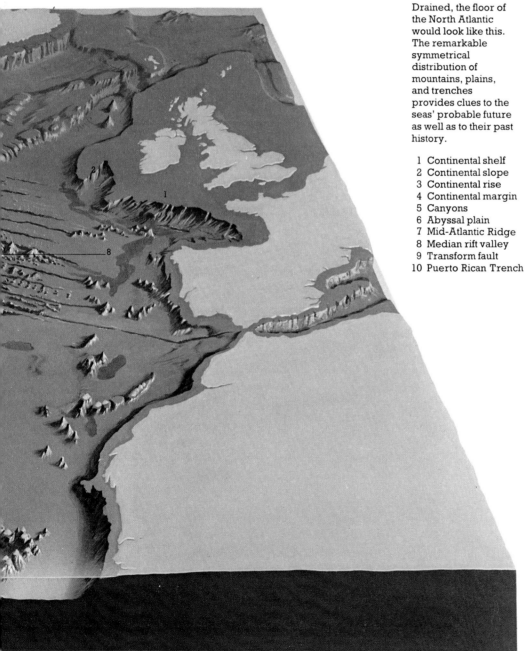

Drained, the floor of the North Atlantic would look like this. The remarkable symmetrical distribution of mountains, plains, and trenches provides clues to the seas' probable future as well as to their past history.

 1 Continental shelf
 2 Continental slope
 3 Continental rise
 4 Continental margin
 5 Canyons
 6 Abyssal plain
 7 Mid-Atlantic Ridge
 8 Median rift valley
 9 Transform fault
10 Puerto Rican Trench

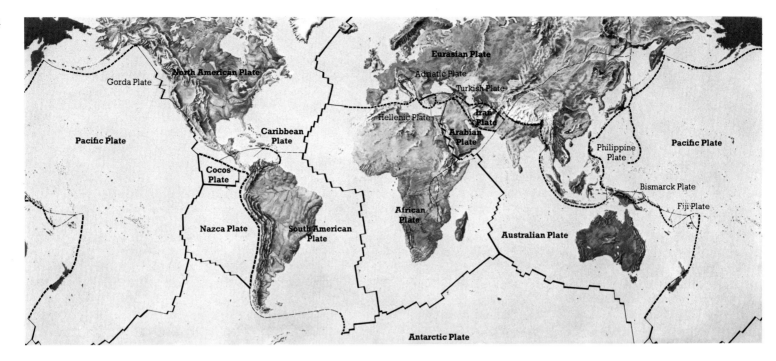

southwest of Guam, the Marianas Trench descends to 11,022 meters, which is enough to drown Mount Everest, with a couple of thousand meters to spare. There is a deep trench right off the west coast of South America; more commonly, though, the side of any such trench that borders a continent is rimmed with a chain of islands, known as an "island arc," and between the island arc and the continent is a marginal sea. The Caribbean is just such an area.

The Active Earth

The movements of the plates that transport the earth's crust are an expression of the internal structure and dynamics of the planet, which has been evolving since it came into being about 5,000 million years ago. We derive our comprehension of the earth's internal structure from our knowledge of its average density, the transmission of earthquake waves through it, and the behavior of its magnetic field. By piecing the evidence together we get an idea of the forces that may be driving plate movements, but we are not sure how they work.

The average density of the entire earth is 5.5 grams per cubic centimeter, as compared with an average density of only 2.7 for continental rocks and 3.0 for the rocks that lie beneath the oceans. Thus, in view of the averages, the central core of the earth must be very dense, about 10 grams per cubic centimeter, and scientists reason that it is probably composed of iron and nickel. Outside the solid inner core lies a 2,000-kilometer-thick liquid layer of these metals, and the liquid, being metallic, conducts electricity. The convection currents generated within this layer create currents of molten metal, which follow complicated paths because of the rotation of the earth;

and since electrical conductors can generate magnetic fields, the interactions of several such currents generate a magnetic field for the entire earth. But because the circulation of the liquid in the outer core is somewhat erratic, the earth's magnetic field is far from constant. Not only does the magnetic pole wander from year to year, but it even flips from north to south about once in every 200,000 years. The results are very important for the plate-tectonic theory, as we shall see.

Above the metallic liquid layer lies earth's mantle, which is nearly 3,000 kilometers thick and consists of more or less mobile rock and dense peridotite composed mainly of the silicate minerals olivine and pyroxene. Solid peridotite of this type is rarely found on the continents but is sometimes found along the 60,000-kilometer-long mid-ocean ridge. This suggests that the mantle material has welled up under the ridges and then solidified. The mantle itself is not homogeneous, and the peridotite gets progressively denser near the core. But there is one region where the rock seems particularly unconsolidated; we call it the "asthenosphere" (from the Greek word for "weakness"), and it lies about 125 kilometers below the earth's surface. Somehow a combination of pressure and temperature in the asthenosphere makes it an exceptionally fluid zone. Below it lies the inner mantle, which we presume to be in a state of very slow convection; and above it, comprising the uppermost layer of the mantle, is the rigid material of the plates. Although we are not yet able to prove our point, we believe that the slow convection of the inner mantle creates the horizontal forces that move the plates.

We call the thick upper layer of the mantle, with its thin topping of crust, the

lithosphere (Greek *lithos* means "stone"). The crust itself has an average thickness on the continents of about 35 kilometers, but is much thinner – around 6 kilometers – under the deep oceans. Almost everywhere but under the mid-ocean ridge there is a fairly sharp boundary line between crust and mantle material. This line is called the Mohorovičić discontinuity (Moho for short), after the Yugoslav seismologist who in 1909 found a way to detect the boundary by means of sound waves from earthquakes.

The Underlying Rocks

We have seen that the earth's crust is much thinner on the ocean floor than on the continents. A close study of the rocks of the ocean floor reveals many other differences between them and continental rocks. The first difference, which has been known for many years, is that oceanic crust is basaltic, with an average density of 3 grams per cubic centimeter, in contrast with the granitic igneous and sedimentary rocks of the continents, with an average density of only 2.7. By visualizing the rocks of the crust as "floating" on the denser rock of the upper mantle (average density 3.4) like a log in water, it becomes easy to understand how the continents maintain their altitude above the oceans. It is a manifestation of the principle of equalization that geologists call "isostasy": because no segment of the earth measured from core to surface can weigh more or less than any other, when heavy rock sinks into the mantle in one place, it causes a compensating elevation of lighter rock somewhere else. Thus, a given block of the crust elevated or depressed by tectonic forces tends to return to isostatic balance level, whether upward or downward, once these forces cease to act upon it.

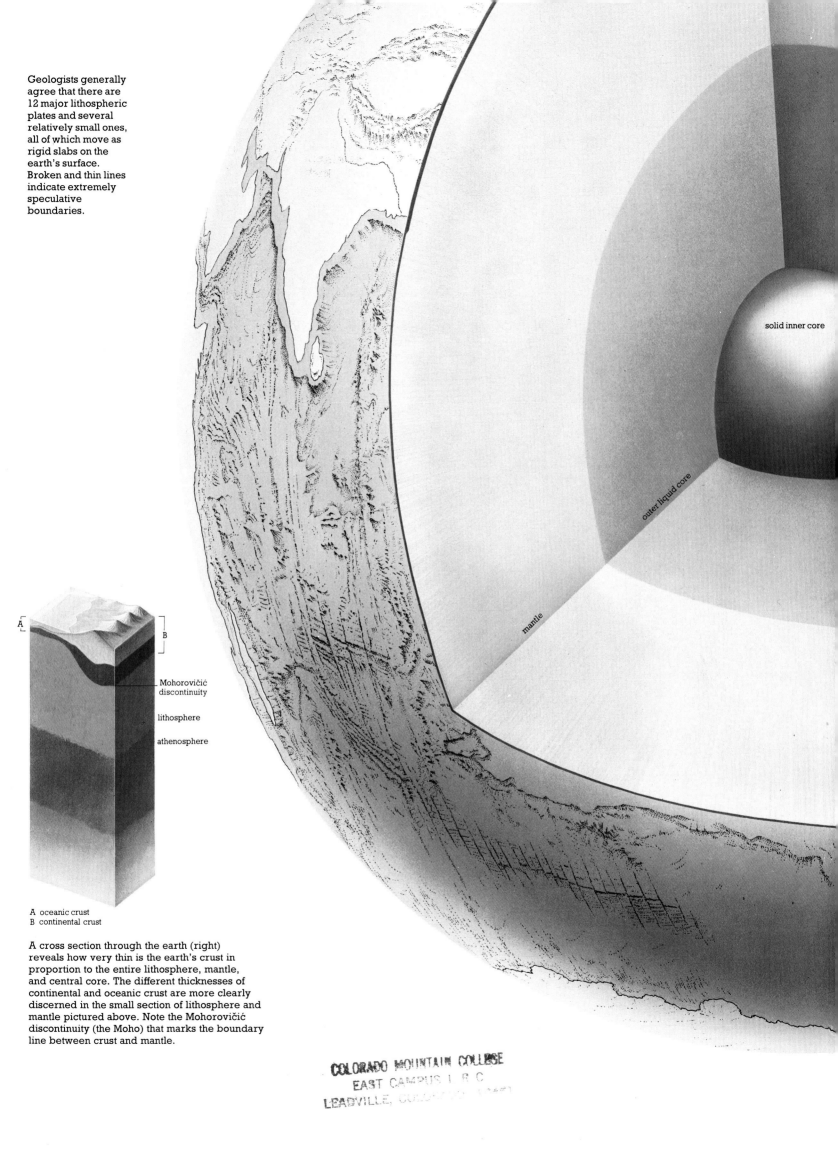

Geologists generally agree that there are 12 major lithospheric plates and several relatively small ones, all of which move as rigid slabs on the earth's surface. Broken and thin lines indicate extremely speculative boundaries.

solid inner core

outer liquid core

mantle

Mohorovičić discontinuity

lithosphere

athenosphere

A oceanic crust
B continental crust

A cross section through the earth (right) reveals how very thin is the earth's crust in proportion to the entire lithosphere, mantle, and central core. The different thicknesses of continental and oceanic crust are more clearly discerned in the small section of lithosphere and mantle pictured above. Note the Mohorovičić discontinuity (the Moho) that marks the boundary line between crust and mantle.

Simple calculations show that a 35-kilometer-thick block of continent should float with some 7 kilometers projecting above the mantle, while an equivalent 6-kilometer-thick layer of denser oceanic crust – made even heavier by the total weight of water upon it – should project only slightly more than 1 kilometer. Thus, the average difference in altitude between the top of the ocean floor and the surface of the continents should be between 5 and 6 kilometers. And those calculations are corroborated by the facts, for the average depth of the ocean is about 4 kilometers, and the average height above sea level of the continents is close to 1 kilometer.

Our use of the word "float" may seem to imply that the earth's crust can drift sideways across the upper mantle; and this was, in fact, the implication of the theory of continental drift put forward by Alfred Wegener, a German meteorologist, early in this century. But any such assumption is incorrect. Both above and below the Mohorovičić discontinuity the lithosphere is extremely rigid, and since the continents are embedded more than 25 kilometers deep in the upper mantle, sideways movement through the mantle is impossible. That is why most geologists rejected the continental-drift theory when Wegener first proposed it in 1912, and continued to reject it right through the 1930s. As was pointed out earlier, we know now that it is not the continents that "drift," but the 100-kilometer-thick lithosphere, which carries the oceanic and continental crust with it as it is moved by forces within the fluid asthenosphere.

The rocks of the ocean floor differ from those of the continents not only in type and density but also in age. A study of the radioactive decay of certain elements in rocks enables scientists to date the period when they solidified, either from magma (molten material that has welled up from the mantle) or from sedimentary particles; and such studies prove that the oldest rocks on land are about 3,500 million years old – almost as old as the earth itself – and that many others are at least 1,500 million years old. No part of the ocean floor, however, appears to be older than 200 million years, and many of the rocks on the mid-ocean ridges are less than 10 million years old! Obviously, faced with the discovery that ocean-floor rocks – those of nearly three-quarters of the earth's surface – have solidified only during the last 5% of the earth's history, geologists want to know why. Why this astonishing difference between the earth's crust on land and undersea?

The theory of plate tectonics provides a credible explanation by assuming that oceanic crust is being formed continuously from the inner mantle at the mid-ocean ridges, thus constantly creating new, young ocean floor. Meanwhile, the older oceanic crust is drawn downward, back into the inner mantle, in the deep ocean trenches. This cannot happen with continental crust; because it is composed of lightweight rock, it cannot be subducted into the mantle, for it simply floats up again. So the continents have been built up for thousands of millions of years while the ocean floor is always in process of being renewed and destroyed.

Although plate tectonics is still only a theory (as is, for instance, the concept of evolution through natural selection), most scientists now accept its validity. There remain, as we shall see, many uncertainties about important aspects of the plates, but few geologists doubt their existence or tectonic importance.

A

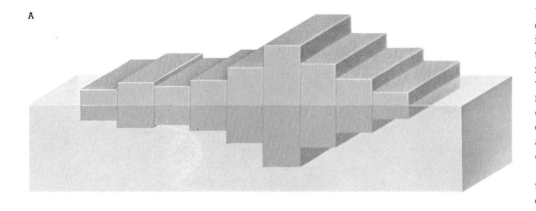

B

C

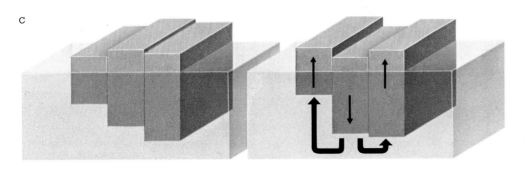

The principle of isostasy: yellow blocks represent segments of the earth composed of lightweight rock, orange of heavier rock, pink of the heaviest. A) Segments of rocks of the same density but differing depths find their own equilibrium levels in the mantle. The higher the "mountain" the deeper its "roots." B) In relation to their lighter neighbors, heavier segments sink deeper into the mantle. C) If further weight is added to a segment, it will gradually sink, and the portion of mantle that the "roots" displace will force its less heavy neighbors to rise.

Pangaea

Throughout geological history continental rocks – which, remember, are lighter than those of the ocean floor – have been accumulating on the surface. Great portions of them, eroded by wind, water, and ice, have been deposited in the oceans as sediments. Meanwhile, vast blocks of continental crust have been shifted to and fro by the moving plates of the lithosphere; and when blocks of continental crust collide, vast mountain ranges like the Himalayas are slowly born. The rocks *must* pile up, since they are not dense enough to be subducted into the mantle.

Successive cycles of continental break-up and collision may have occurred several times in our planet's history, but the last such cycle probably started about 200 million years ago, with the break-up of a single super-continent, to which Alfred Wegener gave the name Pangaea (meaning "all earth"). Pangaea, consisting of all the present continents joined together, apparently formed some 150 million years before that. During the long period of its existence as a unit, the rest of the world's surface must have been one huge, continuous ocean. Then, as Pangaea broke up, creating new ocean floor in the opening rifts, the original oceanic crust was progressively destroyed. Only a few fragments of it are now left; they are found in the northwestern corner of the Pacific. This is because, whereas the other oceans are regions in which new crust is being created, the Pacific is, by and large, an ocean in which crust is being destroyed.

If we are to understand why the Pacific floor is, so to speak, "old," and how the newer ocean basins have been formed, we must begin by tracking the stages of

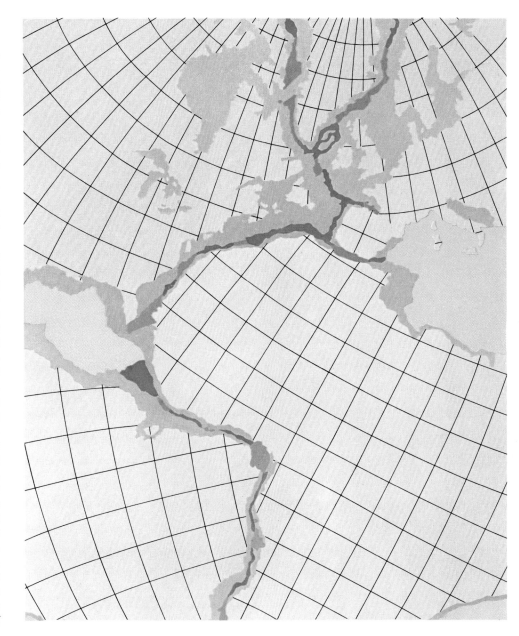

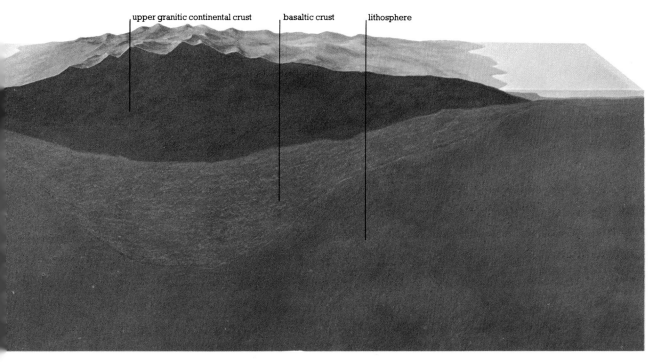

upper granitic continental crust   basaltic crust   lithosphere

Left: Because a continent may be rooted as much as 50 kilometers deep in the upper mantle, its broad edges (the "continental margins") can extend beneath the sea far beyond present shorelines. Thus we can visualize an accurate fitting together of land masses as they probably existed 200 million years ago only by including part of the underwater margins in considering terrestrial topography. This has been done in the above map, with latitude and longitude lines retained to show how much the continents have rotated around since separating.

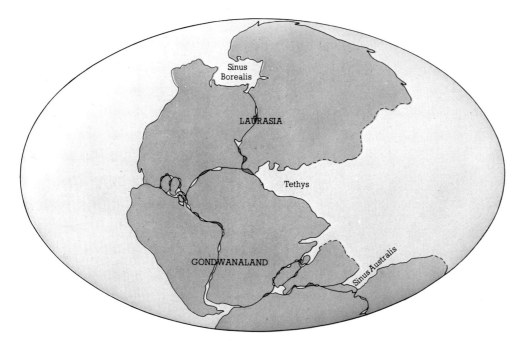

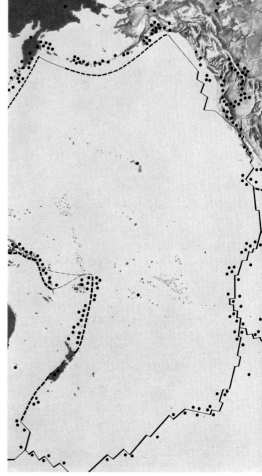

All present land masses were once joined together in a single super-continent. We call it Pangaea; and some of its parts have also been named, as indicated here.

Pangaea's fragmentation. Surprising as it may seem, it is not at all hard to re-create Pangaea and to follow the paths by which the continents reached their present positions. All it really requires, actually, is a precise knowledge of continental-margin contours. For instance, although the rough "fit" between western Africa and eastern South America has been apparent for centuries, an attempt to match the coastlines exactly runs into problems of overlaps and gaps. If, however, they are visually joined at the 1,000-meter contour on the continental slope, the fit is extraordinarily good. The same goes for Europe against North America, and for fitting India, Australia, Africa, and America around Antarctica. A few odd bits, such as Rockall Bank and the Seychelles, are difficult to slot in, but for the most part the form of Pangaea can be seen very clearly.

One further clue to the original position of a given fragment of Pangaea is found in the fact that when sedimentary and igneous rocks solidify, magnetic particles within them tend to become oriented in such a way as to indicate the position of the magnetic north pole at the time. If the continents were fixed on the earth's surface, all rocks of the same age would indicate the same position of the earth's magnetic pole, even though the pole itself moves from age to age. What geologists actually find is that contemporary magnetized rocks from any one continent always point to the same pole – but magnetic particles of the same age in other continents point in different directions. This confirms the relative movement of continents. A thorough understanding of how and why depends on a close study of the ocean floor.

The Earthquake Key

The origin of today's oceans lies in the break-up of Pangaea 200 million years ago and the continuous movements of the plates since then. To visualize what has been happening, we need a way of locating and defining plate boundaries, which are the zones where all the relative movement occurs. Plates may be pressing together, sliding past each other, or tearing apart, and so it is natural to expect a release of energy at the boundaries through friction and breaking of rocks, since it takes a great deal of force to break or deform rock. And it is just such strains near plate boundaries that cause earthquakes, which displace the surface of the earth, leaving cracks or steps. The plane along which any such break tends to occur is called a "fault," and its slope and depth depend on the type of plate boundary – whether the plates at that point are diverging, converging, or shearing past each other.

The fact that earthquakes are not distributed randomly over the globe was already evident more than a century ago, but the meaning of the sinuous chains of earthquake centers was unknown until our own time. We now believe that the earthquake zones extending through the Mediterranean lands, the Middle East, northern India, and all around the margins of the Pacific define some of the principal plate boundaries. An accurate location of the focal point of earthquakes under the ocean has been made possible by the use of increasingly precise measurements from seismological observatories on land, and by the mid-1950s scientists had found a narrow band of earthquakes along the Mid-Atlantic Ridge, in a zone only 160 kilometers wide. Such earthquakes were soon to be associated with the whole length of the mid-ocean-ridge system, and the chain could be traced

into the Gulf of Aden, and on into the East African Rift Valley. This chain of subsea earthquakes indicates the whereabouts of the remaining plate boundaries where plates are presumed to be diverging.

We can now examine how the earthquakes differ on differing types of boundary. Under the median valley of the mid-ocean ridges, where the plates are presumed to be diverging, earthquakes are, in seismological terms, very shallow, just a few tens of kilometers below the surface. Because there are many fissures containing molten magma underneath the valley, the lithosphere is not extremely rigid and resistant to strain, and so the earthquakes are not very violent. Where plates are sliding past each other, as on the western coast of the United States, the north coast of Turkey, and the many offsets in the mid-ocean ridge, earthquakes occur at all depths in the lithosphere – that is, down to about 125 kilometers. It is because the San Andreas Fault in California and the Anatolian Fault in Turkey cut through continental edges that the earthquakes on those faults cause so much damage.

When two plates are convergent, or colliding, one of them must press under the other. Since oceanic crust is nearly as dense as the mantle, it tends to bend down and be subducted back into the mantle. As this hap-

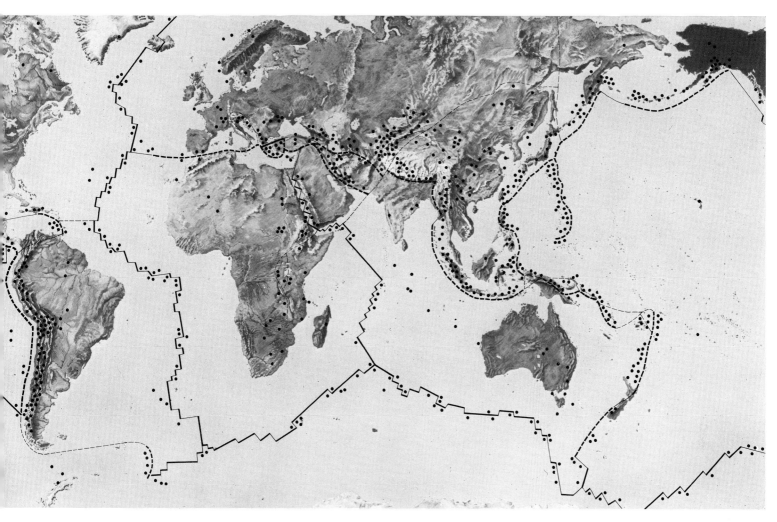

Most earthquakes occur at plate boundaries and appear to be caused by plate movements at those boundaries. Locations of recorded earthquakes are indicated here by dots.

Below: The earthquake zone under an area where two plates collide was first defined and described by the American geophysicist Hugo Benioff. Hence the term "Benioff Zone."

pens, the layer of light sedimentary rock on top of the oceanic crust is scraped off by the overriding plate, and it may accumulate either as an island arc or against the continental margin. The oceanic crust and lithosphere, in a slab about 100 kilometers thick, descends at an angle of some 30°, and the surface of the slab slowly melts as the result of friction and pressure. At a depth of 100 to 300 kilometers – in other words, within the asthenosphere – the lighter molten rocks, forcing their way up behind the subduction zone, form a chain of volcanoes (a pattern that prevails all around the Pacific – the so-called ring of fire – as well as in the Caribbean and the south Aegean). The cold slab of lithosphere of the subducted plate continues downward, causing further earthquakes as it slowly breaks up, until it finally melts at a depth of about 700 kilometers, well within the inner mantle. Over the whole area of the sloping surface of the slab – an area known as the Benioff Zone – deep and powerful earthquakes occur regularly. It is Benioff Zone activity that accounts for the extremely violent earthquakes suffered in such countries as Japan and Chile.

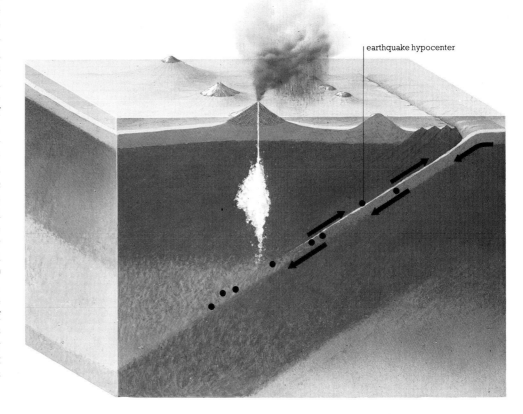

earthquake hypocenter

The 60,000-kilometer-long mountain chain of the
mid-ocean ridge (above) is offset at frequent
intervals by transform faults (below).

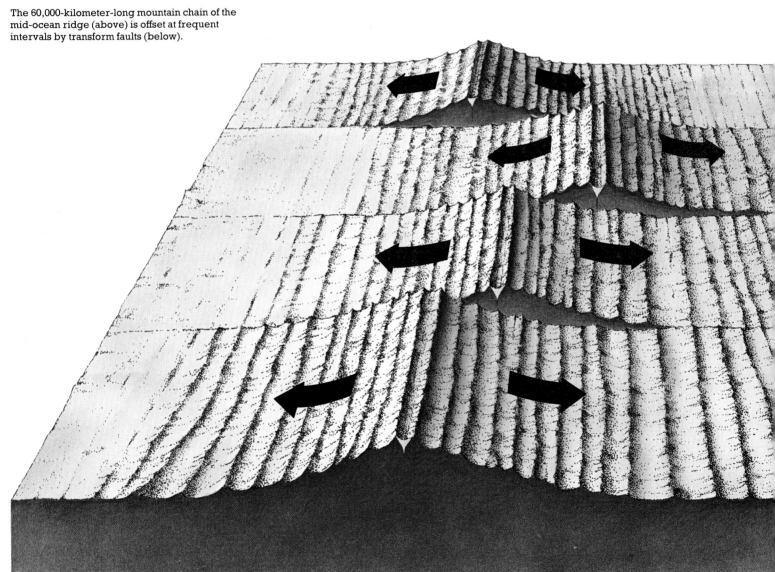

Sea-Floor Spreading

So far we have been assuming that at the mid-ocean ridges the plates move apart, magma wells up, and there are some shallow earthquakes. There is more to the story than that. The pattern of the mid-ocean ridge on a small-scale map of the floor shows it to be a sinuous chain of mountains about 1,000 kilometers in breadth. A close look at the axis of the ridge reveals that there are in fact no curved bends, only sharp angular offsets called "transform faults." The exact trend of the axis is defined by the straight segments of the median valley, which alternate with straight segments of transform faults, usually at right angles to the median valley. Since the initial break-up of Pangaea was not along simple lines, this zig-zag pattern results from the way the mid-ocean spreading process, which we shall now explain, adapts itself to the shapes of the retreating continents.

The splitting of the median valley is probably, though not certainly, related to rising convection in the mantle. At any rate, the following speculative mechanism can be suggested. The walls of the median valley are forced upward and outward by the pressure below, and narrow linear fissures open up in the valley floor. Molten rock (magma) wells up through the fissures and erupts onto the valley floor to form small hills made of rounded masses of lava that are called pillow lava because of their shape. In the crust below the eruption, the stream of molten rock cools and solidifies to form a linear vertical wall, or dike, of volcanic rock. The next split forms on the same line and pulls the dike apart, so that another feeder of magma intrudes along the axis of the previous dike. Each successive dike injection carries older dikes and the overlying piles of lava sideways, away from the injection axis.

The immediate supply of molten magma does not come directly from the asthenosphere (which is more than 100 kilometers below the surface) but from an intermediate chamber, extending along the axis of the median valley, within the thickness of the oceanic crust itself. This chamber is refueled from within the inner mantle. As the plates move apart, the chamber walls are also carried sideways, and molten rock solidifies against the walls as they cool. These rocks cool more slowly than at the surface, and they form coarse-grained crystalline materials such as gabbros and peridotites. The newly created ocean floor is

Pillow lavas in the median rift valley (above) are formed as molten rock cools in contact with the water. The molten rock comes from a magma chamber in the oceanic crust (right), where it accumulates as it wells up from the inner mantle.

32   thus in two distinct layers – an upper one composed of extruded basalts and a lower one formed of the denser gabbros and peridotites. On top of these, sedimentary layers are eventually deposited, with the result that the entire ocean floor (the *true* ocean floor of the abyssal plains) has a uniform three-layer structure. Geophysicists, who can detect the layers by the rate at which sound is transmitted through each, call the sediments layer 1, the basalt layer 2, and the deepest of them, layer 3. All these layers are within the relatively thin oceanic crust, above the Moho.

Between two adjacent segments of the median rift valley, the slabs of lithosphere topped by newly formed crust are sliding past each other at speeds of from 2 to 10 centimeters per year, causing further earthquakes. But beyond the zone of offset the slabs on each side of the transform fault are moving in the same direction, and so the fault becomes dormant – that is, there is no further movement of it. This produces the peculiar effect of a fault that seems to be moving in the middle, but not at either end – a pattern that is, indeed, the defining characteristic of a transform fault.

Below, left: A section through the oceanic crust shows the typical three-layer structure of the true ocean floor.

Below: As molten lavas cool in the median valley and are carried sideways on separating plates, each new slab of rock is magnetized in the direction of the earth's magnetic field at the time when it solidifies.

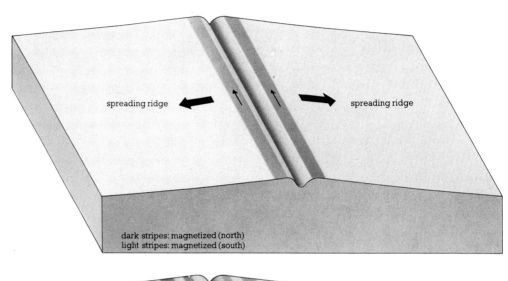

spreading ridge

spreading ridge

dark stripes: magnetized (north)
light stripes: magnetized (south)

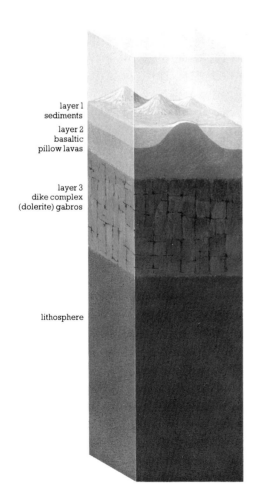

layer 1
sediments

layer 2
basaltic
pillow lavas

layer 3
dike complex
(dolerite) gabbros

lithosphere

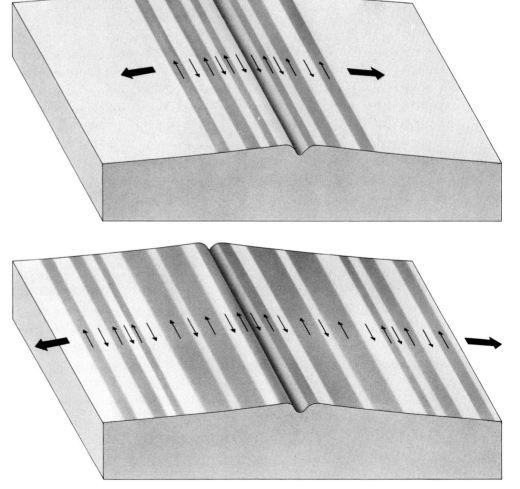

Magnetic Stripes

We have already pointed out that the earth's magnetic poles reverse from time to time because of instability in the circulation of molten iron and nickel in the outer core, and that cooled-down molten rocks preserve a small magnetic field pointing in the direction of the earth's field at the time when they solidified. Thus, igneous rocks that cool when the earth's magnetic poles are in the present direction – so-called normal polarity – have a magnetism pointing northward; but rocks that form when the earth's magnetic field is reversed solidify with their magnetism pointing southward.

The molten magma that is continually being injected at the axis of the mid-ocean ridge can be thought of as a blank magnetic tape, like that in a tape recorder. Actually, as we have seen, a single extruded stream splits so that two "tapes" move away from the axis on opposite paths. As long as the earth's magnetic field remains in one direction, both "tapes" of new oceanic crust will have magnetism pointing that way. Then, when the earth's magnetic field reverses, all new magma cooling in the median valley will also have reversed magnetism. And so, as the millennia go by, two slabs of oceanic crust constantly grow away from opposite sides of the median valley, both with a direction of magnetism that reflects the earth's current magnetic field.

If we assume that the speed of ocean-floor spreading is constant for many millions of years (as it probably is), we would expect the width of each crustal slab with a given polarity, whether normal or reversed, to be proportional to the time for which that polarity lasted. Furthermore, the distance from the median axis to the line in the crust where the magnetism changes direction should be proportional to the total time since that polarity change. We'd expect this to be true of the magnetic "tapes" moving away from both sides of the axis, which should be mirror images of each other. And so it is: a magnetometer towed behind a ship reveals that the magnetism of ocean-floor rocks follows a pattern of stripes. Every stripe is parallel to the mid-ocean ridge; and, in keeping with the fact that the times of reversal of the earth's magnetic field are irregular, the widths of successive stripes with normal and reversed polarity are also irregular, though symmetrically mirrored across the median axis.

The irregularities are useful to the geologist because the magnetic stripes are rather like fingerprints in that certain sequences of broad and narrow bands can be recognized and identified from ocean to ocean, and so sea-floor spreading can be seen and studied as a synchronized process. The rocks associated with each magnetic reversal can also be sampled and

dated by radioactive means. So the magnetic stripes can be accurately dated, enabling us to tell when each stripe was being formed. Furthermore, by selecting equivalent stripes on opposite sides of the ridge, either by matching "fingerprints" or radioactive dating, we can – figuratively speaking – remove all the oceanic crust formed since then and determine the positions of the two underlying plates at the time when those stripes were being formed. Thus the relative positions of the plates and the exact shape of the ocean basin can be recreated for any period in the past.

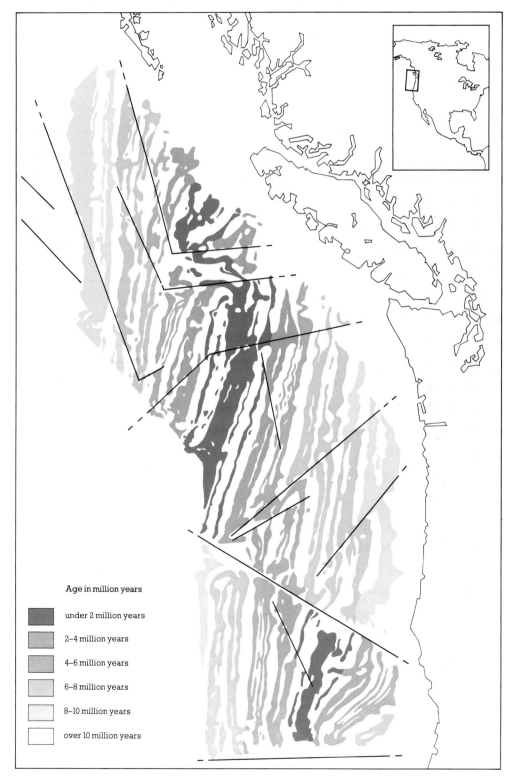

Off the west coast of North America this pattern of magnetic stripes reflects the history of sea-floor spreading in the area.

Age in million years

under 2 million years

2–4 million years

4–6 million years

6–8 million years

8–10 million years

over 10 million years

spreading (passive)

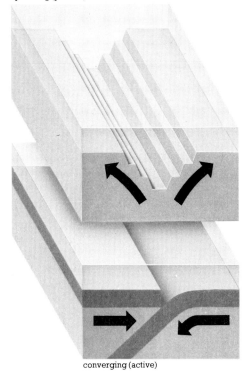

converging (active)

Above: Two types of plate boundary. **Right:** A space photograph of the Red Sea area. The coasts of Africa (lower left) and Saudi Arabia (upper right) were separated about 10 million years ago, and the rift, flooded by the Indian Ocean, eventually became the Gulf of Aden and the Red Sea (top center).

Below: These diagrams show how a rift valley may form within a continent as the result, probably, of mantle upwellings.

Birth of an Ocean

An ocean is born when a great land mass begins to split apart; the fragments of continent-bearing lithosphere move away from each other, and the ocean between them widens as upwelling rock from the mantle creates new lithosphere and oceanic crust as water flows into the new ocean basin. This is what happened at the break-up of Pangaea, which started about 200 million years ago. We can see some early stages of new ocean formation in various parts of the world today.

After the great north-south split through Pangaea had created the Atlantic, there was no further tectonic activity on the diverging margins of the continents. Spreading continues to this day along the median axis of the Atlantic – except between Greenland and Canada – but the zone of the initial rift is marked by the junction between oceanic and continental crust on the split-off margins of each continent. Such margins, where tectonic changes practically ceased once they were well and truly separated, are termed "passive" continental margins. They exist on both sides of the Atlantic, all around southern and eastern Africa, excluding the Red Sea area, and around India and southern Australia. Passive margins tend to be relatively free of large mountains, which, by contrast, are likely to rise on the "leading edge" of continents with "active" margins, (where, in other words, plates are converging instead of drawing apart). This is why there is usually a gradual slope from the great mountain ranges across the continents to the passive margin. The world's biggest rivers – the Mississippi, Amazon, Ganges, Niger, Congo, and Indus – all flow in this way. Over tens of millions of years, many

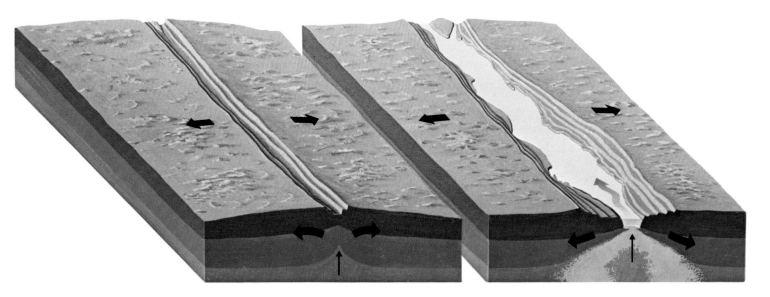

kilometers' thickness of sediments has accumulated on the sea bed around their gigantic deltas. Such enormous deposits are ideal locations for the formation of petroleum.

To see what the first stages of continental rifting look like, we can observe what is happening today in the Red Sea area and eastern Africa, where plate boundaries are diverging. Eastern Africa is bowed upward in a broad arch, about 1,000 kilometers across, which is slowly splitting at the crest to form a linear rift valley, and numerous volcanoes have developed along the line of rifting, presumably because of the molten mantle upwelling beneath. There are also many lakes in the valley floor, and coarse sand and gravel are washing into them from the valley walls. As the rift widens, the floor sinks. Eventually it will sink below sea level, but oceanic crust may begin to form even before this happens – a stage already reached in the Afar region of northern Ethiopia. Then, as the margins spread farther, they will cool and subside, no longer buoyed up by the heated mantle material, permitting the sea to flow in.

The Red Sea itself is an example of the next stage in development of an ocean. During this phase, while the ocean is only 100 kilometers or so wide, it may be blocked by the influx of sediments and by the final stages of volcanism associated with continental splitting and plate divergence. When this happens, the water dries up for a period, precipitating enormous thicknesses of salt, before continued spreading of the plates opens a further gap and the ocean floods in again. This happened in the Red Sea some 10 million years ago (a short time in the geological calendar); and it probably also happened 110 million years ago in the southern Atlantic, where gigantic domes of salt are found on the continental margins. The southern portion of the Atlantic Ocean at that time may well have looked like a sinuous prototype of the modern Red Sea.

Death of an Ocean

Where plates collide instead of diverging, the edge of the lithosphere is swallowed back into the mantle. The cold descending slab slopes down at an angle of about 30° to a depth of as much as 700 kilometers (causing earthquakes over the so-called Benioff Zone region), and the manner of the subduction depends on the way the two plates meet and on the type of material on each side. There are many possible patterns, of course, but the four basic ones are these: oceanic crust can meet oceanic crust; oceanic crust can meet a marginal sea bordering a continent; oceanic crust can meet continental crust; or continental crust can meet continental crust.

Where oceanic crust meets oceanic crust, it seems to be a matter of chance which is subducted and which overrides the other. For example, along the Tonga-Kermadec Trench in the southwestern Pacific the so-called Pacific plate plunges under the Indo-Australian lithosphere; but immediately to the west, in the Solomons and New Hebrides basins, the Indo-Australian plate is subducted under the Pacific lithosphere. In both cases long chains and arcs of volcanic islands are erupting above the descending slab, because the crustal rocks melt as they descend.

The Marianas Trench, with the Philippine Sea between it and Asia, provides an interesting example of the second type of subduction region. Here upwelling magma from the melting oceanic crust is trapped between Asia and the advancing Pacific plate. The floor of the marginal back-arc sea is thus highly active and tends to form local crustal spreading centers that force the crust out again toward the Pacific. This process is not yet fully understood, but it plainly cannot go on for very long in geological terms; sooner or later the advancing oceanic plate must sweep up the island arc and marginal sea floor and crush it against the continent, or else the resistance to ad-

vance will become so great that a new trench and subduction zone will form farther back in the Pacific Ocean. Both processes seem to have occurred in the past, since an observant and informed scanning of a good map suggests that curved chains of islands and mountains have been progressively accreted against the margin of Southeast Asia, while multiple island arcs, marginal seas, and trenches also exist offshore.

The third type of subduction zone is an area where oceanic lithosphere plunges straight under a continent. This occurs along most of the west coast of South America, where what has been named the Nazca plate is subducted, and also along the coast of the northwestern United States where part of the Pacific plate descends. In these circumstances the volcanoes and earthquakes associated with the Benioff Zone are to be found within the mountain ranges of the Andes and the Cascades.

Subduction always destroys oceanic crust, and so it represents a diminution of the area of an ocean. The final extinction of an ocean occurs when one continent meets another. This happened when India collided with Asia 30 million years ago, and it is on the verge of happening with the present approach of Africa to Europe (''verge'' meaning ''within a few million years,'' of course). The ocean that existed along the southern margin of Eurasia at the time of Pangaea is called by geologists the Tethys (for the mythical daughter of Heaven and Earth). India, Saudi Arabia, and Persia eliminated part of the Tethys as they collided with Asia, but remnants of the ancient ocean remain in the Mediterranean, Black Sea, and Caspian. If the present rate of relative movement of Africa toward Europe continues, the Mediterranean too will be gone in less than 50 million years.

An ocean dies when oceanic lithosphere that is being subducted under a continental margin carries a great land mass along with it, so that the two continents collide.

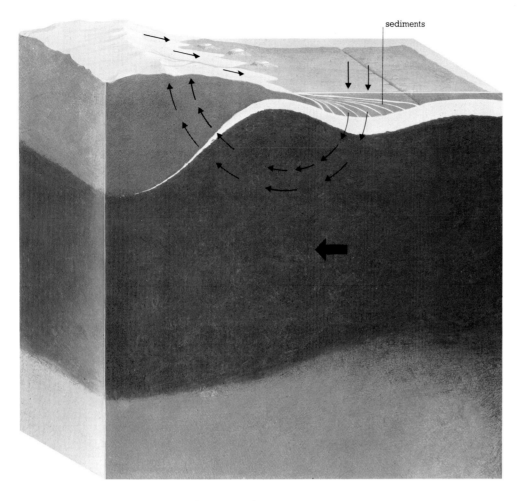

sediments

The Marine Sediment Cycle

Just as there is a cycle of mantle rock becoming oceanic lithosphere and returning once more to the mantle as the plates spread from mid-ocean ridges and their edges are subducted in ocean trenches, so there is a quite different cycle between continental lithosphere and the sediments that cover the ocean floor in most places. The best way to see how this works is to concentrate first on the results of erosion, temporarily ignoring the sediments created by either biological processes (the sinking of billions of skeletons of marine creatures to the sea floor) or the precipitation of salts from sea water by evaporation.

Picture a continent composed of igneous rocks of the lighter, granitic type constantly subjected to the destructive forces of ice, wind, rain, frost, and running streams and rivers. The result is an unending flow of rock particles down to the sea and out to the deep-ocean floor – a process that continues on passive margins for millions of years. As a given continental mass is eroded away, it loses weight, and the thick continental crust of light rock floats upward on the dense rock of the upper mantle (in obedience to what we have already defined as the principle of isostasy). Meanwhile, as the sediments that are washed into the sea pile up on the continental margin where oceanic and continental crust meet, their weight gradually depresses this zone. Thus, although the continents are continuously being eroded away, the process is extremely slow, for the continent keeps rising higher while the neighboring sea floor subsides and new sedimentary rocks are formed.

A different process, but with a similar effect, takes place on active margins. Here the sediments on top of the oceanic crust are continuously carried into the trenches formed by converging plates. But the sediments cannot be subducted into the mantle because they are too light, and so they accumulate on the *underside* of the edge of the continental plate as they are scraped off the descending slab. Because the sediments are lightweight, they serve as a kind of buoy to uplift the leading edge of the plate – a situation that prevails in the region where the Nazca plate goes under the west coast of South America, causing the uplift of the

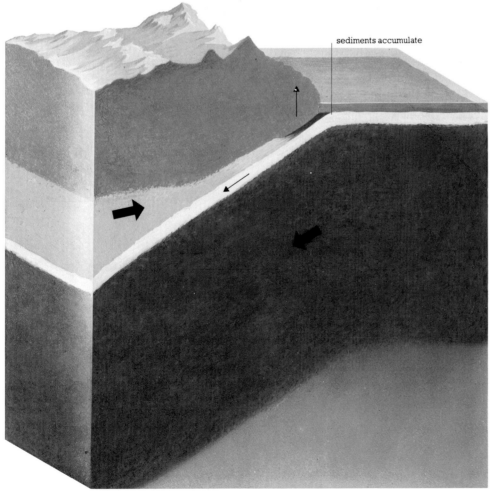

sediments accumulate

Above: On a passive margin, sediments eroded from the continent build up to enormous thicknesses and extend far out into the ocean. The huge weight of such accumulations gradually depresses the underlying oceanic crust, while loss of weight causes the continent to rise.

Left: On an active margin, sediments on the advancing oceanic lithosphere are scraped off on the underside of the continental plate, whose edge is uplifted by the accumulation of lightweight material.

This coarse-grained sedimentary-rock core, showing a matrix of grains of various sizes bonded together by pressure, heat, and chemical action, is about 200 million years old.

Andes Mountains. It is also happening underneath the nonvolcanic islands that form in front of the volcanoes in island arcs. After many millions of years, such islands tend to be swept back against the nearest continent, thus completing the sediment cycle by renewing the land.

In short, rocks eroded from igneous masses in one area are eventually returned by plate movements to another area of the continents as sedimentary rocks. This explains why most continents have a core, or central "shield," of very old igneous rock, surrounded by a broad fringe of younger sedimentary rock and mountain ranges mostly composed of sedimentary rock from the sea floor. Where two continents have collided, as in the collision of India and southern Asia, which created the Himalayas, sedimentary rocks are trapped in the center of the resultant land mass (as, indeed, are small pieces of oceanic crust that were uplifted in the violent convergence).

We can now add a final word about the biological and precipitated sediments. Calcareous and siliceous deposits are formed on the sea floor by the ceaseless rain of skeletons from planktonic animals; and in shallow marginal seas coral reefs and sea shells contribute to the thick beds of chalk and limestone that may eventually be uplifted onto a continent. Then too, when sea water is warmed, calcium carbonate precipitates out and settles onto the bottom, as is happening today in places like the shallow banks of the Bahamas. If an arm of the sea is completely closed off – as the Red Sea and Mediterranean have often been during the last few million years because of movements of the neighboring plates – all the water eventually evaporates, leaving huge thicknesses of solid salt on the floor.

History of a Revolution

The story of the plate-tectonics theory is a tale of hesitantly poised incredulity that has gradually changed to almost universal belief as a result of steadily accumulating evidence. The way that continents seem to dovetail across the Atlantic and the apparently related worldwide chains of mountains and earthquakes have long suggested the possibility of continental shifts in position. But scientists could not be expected to believe something for which there appeared to be no conceivable mechanism and no hard evidence – and it was not until the 1960s that oceanographers and marine geophysicists finally produced an explanatory theory and evidence to support it.

In 1912 the German meteorologist Alfred Wegener published a reasoned case for the movement of continents, basing his argument on an unprecedented reconstruction of Pangaea. As supportive evidence, he discussed the distribution of similar fossils in southern Africa, South America, India, and Australia; an early ice age that demonstrably covered parts of those same land masses; geological similarities between Africa and Brazil; and, of course, the fit of the coastlines. Developing an already ac-

We owe our present comprehension of plate tectonics and sea-floor spreading to the brilliant deductive and exploratory work of a number of geologists and oceanographers. As this abbreviated list of names and achievements indicates, the basic theory was developed in less than two decades, during the 1950s and '60s. Current efforts are largely aimed at testing and refining the theory.

1950s:	Evidence of the movement of the continents relative to one another accumulated from studies of palaeomagnetism, but the mechanism cannot be explained.
1952–60:	Magnetic stripes on the sea floor, with large offset fault patterns, are progressively mapped in the northeast Pacific, but no explanation is proposed.
1953:	Bruce Heezen (U.S.), Lamont-Doherty Geological Observatory, and John Swallow (Britain), National Institute of Oceanography, independently discover sections of the Atlantic Median Rift Valley.
1956:	Maurice Ewing (also of Lamont-Doherty) and Bruce Heezen predict that the whole earth should be encircled by the mid-ocean ridge system passing through every ocean.
1959:	Bruce Heezen and colleagues propose a system of topographic provinces to describe the ocean floor.
1960s:	Sir Edward Bullard (Britain), Cambridge University, and colleagues, produce progressively more refined computer reconstructions of the fit of continents making up Pangaea.
1960:	Bruce Heezen proposes that the mid-ocean ridge is being torn apart.
1962:	Harry Hess (U.S.), Princeton University, and Robert Dietz (U.S.), U.S. Naval Electronic Laboratory, suggest model of mantle upwelling under the median rift valley, with sideways transport of the lithosphere and subduction in ocean trenches.
1963:	Fred Vine and Drummond Matthews (Britain), Cambridge University, suggest that the earth's polar reversals account for the magnetic stripes on either side of the Carlsberg Ridge in the Indian Ocean; and they propose a general model of mantle upwelling under all ocean ridges, sea-floor spreading, and magnetic stripes caused by polar reversals.
1965:	J. Tuzo Wilson (Canada), Toronto University, explains the geometrical nature and movements of transform faults.
1967–68:	Dan McKenzie (Britain), Cambridge University, develops models of plate movements on a sphere.
1968:	Deep-Sea Drilling Project begins to obtain the information from cores that demonstrate the age of the crust at hundreds of locations on the ocean floor.

cepted geological belief that the lightweight continents ''float'' on the denser rock beneath them, Wegener suggested that the continents have drifted through the oceanic crust as discrete entities.

During the 1920s Wegener's work became well known throughout the Western world, but most reputable scientists were unconvinced. By 1928, in fact, the notion of continental drift had been so thoroughly discredited that a symposium of the American Association of Petroleum Geologists formally termed it ''impossible.'' From then on Wegener's suggestion was virtually forgotten except by a few open-minded geologists, particularly in South Africa, where supportive evidence for a former physical connection with South America was especially plentiful. In the 1950s, however, palaeomagnetic studies of the various continents showed that they must somehow have moved in relation to the North Pole, and oceanographic data

bearing on the subject began to pour in.

Wegener's idea of the mechanism of continental drift, with land masses moving through the rock of the upper mantle like floating ships, was incorrect, but he was on the right track. We now know, thanks to the brilliant work of a number of geophysicists and oceanographers, that although the continents themselves are probably embedded in the thick plates of the lithosphere rather than ''floating'' discretely, they have indeed moved according to a pattern very similar to the one Wegener suggested. His work was of crucial importance to the progress of modern geology, for his concept of Pangaea started a revolution in thought that will not reach its climax until we are someday able to explain not only the movements of the plates, but the forces that drive them.

This simplified drawing of a hypothetical area of ocean floor illustrates different ways plates are believed to move in relation to one another.

The Greatest Show on Earth

The most awesome of all worldly spectacles, as we have begun to see it in our mind's eye, is the dynamic activity of the earth's surface as gigantic plates slide and grind against and past one another and the continents embedded in them are torn asunder and collide to the accompanying phenomena of earthquakes, volcanoes, and ''tidal'' waves (tsunamis). Fortunately for us, it is happening so slowly – in spite of an occasional catastrophe – that we have just begun to appreciate that it happens at all, let alone to unravel the details. The world as seen on our time scale is like a single frame of a movie that has played for over 4,000 million years. Only when we imagine the whole film speeded up do we see the wonder of it.

Right: The directions of movement of the plates. Lengths of arrows indicate probable relative speeds.

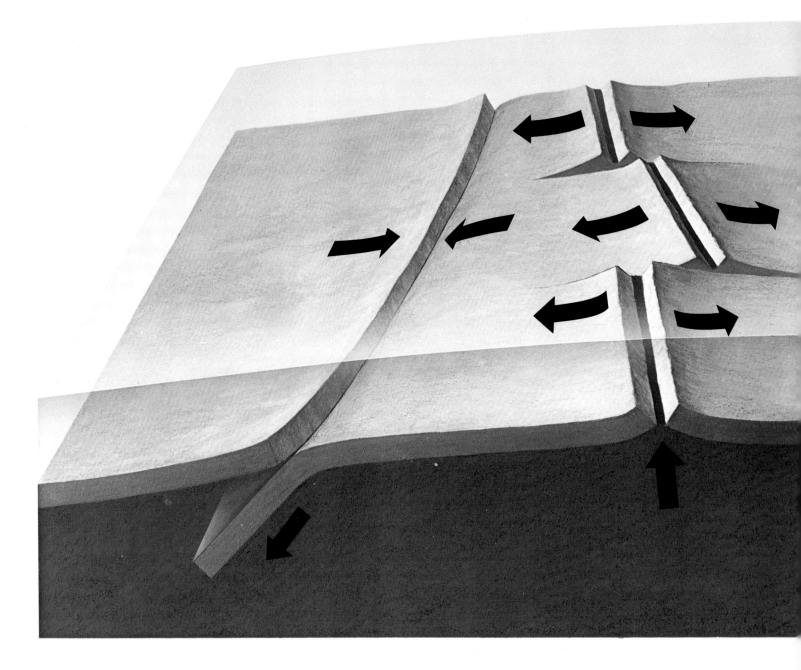

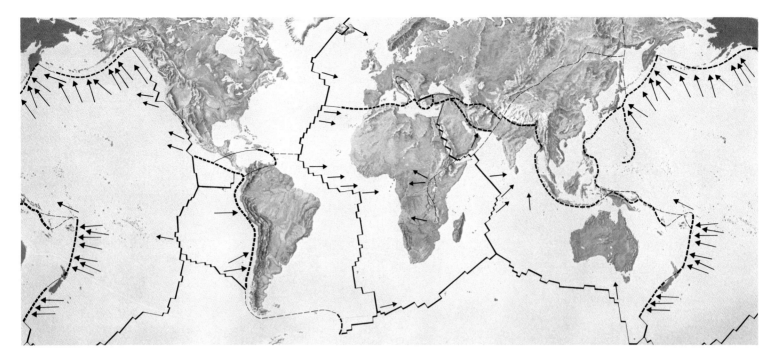

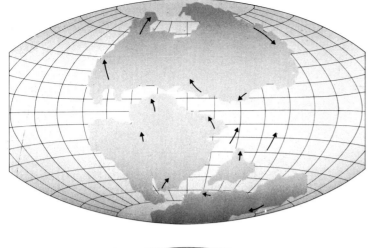

About 200 million years ago Pangaea begins to break up. South America and Africa remain joined, India is separated from Antarctica.

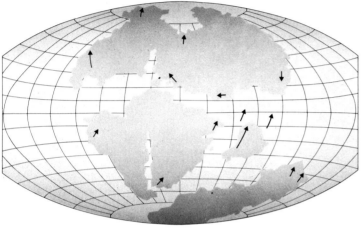

135 million years ago North America, Eurasia, Africa, and South America are beginning to assume familiar shapes. India is drifting toward Asia.

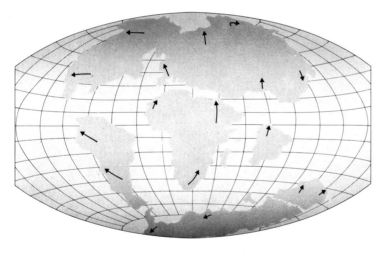

65 million years ago Australia is beginning to detach itself from Antarctica. The gaps between North and South America and India and Asia are closing. Fragments of northern Africa are about to collide with central Europe.

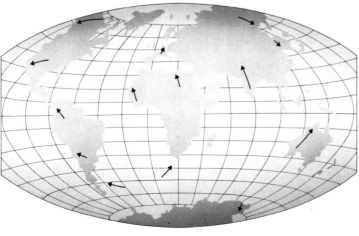

Today's map shows how far the old components of Pangaea have traveled. A few million years from now will see the break-up of eastern Africa.

Geologists use a million years as the basic unit of time, and this has the psychological effect of making the time scale manageable and speeding up the movie. But if we can struggle with a kind of double-think, to appreciate the monumental slowness as well, the show becomes even more impressive. The moving plates have probably been in action ever since the crust first formed, but our part of the performance can be said to open 225 million years (m.y.) ago, with Pangaea all in one piece, surrounded by water. Antarctica stands almost in its present position, with South America, Africa, India, and Australia packed around it, North America crushed in against Brazil and west Africa, Europe and Russia curving northwestward from the Canadian-Greenland coast all the way to the region of the present North Pole. There is a big gap – what geologists now call the Tethys Sea – between the parts of Pangaea that have become Africa and Asia. Several fragments of continental material that now comprise Italy, Greece, Turkey, and parts of the Balkans and Saudi Arabia are on the south side of the Tethys, still joined to Africa.

About 200 m.y. ago the first rifts appear in Pangaea. North America splits away from west Africa and Brazil as the whole North American-Eurasian block (sometimes called Laurasia) begins to rotate clockwise as a unit. A united Antarctica and Australia splits off and moves slightly southeastward, while India, no longer attached to Antarctica, starts to move northward at about 5 centimeters a year. By 135 m.y. ago India is halfway across what we now call the Indian Ocean on the road to collision with Asia. The North Atlantic is widening, and a crack splits Europe from America and Greenland from Canada. South America starts to slide away from Africa, opening a "Red Sea" type of rift in the south but with western Africa and Brazil still attached together.

By 65 m.y. ago Australia is splitting away from Antarctica. The Atlantic has grown to a broad ocean, but northern Europe has still not broken away from Greenland. The northern edge of Africa is beginning to break up, with Saudi Arabia coming into collision with Iran and continental fragments being transported across the remnants of the Tethys – or proto-Mediterranean – to collide with central Europe, where they will ultimately form the Alps, Carpathians, and the Dinaric and Caucasus mountains. About 56 m.y. ago Greenland breaks away from Europe; 20 m.y. later India collides with central Asia to create the highest mountains in the world.

Today the Atlantic is still widening across its entire span from pole to pole; the Pacific is contracting; Africa is rotating counterclockwise, moving northward toward Europe; eastern Africa is splitting up; and Australia is moving northward into the western Pacific.

Acoustics and the Sea Floor

In all ocean-floor experiments we are probing for answers to two basic questions: how deep is the oceanic crust at each point, and what is it made of? Since we cannot go down to see for ourselves, we get many – but by no means all – answers at long distance from instruments and equipment that permit us to identify certain facts about sediments and rocks by means of sound waves projected through the water down to, and through, the floor. The simplest and oldest such tool is the echo-sounder, which projects sound pulses vertically downward and measures the time it takes for the reflected pulse to return to the surface.

Echo-sounders tell us much of what we want to know about the topography of the sea bed – its ups and downs, so to speak. A more complex technique for ascertaining subsurface *structure* is seismic refraction, a process in which sound waves caused by the detonation of several tons of explosive are propagated through water, down through rocks, horizontally along rock layers, and then up to hydrophones at the surface. Because the speed of sound is different in different rocks, the sound waves are bent, or refracted, in various patterns that indicate the density, rigidity, and porosity of the materials through which the man-made earthquake has passed. Such experiments have shown that the average speed of sound in unconsolidated sediments (layer 1) is about 2 kilometers a second; in the lavas originating in the median valley (layer 2), about 5 kilometers a second; and in the dense rocks that cool in the magma chamber under the mid-ocean ridge (layer 3), between 6 and 7. Below layer 3 – that is, in the upper mantle beneath the Moho – the very dense and rigid upper-mantle rocks produce a sound velocity of about 8 kilometers a second. So, realizing that the heavier and less porous a rock the faster sound travels through it, geophysicists can calculate the thickness and probable composition of each layer.

Seismic refraction is a slow process, for it requires the detonation of a number of very large explosive charges to cover only a tiny area of the ocean floor. Obviously, we could learn a great deal more a great deal faster through a continuous repetitive source of moderately high-energy sound, and since the early 1960s geophysicists have been using a number of acoustic devices that produce a steady sequence of high-energy sound pulses to penetrate the sea floor. With such sound sources, a research ship can move continuously on a programmed track, and reflections from different layers in the sea bed can be monitored. Echo-sounding is also being improved, with emphasis on side-scan sonar, which measures distances vertically and obliquely to the sea bed.

Heat, Magnetism, Gravity

Acoustic measurements give us a fairly general "view" of the structure of the oceanic crust. We can fill in some further details by making three other types of measurement: the rate of heat flow from the center of the earth out through the crust, the strength of the magnetic field in a given portion of the sea floor, and the local strength of gravity.

Heat flow up through the crust and the sediments would naturally be greatest where the mantle is hottest, the crust and sediments thinnest, and the rocks composed of the most conductive materials – except for one enormously influential factor: the heat produced in the crust itself by the continuous decay of radioactive minerals. We calculate the flow of heat through the crust at a given point by means of sensitive probes mounted on the side of a core barrel driven down into the sediments, and we have thus learned that average heat flow through the sea floor appears to be about the same as that of the continental crust. This finding provides further support for the plate-tectonics theory. Since oceanic crust is much less radioactive than continental crust, you might expect much less heat to flow from it; extra heat, however, comes from the fact that the oceanic crust is thin and the mantle under the ocean is both hotter and more radioactive. On the mid-ocean ridges the heat flow is particularly great because of the intrusion of hot magma from the asthenosphere.

We have already mentioned the importance of magnetism, but it is worth stressing that the magnetic field due to local surface rocks is very weak compared with the magnetic field generated in the earth's core, and so the geophysicist uses a magnetometer to detect a very small local anomaly – that is, a small difference from the average – whenever it occurs. The banded, striped pattern of anomalies found in the ocean is strictly limited to the oceans; nothing like it has been found on the continents.

Gravity measurements at sea are also concerned with anomalies – with, that is, departures from average worldwide values. All matter, as even the most amateur physicist knows, exerts a gravitational pull proportional to its mass and inversely proportional to the square of its distance from the measuring point, and so any variation in the density of rock from one place to another is reflected in a change of gravi-

There are two ways of using high-energy sounds to penetrate the sea floor and reveal rock strata. *Seismic refraction* (A) uses explosions to project sound into the rock, where it is refracted horizontally and back to the surface. The more rapid *reflection system* (B) projects sound pulses vertically through the strata and records the reflected echoes.

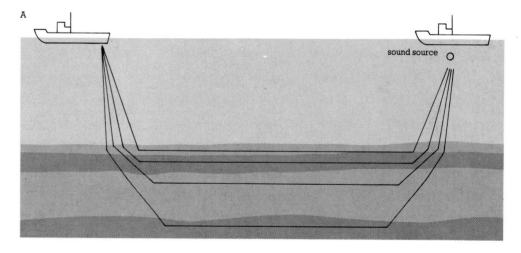

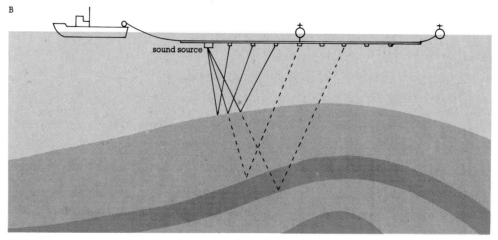

tational pull. Where there is a lot of dense rock near the surface, the gravity anomaly is high (or, in other words, the pull is stronger than the worldwide average); where there is a lot of light rock, the gravity anomaly is low (i.e., the pull is weaker than average). Thus, if the dense rock of the mantle is close to the surface at a given point, the anomaly should be high. And, in general, this is exactly the situation over the deep ocean, where gravity values are indeed higher than on continents. Similarly, high gravity anomalies are found in back-arc spreading areas, such as in the Aegean, where mantle material apparently wells up behind the island arc.

Because the oceanic crust is three-layered, a measurement of local strength of gravity does not in itself indicate which thickness of which layer is producing the anomaly. But gravity meters used in conjunction with acoustic equipment can sometimes do it. The importance of gravity for the student of crustal structure is that it can be measured continuously as a ship steams on a given course, while the thickness of layers can be determined at specific points by seismic refraction to provide a check. Thus, the structure of the layers over a wide area can be mapped quite thoroughly.

Project FAMOUS

Sea-floor spreading at the mid-ocean ridges seems to make sense on the grand scale of global plate tectonics, but how does it work in detail? The concept of a continuously widening crack in the earth's surface, filled with a river of red-hot rock that remains molten although in contact with cold sea water, is impossible to accept. Therefore, the slow yielding and cracking of crustal rocks near the surface must work in such a way that fissures are forced open by flows of magma, and faults slip, in a discontinuous, small-scale, piecemeal manner; yet these generally small events must be so frequent that they add up to the gradual creation of a new ocean floor on a global scale.

The only way to ascertain the probable mechanism of sea-floor spreading on this scale is to descend into the median valley of the mid-ocean ridge and examine the rocks of the floor and walls of the valley meter by meter. This would have been an impossible feat to perform before the advent of small submersible craft capable of descending 2,000 to 3,000 meters below the surface (see chapter 9). But from 1972 to 1975, teams of scientists from French and American research institutes, cooperating in what was appropriately named "Project FAMOUS" (for French-American Mid-Ocean-Ridge Undersea Study), were able to explore large parts of the median valley of the Mid-Atlantic Ridge. The area chosen for exploration, which was mapped out by sup-

The international FAMOUS project, to recover rock samples from the mid-ocean rift valley near the Azores, involved the French research ship *Noroit*, supporting two submersibles – the *Cyana* and the *Archimède* (left).
Below: *Cyana* is prepared for a dive.

The foredeck of the *Glomar Challenger* viewed from the drilling derrick. Several kilometers of drillpipe are laid out ready for use for deep-sea drilling and recovery of rock-core samples.

portive British and Canadian teams, was 400 kilometers southwest of the Azores, where preliminary topographic mapping had been done by precision echo-sounding, while the magnetic patterns were surveyed by means of an airborne magnetometer. A huge side-scan sonar capable of mapping a 15-kilometer-wide band of sea floor was operated from a British research ship, and remote-control cameras were towed backward and forward over the selected area. All the data recorded by such methods were combined with geophysical information to compile special maps so that American and French geologists in submersibles could find their way to the most significant points in the median valley.

During 1975 the undersea craft made numerous dives in the median valley and associated transform faults, enabling the geologists to motor along the foot of the steep scarps that border the median valley and to follow numerous fissures and fault cracks running parallel to the median axis. Using drills and coring machines that could be operated from within the submersibles, the research team obtained rock samples from points at selected distances from the axis and in varying relations to volcanic vents and fissures. Small hills on the floor of the valley turned out to be piles of pillow lava in bulbous tumbled heaps, with fallen rubble around the front of each lava flow. The lava, it was discovered, appears to be less than 10,000 years old and to lie right over the presumed axis of injection of new crust. All signs indicate that as the molten lava reaches the sea floor, it cools in contact with the water, and the skin of rock breaks again to emit another flow. This repetitive process creates the billowing pillow effect and the fallen fragments.

Between the lava hills and the walls of the median valley, the FAMOUS geologists found older lavas, about 100,000 years old — a find that seems to confirm the spreading from the axis. And explorations of the fracture zones of the transform faults revealed much older rocks than those of the median valley, as well as types of rock characteristic of much deeper layers of the earth's crust. Such findings are just what we could expect if the spreading mechanism operates in accordance with plate tectonics, and so the FAMOUS project provides significant corroboration of certain aspects of the theory. Most scientists have little residual doubt that the median valley is truly the place where new oceanic crust is being formed. But the success of Project FAMOUS is only a beginning. A great many questions remain to be answered.

While the theory of plate tectonics was being developed by marine geophysicists, geologists, and oceanographers in the early 1960s, it became increasingly clear that definitive evidence was needed to support it. Magnetic stripes are pretty convincing, but there could be other explanations for them. So scientists agreed on the necessity for conducting an experiment that might prove – or disprove – the supposition that plate movements have been going on for millions of years, and are still going on.

Any such experiment would naturally have to be on a grand scale, but the one proposed was simple in concept, though broader in scope than Project FAMOUS, which was to restrict itself to an examination of the ocean floor in a small area of the Atlantic. This experiment involved drilling many holes in widely separated parts of the floor in order to examine the thickness of sediments and to compare samples of rock from layer 2 (the extruded-basalt layer) of the oceanic crust. If it is true that oceanic crustal rock all over the globe spreads from the mid-ocean ridges, every bit of basalt in layer 2 must be of an age directly related to its distance from the mid-ocean ridge. This must be true not just in general, but, if the plates are rigid, with extreme precision and without exception everywhere on every ocean floor.

The project was certainly ambitious: in the early '60s no holes had ever been drilled in very great water depths and the offshore oil industry had not yet started to develop the technology for drilling beyond the continental shelf. There was some experience to build on, however. In 1957 geologists had worked out a plan to drill a hole through all three layers of the six-kilometer-thick oceanic crust in order to penetrate the Mohorovičić discontinuity and sample the material of the upper mantle, so as to resolve many uncertainties about the nature of the lithosphere. Rising costs caused this experiment – known as the Mohole Project – to be abandoned in 1966, but not before much spadework had been done on the central problem of how to get a drill bit down thousands of meters below the deep-ocean floor (a much less difficult problem, incidentally, for those who wanted to go only as far as layer 2, not all the way to the Moho).

The new project, sponsored by a group of American oceanographic laboratories and conducted from a specially built drilling ship, the *Glomar Challenger*, came to be known as the Deep Sea Drilling Project (DSDP). It has also been termed "the most successful experiment ever done." From 1968 to 1975 the *Glomar Challenger* drilled more than 400 holes during 44 cruises, or "legs," in every ocean. In one such leg, a series of holes were drilled at locations

that we have barely begun our study of the ocean floor and the movement of plates, and that many, many years of intensive research lie ahead. Our ideas about plate tectonics are based on a thin scattering of information. The great importance of the theory is that it can unite many previously unrelated observations – biological, geological, and oceanographic – into a coherent whole. But we still lack knowledge of too many basic elements in the picture; above all, as has already been emphasized in these pages, we have not yet found an adequate explanation for the driving force of the plates. Why has Africa progressively broken up at the front edge of the continent, with the broken pieces apparently moving more speedily than the rest? What is the significance of the so-called hot-spots – such areas of volcanic activity as the Hawaiian Islands, which seem to go on emitting lava even as the plates move over them? Those are just samples of unanswered questions. There are dozens more, and we are still forced to come to conclusions on the strength of only a few drillings and on topographic maps with great gaps in them. For instance, in the North Atlantic, the best-surveyed ocean, there are many areas where we have only one sounding for every 2,700 square kilometers.

In the years to come, improvements in acoustics will speed detailed mapping of the floor and seismic penetration of the sediments and sedimentary rocks. Various kinds of side-scan, multi-beam, and sector-scanning sonar will enable computers to plot out contour maps over a broad swathe on either side of a ship's track. Seismic-reflection techniques developed by the oil industry will be used for research on the continental margins as we try to find the best sites for research drilling in international programs. And we are sure to improve our technology for deep-ocean drilling to a point where we can explore in detail selected problem areas of plate tectonics on ocean ridges and passive margins, and in ocean trenches. All this lies ahead, but it will take time.

Fortunately, there is a strong commercial motive for all attempts to increase our understanding of sea-floor geology. Since so many of the oil and mineral resources that we exploit today, whether on land or under-sea, were originally formed in the sea bed, everything we learn about the conditions of formation can help us in prospecting. When we drill for geophysical research reasons at present, we must be very careful *not* to strike oil in deep water, for a well struck unexpectedly could not be closed off. But the more we learn from such cautious drilling, especially where it penetrates thick sediments, the more capable we become of pinpointing the probable location of untapped oil reserves.

A core of sedimentary rock drilled by the *Glomar Challenger* shows banding due to changing conditions as the rock was gradually consolidated. To guide decisions about further drilling, laboratory work is constantly carried on during the voyage. Here a geologist is analyzing a slice from the core.

whose age had been predicted on the basis of the mid-ocean-spreading theory and the magnetic patterns of the sea floor. Typical of the remarkable findings: on the basis of the plate-tectonics theory, it was predicted that the earliest sediment at the bottom of layer 1 at one site would be 38 million years old; in fact, drilling revealed the correct figure to be 39 million. Indeed, legs in the Atlantic, Pacific, Indian, and Antarctic oceans all confirmed the spreading theory.

In 1975 an International Phase of Ocean Drilling (IPOD), with many participating countries, was initiated. Its objective is to learn more about specific areas of the ocean floor, about continental margins, about sediments and fossils on the deep-sea bed, and, of course, about the many remaining mysteries of plate tectonics. IPOD has already borne fruit. During Leg 48 in 1976, holes were drilled in the region of the Bay of Biscay and Rockall Plateau. The rocks found revealed that swamps and coral reefs had fringed the continent of Europe 130 million years ago during the widening of the North Atlantic. As the margin sank into a depth of 3,000 meters, thick layers of sediment accumulated, which may well contain oil.

Looking to the Future

We can think of the oceans' future in two quite different time scales, geological and human. On the geological scale – in terms of tens of millions of years – if we assume that plate movements will continue in their present directions, we can predict with some assurance that the Mediterranean will close completely, the island arcs of the Far East will continue to be uplifted, Australia will move farther out into the Pacific, and the Atlantic will widen. In perhaps several hundred million years, all the continents will have reassembled, and the Pacific will have been eliminated. Such changes will not only rearrange geography but will undoubtedly cause vast changes in climate and in biological evolution.

On the human time scale, we must admit

2
The Water Itself

Jens Meincke

It was not until the middle of the nineteenth century that a number of bold expeditions into the arctic and antarctic regions enabled us to fill in the last gaps in our knowledge about the global distribution of land and sea. Those expeditions ended the phase of surface mapping that constituted a necessary first step toward a scientific understanding of the oceans. As a result of exploratory probings that had begun in the fifteenth century with the Portuguese and Spanish transoceanic voyages, we knew by the mid-1800s that our planet's surface is mainly oceanic: 361.1 million square kilometers, or 71%, water; and only 147.9 million square kilometers, or 29%, land. In fact, as we have learned in our own century from colored photographs taken in outer space, the earth looks from a distance like a "blue" planet.

Even in the so-called "land hemisphere" – the hemisphere with the highest proportion of land – 51% of the area is covered by water. (If you twirl your globe a bit, you will see why the pole of the land hemisphere is located near the estuary of the river Loire in northwest France, whereas the half of the earth known as the "water hemisphere" has its pole near New Zealand.) Over 90% of the water hemisphere is covered by water; and as for our familiar Northern and Southern Hemispheres, the amounts of water coverage are 61% and 81% respectively. Because the processes that govern the effects of incoming solar radiation are very much different for water and for solid earth, the asymmetric distribution of land and water over the globe is a highly important factor in determining climatic conditions.

The continents and island arcs divide the global water surface into roughly four sub-areas, which, by international agreement, are identified as the Atlantic, Arctic, Pacific, and Indian oceans. Where no natural boundary separates these ocean areas, we distinguish them by means of such obvious geographical lines as meridians, shortest distances between capes, etc. Each of the great oceans is further subdivided into a main part and its adjacent seas, and we classify those small seas according to their most obvious features. For instance, we speak of bodies of water like the North Sea and the Bering Sea, which are located along the margins of continents and have wide connections to the open ocean, as marginal seas. Or we identify seas that are enclosed by different continents as intercontinental – the Mediterranean is an example – and those that are embedded in a continental landmass, such as the Baltic or Hudson Bay, as intracontinental. This method of classifying the many "adjacent" seas helps us to understand some of the reasons for the wide range of such characteristics as temperature and biological productivity that determine their economic potential.

Sea and land are very unevenly distributed on the globe, with over 90% of one hemisphere and about half the other covered by water. The orientation of these so-called water and land hemispheres is clearly visible in this diagram.

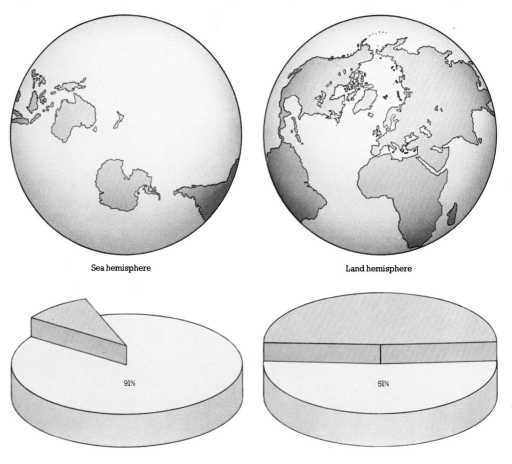

Sea hemisphere · Land hemisphere

91% · 51%

Our present knowledge of the varying depths of water in the world's oceans is based on a vast number of soundings that began to be systematically registered back in 1853 under the direction of an American naval officer, Matthew Fontaine Maury. In those pre-electronic days, when depth was tested by a length of piano wire paid out from a stopped vessel, each sounding took a very long time, and so it is not surprising that by 1900 the total number of deep-ocean soundings did not exceed 6,000. Today a research vessel may produce as many as 20,000 soundings a day by means of echo-sounders, which automatically record how long it takes a signal emitted from a vessel in motion to hit the bottom and be reflected back to the surface. The enormous increase in soundings has made it necessary to divide the oceans into several sounding areas, with various national hydrographic offices responsible for registering incoming depth information for their designated areas.

Even so, however, although we now have detailed depth surveys for most coastal zones, the topography of very few oceanic areas has been fully charted. To realize the immensity of the problem, consider a fact that has already been emphasized in the first chapter of this book: many parts of the ocean floor slope down more steeply than the highest mountains on the continents. To measure depths accurately in such places would require a succession of sounding lines with distances of only a few hundred meters between neighboring lines – an unrealistic objective for the whole world ocean. Still, we do know the *major* topographic features of the sea floor, and there is no reason to believe that surprising discoveries of spectacular new depths lie ahead.

The deepest spot we have found, as the reader surely knows by now, is southwest of Guam in the Marianas Trench – 11,022 meters. But it should be remembered that in only 1% of the total oceanic area is the water deeper than 6,000 meters. Depths between 3,000 and 6,000 meters prevail in 76% of the oceans, and in nearly a third of the remaining 23% the water is no more than 200 meters deep. The mean depth, however, is quite startling: 3,729 meters. Thus, there is so much water in the oceans' basins that if the globe's surface were leveled, the entire earth would be covered with a 2,430-meter-deep layer of water. Yet, although that figure may seem enormous, it becomes nearly negligible on a global scale; assuming, for the sake of relativity, a globe with a diameter of 40 centimeters, the mean ocean depth would be barely one-tenth of a milli-meter – the thickness of a page or two of this book. So the ocean is, in fact, like a thin skin on the earth's heavy body. Moreover, if we take the general horizontal dimension of a "typical" ocean basin as being 20,000 kilometers and its mean depth as 4 kilometers, it is clear that an ocean is about 5,000 times as wide as it is deep. Water movements in such a thin sheet of water must, therefore, be dominantly horizontal, with relatively minimal vertical movements. This does not mean, of course, that vertical currents can be neglected, as we shall see in the following pages.

Three of the four great oceans – the Atlantic, Pacific, and Indian, with a mean depth of some 4,000 meters – account for 89% of total sea area. Smaller seas may be remnants of older oceans (e.g., the Mediterranean), island-arc seas (the Caribbean), or flooded parts of the shelf (the Baltic).

	Area million km²	Mean depth m	Max. depth m
Pacific Ocean	166.24	4,188	11,022
Coral Sea	4.79	2,394	
South China Sea	3.68	1,060	
Bering Sea	2.26	1,492	
Sea of Okhotsk	1.39	973	
Yellow & East China Seas	1.20	272	
Arafura Sea	1.04	197	
Sea of Japan	1.01	1,667	
Atlantic Ocean	85.62	3,736	9,219
Caribbean	2.75	2,491	
Mediterranean Sea	2.51	1,502	
Gulf of Mexico	1.54	1,512	
North Sea	0.58	93	
Black Sea	0.51	1,191	
Baltic Sea	0.38	101	
Irish Sea	0.10	60	
Indian Ocean	73.43	3,872	7,455
Red Sea	0.45	538	
Persian Gulf	0.24	100	
Arctic Ocean	9.49	1,330	5,449
Hudson Bay	1.23	491	
Baffin Bay	0.69	861	
World oceans	361.13	3,729	11,022

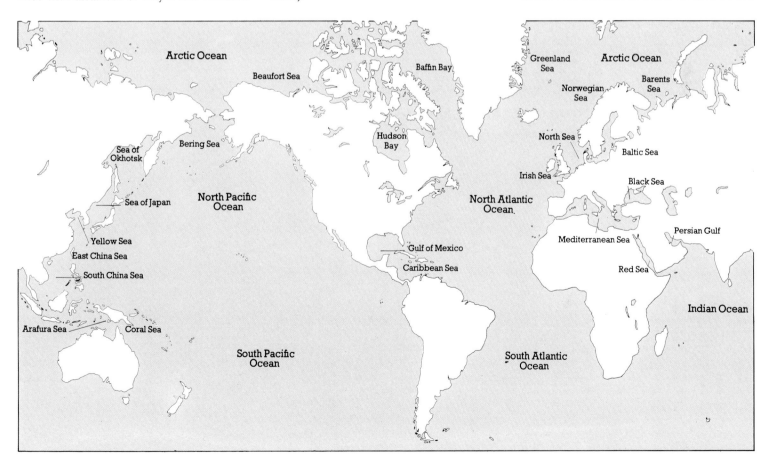

The Ocean Basins

The opening chapter of *The Undersea* has given the reader a detailed view of the form and structure of the ocean floor as we believe it to have evolved through the ages (and to be still evolving in a never-ending process). We cannot understand the behavior of the ocean itself, however, without constant references to the morphology of the basins in which the water lies. And so, at the risk of a small amount of repetition, I shall now summarize very briefly the essential facts about the contours of the floor – but without revisiting the fascinating territory of plate tectonics that has already been fully explored in chapter 1. During the course of the next few paragraphs it should become increasingly evident that just as the form of the ocean basins is not unchanging, so the volume of liquid water in the basins is by no means constant.

There are two dominant levels to the relief of our planet's surface. One, representing the normal surface of the continents, lies within a few hundred meters above sea level; the other, known as the "deep-ocean floor," lies at depths of between 4,000 and 5,000 meters below the surface of the sea. This second level, covering more than half the earth's surface territory, is divided into distinct basins by two major topographic features: continental margins that form its side walls, and mid-oceanic ridges that form partitions within it. These ridges, found in each of the four main oceans, are a kind of submarine mountain range that is very much larger, if not higher, than anything on the continents. Rising two to three kilometers above the deep-ocean floor, they are about 1,000 kilometers wide, and they form a single connected range extending for 60,000 kilometers throughout all oceans. As we discuss deep-sea currents and their effects on the renovation of deep water in the ocean basins, we shall begin to see the oceanographic significance of the ridges.

The continental margins constitute a zone of transition from ocean to continent. The continents do not usually end at the coast; they extend out below sea level, in some places for hundreds of kilometers. Thus, it is not technically correct to speak of the continental shelves and slopes as part of the true ocean floor, even where they lie under many meters of salt water. The true ocean floor is the deep-ocean floor, which, as I have said, is walled in by the continental margins. The margins themselves have the same rocky foundation as the continents, but their rocks are covered over with eroded and weathered material that has been carried down to the sea from the continents, chiefly by rivers.

Continental debris on the shelves may be as thick as six kilometers. And the shelves, which sometimes stretch for up to 300 kilo-

meters from the coast, may drop down to 200 meters below sea level. They tend to be fairly flat and ledge-like because surface waves and shallow-water current activities work toward an even distribution of eroded material. In many places, of course, the shelves are far too deep to be influenced by surface waves. But this was not always so, their relatively flat topography dates back to various glaciation periods, when so much water was imprisoned in glaciers that the sea level was a good deal lower than it is now.

At the seaward boundary of the shelves we find the steeply descending continental slopes, which are cut into at many places by deep, narrow canyons. The canyons have been caused by the frequent avalanche-like slumping of portions of slope material as the result of either earthquakes or top-heavy local accumulations of sediment. When such submarine catastrophes occur, huge amounts of suspended matter are carried far out into the oceans, and these "turbidity currents" can develop speeds of up to 25 meters a second. One of the rare documented cases of the destructive energy of turbidity currents dates back to a November day in 1929, when one such current did great damage to several deep-sea telegraph cables on the continental slope south of Newfoundland.

Beyond the slopes and on either side of the mid-oceanic ridge lies the vast area of the deep-ocean floor, whose most striking feature is the absolutely flat abyssal plain next to the continental margin. The abyssal plains are a direct result of turbidity currents; material that such currents have transported out into the ocean basins has sunk down and gradually formed a "blanket" of sediment that hides the original topography of the floor.

Thus, on a large time scale, the flow of eroded material from the continents into the oceans has drastically changed the surface configuration of the basins. And there is another process that induces locally important changes in the height of the sea floor: because the flow of material from the continents alters the load on the earth's crust, and because the crust is somewhat flexible, it compensates for load changes in one place by either sinking into or rising out of the earth's underlying mantle in another. For example, while the North Sea shelf has been sinking at a rate of 0.7 millimeters a year because of a heavy sediment load, central Scandinavia has been lifted more than 250 meters over the course of 12,000 years as a result of the removal of the weight of ice that melted at the end of the glacial epoch.

Since the coastline is where sea level and land meet, sinking or rising motions of either sea bottom or land cause changes in coastal configurations. But such variations

can also be brought about by changes in the volume of water in the oceans. The melting, for instance, of about 360,000 cubic kilometers of land ice would mean a one-meter rise in mean sea level. Thus, with the present volume of land ice amounting to some 30 million cubic kilometers, there would be a sea-level rise of roughly 85 meters if all the ice were melted. During past ice ages, the volume of land ice has been as high as 70 million cubic kilometers, and at such times the sea level has been around 110 meters lower than it is today. Right now, in fact, land ice is melting at a rate that creates a sea-level rise of 1.1 millimeters a year.

The combination of both types of change – the height of sea floor and land, and the volume of liquid water in the oceans – can lead to considerable withdrawals or advances of the coastline. This may not strike the reader as a very pressing problem, but consider for a moment its effect on the Netherlands. Assuming present trends to remain fairly constant, the Dutch coast will

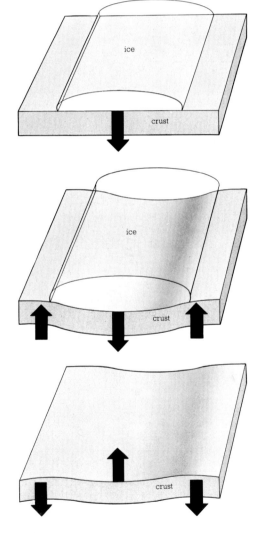

Freezing and melting icecaps, which may be several kilometers thick, not only remove water from and add it to the ocean; they also bring on local changes in the height of the coast through the compensation process of isostasy, as illustrated above.

sink by 18 centimeters and the sea level will rise by 33 centimeters during the next 300 years. That will mean a vertical change of the coastline by 51 centimeters – a substantial and dangerous amount. It is by no means too early for regional authorities everywhere to begin long-term planning for coastal areas that will be affected by this change in the shape of the sea floor. In the Netherlands, extensive areas of cultivated land will be flooded unless the already existing system of dikes and locks is immensely enlarged and improved.

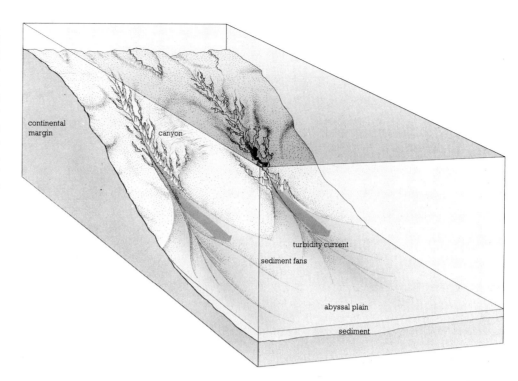

Right: How turbidity currents sweep continental debris out to the deep-ocean floor. Earth tremors or the weight of sediments on the continental slope can cause underwater avalanches, which cut deep canyons in the slope, and the force of the slump sets up a swift current that bears off great quantities of suspended matter. No turbidity current has been photographed, but the picture below, taken at a depth of 40 meters in a canyon off the California coast, shows sand running downward like a waterfall.

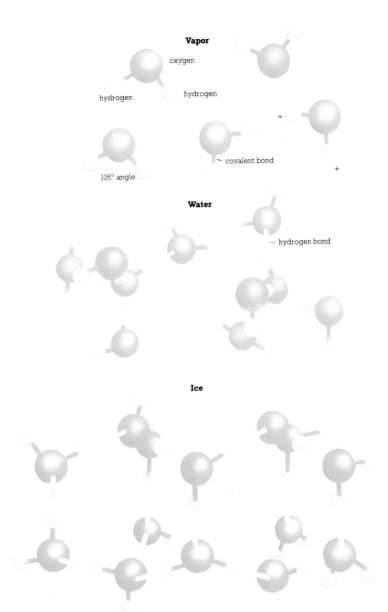

Vapor

oxygen

hydrogen   hydrogen

~ covalent bond

105° angle

Water

— hydrogen bond

Ice

The properties of both sea and air are partly determined by the physical characteristics of water. Left: In the vapor state, molecules of water move about freely and independently; in liquid water they are packed much more closely together and are attracted into groups by their oppositely charged sides; in ice they are set in a rigid framework, which is actually less dense – i.e. weighs less per unit volume – than liquid water. Above: The floating ice of the Arctic Ocean, through which the U.S. nuclear submarine *Whale* is thrusting its bow and conning tower, was formed thousands of years ago at the North Pole. Snow falling on the surface and new sea ice from Canadian and Russian continental shelves maintain the frozen covering. Such ice floats could not form on the open ocean today.

Water – Fresh and Salt

Chemists have discovered that a water molecule is constructed asymmetrically – that is, both of the positively charged hydrogen atoms join the negatively charged oxygen atom on the same side of it instead of on opposite sides. This makes water a "dipole" – that is, negative charges dominate one side of the molecule, positive the other. Consequently, several molecules coming together attract each other and can easily combine their oppositely charged sides to form molecule clusters. Depending on temperature, such clusters, which are not irregular, assume various geometric shapes of different density (i.e., of different weight). These elementary physical facts give us the clue to an understanding of the unique properties of water that lead to its key role in the drama of life on earth.

Consider a freshwater lake in the dead of winter. What property of water keeps the lake from freezing down to its bottom and thus killing all life within it? Answer: the way that its density changes in response to tem-

perature changes. As with most natural substances, water gets heavier and heavier when cooled, but, unlike other substances it does not increase its weight at temperatures below 4°C. Further cooling increases the proportion of less densely packed molecule clusters, with the effect that the weight is diminished again. Thus, when freezing starts (at 0°C), ice is mainly built up by the least densely packed molecule clusters, and, being lighter than water, it floats. This is an uncommon property of water, for most substances are heavier in the solid than in the liquid phase. As a freshwater lake cools in winter, dense surface water sinks to the bottom until the lake as a whole has reached a temperature of 4°. Further cooling results in the formation of a less dense floating surface layer, which, when it finally freezes, cuts off any further heat exchange between the atmosphere and the water below the ice, and inhibits further cooling.

Another unique property of water is due to the unusually strong coherence of its molecule clusters – a coherence that arises from the electrical attraction of one dipole

to another. It takes a lot of energy to heat water, and, therefore, cooling water releases large amounts of heat. That goes, too, for its solid and gaseous phases (ice and steam). In fact, water can store about ten times as much heat as soil or dry air can. And so the vast quantity of water on earth – the only planet that has it – keeps atmospheric temperatures from being killingly variable. The moon, for example, has daily temperature shifts of 280°C, as compared with extreme changes from day to night of only 40° even in such hostile earthly regions as deserts. Moreover, water and water vapor are mobile and can actively distribute heat to even out imbalances. This is particularly important for our climate, since major oceanic and atmospheric flows transport the heat surplus from lower to higher latitudes, and so keep global temperature differences relatively low.

So far I have been discussing pure water, but the properties of salt water are not much different, though it contains on the average 96.5% of pure water and 3.5% (35 parts in a thousand) of salts. There are a few

This diagram shows the concentration of some 40 chemical elements in a given quantity of sea water. Easily the largest component is H_2O itself, but more than 70 different elements have been washed into the ocean by rivers or blown in as dust, or else have been emitted by submarine volcanic action.

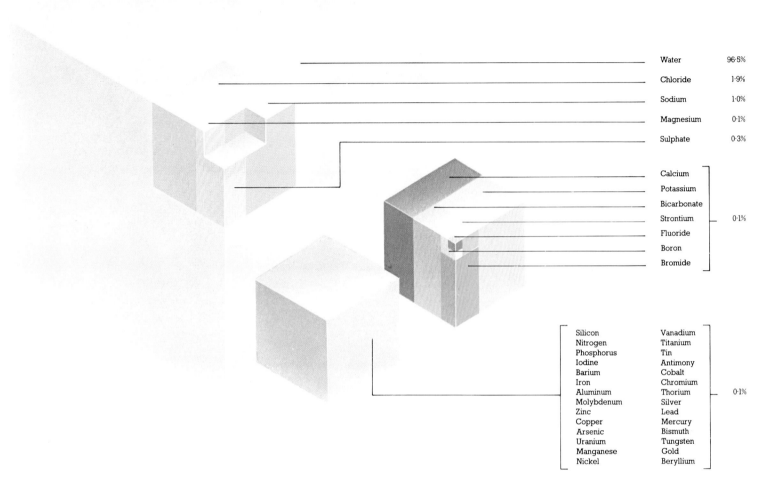

Water	96·5%
Chloride	1·9%
Sodium	1·0%
Magnesium	0·1%
Sulphate	0·3%

Calcium	
Potassium	
Bicarbonate	
Strontium	0·1%
Fluoride	
Boron	
Bromide	

Silicon	Vanadium	
Nitrogen	Titanium	
Phosphorus	Tin	
Iodine	Antimony	
Barium	Cobalt	
Iron	Chromium	
Aluminum	Thorium	0·1%
Molybdenum	Silver	
Zinc	Lead	
Copper	Mercury	
Arsenic	Bismuth	
Uranium	Tungsten	
Manganese	Gold	
Nickel	Beryllium	

differences, though, that increase the critical importance of water to our planet. Take, for instance, the freezing of sea water. The temperature for its maximum density is much lower than 4°C; and the saltier it gets the lower the temperature at which maximum density is reached. Its freezing point is also lower – so low that most sea water actually freezes *before* reaching the maximum density it could attain. Theoretically, then, the ocean is unlike a freshwater lake in that it could freeze – given cold enough weather – from surface to bottom. In reality, however, its large heat content prevents ice formation. When it is being cooled by cold air, sinking surface water is constantly replaced from below by warmer water, which releases its heat into the atmosphere and warms it up. It is merely in shallow areas like the wide shelves of the Arctic Ocean that ice can develop; and once it is there, it floats on the surface and prevents the rest of the water from freezing. Only one deep oceanic area – the central Arctic – is completely covered by ice. The ice there dates back to worldwide glaciation periods and is so thick

that under present conditions it cannot melt during the short arctic summer.

What about the salt itself in sea water? Well, for one thing, there is enough salt in all the oceans to cover the continents with a layer 150 meters thick. Then, too, all the 92 natural chemical elements are probably included in the ocean's salinity (we have so far found about 70 of them). Analysis shows, though, that more than 99.9% of salinity is accounted for by 11 major components, and we group all the others together as "trace elements." Throughout the geological ages, the various salts have been washed into the oceans by rivers that carry sediments from continental rocks. The fact that the 11 major components are found in constant ratio to one another all over the world's oceans indicates a thorough mixing of ocean waters; it is no wonder that the sea has been called "nature's great washbowl." Trace elements, however, do not show a constant ratio of composition, for they are directly involved in the marine life cycle. Phosphate, nitrate, and silicate, for example, are much depleted in sunlit surface water, where they

are used up by floating plants (the phytoplankton). In the deeper layers without any plant productivity, they exist in much higher concentrations.

But although the concentration of trace elements in the water is low, it may be very high in plant and animal tissues – with important consequences for man. For example, a kilogram of dry seaweed contains five grams of medically important iodine, whereas the same amount is normally dissolved in a hundred tons of sea water. There is a dangerous side to this organic enrichment of substances that exist only in low concentrations in the water. In 1974, for instance, several thousand tins of northeast Pacific tuna had to be withdrawn from the market because of extreme concentrations of mercury in the fish. The mercury came from the waste waters of American paper-production plants, which had discharged the wastes at concentrations presumed low enough not to harm the environment. Unfortunately, that presumption had failed to take into account the ability of marine organisms to enrich trace elements in their tissues.

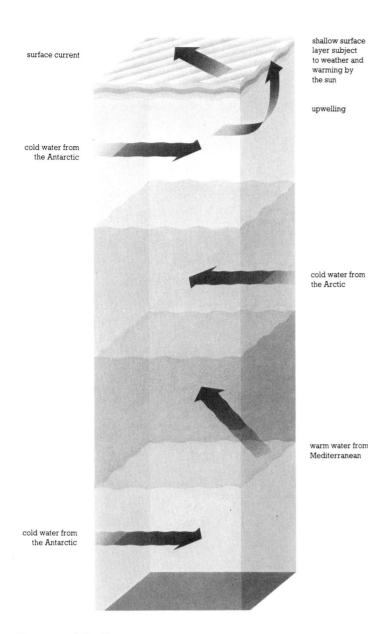

surface current

cold water from
the Antarctic

cold water from
the Antarctic

shallow surface
layer subject
to weather and
warming by
the sun

upwelling

cold water from
the Arctic

warm water from
Mediterranean

Layers of the Sea

If asked to estimate the average temperature of all the water in the sea, most of us would probably suggest a much higher figure than the correct one: 3.8°C. Obviously, what we are familiar with is the distribution of temperatures at the oceans' surface, and it is a fact that very low surface-water temperatures are found in polar regions only. Average readings of less than 3.8° are restricted to latitudes north of 67°N and south of 57°S; the remaining 90% of the oceans' surface has an average of 19.4°. Why, then, does the ocean as a whole maintain an average of only 3.8°?

Knowledge about the temperatures of subsurface ocean waters dates back to the eighteenth century, when scientists brought up deep water in wooden buckets with a lid that closed automatically when the winch shifted from lowering to hoisting. Three deep-sea temperature determinations were made by this means when the *Resolution* under Captain James Cook sailed round the world from 1772 to 1775. More accuracy

Contrasting layers in a column of water can be pictured in such models as this one. Each layer has a different salinity and temperature, and most are moving in different directions. We use various measuring instruments to build up actual profiles (see facing page). The tube above is for

became possible in the late nineteenth century, with the development of a thermometer whose mercury thread could be kept in a fixed position throughout the long haul up from the depths by simply reversing the thermometer. With this mechanical instrument, which is still in use today, a rapid survey of deep waters became possible. It soon revealed a sandwich-like stratification of the water masses. Except in polar and subpolar areas, warm and highly saline surface layers lie on top of various cold layers, which get colder and less saline as depth increases. Since decreasing temperatures make each layer denser than the one above it, the layers rest stably upon one another. Here is an example of typical vertical distribution in the central Atlantic:

From the surface down to 50 meters the temperature is 27°C. Then in successive

launching an expendable bathythermograph, which records depth and temperature continuously while falling on the end of a fine wire launched from a moving ship. At a predetermined depth the wire and the instrument are both jettisoned.

layers there is a sharp fall to about 10° at 400 meters and 5° at 900. Thereafter, the decrease is very slow, although various layers are still identifiable, until near the bottom, at depths greater than 5,000 meters, the temperature sinks to 1°. Salinity ranges from 36.5 parts a thousand at the surface to 34.9 near the bottom, with a layering parallel to temperature. Such stratification is highly interesting, especially in deep, cold waters. You might expect that after a certain time all differences would be eliminated by gradual mixing; instead, stratification is always a feature of the sea – and so, somehow, there must be a constant renewal of the individual layers. Hence the answer to our initial question about the maintenance of generally low temperatures in the oceans must involve an explanation of the formation and persistence of layering.

Temperature

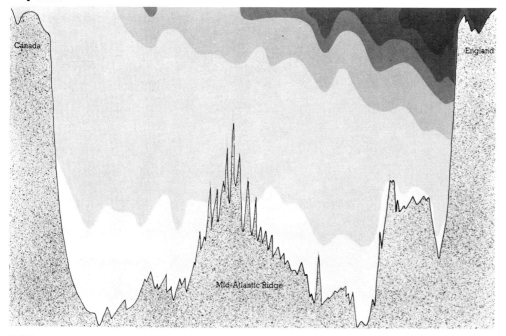

Salinity

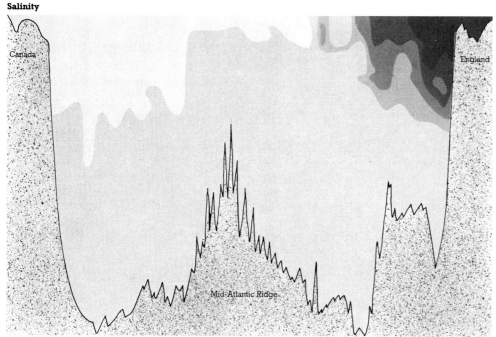

These profiles of temperature and salinity across the Atlantic were built up by means of oceanographic thermometers. Note the low salinity of the cold Labrador Current and the high salinity of the warm North Atlantic Drift.

53

Let us, therefore, look at the distribution of strata along a north-south section through the Atlantic Ocean from the Arctic to the Antarctic. In general the strata are horizontal, but north of 50°N and south of 40°S they bend more and more upward, until they are nearly vertical. The coldest deep-water layers come right to the surface within the polar seas. There, cold as the water is, it is warmer than the air, which cools it further, thus increasing its weight and making it sink before it has a chance to freeze. It is replaced by warmer water rising from below, which is again cooled and sinks, and so on. Since the degree of cooling determines the depth to which the water finally sinks, polar-sea surface waters, which naturally become the heaviest of all water masses, fill the deepest regions of the world's oceans. Water masses formed in subpolar regions

such as those southeast of Greenland are not quite so heavy, and therefore sink only to intermediate depths. In subtropical and tropical zones, where warm air constantly heats the surface layer of water, the warm lightweight water floats atop underlying water masses, where it prevents the atmospheric heat from penetrating to lower strata. Anyone who has swum in warm water on a sunny day knows the sensation of sticking the feet down into what feels like icy depths below the surface. In the open oceans there is a layer of sharp temperature-decrease known as the "thermocline" at depths between 30 and 80 meters.

Thus we have an answer to our question about why the ocean is so cold. It is heated and cooled at its surface only. Since heating creates an effective barrier against the downward propagation of heat, warm

waters remain in surface layers. Cooling at the surface, however, draws more and more water into the cooling process. And so cold water is continuously being produced in polar and subpolar areas, while warm water tends to sit rather statically on the surface in the tropics and subtropics. Finally, the frigid polar water spreads from its northern or southern place of origin into the interior of the oceans at a fast enough rate to preserve the characteristics of deep layers all over the ocean basins.

Not only heating and cooling but also the process of evaporation can change the density of surface waters. The Mediterranean has a salinity of up to 40 parts a thousand as a result of the high rate of evaporation. Because of this, the surface water becomes so dense that it sinks down, fills the deep Mediterranean basin, and finally flows into the Atlantic through the Strait of Gibraltar in the form of a "salt-water injection." It can be traced as a warm, saline layer at a depth of 1,200 to 1,500 meters all the way down to the equatorial Atlantic.

One last question must be answered here: how long does it take for a particle of water to complete the cycle from surface to depth, horizontal spreading, and back again to the surface? Interestingly enough, the answer is different for the Atlantic and Pacific oceans. This is because the Atlantic has cold-water sources at its northern and southern boundaries, whereas the Pacific has cold sources only to the south, as a result of which the Atlantic's production of deep-water masses is significantly greater. This in turn means less oxygen in the Pacific depths than in the Atlantic. Oxygen can be absorbed into the water only during surface contact with the atmosphere, but it is constantly being used up at all depths by sea-dwelling organisms. The relatively low content of oxygen in Pacific deep waters indicates a lack of frequent exchanges of oxygen-depleted and atmospherically replenished waters. It probably takes about 900 years for a water particle to sink, spread horizontally, and mix back up to the surface of the Pacific, as against an estimated 500 for the Atlantic.

Because of the long periods of time between two successive contacts of deep water with the atmosphere, the oceans may be said to serve as a kind of "climatological memory" of the earth. Climatic changes that occur over periods of several hundred years are softened, so to speak, by three closely related big facts about the oceans: the very long time that water particles "reside" in the depths, the extraordinary capacity of water to store heat, and the enormous volume of deep-ocean waters.

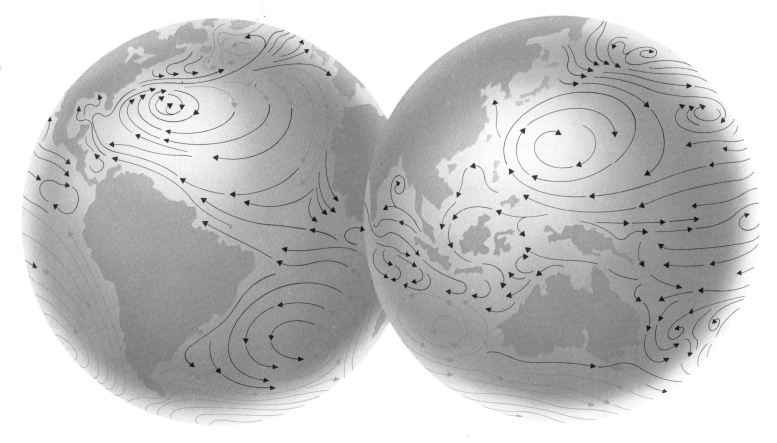

Circulation Systems

There are two main types of current in the world's oceans: surface-layer circulation, which can and must be fairly well understood for the sake of efficient navigation, and circulation within the depths, about which our knowledge is limited. Our present reasonably good knowledge of the distribution of currents in the oceanic surface layer is largely a by-product of the routine determinations of ships' positions by many generations of seafarers who have rightly assumed that the difference between the position predicted from a ship's heading and speed and the actual position fixed astronomically is caused by such currents.

The major part of the surface circulation in each ocean consists of a complex circular pattern of currents extending from the equator, or near it, either northward or southward. The Atlantic provides an outstanding example: the Gulf Stream, which gets its name as well as its deep blue waters from the Gulf of Mexico. After passing through the Straits of Florida, the Gulf Stream flows parallel to the continental slope northward along America's East Coast at a maximum speed of about 9 kilometers an hour, and it transports 50 million cubic meters of water every second – 30 times the total river discharge into the world's oceans. North of Cape Hatteras, it veers away from the continental slope and broadens from a narrow band a few tens of kilometers in width to become the two-to-three-hundred-kilometer-wide, more meandering North Atlantic Current, which crosses the Atlantic in a northeasterly direction. While part of it penetrates into

the Norwegian Sea, the largest portion turns south to become the wide and sluggish Portugal and Canary currents; then, at the latitude of the Cape Verde Islands off the northwestern coast of Africa, the Canary Current joins the broad North Equatorial Current, which crosses the Atlantic toward the coast of South America. There it turns into the Caribbean, and finally closes the current circle by flowing through the Yucatan Channel into the Gulf of Mexico.

Few current circles are as easy to trace as the Gulf Stream, but they exist throughout the world's oceans – in every one, in fact, but the Arctic. The flows are clockwise in the Northern Hemisphere, counterclockwise in the Southern. The most prominent features of the hemispheric circles are the intense poleward flows along the western circumference and the weak flows toward the equator in the east. This west-east asymmetry is brought about through certain basic forces that act on oceanic flows. The driving force of surface currents is, of course, the winds, which create waves that, because of friction, decay into a net water drift. Under steady conditions, as in the trade-wind belts, the speed of a current reaches approximately 1% of wind speed. Once initiated, however, there are secondary forces that may affect the drift in several ways. One is the friction caused by neighboring water masses or the sea floor, which tends to slow the current down. Another is the deflecting force of the earth's rotation.

It was a French mathematician, Gaspard de Coriolis, who, in 1835, first pointed out that any particle moving relative to the earth's surface is deflected toward the right in the Northern Hemisphere and toward the

Surface currents, which are driven by the winds, consist principally of clockwise loops in the Northern Hemisphere and counterclockwise loops in the Southern. As shown in this map, local conditions can change the general pattern considerably, however. Red arrows denote warm currents, blue arrows cold.

left in the Southern Hemisphere. Although, on the average, the magnitude of the deflective force is only one-millionth of the gravitational force of the earth, its effect (which we call the "Coriolis effect") is significant for horizontal atmospheric and oceanic flows. In other words, the force of the earth's rotation deflects oceanic currents, like the winds, from the direction of their driving forces. This can be clearly seen, as it affects the atmosphere, in weather maps that show winds taking spiral paths around the center of low-pressure areas instead of going in directly. In the open ocean it has been found that surface currents in the Northern Hemisphere are deflected by 45 degrees to the right from the wind. The North Equatorial Current is a good example: it flows due west under a southwesterly trade wind.

Careful observations of winds and currents have led to the formulation of a model that quite adequately delineates the main surface currents of the world's oceans. The general picture reveals that surface circulation is chiefly driven by the wind system over the oceans, and that the wind-induced currents extend to a depth of 100 to 200 meters, with the exception of the intense currents along the western circumferences of some circles and around the Antarctic continent, which reach down to 1,000 meters. The hemispheric current circles are due to a combination of the Coriolis effect and the continental boundaries; and the difference between the intensity of eastern and western flows is largely due to the fact that the force of the Coriolis effect increases from the equator toward the poles.

Until the end of the nineteenth century we knew nothing at all about subsurface currents apart from the fact that the deep waters *are* in motion. Now we know that the velocity of such currents is only about a tenth that of surface circulation. And since in most places the wind does not directly affect water at a greater depth than about 200 meters, deep circulation is driven by less obvious forces. These appear to be, broadly, of two kinds, both of which originate on the surface and one of which is simply an extension of wind-driven surface currents. When those are directed against a coastal boundary or even against one another in the open ocean, the water within them may pile up to a point where local pressure anomalies are created in the depths, and the ocean adjusts to such irregularities by setting up compensating currents. The second type of force is due to climatic differences at the surface. When waters are cooled, as in subpolar regions, or evaporated by warmth, as in the Mediterranean, their densities increase and they sink. Thus, waters of different densities are mingled – an unstable situation adjusted by deep-water currents that redistribute the denser water underneath the lighter.

Such measurements of deep circulation as we have been able to gather indicate that the most intense flows have speeds of around half a kilometer per hour, and that these occur underneath the western-boundary currents of the surface. The direction that a deep-water current takes, however, is toward the equator – the reverse of the surface flow – and deep circulation does not generally develop ocean-wide current circles. One reason for this is the submarine ridges whose height serves as a barrier to the deep-water flow.

It would be wrong to think of surface and deep circulations as separate systems. They are, after all, closely related. Surface waters sink into deeper layers after cooling or evaporation. Waters from deeper layers rise to the surface when surface currents veer away from coasts or are forced by the winds to change course in the open ocean. The significance of this vertical type of circulation lies in the transfer of properties. Surface waters are rich in gases, especially oxygen, that are essential for decomposing organic matter that has sunk to the depths; deep waters are rich in the inorganic nutrients needed in the surface layers for sustaining life. So although the upwelling and sinking currents are not strong – their speeds are only a few centimeters a day – they are highly important for the health of the sea.

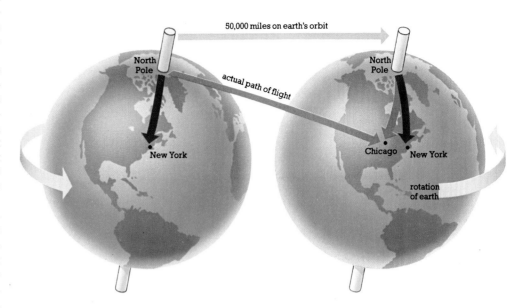

Since the days of sail, seamen have been systematically gathering information on winds and currents, whose curved paths, we now know, are due to a force known as the Coriolis effect. As the earth rotates, points on the equator travel faster than points near the poles, and so an object moving poleward tends to accelerate "ahead" of the earth beneath it, while an object moving from pole to equator gets "left behind." In effect, this means that objects moving in the Northern Hemisphere always curve to the right, and objects moving in the Southern Hemisphere curve to the left. As illustrated above, a rocket fired from the North Pole would have to be aimed at New York in order to hit Chicago. Similarly, a rocket fired from the South Pole would have to be aimed at Montevideo so as to come close to Valparaiso.

Two stages in the formation of an actual eddy in the Gulf Stream. Top: The current has meandered, and a 400-kilometer-long loop of cold water has developed in the current near an already existing eddy. Bottom: Four days later, the end of the loop, now isolated, has become a new eddy while the old one has diminished in size and strength.

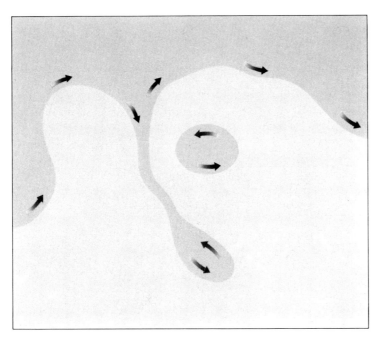

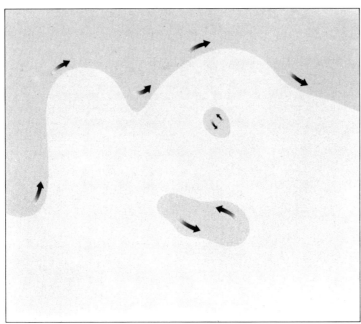

Undersea Weather

Ocean currents as indicated in a map of the seas look as if they are steady flows, with only small variations in space and time. This very wrong impression is inescapable, since the maps are based on average values from thousands of observations routinely recorded as part of the daily positioning of all ships. Any illusion of steadiness is quickly dissipated, though, when we examine an account of current speed and current direction as obtained over a period of months at a fixed position and consisting of hourly observations by electrical recording instruments. Drastic changes can occur within a single hour; moreover, there are daily and twice-a-day variations related to tidal flows, along with a whole class of transient fluctuations enduring within a range of from several days to

weeks. The significant feature of currents, in short, is not their steadiness but their variability, as evidenced in particular by their tendency to meander (i.e., to twist and turn for no apparent reason) and to form eddies (bodies of water that rotate within the main current flow).

Meanders and eddies are especially well documented for an area of the North Atlantic in which there is a "polar front" where warm water meets cold water, much as subtropical air masses meet subpolar air masses in a sharp transition zone at the atmospheric polar front of the middle latitudes. The polar front in the North Atlantic occurs about where the northern part of the Gulf Stream becomes the North Atlantic Current. Mariners recognize it as the "cold wall" on their approach route to the United States from Europe. It can easily be identified even by a landlubber, for the

color of the water suddenly changes from deep blue to greenish blue, and the surface temperature drops several degrees.

Oceanic and atmospheric fronts have two characteristics in common. First, they are maintained only as long as the water or air masses move relative to each other. The northern Gulf Stream, for example, running on the warm side of the "cold wall," moves alongside, though much faster than, the adjacent cold water; but – and this is the second characteristic that it shares with an atmospheric front – it does not hold its course steadily. In fact, meandering is the "normal" state of a front. Such boundaries are controlled by a delicate balance of forces resulting from differences of temperature and velocity between the converging air or water masses, the inertia of the masses involved, and the deflecting force of the earth's rotation. Even a small disturbance can turn the flow from its initial path, and once this happens there follows a serious imbalance, intensified by the fact that the deflecting force of the earth's rotation changes according to different latitudes. The out-of-balance forces then interact like a slowly vibrating spring, and the current begins to meander. People who live in the middle latitudes are familiar with this meandering motion of the atmospheric polar front, which happens whenever a series of low-pressure systems pass over an area. At such times, spells of cold and warm weather alternate swiftly, and individual meanders may grow so large that they are completely cut off from the front and drift independently as warm or cold eddies.

Today, satellites permit us to monitor oceanic eddies and meanders virtually continuously, with infrared sensors automatically mapping temperature contrasts such as those found along the "cold wall." As a result of studies of such maps, we are beginning to learn a lot about the basic features of the "normal" deviations of currents. Meanders, for instance, can be as much as a hundred kilometers wide, with distances of several hundred kilometers from curve to curve, and they can advance at a rate of between 10 and 50 kilometers a day (technically known as their "propagation speed"). As for eddies, many of them rotate in circles with 100-kilometer diameters; and we have discovered that a single eddy can exist for up to several months.

It should not be assumed, because I have been discussing meanders and eddies in relation to oceanic fronts, that they occur only in that context; they occur in all major flows. Just as the difference in velocities along a front will induce a meandering motion, so too will the flow of water across submarine ridges or the influence of rapidly changing weather conditions at the surface. It is because meanders and eddies are an oceanwide phenomenon that current rec-

Horizontal layering of water is most marked near the surface where the sun warms the water from the top, and the wind mixes it. Left: A pellet of falling dye leaves a disturbed trail in the wind-mixed layer, then kinks sharply as it descends into a stiller, cooler layer moving at a different speed.

Modern oceanographic experiments use thousands of readings taken almost simultaneously with calibrated instruments at various locations. The sampling mechanism below contains a modern electronic system for measuring water temperature, pressure, and conductivity and for transmitting findings to a recorder on the ship. Along with the electronic package there is also a set of traditional water-sampling bottles and thermometers, which are used for standardizing electronic results.

ords taken almost anywhere in the world reveal so many apparently inexplicable fluctuations. And a large part of the energy of oceanic flows is actually contained in the meanders and the eddies rather than in the straightforward currents (much as a rope or wire stores most energy when it vibrates). We do not yet fully understand the role of these energetic fluctuations in determining the behavior of the oceans, but they are doubtless a major factor in the process of transferring energy within the oceans' interior. For example, an eddy that drifts off the intense Gulf Stream will gradually transfer both energy of motion and heat content to its new environment.

In fact, meanders and eddies in the sea resemble closely what we speak of as the "weather" in the atmosphere – that is, its constantly changing conditions, with storms succeeded by quiet periods, and vice versa. Day after day, weather patterns in the air around us – patterns just like meanders and eddies in the water – hold and transfer most of the energy contained in atmospheric circulation. This holds equally true for oceanic circulation.

So the oceanic currents are more important for their fluctuations than for their on-course flow. And, in addition, they greatly influence the stratification of the water: fluctuations in the layering are immediately perceptible whenever eddies and meanders pass by. Another process, which causes vertical fluctuations of the layering, is associated with internal waves, which are simply an undulating motion of the subsurface layers. The size of such waves depends on the stability of the stratification – i.e., the vertical density differences; the *greater* the differences, the *smaller* the internal waves, for the layers are less easily

displaced. Thus, in the depths, where density differences are small, the waves tend to be very large, but internal waves become negligible at the surface, where there are great density differences between water and air. Like surface waves, internal waves can become so steep that they curl over and break. This is another way in which waters are mixed from one layer to the next – assisting the vertical transfer of properties.

Internal waves can be so enormous and powerful under certain conditions that they endanger submarine craft. In the Strait of Gibraltar, for example, at the interface between inflowing Atlantic water and outflowing Mediterranean water, both traveling at different speeds, internal waves can be as much as 250 meters high! The loss of more than one German submarine trying to dive through the strait during World War II has been attributed to this situation.

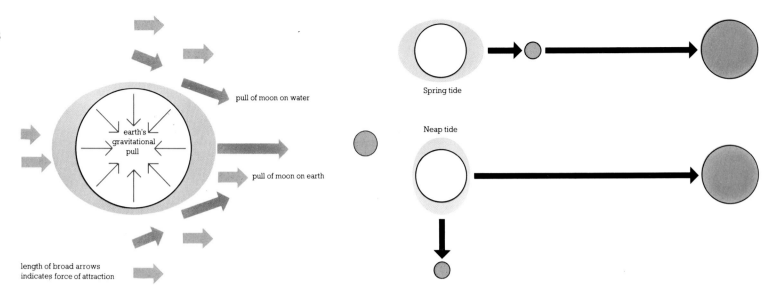

pull of moon on water

earth's gravitational pull

pull of moon on earth

length of broad arrows indicates force of attraction

Spring tide

Neap tide

Tides

Along many coastlines the tides are the most obvious cause of changes in sea level. The rise and fall of the sea and the relationship of the regular change of currents to the moon's orbit are familiar facts to people living along tidal coasts. The origins of our Western culture, however, were centered around the Mediterranean, which has extremely weak tides; and so documentation on tides was scarce until Alexander the Great had to face them in the Indus estuary, and Caesar around the British Isles, in 325 BC and 55 BC respectively, during their conquering campaigns.

The system of tide-generating forces is thoroughly understood – a fact not true for many other aspects of oceanography. Tides are generated by the varying gravitational forces of mass attraction between the earth and other celestial bodies – chiefly the nearby moon and the far-away sun. Astronomers have been able to compute tide-generating forces ever since Isaac Newton discovered the forces of attraction between two bodies. Superficially, the magnitude of the combined tide-generating forces of the moon and the sun would seem infinitesimal – less than a ten-millionth of the earth's own gravitational force. (That corresponds to an addition of about 0.01 gram to 100 kilograms.) The ocean nonetheless reacts because the moon's pull on the water at the side of the globe away from the axis joining the earth and moon is largely horizontal; and small though it may be, it is not counteracted at those points by earth's gravity, so that it does make a difference. The sun, which has less than half the moon's tide-generating force, produces similar horizontal pulls. And it is these horizontal forces that affect the oceans.

The distribution of the moon's tidal forces on a completely water-covered globe would generate one bulge toward the moon and another in the opposite direction. The fact that earth rotates underneath those tidal

Above left: The moon's attraction causes two sea-surface bulges – one facing the moon, the other on the opposite side of the earth – and the far-away sun has a similar, but lesser, effect. Right: How the combined forces of sun and moon produce high (spring) or low (neap) tides.

bulges explains the occurrence of two high and two low waters (semidiurnal tides) per "moon-day" of 24 hours 50 minutes. Beyond that, because the earth's equatorial plane is inclined relative to the plane of the moon's orbit, there is diurnal inequality – i.e., differences in elevation between the two high and two low waters. Finally, the sun's tidal forces enhance the moon's tidal bulges at times when the sun and moon are in line (*spring* tides, which occur at full moon and new moon); and they counteract the moon's tidal bulges during half moon (*neap* tides), when the sun and moon pull at right angles to each other.

All this would happen with perfect regularity if the earth were entirely covered with water. But since this is not so, the tides do not behave as neatly as all that. In various places there are varying delays between the coming of a spring tide and the full moon. There are also big differences in tidal amplitudes and phases over small distances, and there are, too, such things as purely diurnal tides (that is, only one high and one low water a day). All these features have been revealed through years, or even centuries, of continuous observations of tidal heights. For a long time, such observations were still available from a tide gauge established in Amsterdam in 1682. Since then, records have been obtained at approximately 3,000 stations on islands and along coasts; and ever since 1965, when it became possible to use bottom-mounted capsules for recording deep-water pressure changes due to variations in sea level, deep-ocean tidal heights have also been available. So we can now study an extremely complex chart of the oceans' real tides – and what we see is very different from what might be anticipated from a

simplified picture of the uniform global pattern of tide-generating forces.

Numerous models to explain the discrepancies have been developed. What they add up to is a fairly simple general observation: tide-generating forces produce a tidal wave in each ocean basin, and the periodicity of this wave is constant; but its propagation speed (the rate at which it moves), its direction, its amplitude, and its wavelength are influenced by the basin's geometry and the deflecting force of the earth's rotation. (Note my use of the term "tidal wave." This is its correct meaning. The popular use of "tidal wave" to describe a destructive wave caused by an earthquake is frowned on by oceanographers, who call any such wave a "tsunami," the Japanese word for "storm wave.")

The semidiurnal tide in the open ocean, with its period of 12 hours 25 minutes (half the lunar day), usually has a wavelength from crest to crest of several thousand kilometers and a height from trough to crest of only 50 centimeters. The water particles under the wave, moving slowly backward and forward as the tidal wave passes, generate currents that flow at speeds of about 0.4 kilometers an hour. But all these figures alter considerably when the tidal wave travels onto a continental shelf or into an adjacent sea. Since the shelf is relatively shallow, the wave piles up to a much greater height and the current velocity increases. In the adjacent seas there may be an additional important effect: as the tidal wave advances, it is reflected (i.e., forced back toward the entrance) by the inner portion of the large body of water; if it then reaches the entrance just in time to meet the next tidal wave sweeping in from the great ocean, the doubled force results in a much enhanced tide. There are many adjacent seas and bays that, because of their proportions of depth to length, are attuned to the timing of oceanic tides in this way, and we call their tides "resonant." The most striking example is the inner Bay of Fundy, an arm of the Atlantic between New Brunswick and Nova Scotia, where the sea level rises and falls as much as 14 meters at a time. The North Sea and English Channel also provide examples of resonant tides – currents with speeds up to 14 kilometers per hour (Pentland Firth) and rises of up to 10 meters (the Gulf of Saint-Malo).

The extremely weak tides of other adjacent seas such as the Baltic, the Mediterranean, or the Caribbean are explained by the fact that they are not resonant, or else that the connection to the open ocean is too narrow. In other words, because of the shape of their basins and other geographical features, either the natural time that it takes for a tidal wave to be reflected is so different from the oceanic tides that the incoming tide is canceled out, rather than amplified, by the outgoing one, or the energy provided by the oceanic tide is too small to set up co-oscillating motions of the adjacent basin.

Despite the complex tidal pattern, there is no problem in predicting tides. Once such characteristics as times of high and low water are known in relation to the moon's and the sun's phases at a particular location, the complete pattern can be worked out accurately by means of a forecast of the moon's and sun's orbit. This is the procedure followed by national hydrographic services in preparing tide tables.

The importance of a knowledge of tides can hardly be overestimated. Such knowledge is still essential for coastal and estuarine navigation, even though many ports now have lock systems that make them at least partially independent of the tides. In our current hunger for new sources of energy, much thought is being given to the possibilities for using tidal energy; as we shall see in chapter 6, it is far from easy to harness the tides, but it is already being done with some success in the Rance estuary near Saint-Malo in northwest France. But even if we neither know much about nor use the tides, they affect all of us. For instance, they may exert a strong effect upon regional climates. In areas with powerful tidal currents, the waters are continuously forced into motion from top to bottom, as they are around the British Isles. Thus, the summer warmth in such waters is evenly distributed throughout the water column, whereas only the upper layers are heated in more quiet waters. Consequently, water temperatures in regions with strong tides remain low during the summer, and this cools the air along the coast. In the autumn and winter, however, strong tides bring the heat content of deeper waters up to the surface, and so air temperatures stay relatively high. Thus, tidal stirring of a given portion of the sea can contribute appreciably to a softening of the annual variations in temperature.

Below: A small ocean tide interacts with continental shelf and narrow bay to produce a huge tide in Canada's Bay of Fundy. The photograph, taken at low tide, has a high-tide mark superimposed to show the 14-meter rise.

Above: Mont-Saint-Michel at the head of the Gulf of Saint-Malo – where Atlantic tides already amplified by the English Channel are further amplified – looks like this at low tide, but may be encircled by water at high tide.

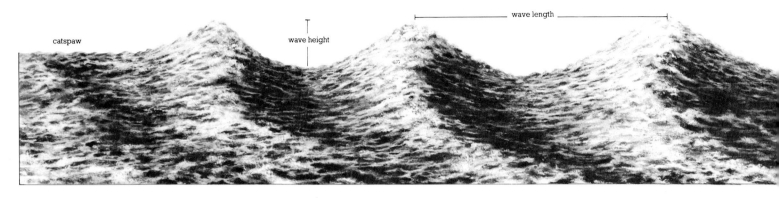

catspaw

wave height

wave length

Wind, Waves, and Storms

The sea's surface is seldom glass-flat. Almost always there are ripples or waves, varying in height from a few millimeters to several tens of meters in a storm. Common sense tells us that they are caused by the wind – which is true – and that the biggest waves are built up by the strongest and most persistent wind – which is also true. But there is another factor (which sailors call the "fetch"): the distance across the water surface over which the wind can blow while building up waves. The greater the fetch, the bigger the waves. And common sense confirms this, for we all know that a gale at sea creates gigantic white-crested breakers, whereas the same gale blowing across a narrow river for several days merely ruffles the water.

When a wind begins to sweep across a great expanse of calm water, the first patches of ripples, called "catspaws," appear as the wind speed reaches 2.5 kilo-

meters an hour. With increasing speed, the catspaws grow into increasingly steep waves, and the first crests start breaking at a wind speed of 13 kilometers an hour. If the wind persists or rises, waves appear that grow constantly higher from trough to crest and longer from crest to crest; they obtain their energy from the smallest waves, which become too steep and break when they get close to the crests of bigger ones. At any one time there is bound to be a mixture of waves of different heights and lengths, and so we speak of a "spectrum" of waves, just as we refer to the spectrum of visible light. After a while, depending on wind strength and fetch, the wind will have created the biggest waves – both highest and longest – that it possibly can, at which point the sea is said to be "fully arisen." In the past 20 years, oceanographers have devoted much effort to measuring the spectrum of developing and fully arisen seas in relation to wind speed, fetch, and wind duration. From an improved understanding of the resulting

The diagram above defines the terms used in studying waves and their characteristics and does not, of course, attempt to give a realistic profile of any portion of the sea. For instance, it may take many hundreds of kilometers for the winds to build a wave to a fully arisen storm state, and swells precede storms, retaining their energy over long distances.

relationships will come greater accuracy in our predictions of wave conditions at sea.

A wave on the surface does not simply raise and lower the surface, nor does the water actually move forward. In fact, as each crest passes, the water particles rise up, advance, descend, and move backward again, describing an almost perfect circle. And particles nearer the surface force deeper particles to move in smaller circles, so that the passing wave sets a whole column of water oscillating. After the wave has passed, only the energy has moved on; the water particles have completed their orbits and returned to their original positions. In practice, the depth to which a passing wave

Left: Underwater seismic activity caused this gigantic wave – a tsunami – which actually engulfed and drowned the unfortunate observer on the left of the picture. Tsunamis frequently follow undersea earthquakes.

Right: As waves approach the shelving shore, orbital motion under each wave becomes flattened to an ellipse, and the wave slows down. Crests pile up and eventually break to form surf.

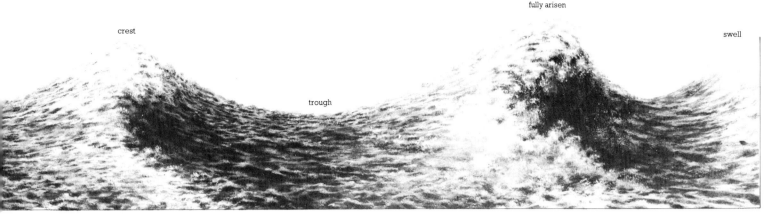

fully arisen

crest

swell

trough

sets water particles oscillating is taken to be half the distance from crest to crest at the surface. Thus there can be an interaction between waves and sea floor in water that is shallow in relation to wavelength. For waves traveling from deep water shoreward this has the following effect:

As long as the water depth is greater than half a wavelength, the oscillating movement of the water particles dies out in midwater and there is no friction from the bottom. We call such waves "deep-water waves." They turn into shallow-water waves when the water depth becomes less than half the wavelength. Then the bottom particle orbits are forced into a flattened elliptical shape, and the wave is therefore slowed down. As the wave runs toward land, its increasing interaction with the bottom brings the crests closer together; they pile up and finally break.

Even in deep water, the speed with which the crest of a wave moves – its "propagation speed" – depends on the distance between crests. Regardless of their height, waves with the longest wavelengths travel fastest. The longest waves move at about 55 kilometers an hour and thus can move out ahead of a rotating pattern of storm winds, for storm systems usually travel forward at only about 35 kilometers an hour. These long waves moving without a wind blowing over them are known as "swells," and the arrival of an onshore swell is often the sign of an approaching storm.

Apart from creating what we think of as "normal" waves, wind disturbs the water surface in two other ways, both of which are important, especially near coasts. First, a moving storm system is usually associated with low atmospheric pressure, which causes a slight rise in sea level, and the resultant widespread bulge in the water surface – known as a "surge" – moves with the storm. The length of such a surge may be several hundred kilometers; it may grow to a crest-to-trough height of two meters as it crowds into a shallow coastal sea; it may

take several hours to pass; and if it coincides with a high tide and rough sea, it can result in dramatic flooding of the coast. An even more important type of disturbance, called "wind set-up," is the direct effect of waves' being continuously wind-driven against the shore – a situation that can raise the sea level by as much as three meters.

Dependable prediction of wave energy and travel direction and of storm surges and wind set-up can make the difference between life and death for people working at sea or living near the coast. Only if we know the average and worst probable conditions in a given region can we build maximum safety into ship and oil-rig design; only if we can anticipate the arrival of particular storms can we prepare emergency defenses. Coming into increasing use in recent years have been automatic wind- and wave-recording buoys. Extensive networks of these buoys are already being planned for several intensively used stormy areas such as the North Sea.

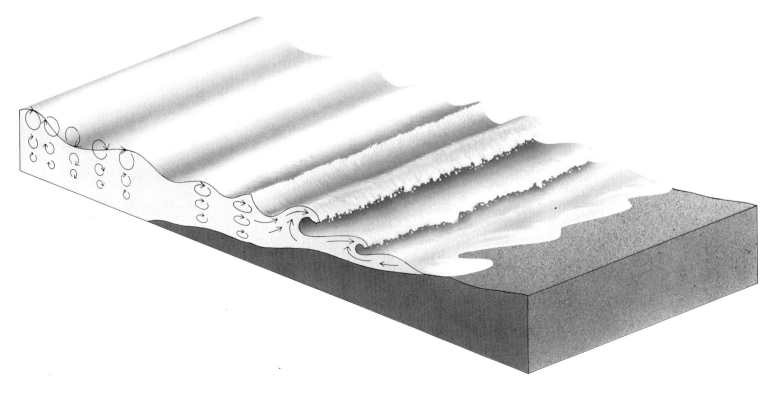

Sound in the Sea

Light and radio waves can penetrate only a few hundred meters into sea water, whereas sound penetrates the water very effectively and can be bounced off the floor and the underlying layers of rock. The sound may, however, be bent and distorted not only as it passes through different layers of rock (as we have seen in chapter 1) but also when it passes obliquely through the stratified layers of the water itself. This means that if you want to use sound for navigating or for locating a submarine, you must first calculate how the path of the sound will be warped by the layered structure of the water; on the other hand, if you are primarily interested in studying the oceanic layers, the warping will help you to detect them. Thus, a knowledge of how sound travels through water is essential for oceanographic research as well as military and commercial activities.

Whether in the atmosphere or in water, sound is not transmitted as a wave vibrating from side to side. Instead, it travels as a compressive wave, much as the successive phases of extension and compression travel back through a traffic jam as cars move forward and close up the gaps in groups. Because a compressive wave is transmitted most efficiently in a relatively incompressible medium, water transmits sound better than air. And because the properties of sea water – especially its temperature and thus density – vary from place to place and depth to depth, the velocity of sound in it varies accordingly. In other words, sound waves radiating from a given source accelerate in some layers and slow down in others, so that the beam or ray path of the sound may seem to wander erratically. Actually, though, the path, far from wandering, obeys clearly defined acoustic laws.

As water deepens, it is, of course, compressed by the weight of water lying on top

of it, and the speed of sound would normally increase slightly with each slight increase in pressure. Near the surface, however, this effect is counteracted by the progressive decrease in temperature, which causes the sound to slow down as depth increases. The ultimate effect of the two phenomena working in opposition is that a layer with minimum sound velocity usually occurs at a depth of between 700 and 1,500 meters. In greater depths, the effect of pressure dominates, since the temperature changes only slightly, and the speed of sound increases again.

When we use sound in water, we sometimes want to transmit the pulses vertically up and down, as in echo-sounding off the sea floor; sometimes horizontally, as in communicating between a submarine and a beacon at the same level; and sometimes obliquely, as in side-scan sonar, searching for fish or for lost objects on the sea floor, or communicating from ship to submarine. In a directly vertical path the speed of the sound varies, but the path does not bend, and so the corrections to be applied to echo-soundings are fairly simple. In a horizontal path there is some refraction, or bending, of the sound waves wherever there are strong gradients of temperature or sloping interfaces between bodies of water of different densities, and some of the transmitted energy is lost from the direct path. But at the depth at which sound travels most slowly – 700 to 1,500 meters – we get the paradoxical result that it carries over enormous distances (up to 10,000 kilometers) as if trapped in a pipe or channel. In general, then, both vertical and horizontal directions provide relatively efficient paths for carrying sound signals.

In many situations, though, we need to send signals obliquely through the water; and this is very difficult, for sound rays that continuously pass at an angle between zones of different sound velocity are refracted again and again. Thus, although the energy can be transmitted long distances, any signal becomes garbled, waves arrive out of step because they have traveled by different paths, and certain areas may get no sound at all and hence be "invisible" from the transmitting station. Since the most rapid variations in speed of sound occur in the layers of great temperature change near the surface, the best way to improve the performance of, for example, side-scan sonar or submarine detection equipment is to lower the instrument a hundred meters or so below the surface.

In order to predict the path of a beam of sound through the water, we need to have its "velocity profile" – i.e., we must know its speed at every depth in that area. We can get the necessary information either by means of instruments specially made for this purpose or by computing the velocity

The speed of sound waves in water varies with temperature, salinity, and depth. Most of the sound generated in surface layers (right) is refracted back to the surface, and a shadow zone is created where almost no sound penetrates. Below: Sound waves do not travel as side-to-side undulations but as pulses of compression and expansion – suggested here by the progressive contraction and expansion of sound-wave symbols at successive moments.

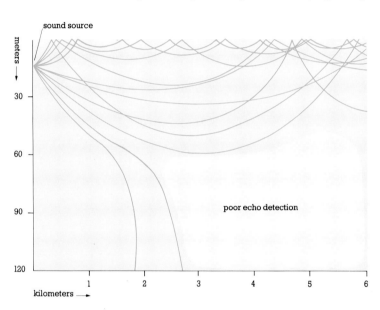

profile on the basis of measurements of temperature, depth, and also salinity, which has a small effect of its own.

The sound waves used by instrument manufacturers generally have a frequency of between 7,000 and 100,000 cycles a second. The human ear can just about hear 10,000 as a high-pitched squeak, but higher frequencies are ultrasonic – that is, cannot be heard. These high frequencies can be modulated like radio waves, varied in strength or frequency, or chopped into short bursts, to provide all sorts of codes for carrying information through the water. This may be used for identifying navigational beacons, relocating oil wells, transmitting numbers or instructions to switch automatic devices on or off, keeping track of submarine craft, conveying speech by voice or morse code, finding and identifying shoals of fish, tracking oceanographic instruments programed to float within the deep ocean currents, etc.

The apparent vagaries of sound paths in water often frustrate military attempts to track enemy submarines, and trawlers in search of rich hauls are frequently led astray by promising signals from the echo-sounder that turn out not to be the hoped-for fish. Nevertheless, sound is our only means of "seeing" underwater. That is why marine scientists and technologists never relax their efforts to understand it better and to improve our tools for utilizing it.

Right: One way to use sound for mapping the sea floor. After recording powerful sound pulses refracted through subsea rock layers, the instrument sphere being recovered here has been released from a bottom anchor by acoustic command from the mother ship.

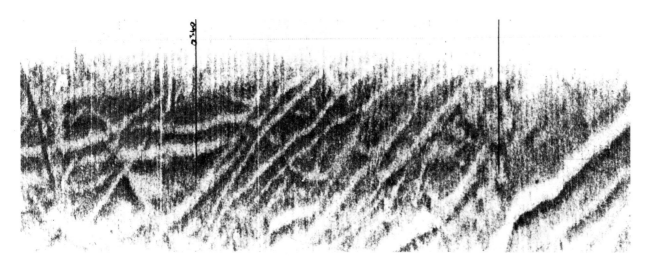

Above: Britain's Geological Long-Range Inclined Asdic – GLORIA – is a towed "fish" that bounces sound waves off the sea floor to produce shadowy, oblique pictures of the topography. Left: A smaller kind of side-scan sonar was used to make this picture of huge plow marks – each about 20 meters wide and 2 meters deep – in the sea bed off Norway. They were made by ancient icebergs.

How Ocean and Air Interlock

The activities of productive industries such as fishing and offshore-oil drilling are limited by weather conditions below the water surface as well as above it. For this reason, government-sponsored meteorological and oceanographic laboratories have for years been collecting information on winds, waves, currents, temperature, and salinity; and since the advent of submarine craft and acoustic devices for use in war and peace, there has been an added incentive to better understanding and prediction of undersea weather. Even more important, we now know that undersea weather and atmospheric weather are inextricably entwined. Within the last decade, meteorologists have been making vast computer models of global atmosphere to improve their ability to make long-range forecasts. And the conviction has grown that we cannot begin to understand the behavior of the atmosphere over a time span of more than a day or so without taking into account the complexities of the interchange of heat and moisture between ocean and air. The longer the time span, the more dominant the role played by the water, because, as we have seen, the ocean acts as a sort of memory, storing heat or transferring it from one part of the world to another, but only after long delays.

The broadly global significance of the ocean in transferring heat, and thus controlling weather and climate, can easily be demonstrated. Let's begin by considering the total balance of heat and light energy that strikes the earth and is radiated back into space. Looked at from outside the atmosphere, the earth-plus-ocean-plus-air system neither gains nor loses heat on balance, since the same amounts that are absorbed are eventually lost. But there is a catch from the earth-dweller's standpoint in that the heat radiated back to space is lost evenly from all over the globe, whereas most of the incoming heat is absorbed in the tropics, where the sun's rays are most direct. Thus, while more heat is gained than lost in the tropics, more is lost than gained in the regions north and south of 38° latitude. Yet, despite this imbalance, equatorial areas do not get hotter and hotter, and polar areas do not get colder and colder. Instead, more or less constant climatic conditions are maintained on earth because heat is continuously transported poleward by ocean currents and atmospheric winds. Slight changes in the pattern of heat transport via the sea can produce dramatic changes in short-term weather and longer-term climate.

The ocean and air are interlocked not only in the constant exchange of heat by radiation or convection, but also because so much heat comes into the air from the evaporation and condensation of the water. A calorie of heat (the same calorie that weight-watchers worry about) is defined as the amount needed to raise one gram of water through one degree centigrade. As you warm a gram of water, then, each degree of increase in temperature requires another calorie of heat – until, if you heat the water to the evaporation point, it suddenly absorbs 750 calories in making the transition from liquid to vapor. It takes that much energy, in other words, to turn water from a liquid into a gas; and when the vapor condenses again, the same amount of energy – we call it "latent heat" – is released. Air that is carrying vapor is therefore transporting a lot of latent heat.

Since 88% of the water vapor in the atmosphere comes from the ocean, and 70% of this comes from evaporation in the tropics and subtropics, it is clear that the sea in

International oceanographic experiments require ships, buoys, aircraft, and submerged instruments, all working together. Some of this equipment is seen below. Left: An oceanographic winch is poised to lower the water-sampling bottle in the background to a precise depth. Right: This free-falling device is sensitive enough to measure even the smallest changes of velocity in many layers of water. At a given depth it will release its ballast and bring its measurement recordings back to the surface.

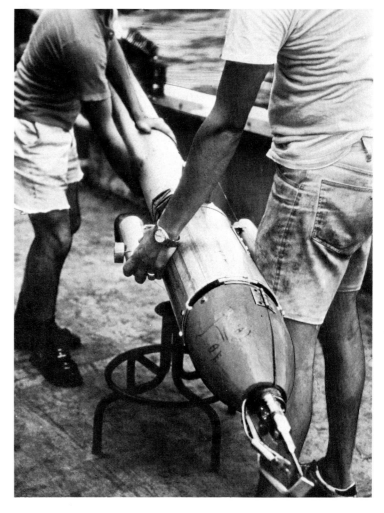

these regions is in effect pumping heat into the air. That heat is then released elsewhere when clouds form and rain falls. Warm air on its own could never carry heat effectively into the mid-latitudes to produce the climate we enjoy. It is the latent heat of water vapor picked up from the tropical ocean that does it.

I have already discussed the importance of horizontal and vertical oceanic currents in transporting warm and cold water. Here I need only add that about 30% of the total global heat transferred in poleward directions is in the form of warm ocean currents. The remainder travels in the atmosphere, most of it in the form of latent heat. And there is a further interlocking of ocean and atmosphere, of course, in that the oceans' currents are driven by the winds. The system is so complex, indeed, that we are nowhere near being able to devise a complete climatic model for it. What is still needed is a staggering amount of factual information about the sea itself and about the myriad probable interactions of sea and air.

The study of such large-scale phenomena can be undertaken only through huge observation programs and experiments. A recent example of this kind of highly organized and widely sponsored project was labeled the Global Atmospheric Research Program: Atlantic Tropical Experiment. It involved a period of field study in the summer of 1974 – study carried on by 39 research ships, 13 aircraft, several satellites, and 4,000 scientists and technicians in an oceanic area stretching from the Cape Verde Islands to the equator and from Africa to South America. The operation was coordinated by an international scientific-management group set up by the Intergovernmental Oceanographic Commission and the World Meteorological Organization, which are specialist groups of the United Nations. And it has provided us with our first opportunity to arrive at an accurate measurement of the total flow of energy between sun, air, and sea.

This project is only one of several being conducted or planned by a new generation of globally cooperating oceanographers, who feed data into world data centers maintained in Washington and Moscow. Meanwhile, more routine work goes on in universities and state-subsidized oceanographic institutes all over the world, and in the several hundred research ships now maintained by many nations and equipped with a whole new range of moored instrument buoys, submerged instrument platforms, and free-drifting instruments. It is a pleasant reflection on oceanography that it is still such a pioneering activity that oceanographers from all over the world keep in personal touch with one another regardless of political divisions and power-bloc rivalries.

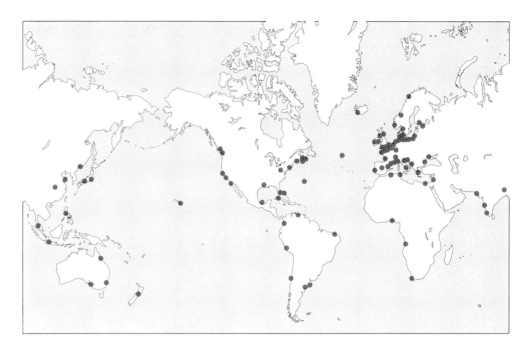

The dots on the above map show the location of only a small number of the nearly 2,000 oceanographic institutes and university departments currently specializing in marine studies throughout the world. Note that although most of them are in the developed countries, the third world is by no means inactive in this field.

Below: In 1974 the U.S. National Oceanic and Atmospheric Administration's research ship *Discoverer* – here seen towing a meterological recording balloon – participated in the international Atlantic Tropical Experiment that studied heat-exchange characteristics of the ocean as they affect the earth's weather.

3
Plant Life

Charles A. Yentsch

In most of the ocean the natural rate of production of living matter is equivalent to that of a terrestrial desert. But in some areas, such as parts of the continental shelves and in coastal waters, estuaries, and coral reefs, local productivity is high. Thus, because the oceans are very large in relation to land masses, about half the living matter created in the world each year is generated in the sea. This high local richness but low average of life means that although there are bountiful marine supplies of food, we can find it only in limited areas, which vary from year to year with changing weather and ocean currents.

To better our understanding of such variations, biologists and ecologists analyze thousands upon thousands of samples of the floating plant life of the oceans. The scientists constantly study the chemicals and nutrients dissolved in seawater that are essential for plants, and they make frequent measurements of the amount of the sun's energy that reaches the water's surface and the amount that penetrates to different depths. They also learn all they can about how plant cells trap sunlight and chemicals, how the individual cell grows, reproduces, and dies – or is eaten – and how the materials of the dead plant are returned to the cycle of life. Such detailed study of the sea's plants (which are not generally eaten by man) is based on the obvious fact that they are the primary producers of food; only if marine vegetation thrives will the oceans continue to teem with animal flesh. And the type of vegetation that matters is mostly found in plankton – the extensive communities of living organisms, both animal and vegetable, that float or drift with the water, having only limited means, if any, of propelling themselves along. In this chapter, therefore, we are going to investigate only the phytoplankton (the general term for planktonic plant life), with virtually no reference to plants that grow fixed to the seashore or sea bed. Although harvested in small quantities for human use, fixed plants contribute little to the chain of eat-and-be-eaten that produces valuable commercial fish.

The earliest studies of plankton in the deep ocean were made in the 1870s during the world cruise of the British research ship *Challenger*. The observations of this expedition confirmed the existence of planktonic life in all the open oceans (and, incidentally, of animal life in the greatest depths). Two basic tools, the plankton trawl net and the microscope, were essential for the *Challenger*'s great scientific achievements. The microscope had been in use for centuries, of course, but the plankton net was an invention that antedated the *Challenger* voyage by only a few decades. In its simplest form, such a net is merely a fine mesh bag, roughly conical in shape, tapering

down to a small bucket or trap, that can be towed by a vessel at varying depths; but countless improvements on the basic design have been made since the mid-nineteenth century. A famous twentieth-century modification is the continuous plankton recorder – developed by Alister Hardy, a pioneering British marine biologist – which traps plankton on a roll of cloth wound past an orifice by a small propeller. In the most recent version of this device, now being used by the U.S. National Marine Fisheries Service and the Scottish Marine Biological Association, the continuous plankton sampler is mounted in a large towed body (or "fish"), and the water flowing past the "fish" spins a propeller that both winds up the roll of cloth and alters the pitch angle on a set of vanes or stubby wings, so that the "fish" alternately dives and rises again. This undulating recorder can be programed to dive at different speeds and different depths, so as to sample a whole water column along the ship's track.

It was only after marine biologists had gained some knowledge of the distribution of different species of phytoplankton that they also began to concentrate on what goes on inside phytoplanktonic cells. Unless we know how the cells function, why they differ, what advantage one type has over another, and why different types grow best under differing conditions, we cannot understand the reasons for the growth or death of the billions of cells that make up the total planktonic crop. Since about 1945, microbiologists who study phytoplanktonic cells have profited from the introduction of such advanced laboratory tools as radioactive tracers to unravel chemical and biochemical reactions, and complex chemical analytic techniques to detect minute quantities of different organic molecules. Shipboard computers, too, are now used for processing much of the data acquired in research

ships on the distribution of plankton and the workings of phytoplanktonic cells.

Although some biologists are necessarily concerned with identifying and naming the numerous different species of phytoplankton, I shall say little about taxonomic matters in the following pages. The really important area of interest for phytoplankton studies today involves our effort to comprehend the total relationship of microscopic plant life to both the nonliving environment and animal life. As the world's populations multiply their efforts to take food from the sea, it becomes more and more desirable to

have informed estimates of the yearly quantity of living matter generated in the sea. How we determine the most effective level of fishing in a given area must depend largely on what we know about primary productivity in that area for the previous few years. Thus, if phytoplanktonic productivity can be understood to a point where, by measuring oceanographic and meteorological conditions, we can predict levels of plant growth, we can try to plan fishing for the maximum benefit to humanity, with minimum damage to the ocean. That is a goal worth working toward.

A hundred years ago, when the British research ship *Challenger* (top left) circumnavigated the globe to make the first truly scientific survey of the world's oceans, specimens of plankton were collected in trawl nets towed by the vessel; and present-day scientists use various modifications of this equipment. One such is the continuous plankton sampler, housed in a "fish" (above) that alternately dives and rises as it picks up samples. Close study of the distribution of phytoplankton, as indicated on the map below, shows that production is greatest where vertical mixing returns vital nutrients to sunlit layers of water. Such areas (darker blue) are most often found in coastal waters, shallow seas mixed by the tides, and deeper waters with upwelling currents.

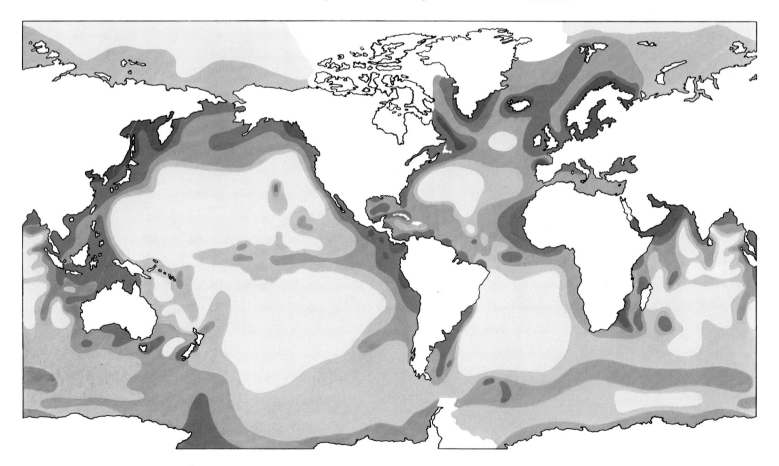

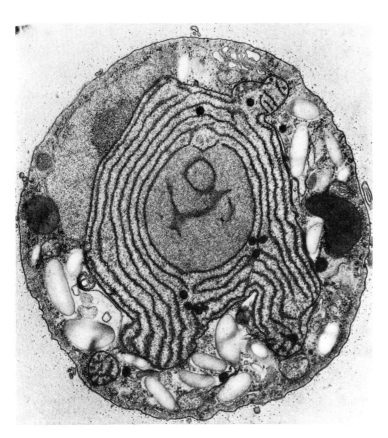 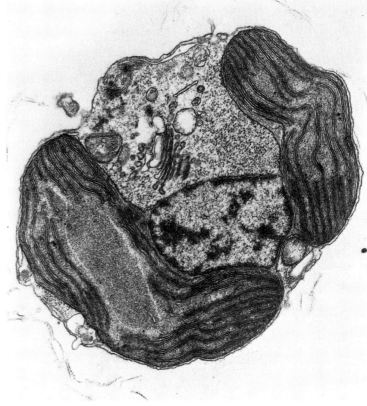

From Light to Life

The fundamental chemical in living things is carbon. Atoms of carbon are joined together in rings and chains and loops and branched chains, along with other chemicals such as nitrogen, phosphorus, oxygen, and hydrogen, to form all the proteins, carbohydrates, enzymes, and other materials essential to life. The only entities that can perform the natural miracle of constructing these complex organic molecules without being nourished by the bodies of other organisms are plants. Marine plants, which lack the variety and security of land-living trees and flowers, are classed (with few exceptions) as algae. The algae that are attached to the bottom of the sea or to the shore live in a very narrow zone, defined by the depth to which light penetrates for growth; in spite of their conspicuous bulk, it is only in near-shore waters that they contribute much to the quantity of carbon built into marine life. Most of the carbon fixed into the creatures of the ocean is fixed by the phytoplankton, which can live down to about 100 meters everywhere in the vast expanse of the sea.

It takes energy to build living cells and organic molecules, and the direct source of this energy is sunlight for the primary producers, which absorb light and convert chemicals into living matter by the process we call photosynthesis. A planktonic algal cell can be thought of as a kind of transparent bag filled with light-absorbing pigment, chlorophyll. In the water that surrounds the cell-bag there are molecules of carbon dioxide gas, nitrates, phosphates, and other chemicals, which have been either washed down from the land or dissolved from contact of the surface water with the air above it. Given ample quantities of such nutrients along with sufficient light energy, the pigment-filled cell will grow until it divides in two. Under ideal conditions, this doubling process can occur several times a day, producing a "bloom" (the biologists' word for an explosive increase in the quantity of phytoplankton within a specified area).

In order to absorb the sun's energy efficiently, the cell must contain a special molecule that can be raised to a high energy level by the light, somewhat as a battery is charged. This stored energy is released in a controlled way to join atoms together and build other molecules. The incoming light from the sun is more or less white – that is, it contains light of all the different colors, or wavelengths. Because it is not possible for the special molecule to operate efficiently at all wavelengths, different species of alga contain pigments that work best at different wavelengths – in other words, at different parts of the spectrum. The active absorbing chemical (i.e., the pigment) absorbs only a narrow band of light, reflecting the rest, and it is this reflected light that gives plants their color. The most common pigment, of course, is green chlorophyll, but we also find a wide range of the orange-red pigments known as carotenoids among the phytoplankton, and the blue of the so-called blue-green algae and red of red algae come from a pigment called pycobilin. The white sunlight that arrives at the ocean surface is

Different algae absorb different wavelengths of light, and their color comes from the parts of the spectrum they reflect. On the facing page and above are electron micrographs – enlarged 30,000 times – of sections (from left to right) through a red, a golden-brown, and a brown algal cell. (Unfortunately, electron beams, like X-rays, do not produce color.) The maze-like lines are light-receiving pigments; the rounded or elliptic blobs – most clearly visible in the red cell – are energy-storing starch grains.

filtered by the water and changes color as it penetrates more deeply. Obviously, a given species of plant will thrive best in water that has the right color for its particular photosynthetic pigment or combination of pigments. When, therefore, we broadly classify the various planktonic plants as green, brown, red, or blue-green algae, or as purple or green bacteria, we are implying that each of them preferentially absorbs certain wavelengths from the surrounding water and reflects those that give it its identifying coloration.

To learn how photosynthesis and the fixation of carbon work, or to study phytoplanktonic productivity in a given area of the sea, biologists have to measure changes in the amount of carbon dioxide or oxygen in the water around the cells, as well as the increase in carbon within the cells. This is more easily said than done. In the laboratory, algae can be observed in transparent jars illuminated with precise intensities of light, and exact measurements can be made, but laboratory-scale experiments, while always valuable, are obviously restricted in scope. Recently, divers have been employed to install arrays of trans-

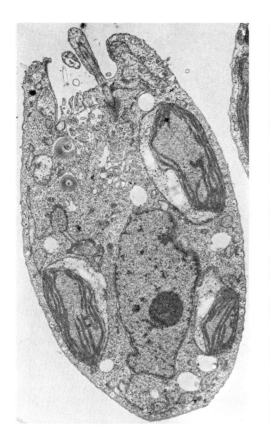

Left: A micrograph of a double-celled rigid diatom. The radiating pigment lines are a mixture of chlorophyll and carotenoids. Oil droplets in the cell walls aid buoyancy.

Algae can be studied in labs, but their reactions are best observed in the open sea. Below left: Filters over specimen tanks help simulate different intensities of light at various depths. Right: At sea, glass jars suspended in a protective frame trap plankton in their natural environment, where divers can measure their growth rate with a minimum of experimental interference.

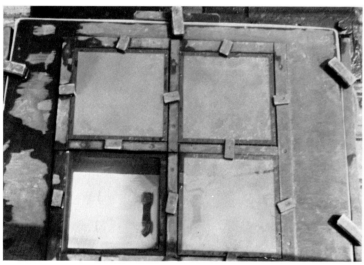

parent jars or canisters at different depths in the sea. If you want to study free-drifting phytoplankton in actual underwater conditions, it is difficult to discover the amount of photosynthetic activity by measuring changes in amounts of carbon dioxide and oxygen in the surrounding water, since the percentage changes are imperceptible. So the divers have to trap the plankton inside a glass jar underwater, inject a measured quantity of carbon dioxide made from radioactive carbon-14, and then measure the rate at which radioactivity appears in the algal cells. A natural algal population can be studied in this way in the exact water and light conditions of the open ocean, with just the minimal experimental interference caused by the presence of the glass side of the jar. This very sensitive technique has

increased our ability to calculate the rate at which carbon is being fixed into living matter at selected depths in small portions of the sea. To study the variations, a string of transparent bottles containing algae in a radioactive solution can be suspended at intervals down to 100 meters; by adding up the various rates of carbon fixation, you get an idea of the total productivity under, say, a square meter of ocean surface.

Because it would be impossibly expensive and time-consuming to repeat such radioactive-tracer experiments all over the enormous expanse of the oceans, even at large intervals, many researchers have explored the possibility of surveying great patches of water from space satellites. This seems a good idea because much of the variation in color of the open ocean is

probably due to different concentrations of chlorophyll and other pigments in the phytoplankton. If satellite observation were to show a clear correlation among color, primary productivity, and concentrations of feeding fish, fishing fleets would directly benefit from the results, for they would need to spend far less time searching for large shoals. Recent experiments in the Atlantic indicate that the correlation may well be good enough to prove useful.

The study of phytoplanktonic photosynthesis is more complicated than an equivalent study of land plants because the intensity and color of undersea light vary so considerably as depth and water transparency change. We'll now take a closer look at what happens to sunlight as it passes down into the sea.

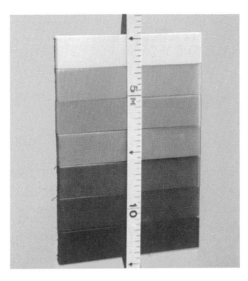

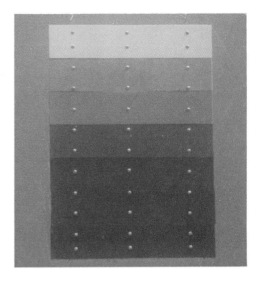

Light in Depth

The photosynthetic pigments of a phyto-plankton cell trap solar energy and store it in complex chemical molecules. For the survival and growth of the cell, this energy is gradually released in the process known as "respiration." How a particular cell reacts to light and how well it grows and repro-duces depend on the efficiency with which both photosynthesis and respiration can proceed under different light conditions.

When sunlight – which is almost pure white – strikes the surface of the sea, several factors govern the depth to which it pene-trates. Latitude matters, for example, since the sun's rays are brightest at the equator and weakest near the poles. Wind matters, too; bubbles scatter and reflect the rays back into the air, and waves cause under-water illumination to flicker because their convex surfaces may focus the light like a lens, while the troughs disperse it. The flickering can produce peaks of brightness that are a hundred times stronger than the average – which suggests that we may not

be able to rely on calculations of photo-synthetic reactions that are based on "aver-age" brightness in a given area of the ocean. Finally, about half the earth's surface is likely to be covered by cloud at any one time, and the pattern of cloud cover nat-urally affects the brightness of any light that reaches the ocean.

Once through the sea's surface, the white light is absorbed selectively. That is, the spectrum of rainbow colors – red, orange, yellow, green, blue, indigo, violet – is ab-sorbed preferentially at both ends until only a narrow bank of blue-green color is left; at a certain depth this too is absorbed and there is total darkness. The light rays are absorbed not only by the water itself but also by particles containing pigments, such as algae, or by organic dyes dissolved in the water. Moreover, light doesn't just shine straight down into the water and fade out through absorption but is scattered side-ways and upward. Pure water absorbs red and scatters the blue diffusely, producing a pervasive blue glow. Water that contains particles of clay, silt, or algae also scatters

The pigments of a plant absorb wavelengths it can use for photosynthesis and reflect those that we see as its "color." Plant coloration in water is much affected, too, by the depth and color of the water itself. Top left: A many-hued panel as we would see it on land. Center: The panel 5 meters down in clear blue water has a bluish tinge; but color distinctions, though dimmed, remain visible. Above: At 20 meters the water has filtered out most of the red band of the spectrum. Facing page, top left: In murky coastal waters, only red light can penetrate to a depth of a single meter. Note the absence here of blue and green light. Thus, in coastal waters a plant is likely to photosynthesize most efficiently with a very sensitive pigment absorbing yellow-red light.

red, yellow, and green light in shallow depths where these wavelengths have not yet been absorbed, making the ocean look green, yellow-green, or brown. In all cases, the plankton cell is irradiated by light of a different brightness and color from every direction.

The maximum depth at which photo-synthesis can function depends on the spe-cies of phytoplankton, the intensity of light at the surface, the active photosynthetic pig-ment of the plant, the reduction of light in-tensity as it is absorbed by the water, and the color spectrum of the light at depth. In the clear waters of the deep ocean some plants can photosynthesize as far down as about 100 meters, but the maximum is only about 30 meters in coastal and continental-shelf water, which is generally less trans-

To pinpoint correlations between algal pigments and light is difficult; the spectrum is more complex than it seems in color charts. Even sunlight is distorted at water surface, for various molecules of gas in space and air absorb parts of it. This diagram shows amounts of light intensity absorbed by four kinds of molecule before light strikes the sea.

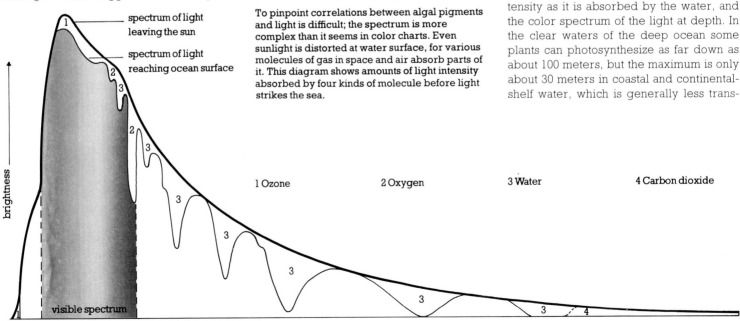

spectrum of light leaving the sun

spectrum of light reaching ocean surface

brightness

visible spectrum

1 Ozone 2 Oxygen 3 Water 4 Carbon dioxide

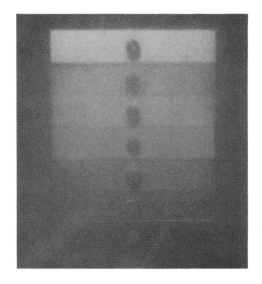

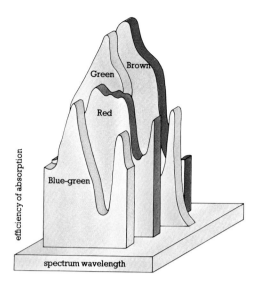

Because of different types of pigmentation, each kind of alga absorbs certain parts of the spectrum more efficiently than it does the rest. For example, as illustrated here, red algae absorb best through the blue, green, and yellow bands, whereas blue-green algae absorb light in two sharp peaks in the blue and yellow bands.

parent than the deep ocean. Near the top of the euphotic zone – the depth zone within which photosynthesis is possible – the algae are bathed in almost pure white light, while at 50 to 100 meters they can absorb only a narrow band of blue-green. To estimate the quantity of plant life that can grow in a one-meter-square column of water, we'd need to know not only the intensity and spectrum of the light at each depth, but also the way in which the pigments and cells of any algae present in the column will respond to these factors. An ability to calculate plant productivity within a given portion of the sea would obviously help to advance our knowledge of marine protein resources, but the factors that I have just mentioned are exceedingly difficult to come by. Rates of photosynthesis do not respond in a simple manner to intensities of light, and rates of respiration behave differently from rates of photosynthesis.

As I have pointed out, photosynthesis stores energy in complex molecules, which are broken down in minute stages to release the energy required for manufacturing the different chemicals required by the cell. This process of respiration can be viewed as a kind of burning up of the material accumulated during photosynthesis, and its waste products are water and carbon dioxide, which are excreted from the cell. For all practical purposes, the rate of respiration can be considered as independent of surrounding light conditions.

When there is a very low intensity of light, the rate of photosynthesis lags behind that of respiration, and so the algae virtually burn themselves up, for they cannot replace the respired carbon fast enough to grow. When light intensity is increased, the rate of photosynthesis speeds up. It does this at first in proportion to the increased amounts of light; but after the photosynthetic rate reaches what's known as a "compensation point," where photosynthesis and respiration are equal, it exceeds the rate of respiration. Increasing light intensity produces more rapid growth or cell division, until eventually the various enzymes involved in the carbon-fixation process become saturated; and from the saturation point on, very little extra photosynthesis can occur, no matter how much increase of light there may be. Once the enzymes are saturated, the rate of fixation of carbon actually drops with brighter sunlight.

We know that light conditions at a given depth cannot be described simply in terms of intensity, since both intensity and color vary with direction even at a single point. And so, once more, it is easy to see why marine biologists face a practically insoluble problem whenever they try to make a dependable estimate of productivity. They can, to be sure, base *rough* estimates on simple assumptions that the rate of photosynthesis depends on the total available light intensity at each depth, but at times simple assumptions are not trustworthy. Nevertheless, we can say that near the surface the light may supersaturate an algal cell so that photosynthesis is only moderate. At a certain depth, there will be a maximum rate, but it will again diminish with greater depths until the light intensity drops to just the amount needed for matching respiration. Below this, the compensation point, the cell cannot grow and reproduce, even though, strictly speaking, it can continue to photosynthesize at a slow rate.

Modern scientists are trying hard to build sophisticated models of the interrelationships of various types of marine plant with light intensity at various depths and individual rates of photosynthesis and respiration. But, as you can see, there are complexities here that continue to thwart even the most advanced of computer technologies.

As shown below, different parts of the spectrum reach algae in varying kinds of water. Here we compare transmissions of light through representative bits of oceanic and coastal waters. Clear deep water transmits light on a broad spectrum from blue to red; if water is murky, light loses brightness, though with little change of color. But sediments and plant life in shallow coastal water shift the spectral-curve peak to the red end.

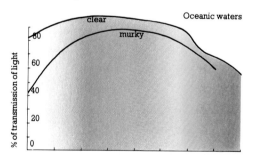

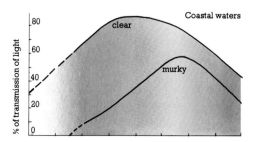

The diagram, right, illustrates the relationship between respiration – the rate at which an algal cell burns up carbon – and photosynthesis. Although respiration itself is independent of light intensity and does not change with depth, the rate of photosynthesis is inadequate to sustain continuous respiration below a certain depth (the compensation point). Above that depth photosynthesis begins to exceed respiration, eventually reaching a saturation point near the surface. Above that point, although respiration remains unchanged, plant growth may actually decrease.

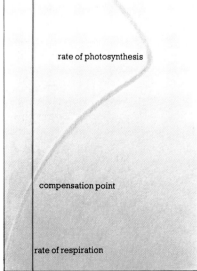

"Watering" the Plants

Primary production in the open oceans tends to be low, because the light is at the top of the water but most of the available nutrients are near the bottom. Productivity can be really good only when these two essentials for plant life come together, and this can generally happen only when wind or temperature changes stir up the water and mix it from top to bottom. Such vertical mixing brings the nutrients toward the surface and carries the plankton down in turbulent eddies, spreading both nutrients and plants throughout the upper layers of the water column. In fact, a severe storm over deep water can force the mixing depth down so far that much of the phytoplankton is carried down to where, because of low light intensity – or no light at all – photosynthesis becomes impossible. When this happens, total productivity may drop instead of rising, even though the plants passing through the upper layers are richly supplied with both light and chemicals. So in the summertime there may often be too little mixing for the good of the phytoplankton, and in the winter too much.

Oligotrophic surface water (water that is low in nutrients) is apt to occur in summer for two fairly obvious reasons. First, the strong sunshine warms the ocean's upper layers, and each layer expands slightly, and becomes less dense than the layer on which it rests. Result: a stabilized water column whose layers neither rise nor sink. Secondly, since summer winds normally tend to be gentle, the stability can continue so long that the algae near the surface use up all the local nutrients, with no replacements from below. And such warm-weather oligotrophy can happen not only in the open ocean, but also in coastal waters.

In the autumn months, as surface waters cool, grow denser, and start to sink, wind strength increases and mixing begins. It is when the critical mixing depth exceeds about 100 meters – below which the light is too weak for photosynthesis – that total plant productivity in the open ocean plummets. On the continental shelf and in coastal waters, however, winter storms often mean better, rather than worse, productivity, for vertical mixing brings up nutrients in relatively shallow waters where the plankton cannot be carried down too deep for the sun's rays to reach. But the best conditions for plankton growth everywhere are likely to occur in the spring and autumn. Springtime bloom is usually the year's largest, since winter storms have already enriched the surface water, which strengthening sunlight warms up while there is still enough wind to prevent stratification of the water column. In the fall, as rising winds mix the water and bring up nutrients before light intensity weakens, another bloom can occur.

Seasonal vertical mixing is not, of course, the only way in which the necessary mingling of nutrients and light gets through to the phytoplankton. A second process – known as upwelling – brings nutrient-rich bottom waters to the surface in a steady upward current instead of mingling top and bottom water in particular water columns. Upwelling is caused by the interaction of oceanic currents, the shape of the sea floor and continental slope, and alongshore winds; and upwelling waters move upward on a slant rather than vertically. Although such movements also tend to be seasonal, since they depend to a large extent on wind strength and direction, they are more predictable than storm mixing. For instance, we find reliably strong areas of upwelling off the coasts of Peru, southwest Africa, and California, and the dependably high plankton productivity in such places makes them extremely important fishing centers.

There is a less well-known process that results in rich areas of primary production – and hence the potential for good fishing – along one side of a major current. This stems from the patterns of pressure and density that build up around such currents as they flow hundreds or thousands of kilometers on the spinning earth. In the Northern Hemisphere, for example, a northward current such as the Gulf Stream or the Kuroshio produces uplifted fertile water on its western edges because the flowing mass of water experiences a force at right angles to the direction of flow (pretty much like the atmospheric force that makes winds blow in circular patterns). Because of this, the water level at the surface is raised on the eastward side of the current, with a consequent slight slope all the way across, and higher-density nutrient-rich water then rises to the surface in the somewhat lower western waters.

Finally, conditions for concentrated phytoplankton production may occur when two water masses are convergent, so that one has to slide over the other. Wherever this happens, the plankton population in the surface waters along the interface increases, because many buoyant organisms in the downward-moving water mass float upward and accumulate in ribbons. We can see the results on a grand scale in the Southern Hemisphere, where the waters of the Antarctic meet the Pacific and Atlantic, and on a much smaller scale along many coasts where estuarine water flows over more saline seawater.

By studying natural phenomena like those I have been briefly reviewing, we've learned a lot in recent years about the peculiar conditions that favor marine plants. Let's turn now from a consideration of their need for nutrients in general to an examination of the specific chemicals required by the ocean's masses of growing algal cells.

Why winter is a better "growing" season than summer for floating plants. Below left: Low surface-water temperature in winter combines with winds and waves to help circulate the water, bringing nutrients into the euphotic zone. Below: Although summer sunlight deepens the euphotic zone, it also warms the surface water, and the warmth and gentle winds result in stabilization of the water column. Nutrients, therefore, circulate less widely, impoverishing the upper layers of water, and so the phytoplankton starve.

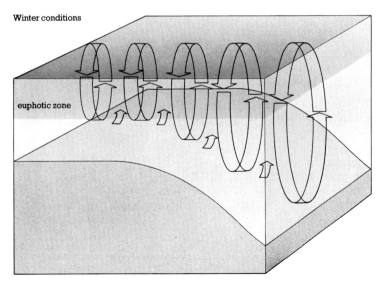

Winter conditions

euphotic zone

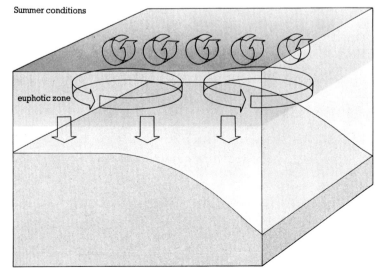

Summer conditions

euphotic zone

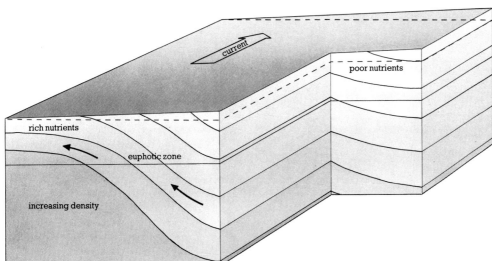

current

poor nutrients

rich nutrients

euphotic zone

increasing density

Wave action is a visible indication of the vertical mixing of phytoplankton with nutrient chemicals. But too much turbulence can decrease productivity by carrying algae down below the euphotic zone – the zone within which enough light for photosynthesis penetrates.

The patterns of pressure and density built up by major oceanic currents affect phytoplankton productivity along their flanks. For example, plants are much more likely to flourish in the euphotic zone on the western edge of a strong northward current in the Northern Hemisphere than in either the center or the eastern side. The reason: rotation of the earth produces a deflective force (known as the "Coriolis effect"), and nutrient-rich water on the left-hand edge of the current rises as illustrated left.

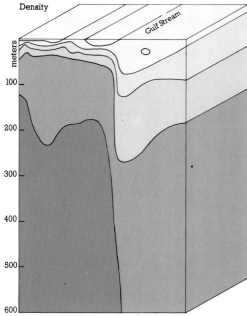

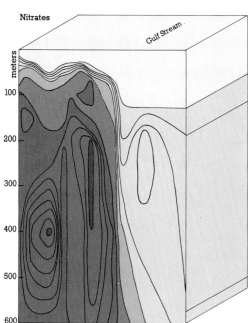

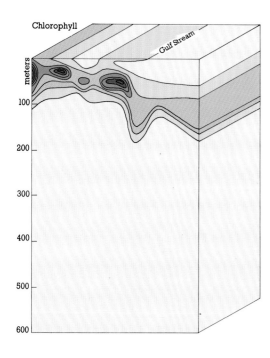

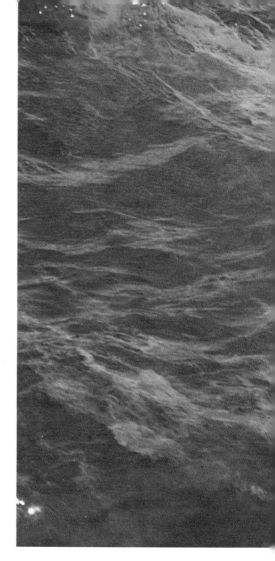

Phytoplankton Food

In most of the world's oceans, the abundance of phytoplankton is controlled – apart from such obvious limiting factors as light and water temperature – by four major nutrients: carbon dioxide, nitrogen, phosphorus, and silica. Marine biologists have named a number of trace metals and vitamins that also seem to be growth requirements or stimulators. But the role these play in determining which species of plant life flourish best is probably a minor one. It is the four big nutrients that count.

Seawater contains a large amount of carbon dioxide, some dissolved directly from the atmosphere, more washed down from carbonate rocks in streams and rivers. It is essential for plant growth. Without vast supplies of it, the phytoplankton would die; without the phytoplankton the zooplankton would die; and so on up the food chain. Fortunately, there is little danger of a lack of carbon dioxide in the world's oceans. Most of the sea's inorganic carbon occurs in carbonate and bicarbonate compounds that serve as a kind of bottomless reservoir of carbon dioxide, since they are converted into carbon dioxide when the concentration of CO_2 in the water drops.

Where nitrogen, phosphorus, and silica are concerned, it is a different matter, for they are in comparatively limited supply. All three of these nutrients are quickly depleted during the springtime and autumn phytoplankton blooms. Nitrogen is not directly available as a nutrient to most plants. True, there are blue-green algae in some parts of the ocean that can absorb pure nitrogen dissolved from the atmosphere, but they are somewhat rare. In general, it must be absorbed by phytoplankton in the form of ammonia (that is, combined with hydrogen) or as a nitrate (an oxygen compound). Similarly, the algae can get their phosphorus only from mineral phosphates (phosphorus with oxygen) dissolved in the water or from phosphate combined with complex organic molecules from dead animal and vegetable matter. Organic phosphate cannot be absorbed directly by phytoplankton cells but must first be broken down by algal enzymes.

Silica is a less universal marine-plant requirement than the three others. Like the nitrates and phosphates, it is not itself an element but is an oxygenated form of silicon, which is a major constituent of most

These three diagrams show different, but profoundly related, aspects of a cross-section of the sea between Cape Cod and Bermuda, with the direction of the Gulf Stream shown by an arrow. It is immediately apparent that dense water has risen on the western rim of the current (top) bringing nitrates up toward the surface (middle) there, and that this is where we find the thickest concentration of chlorophyll – i.e., of algae.

rocks. Though extremely common, however, it is almost insoluble, and so its concentration in seawater is limited. Yet a major component of the phytoplankton, the diatoms, need it for the stiff structure of the cell wall. Although, as far as we know, there is no other involvement of silica in the metabolic processes of algae, this nutrient is so vital for the diatoms – and there are so many of them in the plankton – that the limited concentration of silica in seawater drops drastically at the time of a major bloom.

Clearly, then, nitrogen and phosphorus are the chemicals whose presence or absence in a given area has most to do with whether or not plant growth thrives. As we've seen, carbon dioxide is in limitless supply everywhere, and many algal cells do not require silica at all. Indeed, marine biologists now suspect that even phosphorus is of secondary importance and that nitrogen is the primary substance that either induces or limits growth. The variations in concentrations of phosphorus and nitrogen have been studied intensively for many years, particularly in coastal waters. Typical concentrations that scientists have measured in waters of the continental shelf of the U.S. Eastern seaboard show significant seasonal variations as the result of upwelling, vertical mixing, and the intake of nutrients by the phytoplankton. During the summer months the surface waters down to 50

meters – the average depth of the most fertile zone in this region – are gradually depleted of nitrogen and phosphorus because of the removal of nutrients by the spring bloom and the subsequent absence of vertical mixing. By November, however, autumnal storms have stirred up the water, and the concentration of both nutrients has begun to increase; and in January there is likely to be 10 times as much nitrate, and two to three times as much phosphate, in the top 50 meters as in July. Thereafter there may be a slow drop until May, when concentrations of both nutrients fall dramatically as, with decreased vertical mixing, the algae begin to use up these vital chemicals in the upper layers.

Since living organisms contain about 15 times as much nitrogen as phosphorus, you would expect the nitrate content to drop more rapidly than the phosphate – and that is exactly what happens. There is almost always 10 to 15 times as much nitrogen as phosphorus in the waters below a depth of 100 meters, where plants cannot use it because of the absence of light. In the upper levels, though (particularly toward the end of summer), the algae may totally exhaust the nitrate supply even while some phosphate still remains in the water. Thus we see that the lack of nitrogen can be justifiably viewed as the ultimate limitation on the growth of phytoplankton.

Above: A springtime bloom of phytoplankton. Of the chemicals most essential for algal growth – phosphorus and nitrogen – nitrogen is the more likely to be in short supply by the end of bloom periods. These diagrams of a cross-section of water off the East Coast of the U.S.A. show nitrate-phosphate ratios in January and September of a recent year. Below 100 meters, where light does not penetrate, there is always 10 to 15 times as much nitrate as phosphate, and in January (just before the spring bloom) the ratio in much of the surface water is also close to 10. In September (before autumnal storms) surface-water nitrate has been depleted much more than has the phosphate.

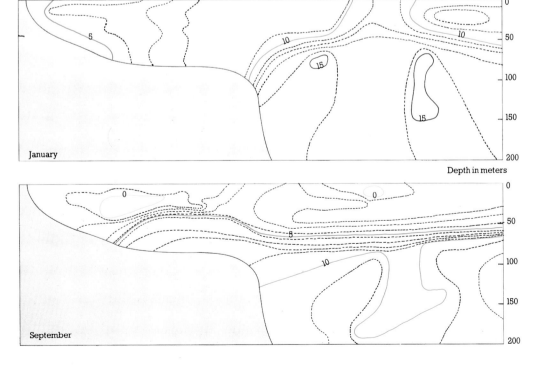

Depth in meters

Diatoms

Dino flagellates

Coccolithophores

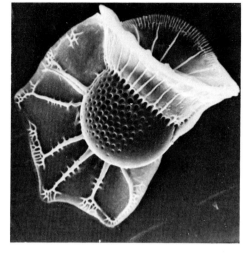

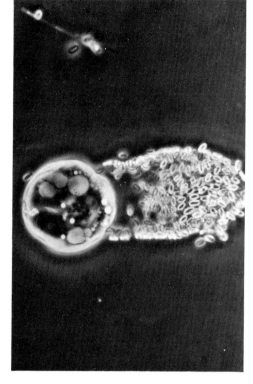

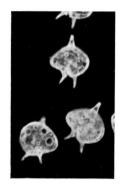

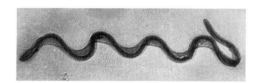

There are three main groups of planktonic alga, with a variety of shapes within each. Far left: Two diatoms, which have rigid silica-impregnated cell walls. Center: Three dinoflagellates, whose cells are composed of cellulose. Note the protruding flagella (a vastly enlarged example is shown left), which, when vibrated, propel the plant. Above: Coccolithophores, which have calcium carbonate cell walls; in this micrograph the cell is shown breaking through its protective shell.

The Many Kinds of Alga

The tiny cells of marine algae are astonishingly varied in shape and size. Indeed, there are over 1,000 species in the phytoplankton – not a great number, certainly, as compared with the hundreds of thousands of species of terrestrial plant, but still quite a lot if you think of them as all living in much the same environment. In fact, though, they do *not* inhabit a single environmental niche. Phytoplanktonic plants are so completely at the mercy of the winds and waves that slight percentage changes in buoyancy or nutrient availability must be regarded as major factors in determining the success of one species or another. Obviously, the many shapes and structures of these unicellular plants have evolved for specific reasons, and each must confer a special advantage within a given oceanic condition.

To begin with, it is possible to group the planktonic algae into three main classes: the diatoms, the dinoflagellates, and the coccolithophores. The main distinction among the three is in the structure of the cell wall, which is made of silicate in the diatoms, cel-

lulose in the dinoflagellates, and calcium carbonate in the coccolithophores. Within each of these general groups are many different species with differing metabolic requirements, and researchers today are deeply interested in the question of how they have evolved and why there is high or low species diversity in a given area at any particular time.

"Species diversity" is the technical term for the number of species found in a volume of water. If, for instance, we find 50 or 60 different kinds of alga in a liter of ocean water, we say it has high species diversity even when the total quantity of phytoplankton is low; but we say it has low species diversity if only 5 or 10 kinds are included among the cells, even when there are millions of them. It may seem odd that we usually find low diversity in conjunction with high productive blooms, and high diversity in poorly nutrient waters, but there is a good explanation for this apparent paradox – an explanation based on the reasonable assumption that nearly every species is adapted to a precise range of environmental conditions. If those conditions

suddenly changed – as they might during an ice age, a volcanic eruption, or a submarine landslide – most species would be wiped out, unable to cope with the sharp changes. The only ones to prosper would be the few that had either specialized in being highly flexible or were already adapted to the new conditions. Thus, wherever the environment changes rapidly or often, we can expect low species diversity; but where conditions remain constant, hundreds of species can develop, each to exploit some special niche. Since the deep-ocean environment remains more or less constant, it favors species diversity although nutrients in the sunlit layers of water are relatively sparse; but fewer species can survive in waters where rapidly changing conditions result in occasional explosions of the plankton population.

The variation of species diversity from season to season in the same area is a consequence of the fact that the conditions for growth sometimes prove to be ideal for certain species. Those algae then proliferate at the expense of competing plants, which are deprived of nutrients by the swiftly multiplying few. Thus, a dense spring bloom tends to have low diversity. On the other hand, in seasons when conditions are generally poor for all species, many kinds of alga tend to be present in low concentrations, waiting, as it were, for the pendulum to swing in favor of one or a few of them. There are four factors that determine the success or failure of every such species. First, is it buoyant enough to stay up in the sunlit layers of the sea? Next, how well is it adapted to the intensity and color of the available light? Thirdly, can it grow quickly in the concentrations of nutrients present in the surrounding water? Finally, how good are its chances of not being eaten too fast by the zooplankton?

The buoyancy of most algae is affected by their growth rate. In periods of rapid growth they float easily in the sunlit upper waters, but they sink when they become deficient in nutrients. You can actually see the effects of this gain and loss of buoyancy during a plankton bloom. At the beginning the population is concentrated near the surface; as the nutrients are exhausted, it slowly sinks. Eventually, the nutrients in the euphotic zone (below which not enough light penetrates for photosynthesis) have been used up, but because they are still abundant below the euphotic zone, the water there is denser – and so the sinking, dying algae tend to "float" near the bottom of the zone. Laboratory studies show that different species sink at very different rates. Some of the dinoflagellates and coccolithophores, for example, have whiplike flagella, which beat rhythmically, providing a crude form of locomotion that keeps them afloat for a while. But as the nutrients grow scarcer, the flagella stop beating, and they too fall to their death.

You might expect the adaptation of various species to light intensity and color to be their most obvious distinguishing feature. Since red pigment absorbs blue light and reflects red, it would appear that red algae should thrive in the deepest water, where only blue light is present; and, logically, since green algae absorb red light and reflect the blue and yellow rays, they ought to do best near the surface, where there is plenty of red light in the white rays of the sun. To a certain extent, this is true of fixed seaweeds: you find bright green fronds in tidal rockpools, brown laminaria in 10 to 20 meters of water, and small, dark-red algae at the limit of light penetration. Even with the fixed seaweeds, though, there are many exceptions; some red seaweeds, for instance, do best near the surface. And, surprisingly, the actual color of the photosynthetic pigment in phytoplanktonic cells seems much less important than other factors for most species. So far, indeed, scientists have been unable to discover any sort of tidy correlation between color and habitat among

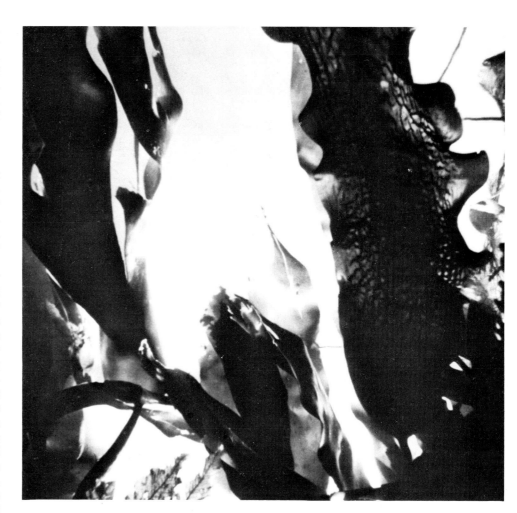

Laminaria kelp (above) is a plant that most swimmers know well. But such fixed seaweeds comprise only a tiny percentage of salt-water vegetation, which mainly consists of the barely visible phytoplankton shown on the facing page.

phytoplanktonic species. Although we know that some pigments must be much more efficient than others under certain light conditions, such other matters as nutrient supply seem to have more effect on species' abundance and level in the water.

Why, you may ask, should the availability of nutrients favor some species over others? Surely, the more food the better for every type of alga? But this is no truer at sea than on land. Cactus and scrub survive in soil that could not support an apple tree. If the soil and weather conditions were more favorable, trees would take over and cactus would not have a chance. In short, some species are experts at thriving in bad conditions. The truth of this statement as it applies to marine plants can be tested by studying the response of various species to different concentrations of a nutrient such as nitrogen. If you start with a very low concentration, all species will grow slowly, if at all. When you increase the concentration, some species respond at once, rapidly soaking up the nutrients, while others still cannot grow successfully on so little. As you further increase the supply, however, the species that responded first begin to show

less and less improvement, whereas the slow starters grow faster and faster. This variability means that in the changing conditions of the real ocean there is always some species that can exploit the grimmest situation and continue to grow, even if only slowly, as long as the concentration of nutrients is not absolutely zero. As concentration levels range upward from oligotrophic (poor) to eutrophic (rich) so the phytoplankton species appear and disappear according to their ability to flourish under different conditions.

The fourth factor affecting the success or failure of a given species is its "tastiness." Not all plants of the phytoplankton are suitable food for all herbivorous members of the zooplankton. The herbivores have preferences, just as cows and goats do; and, to continue the analogy, goats can eat plants that cows cannot. Even a brief discussion of what eats what in the plankton is out of the question here. Clearly, much depends on the size and maturity of the herbivore, the total amount of phytoplankton food present, the ease or difficulty with which a species of alga can be eaten, the temperature of the water, and so on. Beyond that generalization, what matters from the standpoint of our limited study of diversity is that this is one more reason why a thousand species of phytoplankton have evolved in the course of hundreds of millions of years.

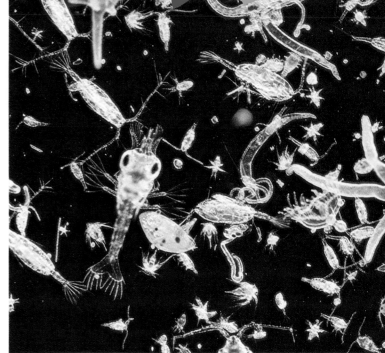

Who Eats the Plants?

I have already alluded to the fact that the phytoplankton blooms of spring and autumn are inevitably followed by a sharp decline. This is due not only to seasonal and nutritional variations discussed in earlier sections of this chapter, but also to the fact that physical removal of plant life from the upper layers of the ocean results from natural death and – more especially – grazing by zooplankton. The first step in the transfer of energy from primary producers (plants) to top consumers (fish, sea-dwelling mammals, and, ultimately, man) is the herbivorous grazing process.

Marine biologists are profoundly interested, of course, in the processes of grazing, but our understanding of this complex subject is still very limited. You might suppose that answers to some of the key questions could be arrived at by means of analogy. Could we not learn something about herbivore relationships with algae, for example, by comparing the zooplankton with cattle eating grass or insects feeding on leaves? The answer is no. Land plants have to stand up against gravity, and so they have hard roots and many woody or fibrous parts that grazing or browsing animals do not eat. By contrast, plants of the plankton are entirely digestible, and they are gobbled up whole. Grass and trees can be grazed continuously, yet go on growing and producing seeds. An algal cell is either eaten or not eaten; if eaten, it is dead and cannot reproduce. Thus, the effects of herbivorous feeding are much more devastating in the sea than on land.

For example, the total mass of new plant growth each year in the United States has been estimated at 5,000 million tons. Farm animals eat just over 10% – about 515 million tons – of this total, and the annual net growth of the animals is only 25 million tons. Thus, in the most efficient agricultural system in the

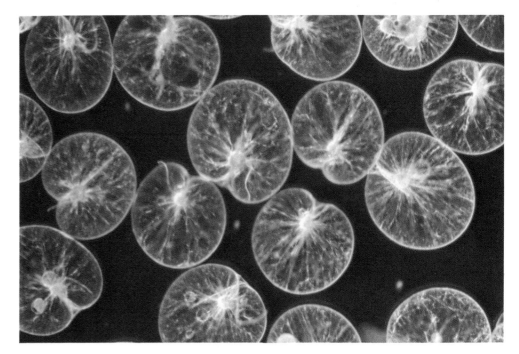

Semiannual phytoplankton blooms, producing growths like this thick crop of luminescent dinoflagellates, are the vital first link in the marine food chain. Unlike terrestrial plants, nearly all of these will be eaten by herbivores.

world, only about 0.005% of the annual total growth of plant matter is converted directly into animal flesh. Since only 1.5% of the annual vegetable growth goes directly for human consumption, and since the amount of plant food eaten by wild animals and insects does not appreciably alter the balance, nearly 90% of all new terrestrial vegetation is left uneaten. As a whole, of course, the United States is outstandingly rich in plant life, and there are many desert areas on earth where the amount of vegetation can barely support even sparse animal populations. But the sea is, so to speak, *all* desert; most of the time, the total mass of marine animals actually exceeds the total mass of plants!

It is only the phenomenal burst of phytoplankton production in bloom periods that fuels the marine food chain. Because each plant cell divides again and again, the quantity of plant life produced and eaten in the course of a year is many times greater than the quantity you would find in the ocean at any one time. Herbivorous animals must therefore depend entirely on current phytoplankton production, and if a bloom fails there is no stored capital of vegetation from previous years. At such times, larger animals can still feed on smaller ones and are not immediately affected by poor phytoplankton production, but the effects of a bad grazing year will eventually work their way up the food chain until they cause a bad year for fisheries.

The relationship between the phytoplankton and the marine animals that feed on it is delicately balanced, for not only is there a severely limited supply of vegetation, but

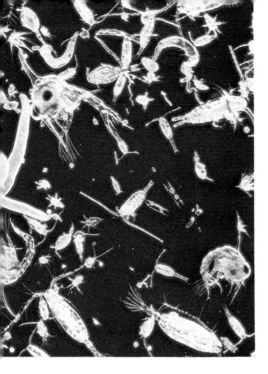

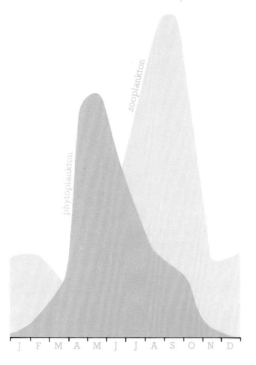

As this diagram indicates, there is a surprisingly long lag between the peak production period of Northern Hemisphere planktonic algae (early spring) and maximum zooplankton growth (late summer). For some reason the herbivores do not begin to multiply rapidly until some time after conditions for algal growth have worsened.

Above and top left: Two views of zooplankton (highly magnified, of course) grazing during an autumnal phytoplankton bloom. The plants are mostly diatoms; the animals are various types of copepod and crustacean larva.

different kinds of herbivore feed at very different rates. It would, perhaps, be fairly easy to study the plant-animal relationship if the zooplankton could be accurately categorized as herbivores, carnivores, and omnivores. But we cannot yet do this: a given omnivore sometimes eats planktonic algae, and at other times eats herbivores, carnivores, omnivores, or even the larvae of its own species. Thus the food chain gets tangled up, so that marine biologists prefer to speak of food "webs" rather than "chains." And – not surprisingly, considering the complexities – biologists are now enlisting the aid of computers for making models of the flow of energy through such webs.

Most of our estimates of the rate at which zooplankton consume phytoplankton are based on the quantities of zooplankton caught in towed nets at different seasons. But it is not possible to design a net that traps only herbivores or only carnivores, or to try to identify under a microscope the species of all the myriad zooplankton in, say, a square kilometer of seawater and sort them out quantitatively. So we have to make simplifying assumptions, none of which is entirely dependable. For instance, although we can draw certain conclusions from comparing the total abundance of zooplankton with the general availability of phytoplankton at different times and places, those conclusions become suspect when we discover the extreme variations in plant and animal populations over remarkably short distances. Such local variations remain largely inexplicable.

Even the timing of maximum zooplankton growth in relation to phytoplankton blooms is rather puzzling. In the Northern Hemi-

sphere, the zooplankton is generally most abundant in July, August, and September. Yet the greatest bloom of algae occurs typically as early as March inshore, and in May and June offshore. Moreover, zooplankton stocks tend to be low in the winter, although phytoplankton abundance near the coast may be moderately high. Naturally, we'd expect to find the peak of zooplankton growth following on the heels of that of the phytoplankton, but, surprisingly, Northern Hemisphere zooplankton is ordinarily least plentiful near the coast in April after the spring bloom has begun, and most plentiful in midsummer when phytoplankton growth has been at its lowest for quite a while. Clearly, planktonic animals do not start to multiply rapidly until some time after the plants do, and they continue to increase – grazing, cropping, and reducing the vast population of algae – long after ideal conditions for algal growth have de-

Sedentary herbivores can be as hard on fixed algal species as the zooplankton can be on the phytoplankton. The rock under this limpet is bare because the limpet has grazed it clean of green algae.

teriorated as a result of the lack of nutrients.

To return to our earlier comparison of the marine situation with the terrestrial: the spike of spring growth in the ocean is not unlike what would happen if trees and flowers and grass and shrubs pushed up from a few seeds every spring, proliferated wildly for a couple of months, and were then stripped bare and gnawed to nothingness by pullulating swarms of locusts, other insects, mice, and birds, within the next two or three months. In this process, almost the entire production of plant life would be consumed and relatively little would rot away. Even a little is *something*, however. Some of the planktonic vegetation does die a natural death, as we shall see in the following pages.

Death and Decay

As the growing phytoplanktonic algae absorb mineral elements and other nutrients from the sea, where such chemicals are in low concentration, they gain weight, as do all living things that are eating well. In other words, by drawing the nutrients inside their cell walls, the plants achieve concentrations many times those in the surrounding water; and as a result of their increased density they tend to sink. While they are alive, they secrete fats and oils that keep them buoyant, but when they die they inevitably sink (assuming, of course, that they have not already been eaten). Since, as we have seen, the availability of nutrients in the top 50 to 100 meters of water is critical for the plant's well-being, any transport downward of such concentrated nutrients by dead or dying algae is a matter of great importance: how can these nutrients be recovered from the algae and recycled into the upper layers of the ocean? Luckily, much of the necessary work is constantly being done by herbivorous members of the zooplankton. And bacterial decomposition does the rest.

The grazing by zooplankton achieves recycling in two ways. First, the plant-eating animal breaks open the algal cells as it eats, and thus releases internal fluids containing complex organic molecules and nutrients, which can then be broken down by bacteria. Secondly, after digesting its meal, the herbivore excretes fecal pellets that contain ammonia, phosphorus, and other organic debris. Some of this material, especially the ammonia and phosphorus, is dissolved by the seawater and becomes directly available as a nutrient for the phytoplankton; the rest is eventually broken down by bacteria. Sometimes, in fact, the fecal pellets are so dense that they sink rapidly to the bottom without disintegrating. In such cases, all the material in them becomes subject to bacterial decomposition.

Recycling is also accomplished by direct bacterial action on the dead or dying algae. We do not entirely understand the mechanics of this process. If, for example, you add phytoplankton cells to a mass of seawater, there is a measurable consumption of oxygen (known as "oxygen demand") even when there is no photosynthesis going on. We can infer from this phenomenon that something in the water is breaking down the chemicals of the phytoplankton and using oxygen in the process. The "something" is bacterial action, of course, but just how the bacteria attack the organic molecules of the cell, and what sequence of changes takes place before basic nutrients are released back into the water, we do not know.

Evidence for what happens comes mostly from analogy with the biochemical processes associated with the soil bacteria that break down vegetable wastes and manure so that such materials are decomposed into the best of all possible fertilizers. The stages of decomposition in the sea must surely be similar to those on land. Let's take nitrogen as an example. When bacteria attack a dead algal cell, they gradually break down its nitrogen-rich molecules, which are in the form of proteins and amino acids. In the normal course of events, the bacteria would decompose these to ammonia – that is, to nitrogen combined with hydrogen – which is a perfectly accessible source of nutrient nitrogen for phytoplankton; but this does not seem to be what happens, since most of the nitrogen dissolved in the ocean is not present as ammonia but is in the form of nitrate or nitrite, which are oxygenated compounds that phytoplanktonic plants can also readily absorb as nutrients. Obviously,

there must be a force at work that adds oxygen to the nitrogen at some stage in the breakdown, and in the soil this force consists of a further group of microorganisms called "nitrifying" bacteria. Such bacteria, then, are most probably present in water as well as in soil.

The phosphorus released from dead algal cells is bound up in large organic mol-

Although the vast majority of planktonic algae are eaten, some do decay and die. In the electromicrograph below we see a number of diatoms, one of which is reproducing itself by dividing while others (in the bottom corners) are dying. The siliceous skeletons of the dead diatoms will slowly sink to the ocean floor to join the diatomaceous ooze (pictured at the bottom of the page). Before reaching the floor, the nutrient chemicals of the cells will have been released and recycled to the phytoplankton.

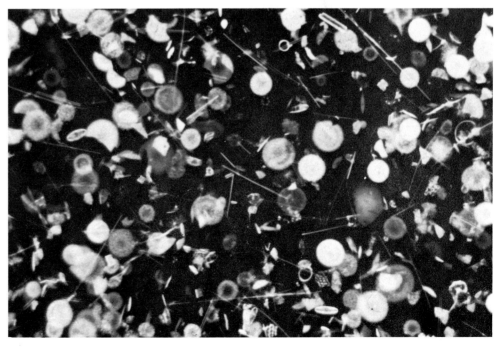

ecules that cannot be reabsorbed directly by living cells. Thus, since the algae have no way of extracting the phosphorus from the molecule, bacteria – or possibly enzymes – must once again come to the rescue. They attack the big molecules, cleave the phosphorus from its chemical bonds, and release it as easily absorbable phosphate.

Because most of the direct decomposition of planktonic algae in the deep ocean occurs in the upper few hundred meters, only a very small amount of the surface production from photosynthesis sinks all the way down to the ocean floor. Quite a lot, though, drops well below the lower limits of photosynthesis, and the broken-down nutrients remain in the deeper levels outside the cycle of primary production until storm winds cause them to be brought up through vertical mixing, or else an upwelling current turns the water over. On the continental shelf and in shallow coastal waters, however, a considerable quantity of organic debris, dead algal cells, and fecal pellets from zooplankton falls to the bottom, where it is trapped in the sediments. But it is only temporarily trapped; further decomposition takes place within the sediments, through the action not only of bacteria but of the numerous bottom-dwelling animals that ingest the sediments in order to filter out and digest the organic material. The end of the story never varies: final waste products are returned to the water as dissolved chemicals, which can once again be absorbed by the phytoplankton. And, in this way the cycle of life, death, decay, and life again goes on and on.

Below: How the endless cycle of life, death, decay, and life progresses in the world's oceans. Note that the chemicals in a dead cell are generally broken down and recycled within the top few hundred meters of the water. The only nutrient-rich material that ordinarily sinks down to the depths or to the bottom gets there in the form of dense fecal pellets from the zooplankton; after bacterial action decomposes these, the nutrients are recycled upward through vertical mixing or upwelling currents.

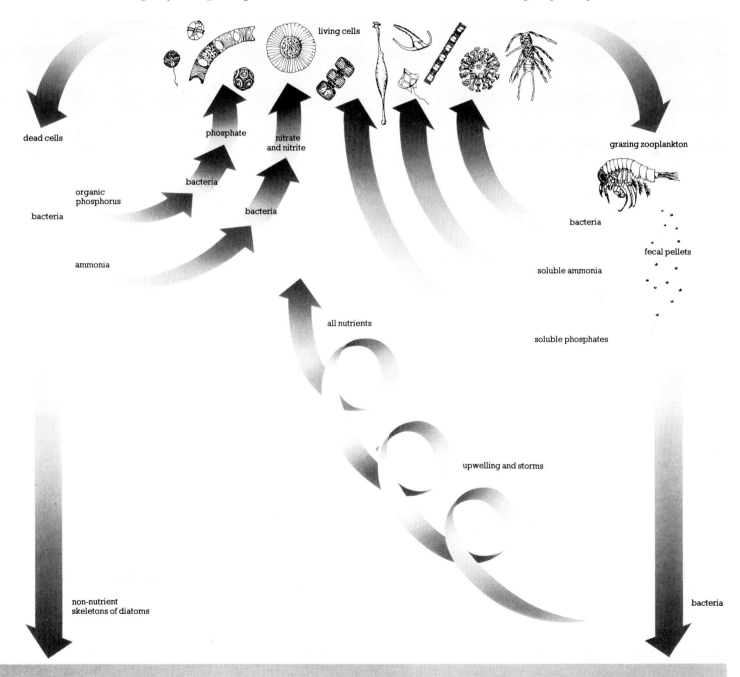

living cells

dead cells

phosphate

nitrate
and nitrite

grazing zooplankton

organic
phosphorus

bacteria

bacteria

bacteria

bacteria

ammonia

fecal pellets

soluble ammonia

all nutrients

soluble phosphates

upwelling and storms

non-nutrient
skeletons of diatoms

bacteria

Two stages of the semiannual boom and slump of phytoplankton. During the winter only a scattering of diatoms are available to such zooplankton grazers as the barnacle larva distinguishable here. Spring produces a phytoplankton bloom, with diatoms and dinoflagellates in profusion; swift growth has been made possible by the presence in upper layers of nutrients brought up by winter storms.

The Odds for Survival

The boom-and-slump relationships between nutrients and phytoplankton and between phytoplankton and herbivores appear at first sight to be very unstable. Yet the system as a whole continues from year to year to provide a more or less steady input of biological energy for the food web. Ecologists would like to be able to make a predictive model of the food web that would take into account all the factors I have been discussing in this chapter, including the possible effects of pollution or the removal of predators by overfishing. But one of the most difficult things for any such model to explain would be how, in the first stages of production, the extreme oscillations of nutrients, phytoplankton, and zooplankton respond to variations in the environment without getting out of control. The system as a whole must be much more stable than any of its parts. If we are to understand how productivity is maintained at a viable level, we need to know more about what lies behind the stability and capacity for self-correction of the phytoplankton population.

There are two extremes at which primary productivity does come to a stop: (1) when there are no nutrients at all or the water becomes extremely toxic; (2) when grossly excessive nutrients are added to the water, so that an overproduction of algae at the surface cuts off light penetration. We shall examine this second phenomenon later on. But both phenomena can occur only in severely disturbed situations. In investigating the stability and productivity of the system, we are more interested in how it functions normally, in response to the ordinary random variations in nutrients and weather conditions.

Marine biologists have constructed simple models designed to show the relationship between the growth of a population of herbivores and their food, but even the simplest such model needs the complex support of a computer to reproduce the "typical" boom-and-slump oscillation. When random variations in conditions are introduced, the model suggests that the food – that is, the algae – would often have to be completely consumed by the herbivores. If it *were* completely consumed, of course, there would remain no algal cells to take advantage of the next appearance of conditions suitable for a bloom. An analogous situation in human society might be the almost inevitable fate of a gambler who keeps betting on some random event: sooner or later there will be such a long run against him that he will lose all his money. Thus, even in extremely subtle models that take into consideration several species of alga, one species or another always gets eliminated by chance.

In the real ocean, however, the pattern of nutrients, phytoplankton, and zooplankton grazers seems remarkably stable in the long run. As far as we can tell, marine plant species simply don't go extinct. There are big· variations in total and regional productivity from year to year and place to place, of course. But in general the system appears to be extraordinarily self-perpetuating – and we can only theorize about the reasons why.

I have pointed out that stable en-

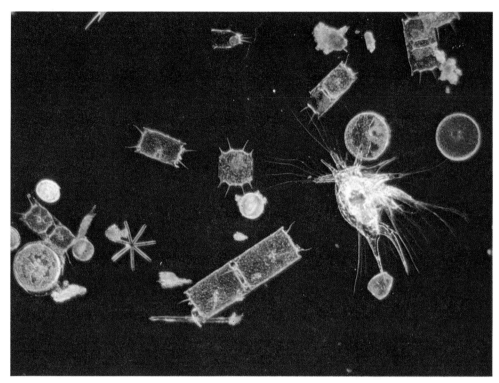

Winter

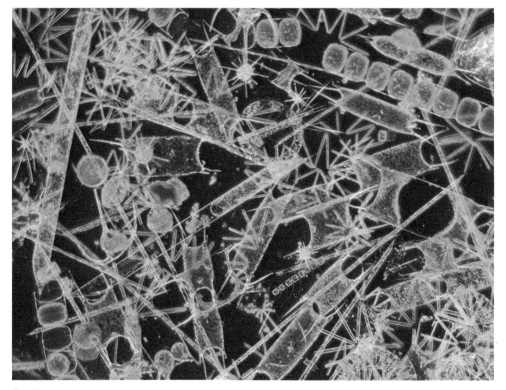

Spring

vironmental conditions tend to produce a great diversity of phytoplankton species, whereas variable or catastrophic conditions encourage only a small number of species. It is therefore tempting to suspect that the reverse might be true – in other words, that a region with a set of widely varying species, each relying in different ways on a variety of nutrients, and each being grazed by differing types of herbivore, might be more resistant to change than is a more homogeneously populated region. For example, highly diverse tropical plant populations might be more resistant to change and disturbance than are the less diverse populations of temperate waters. But this argument cuts both ways, since a complex food web is nonetheless a web, and the disruption of one strand may rupture scores of others, thus eliminating species one by one and perhaps causing the whole system to collapse. Computer models indicate that species diversity does not of itself guarantee stability.

Another possibility is that the "gambler's ruin" type of extinction does occur among algal species, but only on a very local scale, leaving nearby patches or islands of continuing productivity; later on, when conditions are better, the temporarily lost species can spread back into the gaps. Patchiness of plankton distribution in the ocean is often observed on a scale of a few tens to a few hundreds of meters. But for this "island" patchiness to protect the threatened species of phytoplankton from being totally consumed, we would have to assume that currents and waves and turbulence did not also carry the herbivorous zooplankton from one patch to the next. That is an obviously untenable assumption. So the patchiness theory is no more convincing than the other one.

In all probability, then, the missing element in the model – and the least easily predictable one – is the response of zooplankton to variations in the concentration of the algae on which they feed. Within certain limits, the more food there is the faster the zooplankton eat and reproduce. But definitions of these limits have only recently been discovered through laboratory experiments that have pretty well established the maximum rate at which zooplankton can eat, no matter how much plant life is present in the water. If you dilute a suspension of algae and zooplankton in natural seawater, there comes a point when the herbivores simply stop grazing. Thus there seems to be a low limit of phytoplankton concentration – we call it the "threshold" – below which the plants are not eaten. Somehow, when one species is being grazed down to the threshold concentration, the herbivores transfer their attention to another species. This means that when conditions are ideal for the phytoplankton, the rate of production

gets well ahead of the rate of consumption; and when conditions are bad, the grazers can never quite eliminate the last few algal cells.

This, then, is what has almost certainly assured the survival of the phytoplankton in all its amazing variety. Is man capable of destroying the wonderful balance of nature? That question, so often and necessarily asked in our day, will be considered in the next section.

By midsummer the numbers of phytoplankton have diminished again, and there remain just a few dinoflagellates and diatoms still alive and floating in the calm water. But this slack period is followed by an autumnal bloom (which is somewhat less profuse than the earlier one). Grazing and an increasing scarcity of nutrients will complete the annual cycle by bringing on next year's winter slump.

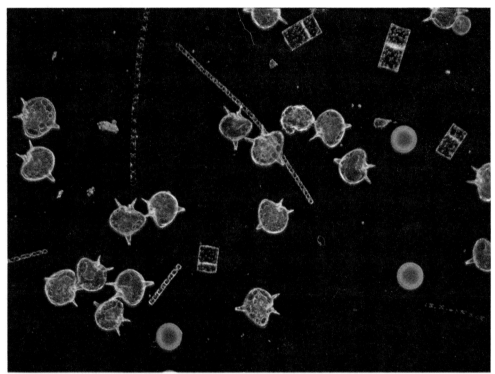

Summer

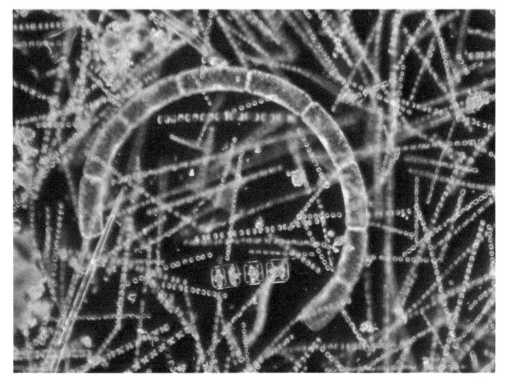

Autumn

Plants and Pollution

Pollutants such as sewage, chemical fertilizers and insecticides, or oil can affect the phytoplankton in several ways; they can kill the plants, can reduce the rate of growth or ability to reproduce, or can contaminate the cells in such a way as to affect the creatures that eat them – all of which, of course, may either decrease or harm our own marine food supplies. Without, at this point, delving into the larger implications for the food web, let's consider the direct interactions between pollutants and the phytoplankton.

The most common killers of oceanic plants are excessive sewage and industrial waste. These begin by actually enriching rivers, estuaries, and coastal waters, but in excess they create a condition called "eutrophication" (literally "healthy nourishment," but in fact *over*-nourishment). The nutrient phosphates and nitrates cause an algal bloom at first, but this soon cuts off the light to deeper layers of algae and restricts photosynthesis. As the bloom thickens, the euphotic zone within which fruitful photosynthesis is possible becomes very shallow. Below it lie large quantities of either dead and decaying or respiring, but not photosynthesizing, algae. Thus there is a growing consumption of oxygen, along with an increasing production of carbon dioxide, throughout the water below the ever-narrowing euphotic zone. Eventually, less oxygen is produced in the euphotic zone than is used up below it; the lower parts of the water column become anaerobic (i.e., totally stripped of oxygen), and, in their search for oxygen, bacteria start to break down natural sulphates and to release hydrogen sulphide gas. This ultimately destroys all remaining algae and other life, and the water becomes a stinking sulphurous mess.

It doesn't always go that far, of course. Prior to the "terminal" condition, however, the species composition of the phytoplankton population keeps changing as eutrophication progresses. The initial enrichment tends to favor a few species, and their proliferation may be considered temporarily beneficial, since an increase in total primary production means more food for herbivores. However, in such abnormal nutritional conditions one or more species of nonbeneficial, or even harmful, plant may become dominant. What happens then is well illustrated by what happened, a few years ago, as the result of the conflicting interests of the duck-rearing and oyster-and-clam industries on the shores of Long Island Sound. As increasing quantities of excreta from domestic ducks in the creeks and shallow waters of Great South Bay washed into the sound, eutrophication brought on a rich bloom of algae – but, unfortunately, of species that oysters and clams could not eat. So, as the bloom continued, the water was depleted of oxygen, and the lack of food and oxygen almost entirely destroyed the mollusk population. Strict control of the levels of nitrogen-rich sewage dumped in the sea is the only way to prevent such occurrences.

Even more potentially dangerous is the widespread dispersal of chemical fertilizers and insecticides. How DDT has been spread all over the world from farm spraying, through the atmosphere and rain, into the oceans and even the ice of the Antarctic, is now well known. The solubility of DDT in water is very low – only 1.2 parts per thousand million – and so you might expect that the phytoplankton could hardly absorb enough to harm their metabolic processes, since the present concentration in seawater is much less than this. DDT, however, is much more soluble in fatty tissues than in water, and so there is some possibility that it could become concentrated in algal cells. To find out what would happen if such plants were exposed to fairly high concentrations under laboratory conditions, cells have been grown in seawater containing 10 parts per thousand million of DDT. The result: a drop of over 20% in rate of photosynthesis. True, the use of DDT is now restricted in most industrial countries, and the quantity in

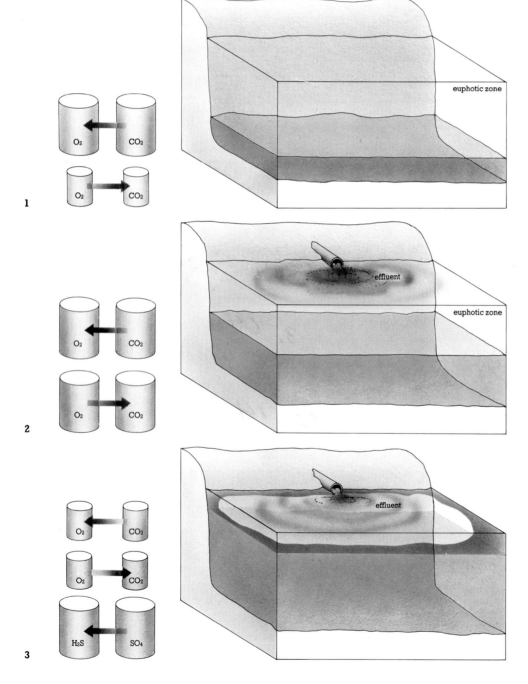

In a healthy body of seawater (1) light penetrates quite deep, enabling plants to consume carbon dioxide and produce oxygen; and bacteria break down dead cells and release carbon dioxide. If an effluent containing nutrient chemicals is added (2), overstimulated algal growth reduces light penetration and the euphotic zone narrows until, in a state of eutrophication (3), nothing can remain alive except the bacteria that produce stinking hydrogen sulphide by breaking down sulphates in their search for oxygen.

the ocean is therefore never likely to reach such a high level. But we cannot be too complacent about prolonged exposure even to low concentrations of various kinds of chemical insecticide.

Most of the effects of a third type of pollution – oil spills – have been studied only on or near the shore, where they involve shellfish, birds, seaweed, and tourists. Since oil seems to do greater harm to herbivorous animals than to the phytoplankton, one side effect of a spill can actually be to the advantage of the plants – and it has sometimes been observed that they may flourish in oil-damaged seas where there are suddenly fewer animals to eat them. But reports are very confused, since tankers carry many kinds of refined and unrefined oils, and oil itself contains a number of aromatic carbon compounds of varying solubility, volatility, and toxicity. The chemicals that would do most damage to phytoplankton are those that dissolve most readily in water but are relatively nonvolatile. Laboratory experiments indicate that salt-water concentrations of 20 to 100 parts per million of such substances would significantly reduce the growth rate of phytoplankton. Though concentrations of such magnitude can occur only in relatively small patches as a result of oil spillage, there is nonetheless cause for concern in the fact that they *can* occur. Toxins absorbed by phytoplankton at levels that do not significantly inhibit algal growth are likely to be ingested by herbivorous zooplankton and shellfish – and some soluble aromatic carbon compounds are carcinogenic. Thus, even very low concentrations in the phytoplankton can result in an accumulation of cancer-producing substances in the herbivores and on up the food chain.

Because of the constantly moving, drifting nature of the phytoplankton population, and because of its rapid rate of growth and destruction, it is very much more difficult to determine how algae are reacting to pollutants in the open sea than in coastal waters. On a worldwide scale, we can probably assume that the concentration of pollutants in the open ocean is still so low as to be unimportant, but it could already have reached dangerous levels in such closed seas as the Mediterranean. Fortunately, the states that border the Mediterranean took a step in the right direction in 1976 when they agreed upon a coordinated effort to study the problem and institute controls.

Three scenes of desolation caused by pollutants. Top: Fixed algae bleached and bearing the marks of limpets killed by an oil spill. But the cure can be as bad as the original evil. Middle: Here detergents used for cleaning up a spill are killing seaweed. Bottom: In this mass of fixed seaweed (known as a kelp forest) growing near a chemical-plant outlet, few of the smaller, more delicate algae can live – a typical lack in polluted waters.

Imprisoned Plankton

Algae account for more than two-thirds of the weight of living tissue in a coral reef. The corals themselves are, of course, animals and will be discussed in the next chapter. But because they depend for their survival on the algae that live with them, the primary productivity of this unique system warrants special attention here.

Algal photosynthesis produces energy and chemicals for reef building in four different ways. First, there are certain species of alga that actually secrete their own stony crust; these grow very strongly in the surf zone, where they continuously reinforce the coral reef at its most vulnerable point. Secondly, there are true planktonic algae which, together with the zooplankton that feed on them, are filtered out of the water and eaten by coral polyps. Since reef-building corals can grow only in waters warmer than 23°C, and since such tropic waters are low in nutrients, it is something of a mystery how the polyps manage to filter out enough food to grow on. Corals, in fact, thrive in some of the world's clearest waters, which are far from rich in phytoplankton. Strong currents and heavy surf may transport large volumes of water over a reef, thus increasing the food supply; but the efficiency of coral growth that depends on planktonic algae for nourishment is still surprising.

There are two kinds of alga, however, that actually live inside the coral – one within the cells of the coral tissue, the other within the stony skeleton – and these are the types that I want to discuss in rather more detail. Those that inhabit the tissues of the coral animal are small, brown, single-celled plants, which the famous British marine biologist C. M. Yonge has dubbed "imprisoned plankton," but which are technically called zooxanthellae. They are always found in coral reefs and never float freely in the water – a clear indication that the algae get an enormous advantage from living in coral tissues and cannot survive outside. Conversely, natural reefs are never found without zooxanthellae in the corals, and so biologists used to believe that the relationship was symbiotic (that is, of mutual benefit to the partners). It was thought that the algae benefited from the physical protection of the corals and the nourishing supply of animal waste products, while the corals could, if necessary, eat and digest their tenants.

We now know that the situation is not that simple; it is possible, indeed, that the algae are more nearly parasitical than symbiotic. The advantage to the plants is that they have immediate access to metabolic waste from the corals – to carbon dioxide, phosphates, and nitrates – and thus do not need to rely on the low concentrations of these essential nutrients in seawater. Laboratory tests have shown that the algae not only remove all the phosphate produced by the coral, but will also use up any extra phosphate that might be added to the water. It seems, therefore, that the concentration of algae in coral tissue is limited only by the supply of nutrients. In practice, such intensive recycling of nutrients ensures that there is a minimum wastage from the entire coral-reef system. Meanwhile, photosynthesis by the zooxanthellae produces large quantities of oxygen within the corals; and since all animals need oxygen, this might seem to be the major advantage accruing to the coral from the intimate alga-coral relationship.

It only *looks* that way, however. In fact, the concentration of oxygen in salt water is nearly always enough to keep corals alive, and so the oxygen from resident algae could be almost superfluous. It's possible that some of the extra oxygen assists the formation of the calcareous coral skeleton. But that is merely a possibility. Similarly, though swift removal of waste excreta from the coral undoubtedly improves the efficiency of coral growth, such a benefit seems at best marginal. As for nourishment for the animals – well, you might think that at least in bad times, when planktonic food is lacking or there is a drop in temperature, the coral would digest away the zooxanthellae, but this doesn't happen. The algae are adapted to live inside the cells of the coral, where they have complete immunity from enzymes of digestion; what actually happens when the coral is short of food is that the algae are ejected from the host cells.

If corals in a laboratory are kept for many days in darkness, but supplied with food, the zooxanthellae die because they cannot photosynthesize, but the coral goes on living. Thus, although the plants depend absolutely on the animals, the animals can apparently do without the plants. This is why it is a temptation to class the zooxanthellae as highly successful parasites. But we are

Below: An electron micrograph of a zooxanthella, with its fingerprint-like pigment lines, black central chromosomes, and white starch grains. These algae live inside corals – but whether as symbionts or parasites is a moot point. Algae of various types make up more than two-thirds of the living-tissue weight of the Great Barrier Reef (facing page) and other coral reefs.

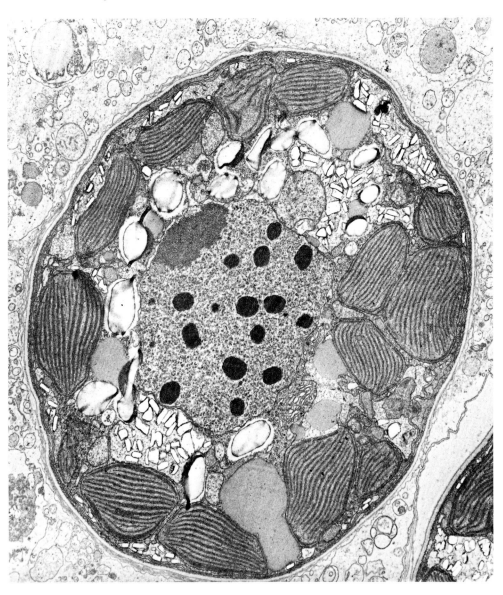

forced back to the fact that not only are reef corals never found without zooxanthellae, but they are never found growing deeper than algae can photosynthesize. So it does look as if, on the scale of a whole reef, the coral somehow needs the algae. We are still trying to discover what that need is.

If the role of the zooxanthellae is ambiguous, the fourth class of algae is even more mysterious. Recent research has shown that two-thirds of the living tissue in a coral head is plant life, and of this only 6% consists of zooxanthellae. The remainder is made up of filaments of green algae that live inside the porous stone skeleton of the coral. But how can this be? How can plants grow inside stone where they are cut off from life-giving light? The explanation is a matter of great controversy. Even if parts of the green filaments may be exposed to light on the surface of the skeleton, the resultant photosynthesis could not be very efficient – certainly not sufficiently efficient to support the rest of the algae inside the stone. Perhaps, instead, these green algae are something like saprophytic fungi, living off dead organic remains (and, indeed, other algae are known to be of this sort). We are still, however, far from a solution to the mystery.

At any rate, there is no mystery about the fact that algae have successfully exploited the coral-reef environment. The zooxanthellae seem to live with the corals in a balance somewhere between symbiosis and parasitism, and the filamentous green algae are probably pure parasites. We do not know how much their productivity actually contributes to the fertility of the reef, but we do know that the reef provides them with a very good home.

Food from Phytoplankton

Annual primary production of phytoplankton growth totals about 15 thousand million tons of fixed carbon. This is the base of the food web on which we rely when we catch mackerel, or tuna, or whales. But fisheries in many parts of the ocean are already exploited up to, or beyond, a sustainable limit, and catches are dropping. There are several possible ways to improve the situation. Those that involve fisheries management and fish farming will be discussed in a later chapter; the broadest possible understanding of the phytoplankton may enable us to plan two other ways of increasing salt-water food production.

In the first place, accurate measurements of primary productivity in the sea, along with an increasing knowledge of marine food chains, should allow us to predict the maximum safe level of fishing effort for every edible species in every sea area. Then, secondly, we can eventually try to farm algae – whether in natural lagoons or in controlled tanks – so as to produce food for farm animals, for man, and for farmed fish and shellfish.

It is no good pretending that we can already predict marine productivity accurately, and the reasons should be obvious. The vagaries of water movement,

nutrient supply, phytoplankton blooms, and herbivore grazing make it very difficult to predict the first two stages, or trophic levels, of the food web. Most commercial fisheries, of course, exploit the fourth level: i.e. the carnivores that eat the carnivores that eat the herbivorous plankton. At each level there is a delay for growing, hunting, and eating, and there is loss of energy representing the energy used up in swimming, metabolism, reproduction, etc. before a creature gets eaten by the creatures of the next trophic level. Any attempt, therefore, to predict fisheries production based on primary productivity by phytoplankton must involve a model that includes all the delays from one trophic level to the next and all the complexities of diminishing energy transfer.

Estimates of the efficiency of energy transfer between trophic levels vary, but scientists generally agree that the productive growth at any trophic level is only from about 10 to 15% of the weight of food eaten. And this loss of energy, remember, applies four times over before we get to edible fish. Thus, a diminution of only 1% in the efficiency of energy transfer at the lowest level results in a 30% diminution in the production of fish! Since an over-all estimate of efficiency is really only a simplification of the network of interactions

among the many species of herbivores, omnivores, and carnivores, any successful predictions of maximum safe levels for fishing efforts must depend on improved models of the grazing process as a first step.

Present world fish catch is about 65 million tons a year, and rough calculations suggest that we could probably raise it to from 120 to 150 million tons by spreading the effort over all the world's oceans. But even that much is a small catch compared with what might be obtained by utilizing, at a trophic level where far less energy has been dissipated, either zooplankton or phytoplankton. In theory, if we were to find our marine food at even the second or third trophic levels – that is, herbivorous zooplankton and first-stage carnivores – we could have an annual harvest of 1,000 million tons. But phytoplankton and herbivorous zooplankton are much too small and thinly scattered to be caught economically. One kind of herbivorous zooplankton, the krill, comes close to being economically

For cultivating experimental strains of different types of alga, small quantities are grown in glass jars under artificial illumination. Nutrients are pumped in through tubes, and the liquid is stirred by an electric paddle to provide equal light for all the algae. This scientist is siphoning off a number of dinoflagellates to establish ideal types for harvesting.

harvestable. But, in any case, to take plankton from the sea would simply be to remove the food of the commercial fisheries that we now enjoy. That would scarcely be a wise move.

What we could perhaps do would be to farm algae as we do wheat or barley, selecting ideal strains, giving them ideal conditions, and providing extra nutrients for them. Algae have already been cultivated experimentally as food for clams and oysters in commercial seafood farms, and this raises the possibility that mass cultivation might be possible on a scale that would produce significant extra quantities of protein for human or domestic-animal consumption. Controlled provision of light and nutrients would enable year-round production instead of the short seasonal blooms of the natural marine habitat.

Experimental cultivation shows that it isn't easy to provide the necessary light for photosynthesis in artificial algal cultures. The tanks of seawater cannot be very large, or they would cover too much land to be practical. So each tank must have considerable depth in order to hold worthwhile quantities of algae; then, if a really strong light blazes over the surface, the resultant dense growth prevents light penetration into the depths, and the phytoplankton farmer is faced with all the problems I have described in an earlier section dealing with pollution and eutrophication. The consumption of power for lights is also expensive. Moreover, to achieve an even distribution of nutrients and the removal of waste gases, the water in the tanks must be circulated either by pumps or by air jets – another costly item. Nutrients and fertilizers have to be purchased in quantities comparable to those required for a large agricultural acreage. And, finally, there's the problem of harvesting commercial quantities of algae, for this can only be done by mechanical centrifuging.

Because of such difficulties, most marine biologists doubt the value of mass culture of algae in totally artificial environments. There is another possible technique, though, which involves farming algae in open pools of large acreage under natural sunlight with domestic sewage (which must be free of industrial poisons and trace metals) used as a nutrient. It should be emphasized that large acreages are needed for an economically feasible farming operation even if the rate of algal production is improved ten times over the rate in the ocean. Again, the process is costly. Shallow natural-water bodies with the right kind of water circulation on the coast are hard to come by, and suitable land for setting up tanks or pools close to the shore is usually expensive. So is the construction of the tanks or of dikes or pits, if necessary. And there is always the problem of how to handle the circulation of enormous quantities of seawater without using extremely costly machinery. Nevertheless, the technique looks promising. If algae are cultivated in this way and fed to clams or oysters, the over-all efficiency is comparable to what we accomplish with cattle when we put them out to pasture in an area where one cow per acre can add from 100 to 150 kilograms to its weight in a year.

During the 1960s it was a popular assumption that the sea was a virtually limitless storehouse of natural resources. The pendulum has now swung the other way, and there is a strong tendency to minimize the role of the oceans as food producers. In fact, each nation tends to emphasize the importance of food from the ocean in direct proportion to the amount of marine food that it already includes in its diet. No knowledgeable person today claims that the ocean contains limitless wealth, but as I have tried to suggest in these pages, the plants of the plankton are remarkably productive and hardy organisms. As we gain an ever clearer knowledge of how they function, we should be able to put them to work for the betterment of mankind by helping them to multiply.

In spite of the obstacles to the successful commercial farming of algae, it can be done, as evidenced in these small tanks of marine plants under cultivation for feeding to oysters at a British oyster farm. Above: Water in the tanks must be artificially circulated to distribute nutrients and remove waste gases.

4
Salt-Water Animals

John Gage

To human beings, adapted as they are to life on land, the ocean may seem a hostile and dangerous environment. But earthly life originated in the ancient seas, and the bodies of all terrestrial animals retain legacies of the millions of years of evolution before living creatures made their first tentative moves to colonize the land. The blood that bathes our cells and tissues has a remarkably similar concentration of salt to that of seawater; and each human life begins in the miniature ocean of the womb, to repeat during its embryonic period many of the evolutionary steps taken by our marine ancestors. Air, in fact, is a less favorable medium for life than is salt water, and terrestrial animals face many difficulties that never confront those that live in the sea.

Since the marine inhabitant lives in a liquid that is chemically almost identical with its internal body fluids, it has no need for a tough, impervious skin to prevent loss of water or life-sustaining salts. Nor does it require a mechanism for controlling body temperature, since most surface waters have a seasonal temperature range of less than 4.5°C, and in the depths of the open ocean the temperature, though cold, is virtually constant. Indeed, only the mammals that evolved on land and later returned to water, such as whales and seals, are warm-blooded. Then, too, because seawater is much more buoyant than the atmosphere, marine animals are less affected than their landlocked cousins by the force of gravity; thus, strong skeletons and muscles are not essential for their support. That is why many of them have been able to develop strange and delicate shapes, often with soft and gelatinous bodies that would be quite impractical in air. Nor is it an accident that the sea is the home of the largest animal that has ever lived, for even the huge bones and muscles of the blue whale would be incapable of supporting its hundred-ton bulk if it were not buoyed up by water. The buoyancy, finally, is responsible for the existence of the strange free-floating planktonic community, which has no true counterpart in the atmosphere, and which makes it possible for many sea-dwelling animals – sea anemones and sponges, for example – to lead a sedentary existence firmly anchored to rocks or shells, where they can rely on the flowing water to act as a generous food-conveyor belt.

Is it any wonder, then, that the shapes and sizes of salt-water animals are so fantastically varied that a casual observer of marine life might well assume that there are a far greater number of animal species in the sea than on land?

The casual observer would be wrong, however. It is true that of the 22 major groups (or ''phyla'') into which zoologists divide the animal kingdom, all have salt-water representatives, whereas only nine

A

are represented on land. It is also true that of the 39,000 or so species comprising the phylum to which vertebrates belong – a phylum that includes all the mammals, birds, amphibians, and reptiles – more than half (roughly 20,000) are marine fish. Furthermore, most of the 21 invertebrate phyla, which all together include more than 97% of the total number of known animal species, have many more species that live in the ocean than on land or in freshwater, Yet, paradoxically, fully 85% of the million or so kinds of animal that zoologists have classified are land dwellers. The reason is that the eight invertebrate phyla that have managed to make a go of it on land have had to evolve into large numbers of closely related species, all of which differ only slightly from other members of the immediate family in their adaptations to the vast patchwork of available habitats.

To understand why this is so, imagine a 10-meter-square patch of Vermont meadow, and compare it with similar areas of Arizona desert, Scottish highland, tropical African rain forest, Siberian tundra, or even a nearby meadow on a different soil. Multiply such differences many times over and you will begin to see why a given family of land animals is bound to have evolved seemingly endless, if relatively minor, variations in physical attributes and life styles. By comparison, habitat diversity is not a major feature of the seas, where a patch of open water differs much less from other patches. Even on the ocean floor, areas where there are many kinds of habitat – coral reefs, for instance – are very few in relation to the vast stretches of muddy bottom in the deep ocean basins. On land and in freshwater, new species can evolve even in similar habitats if, for example, populations become

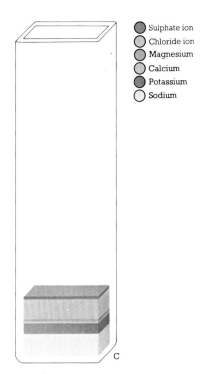

Sulphate ion
Chloride ion
Magnesium
Calcium
Potassium
Sodium

Above: A comparison of the concentrations of ionic salts in seawater (A), in the body fluid of a marine worm (B), and in human blood (C). Though similar salts are present in all creatures, concentration in human blood is notably weak.

isolated from one another by such barriers as islands and mountain ranges; but barriers to interbreeding in the sea are rare, for the continual mixing of the ocean waters into which most marine animals shed their eggs and sperm prevents such isolation. Instead, therefore, of large numbers of fundamentally similar species, such as the 25,000 kinds of beetle or the 15,000 different pulmonate land snails, we find enormous diversity of basic organization, great numerical abundance, and an extremely wide geographical range among the sea dwellers.

Their diversity, abundance, and range will become apparent as we trace the path of increasing complexity from tiny single-celled protozoans, so simple in structure that they can almost be mistaken for planktonic plants, to large squids, fishes, and marine mammals. But as we consider some of the ways in which these different animal groups organize their social life, how they reproduce, and how they compete for the sea's riches in order to survive, we must not lose sight of our own terrestrial world. Without the oceans there could be no life on land, and man's ultimate survival may well depend on how intelligently he makes use of the living resources that we shall be discussing in this chapter.

Illustrated here are the 22 major animal phyla. Note that all are represented in the oceans, as against only 9 on land. Within the 9, however, there are many more species because of the greater diversity of terrestrial habitats.

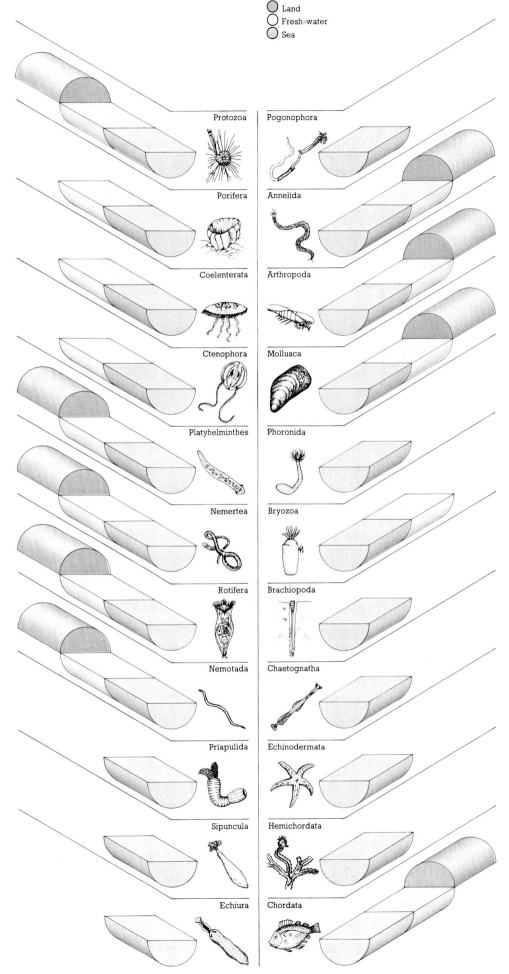

Land
Fresh-water
Sea

Protozoa
Porifera
Coelenterata
Ctenophora
Platyhelminthes
Nemertea
Rotifera
Nemotada
Priapulida
Sipuncula
Echiura

Pogonophora
Annelida
Arthropoda
Mollusca
Phoronida
Bryozoa
Brachiopoda
Chaetognatha
Echinodermata
Hemichordata
Chordata

The Invisible Billions

We take it for granted that the complex functioning of our bodies depends on a vast number of intimately joined cells working together. Yet the smallest marine animals, the protozoans, consist of only one cell, and they can nevertheless swim, crawl, eat, reproduce, and react effectively to external stimuli. Though almost all of them are microscopic, different species range in size from less than a hundredth of a millimeter up to a few centimeters – the size of certain foraminifers. In spite of being rather bigger than most other protozoans, the amoeba is one of the least complex; it is just a shapeless blob of protoplasm that flows within the limits of a semirigid cell wall, feeds by flowing around and engulfing bacteria and unicellular plants, and usually reproduces by splitting in two. So amoebas are not typical of the protozoans as a whole. More typical are the highly organized protozoans that belong to any of the three main marine groups: the ciliates, the foraminifers, and the radiolarians.

Of these, only the ciliates are accomplished swimmers. Most ciliates propel themselves forward by means of hundreds, or even thousands, of hairlike appendages (cilia), which beat rhythmically in a wavy motion, like a sudden breeze over a field of wheat, strong enough to move the tiny creatures – only a fraction of a millimeter long – at the relatively fast speed of a millimeter a second. At that rate, a six-foot-tall human being could swim a kilometer in one minute! What's more, the ciliates seem to swim purposefully, searching for and eating bacteria

A typical protozoan is propelled through the water by its rows of tiny hairs (cilia) all over the body surface. As illustrated below, these have a whiplike action characterized by a rigid downstroke with a relaxed return that progresses in rhythmic waves along each row.

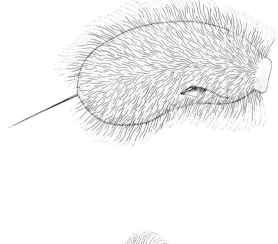

and minute bits of detritus. Some species have only a few large cilia instead of a great many tiny ones. All of them, though, are wonderfully well-coordinated creatures. They can even maneuver to avoid obstacles or unpleasant stimuli – which means that the coordinated action of the cilia can be readjusted in an instant.

It appears probable that an intricate system of fibers that connect the bases of adjoining cilia functions as a kind of nervous system. Scientists who have managed to cut through the fibers have discovered that the coordinated beating of the cilia stops when this has been done. Because it is hard to believe that such functional complexity can be contained in a single cell, many zoologists regard the protozoans in general not as unicellular, but as microscopic creatures *without* cells.

Whereas ciliates tend to swim in shallow coastal waters, especially estuaries and inlets where there are plenty of bacteria to feed on, the other major protozoan groups generally float along with the surface plankton or, as in the case of many species of foraminifer, live down on the ocean floor. Most foraminifers, wherever their habitat, have complex shells made of calcium carbonate, and these are marvelously varied in shape and form. A typical shell consists of a spiral chain of compartments, which are added at successive stages of the one-celled animal's growth. Scientists derived the name "foraminifer" from the Latin word for "hole" because these creatures capture their food through numerous holes in the shell: they put forth sticky strands of protoplasm to ensnare smaller floating animals, which are somehow digested outside the perforated shell (we still do not quite understand how).

The radiolarians are entirely planktonic, and their shells are unlike those of the foraminifers in that they are *internal*. Some of these supporting skeletons are even more intricate than the external "houses" of the foraminifers, but they are normally covered by frothy protoplasm, from which the strands for trapping prey are likely to project. One of the most remarkable – indeed incredible! – feats of the radiolarians is that most of them construct their exquisite skeletons out of either silica or strontium sulphate. Silica is a common land mineral, but there is little of it in seawater, for it is not particularly soluble; and there are only minute amounts of strontium in the world's oceans. Yet there are billions upon billions of radiolarians in the sea, each with its superb variety of silica- or strontium-made shell. How they manage to extract enough of these rare minerals from the water to serve their needs remains a mystery.

So vast are the numbers of foraminifers and radiolarians that their minuscule skeletons are the chief source of the muddy sediments

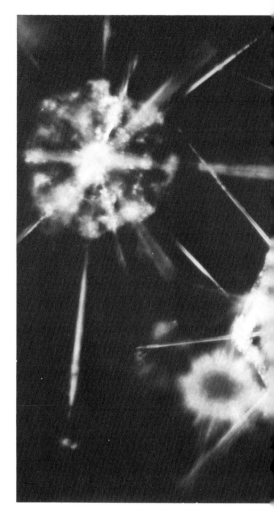

ment or "ooze" that carpets the floor of the deep ocean basins. In some places, the ooze may lie hundreds of meters thick despite the fact that the rate of sedimentation has been no more than a few centimeters every thousand years. Under the weight of all the overlying layers, lower layers have gradually become consolidated into rocks; and in those rocks we can read a petrified history of the ancient seas beneath which they were formed. A century or so ago, when foraminiferan skeletons were discovered in such high ground as the famous White Cliffs of Dover, it first became clear that many high places like this were once at the bottom of the ocean. Today, identification of the protozoan shells can tell us something that is perhaps more important for our generation: the presence or absence of certain kinds of fossilized foraminifer or radiolarian in rock samples brought up in deep drillings into the sea bed helps geologists to decide whether or not offshore-oil deposits are likely to be found in a given area.

Right: This is how the skeletons of billions of planktonic protozoans that have sunk to the sea bed form characteristic deep-sea oozes. Unlike the very deep-lying siliceous radiolarians, the calcareous skeletons of globigerine foraminifers cannot form deposits below 4,000–6,000 meters, because extreme hydrostatic pressure makes calcium càrbonate redissolve in seawater.

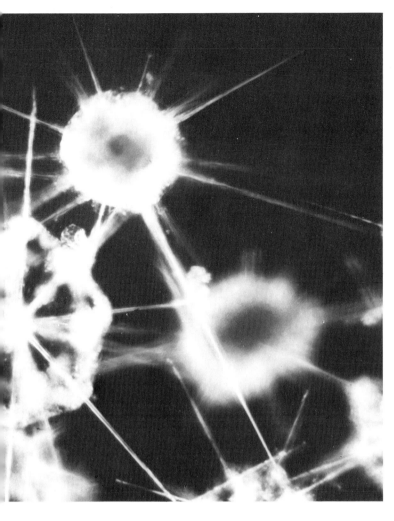

In the plankton of many parts of the world, microscopic foraminifers (above) and radiolarians (left) such as these occur in astronomical numbers. Their long spines and cellular cavities containing fluid with a lower salt content than that of the ambient water help them to float.

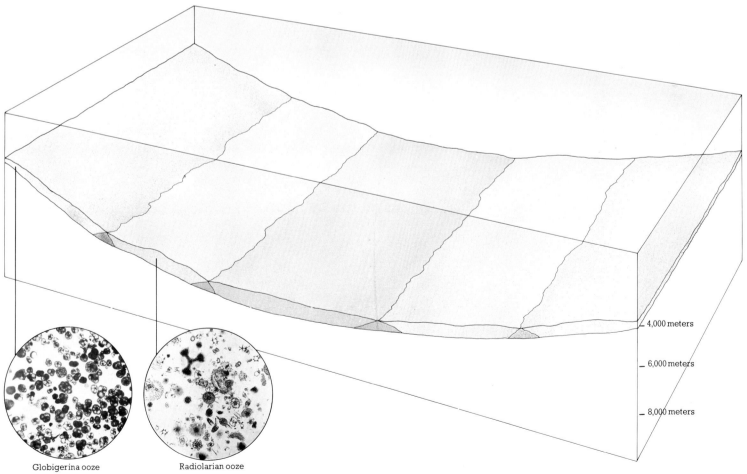

Globigerina ooze

Radiolarian ooze

4,000 meters

6,000 meters

8,000 meters

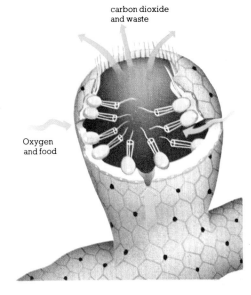

carbon dioxide
and waste

Oxygen
and food

Sponges, Jellyfish, Corals

Among the marine animals that probably evolved from unicellular protozoans, perhaps the least complex are sponges and coelenterates (a large group of multicellular but extremely simple creatures, such as jellyfish, sea anemones, and corals), for their separate lines of evolution did not get far before stabilizing on a more or less constant form. Though the cells of sponges, for example, are specialized for a few different functions, they are neither closely organized nor interdependent. Thus, if a living sponge is torn into pieces, each part eventually forms a new sponge. Even if the individual cells are separated, they wriggle along like amoebas until they come together and build up into whole sponges again.

In the marine world, though, simplicity of structure does not mean unvarying sameness of form and behavior. There are nearly 5,000 species of sponge, with a splendid diversity of shape and color, that inhabit the waters of every ocean from the tropics to the poles. Most sponges live attached to rocks or shells on the sea bed and are plantlike in their apparent inactivity. But they are very definitely animals; by means of the steady beating of little cilia, they draw in water through countless tiny pores, drive it through a meshwork of interior spaces, and, after filtering off bits of organic food and taking out oxygen, disgorge it, along with dissolved wastes, through one or more large openings ("oscula") in the body. The porous body is supported by a delicate scaffolding of criss-crossing, rigid spicules that may retain its form long after the living tissues have disintegrated. Household sponges are gathered from warm-water species whose durable skeletons are made of a fibrous protein known as "spongin." (The spicules of other species are usually composed of silica or calcium carbonate, neither of which would provide comfortable and efficient sponging power.)

The basic structure of the coelenterates is different, but also very simple – just two tissue-forming layers of cells separated by a noncellular gelatinous substance. The cells are more differentiated, however, than those of the sponges; for instance, every coelenterate has a body cavity that functions as a stomach and a sort of mouth surrounded by flexible tentacles. Each individual is technically known as a polyp (from a Greek word meaning "many-footed") because of those tentacles, which in a number of species have singularly effective stinging cells. The sting helps to catch smaller animals for food and protects the coelenterate from possible enemies, but it is seldom powerful enough to penetrate human skin. In some cases, though, the bad reputation of jellyfish is deserved; the notorious sea wasp of northeastern Australia's coastal waters, for example, secretes such a venomous poison that bathers stung by this creature have been known to die within minutes.

There are three main classes of coelenterate: scyphozoans, hydrozoans, and anthozoans. Only a few species are able to swim; most either float or, like the sponges, remain attached to something on the sea bed. Jellyfish, which constitute the scyphozoan group, are notably jellylike because they have much greater amounts of gelatinous substance between their two body layers than do most members of the other classes. And, as every salt-water bather knows, they float about in the water apparently aimlessly – but they are actually searching for food, with trailing tentacles ready to capture the tiny planktonic animals on which they feed. If you have seen hydrozoans, on the other hand, you may have mistaken them for seaweed, for hydrozoan species often form mosslike growths on rocks and seaweed fronds. These growths consist of many polyps, for it is characteristic of hydrozoans to colonize. Most of the individuals in a colony capture and ingest food, but some are specialized for repro-

Above, left: In a typical sponge, waving cilia draw oxygenated water and food particles in through small body pores, and wastes are expelled through larger openings. The early growth phase of the harmless-looking sponge in the photograph can do great damage to shellfish by boring into the shells in order to insert its anchoring system of rootlets.

duction rather than eating. Specialization goes even farther in such free-floating colonies as the Portuguese man-of-war, perhaps the best known of all hydrozoans. Each individual within the colony, which is easy to mistake for a single brilliantly colored jellyfish, has a special function; some, for example, form buoyant gas-filled floats, while others capture prey, still others digest it, and so on.

The third class of coelenterates, the anthozoans, includes both individual and colony-forming species. Outstanding for their beauty are the noncolonizing sea anemones, which live on the sea bed or attached to some other solid surface. Their expanded tentacles constitute one of the loveliest of undersea sights, but the flowerlike beauty is deceptive, for any small creature that blunders into the tentacles is swiftly paralyzed by the stinging cells and dragged down into the stomach. One reason, in fact, why sea anemones and their colonizing close relatives the corals flourish so successfully in the shallow ocean environment may well be their sting, which keeps many predators at a distance.

When we talk about corals, we are likely to be referring to the reef-building species. Actually, there are many kinds, several of which grow in quite cool and deep waters. But because of their persistent collective growth through millions of years, it is the tropical reef-building stony corals that have made the greatest impact on the marine environment. What chiefly distinguishes these from other types is that minute algae live within their tissues in an intimate state of symbiosis. As plants, the algae liberate

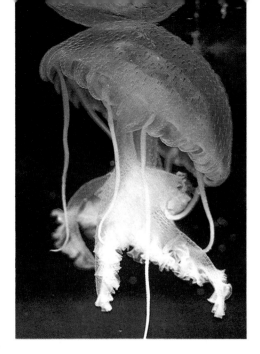

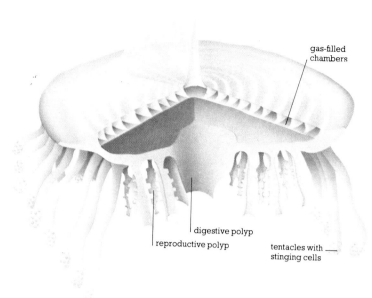

gas-filled chambers

digestive polyp

reproductive polyp

tentacles with stinging cells

Above: This is how the stinging cell that coelenterates use for catching food works. Contact with the prey produces a swift outward thrust of the venomous threads.

Above: Tentacles hanging from the edge of its supporting bell contain the stinging cells of this jellyfish. Captured prey is drawn into a centrally placed mouth, which is surrounded by four frilled arms. Above, right: More complex is the velella – a floating colony of individuals adapted for special functions – whose outer ring of polyps containing stinging cells conveys food past rings of reproductive individuals to a central digestive polyp.

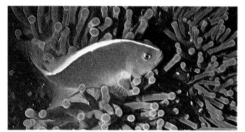

Above: Beauty can be beastly, as this prawn, held captive in the tentacles of a sea anemone, has discovered. But the predator of some creatures may serve as the protector of others. Right: A clown fish shelters among an anemone's tentacles, shielded from stinging cells by its coating of mucus.

This Caribbean coral reef below has several species of the lovely anthozoans. In the foreground massive pillar corals encroach on a growth of fire coral, whose sting has burned many a swimmer. Behind them a flexible gorgonian sways gently alongside towering forms of elkhorn coral.

oxygen, which appears to play an important role in forming the calcium carbonate skeleton on the stony coral. Since algae cannot live without light, coral reefs can be built up only in shallow water, where photosynthesis remains possible. A reef is nothing more than an enormous pile of skeletons, only the top layer of which contains living colonies, which multiply by budding off new polyps. As the little creatures die, new coral growths settle on top of the old, and the level of the reef gradually rises.

Deep drillings have shown some warmwater reefs to be nearly 1,300 meters thick – which raises an obvious question: how did such deep-water reefs begin if coral growth is restricted to shallow water? The answer, implied by the question itself, is that the water in which the reef has grown has not always been so deep. Through the ages, as we now know, there have been sea-level changes caused by the changes in total ocean volume brought about by two factors: the slow release of water that was locked up in ice during glacial periods, and the slow spreading of the sea floor as a result of geological activity. Such changes have been gradual enough for the growth of coral to keep pace with them over the course of millions of years.

The undoubted success of such simple animals as sponges and coelenterates is worth a moment's reflection. They *are* successful – remarkably so – in that they have multiplied and survived almost unaltered despite enormous competition from more highly evolved marine creatures. Yet in evolutionary terms they are certainly "backward," for they seem to have made no biological advances for thousands of millions of years, during which some more recently evolved groups have actually become extinct. Perhaps this illustrates an important but often forgotten truth about evolution: simplicity in the organization of cells does not always give way to increasing complexity. If a simple organism survives and reproduces its kind, what would be gained by evolving further? Because the sponges and coelenterates are successful as they are, no change has been necessary.

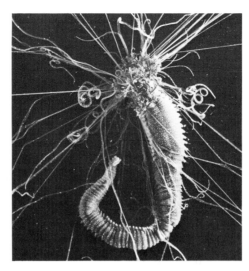

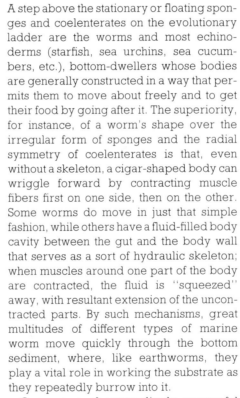

Top left: The ragworm uses its many lateral body extensions for locomotion. By contrast, a sedentary worm (above) need not crawl about in search of food, and so the locomotory extensions of its body are poorly developed.

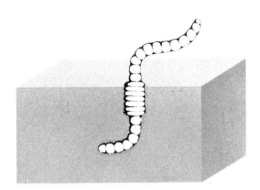

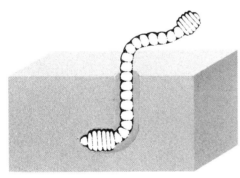

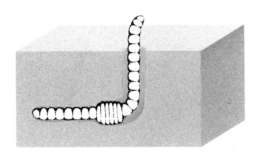

Above: Three diagrams showing how a sedentary worm burrows into the sand. The head is pushed forward and swells to form an "anchor" as the tail draws in. The swelling then passes to the rear of the body, and the process is repeated.

Built for Mobility

A step above the stationary or floating sponges and coelenterates on the evolutionary ladder are the worms and most echinoderms (starfish, sea urchins, sea cucumbers, etc.), bottom-dwellers whose bodies are generally constructed in a way that permits them to move about freely and to get their food by going after it. The superiority, for instance, of a worm's shape over the irregular form of sponges and the radial symmetry of coelenterates is that, even without a skeleton, a cigar-shaped body can wriggle forward by contracting muscle fibers first on one side, then on the other. Some worms do move in just that simple fashion, while others have a fluid-filled body cavity between the gut and the body wall that serves as a sort of hydraulic skeleton; when muscles around one part of the body are contracted, the fluid is "squeezed" away, with resultant extension of the uncontracted parts. By such mechanisms, great multitudes of different types of marine worm move quickly through the bottom sediment, where, like earthworms, they play a vital role in working the substrate as they repeatedly burrow into it.

One group of outstandingly successful ocean-floor worms are the bristly polychaetes (Greek for "many-haired"). The bodies of these marine cousins of the earthworm are divided into a number of segments in which certain vital organs are duplicated. New segments are added as the animal grows, but all of them remain under the general control of nerve centers contained in the head. There are two major kinds of polychaete, a free-living worm and a more sedentary tube-dwelling type. The development of the front end has proceeded farthest in the free-livers, whose heads may have many pairs of eyes, long sensory tentacles, and fierce-looking jaws. As active predators, some of them are excellent swimmers, too; they row themselves through the water with paddlelike extensions of the body wall.

Tube-dwelling worms are much less active. They rarely venture more than a small segment of the body outside the tube or burrow, and so they have little need for a highly developed head to control their movements. Nevertheless, they usually have a sensitive crown of tentacles on the head end. These may be long and sinuous, so as to creep over the mud in search of food, or the crown may be shaped like a beautiful fan through which suspended food particles are filtered into the mouth.

All the polychaetes have evolved primitive excretory and blood-circulation systems, and some of them even have little muscular structures that function as hearts. But their anatomies are much less complex than those of the sea urchins, starfish, and other echinoderms, which are built on a fivefold radial plan. Because of this shape, they lack the definite head end of worms; but to make

Below: Whether walking or swimming, the ragworm's natural body extensions act in succession down its undulating sides; when walking, the bristles help grip the surface.

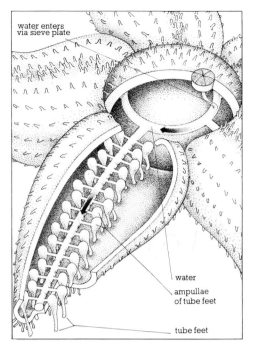

The above diagram indicates how a starfish moves by means of fluid-filling tube feet on the underside of its arms. But, as shown in the photograph (left), suckered feet can do more than just walk. Here they are pulling apart the shell valves of a mussel; once the valves are open, the starfish will thrust out its stomach and devour the soft tissue.

Above: A sea cucumber propels itself through seaweed as the branched tentacles around its mouth collect food particles.
Below: Although tube feet are the echinoderms' major means of propulsion, sea urchins can also use their spines like stilts on level ground.

up for the lack of a head, they have amazingly complicated nervous systems. The "brain" of a starfish, for example, enables any one of the five arms to assume the function of leader and take command of the others' activities. Such a system is essential for the control of movements. Without it, different arms might try to crawl in different directions – and this does, indeed, happen when the nervous system is interfered with, either experimentally or by accident.

The echinoderms (scientists coined a Greek word meaning "hedgehog-skins" for the phylum – and if you step on one, you'll know why) achieve mobility in various ways. The most active are the brittle stars, which move rapidly across the bottom by bending their slender, whiplike arms. In sea urchins, the surface spines have ball-and-socket joints at their bases and can be swung back and forth to aid movement of the radial creatures across the sea bed. But the most remarkable locomotive system, best developed in starfish and soft-bodied sea cucumbers, involves rows of little tube-feet on the underside of the body. These are extensions of a complex internal hydraulic system, and each has at its base a muscular bulb that contracts to extend the foot – rather like the result of squeezing one end of a sausage-shaped balloon. Sucking-disks on the end of the feet help some species to walk up vertical surfaces and to cling to rocks. A single tube-foot is flimsy and weak, but since echinoderms have many hundreds, which work in concert, the animals move along quite efficiently, if slowly Some starfish use their tube-feet not only for walking, but also for opening the tightly closed shells of the mollusks on which they prey. The feet pull in sequence, like a well-disciplined team, until they have gradually forced the shells apart; the starfish then

pushes its saclike stomach out of its mouth, which opens at the center of its lower surface, and the stomach envelops and digests the prey. Other echinoderms have equally distinctive, though very different, feeding techniques. Sea urchins feed mainly on the plentiful mosslike animals, such as hydroids, that encrust rocks and seaweeds; these are bitten off and ground up in an ingenious array of hinged bones and muscles situated just inside the urchin's mouth. Sea cucumbers ingest large quantities of mud and sand from the sea bed and extract small food particles, much as an earthworm does. Sea lilies (crinoids) spread their branched arms in the water to capture suspended food material.

Shallow-water crinoids, incidentally, are the only echinoderms that swim, for although they usually stand mouth upward, clinging to rocks by rootlike extensions from the central disk, they can move in a rather ungainly fashion by waving the branched arms up and down. Some deep-water species, however, are not at all mobile but are attached to the sea bed by long stalks. These flowerlike animals are really living fossils – the remnants of a group of ancient sea lilies that were probably the forebears of all modern echinoderms.

Such gourmets as the French and Chinese look upon certain kinds of sea urchin and cucumber as supremely edible delicacies (the sea cucumber is known, appetizingly, as *bêche-de-mer,* which is a French rendering of a Portuguese name meaning "cater-

pillar of the sea"). Most people, however – and, indeed, most marine animals – are repelled by the echinoderms' prickly skeletons. And there is one species, the crown-of-thorns starfish, that has become especially unpopular among naturalists in recent years because it has been devastating the coral reefs on which it feeds in the Red Sea and the Indian and Pacific oceans. These reefs are, of course, important assets to the areas involved, and it is right that we should be concerned at the damage. But we are by no means certain that the damage is really widespread, or that it is a new and abnormal phenomenon. Such occasional depredations may simply be part of a natural cycle with which the reefs have managed to cope without man's interference for millions of years. It is quite possible that if left alone they will continue to do so.

The Crustaceans

Such familiar marine crustaceans as lobsters, shrimps, and crabs are, so to speak, only the tip of the iceberg, for this immense salt-water branch of the arthropod family includes a great range of forms that inhabit almost every part of the ocean. Some of them are quite sizable – impressively so in the case of the giant Japanese spider crab, which has a leg span of three meters – but most are small, ranging in length from a few millimeters to a few centimeters. The main reason for this is the external skeleton, or crust, that gives the group its name, for the shell of a large crustacean has to be extremely thick in order to prevent buckling and collapse, and so the bigger forms are clumsy and sluggish. The shell or outer casing of an echinoderm is rigid, or nearly so, and the rigidity makes the echinoderms very slow-moving; the arthropods – members of a phylum that includes not only the marine crustaceans but also terrestrial insects and spiders – are much more mobile because the shell consists of hard pieces connected by softer, more pliable tissues that enable the body and limbs to bend. But the thicker the shell, the more weight the animal carries, and the less mobile it becomes. ·

In exchange for the advantage of mobility, arthropods need to molt the protective shell periodically so as to permit growth. Many times during its life, therefore, every crustacean sheds its entire crust, including each hair and spine and even the lining of most of its gut. Before the new shell hardens, the soft body swells by absorbing water, which is gradually replaced by tissues until the animal is once more too tightly corseted and must molt again.

Under its nonliving shell, the crustacean's body is basically composed of a series of hinged segments, each of which has a pair of jointed appendages. In the most primitive members of the group, such as the brine shrimps, the segments, as well as the limbs on each segment, tend to be nearly identical, without clearly distinguishable separate features. But in the more advanced types, the appendages of particular segments have specialist functions and are modified accordingly. A lobster, for instance, has sensory antennae, an impressive arrangement of jaws around the mouth, enormously enlarged claws, four pairs of walking legs, paddles (or "swimmerets") beneath the abdomen, and a tail fan, which is used for flicking the body backward and away from predators.

The bodies of some species are so modified that they don't really resemble the general shrimplike form. The hermit crab, for example, takes up residence in an empty mollusk shell; its abdomen has lost its hard plates and is coiled to fit into the foreign shell, which the hermit carries around as a

Copepods, most of which look like this deep-water species, are the most numerous of marine animals. The bright red coloring is common to many crustaceans in the depths; it is not a disadvantage where no natural light can advertise their presence to predators.

These Japanese fishermen are sorting through a mound of spider crabs for processing at sea.

protective covering. In true crabs, too, the abdomen has evolved differently from those of shrimplike crustaceans: it is reduced to a tiny flap folded beneath the main shell, or carapace. No matter how modified an adult crustacean may be, however, it almost invariably produces a typical shrimplike larva, which returns for a few days or weeks to the ancestral mid-water planktonic environment even when, as with lobsters and crabs, the adult stage has forsaken this for a sedentary life style at the bottom.

It was the planktonic larval stage that provided the vital clue, a century and a half ago, to the identity of a particularly puzzling crustacean, the barnacle. On many coasts, barnacles are the most common of all animals; it has been estimated that they live at a density of over a thousand million per kilometer on some rocky stretches of Atlantic Ocean coastline, where, at high tide, they extend feathery jointed legs to sweep suspended food particles into their mouths. The conical shells of these strange creatures – which have been aptly described as crustaceans that stand on the back of their necks and kick food into their mouths with their feet – make them look more like limpets than crustaceans, and early marine biologists thought that they were, indeed, some sort of aberrant mollusk. Not until 1820, when a British army surgeon collected some tiny, obviously crustacean, larvae floating in vast quantities in the inshore plankton and watched them attach themselves to stones and gradually assume adult form, that it became clear that barnacles are true crustaceans.

Unfortunately, it is also clear, and always has been, that they are an expensive nuisance to shipping, for the moving hull of a ship offers them an environment quite similar to that of the wave-lashed rocky shore. Along with other invertebrates and seaweeds that quickly become attached to the hull, the tenacious barnacles can impede a ship's passage through the water to such an

extent that its fuel consumption may increase by 40% through the course of a single year. So widespread is this fouling of hulls that the world's shipping companies spend millions of dollars a year on costly hull paints that repel barnacle larvae.

Another serious crustacean pest is the gribble, a tiny relative of terrestrial wood lice, which will chew its way through any wooden structure in the sea and has been known to cause the collapse of pilings and jetties. In general, though, crustaceans are more beneficial than harmful to man. Apart from the obvious fact that we eat such relative giants as lobsters, shrimps, and crabs, we benefit indirectly from the much smaller planktonic forms, for these ubiquitous members of the zooplankton graze on floating plant life (the phytoplankton) and are therefore the second link in food chains that ultimately provide an essential source of protein for mankind. The title of Chief Planktonic Grazer undoubtedly belongs to the minuscule copepods. These crustaceans are such dominant constituents of the zooplanktonic community that they probably outnumber all the other animals of the earth put together! Because the copepods swarm in the water much as midges and mosquitoes do in the air (only much, much more abundantly), they are often called the "insects of the sea." Let us turn now to an examination of their very special – and supremely important – world: that of the plankton.

Lobsterling

Lobster larva 6 weeks old

Above: A fine example of a commensal relationship. The hermit crab's soft abdomen is protected by the shell of a dead whelk and the stinging cells of anemones that cling to the shell. When moving to a new home, the crab encourages the anemones to come along by stroking them.

Left: The acorn barnacle, whose calcareous plates shield it from heavy surf, feeds by sweeping particles into its mouth with its legs.

Below: Various stages in the growth of a lobster: the eggs are carried under the female for 9 to 10 months before hatching into shrimplike creatures about a centimeter long; after 4 to 5 weeks in the plankton, during which they molt twice, the lobsterlings sink to the bottom, where they mature in about 7 years.

Above: The translucent female pea crab lives as a parasite within a mussel, and the much smaller male crabs must pass from one mussel to another to fertilize the females. Pea-crab larvae are expelled from the mussel in its exhalant water current.

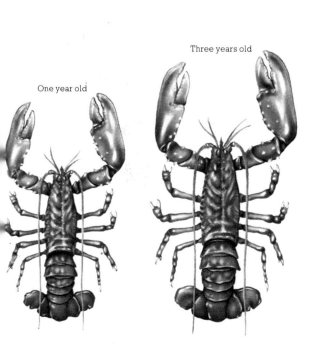

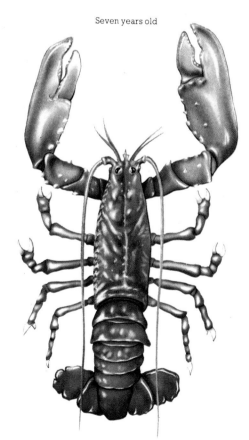

One year old

Three years old

Five years old

Seven years old

Animals of the Plankton

The mid-water world far above the ocean floor is the largest single environment on earth. Here naturalists make a distinction between two types of animal: swimmers, or *nekton*, including most fishes, squids, and whales; and drifters, or *plankton*. This is a man-made and artificial distinction, but a working definition of the zooplankton is that it consists of animals that cannot swim against strong ocean currents, although they may make extensive movements within the planktonic community as they are carried along with it. A few planktonic animals are quite large. The lion's mane jellyfish, for instance, has a body up to a meter across and tentacles 20 meters long; yet because it is a weak swimmer, it must drift with the plankton. For the most part, though, the zooplankton consists of countless numbers of very minute creatures.

Because planktonic plants can survive only in the sunlit upper layers of water, the herbivorous members of the zooplankton are mainly restricted to those same layers, but the carnivores can and do penetrate much farther down. Wherever they live in the water column, however, all planktonic organisms must be able to float. Smallness is one way of maintaining buoyancy, since the smaller the animal the greater the surface area in relation to mass, and therefore the greater the resistance to sinking. Surface area can also be increased without significantly increasing weight by convolutions of the body surface or by slender, fingerlike extensions of the body. This is why bizarre shapes, long spines, and feathery hairs are common features of the drifting animals. In addition, though, many of them are endowed with special buoyancy mechanisms. Among the coelenterates, for instance, such surface-dwelling hydrozoans as the Portuguese man-of-war are generally supported by gas-filled floats; and one of the most beautiful of mollusks, the pelagic snail *Ianthina*, suspends from a self-made raft of bubbles. In some cases, too, the planktonic creature's buoyancy is increased by oil globules enclosed within its body.

All major phyla, from the simplest protozoans to the vertebrates, are represented in the plankton, either by permanent residents or by larval stages that may spend only a few days or weeks in the community before swimming away as nonplanktonic adults. But almost any fine-meshed tow-net sample taken in the ocean is likely to be dominated by the shrimplike copepods that were mentioned in the preceding section of this chapter. Most copepods, which range in size from a pinhead to a grain of rice, are herbivores and filter planktonic plants from the water through which they swim – a vitally important operation, for the copepods convert the phytoplankton into a highly palatable source of food for larger animals. Even man would find copepods good to eat if he could find a practical way to get at them. Unfortunately, the technological problem of filtering sufficient quantities for human consumption out of huge volumes of water seems insoluble. As we shall see in the chapter on the resources of the sea, there is a much larger – and therefore more easily harvested – planktonic herbivore, the krill, which is the chief food of baleen whales and of many open-ocean fishes, and which the Japanese and Russians are currently trying to exploit. Some day it, if not the copepods, may become an important new source of protein for humanity.

Although the zooplankton spend most of their time passively drifting and feeding at a given level of water – either near the surface if they are herbivores or farther down if they are predators – they do not, strangely enough, remain there all the time. Instead, a general vertical migration occurs every day, from a daytime depth well below the surface up toward the surface at night. The largest species, which are usually the best equipped for swimming, seem to make the most extensive migrations; some go downward and upward as much as 600-odd meters. Marine scientists are still puzzled over the significance of this energy-consuming behavior. It appears to be markedly affected by changing light intensity in the water, since the animals tend to go lower on sunny days than on overcast ones. Yet the timing and depth of the migrations, which vary enormously, are by no means completely controlled by this factor. The noted marine zoologist Sir Alister Hardy has suggested that there may be a navigational advantage to the drifters in migrating to different layers of water that are moving in different directions at different speeds: by varying the time spent in the various layers, members of the zooplankton may achieve a measure of control over where they are carried, so that, for instance, they can swarm in areas where the phyto-

This diver is using a fine-mesh net fixed beneath his battery-driven underwater "scooter" for taking a sample of reef plankton. Any such sample contains many thousands of microscopic organisms, both plant and animal.

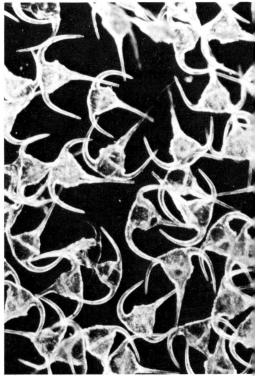

plankton happens to be especially thick.

When we speak of the zooplankton, of course, we are not referring to a single community but to innumerable drifting communities scattered throughout the oceans. There is one thickly inhabited layer of water about which we know very little: the "deep scattering layer" (generally spoken of as the DSL). During World War II, a sound-reflecting layer of the sea, down deep but far above the bottom, often produced a smudge on echo-sounder records, and American Navy scientists studied this curious phenomenon with great interest; the sonar aspect was particularly important from the military standpoint because submarines, it was discovered, could sometimes avoid detection by diving under the DSL, whatever it was. A zoologist at the Scripps Institution of Oceanography in La Jolla, California, soon discovered that the DSL behaved much like the zooplankton, ascending at night and descending during the day. It now seems certain that the DSL is caused mostly by small mid-water fish, whose swim bladders act as excellent acoustic reflectors. Thus, there are probably vast stocks of these and other vertically migrating mid-water animals about which we have little real knowledge. Even if they are not themselves planktonic, these inhabitants of deep water must surely be dependent on the plankton for food since they follow its vertical migrations.

Below: A photomicrograph of a swarm of dinoflagellates. These planktonic organisms – not quite animals, yet not really plants – are sometimes so numerous that they discolor the water, and their toxic excretions can kill other creatures.

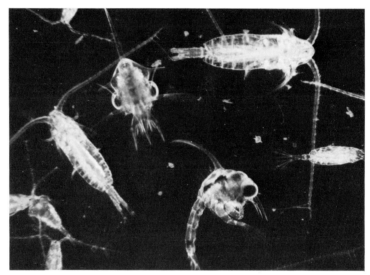

A velella is kept afloat by its gas-filled bladder, but the one at the left won't float much longer; it's being devoured by a snail, which maintains its own buoyancy by secreting bubbles. Left, below: A planktonic sample containing crab larvae as well as a considerable number of copepods (the most common of all planktonic animals).

The diagram below shows the change in depth between daytime and nighttime plankton. This vertical migration, which takes billions of creatures up toward the surface at dusk, appears to be a response to the reduction in light intensity.

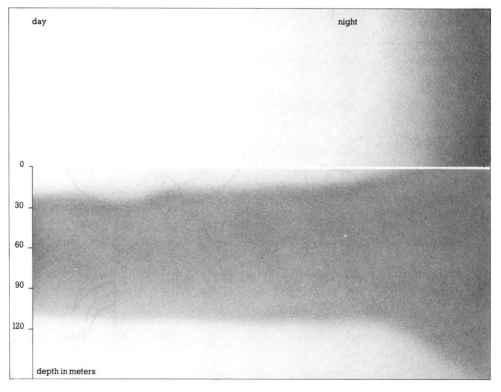

day   night

0

30

60

90

120

depth in meters

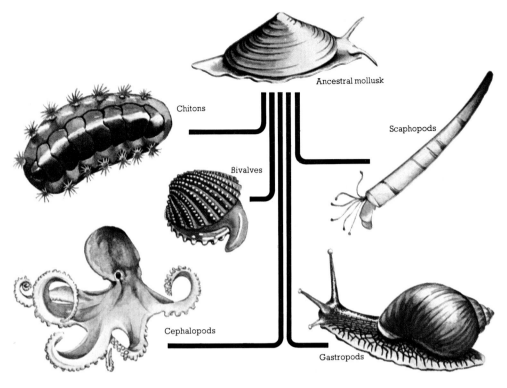

Chitons

Ancestral mollusk

Scaphopods

Bivalves

Cephalopods

Gastropods

The Mollusks

Mollusks are soft-bodied creatures. The few kinds that we eat – such bivalves as clams, mussels, oysters, and scallops – represent only a fraction of the more than one hundred thousand known species. Scientists recognize five main molluscan classes: the bivalves, with their hinged double shells; the gastropods (snails and limpets), which are much the largest group, containing 80% of all molluscan species; the cephalopods (squids and octopuses); the scaphopods, or "tooth snails," which look like miniature elephant's tusks; and the chitons (flattened little animals that cling like limpets to rocks and seaweed on the shore). The last two groups contain relatively few species and are not very significant in the general economy of the oceans. The bivalves, gastropods, and cephalopods, on the other hand, are of tremendous importance.

We find mollusks everywhere in the sea. Some swim, others do not. Some crawl or attach themselves, as do the limpets, to wave-battered rocks; others burrow into sand or mud or even into rock or the wood of harbor installations. And their eating habits are similarly diverse. Some creeping sea snails, for instance, have catholic tastes and are thoroughly omnivorous, whereas other gastropods, such as certain sea slugs (snails that have completely lost their shells), feed only on one species of sponge. Despite their great diversity, though, the bodies of all mollusks have a fundamental similarity in structure based on the "mantle," which is a flap tissue overlying the soft body next to the protective shell, which is actually secreted by the mantle itself. In the cavity between the mantle and the body wall are the gills. The mantle creates more shell along its outer edge as the animal grows, and it is this secretion in thin layers that gives the inner surface of the shell its iridescence. When a foreign body, such as a sand grain, is caught between the mantle and shell, it forms a pearl as it is gradually covered by layer after layer of shell – but valuable pearls come chiefly from the so-called pearl oyster; pearls formed by other mollusks generally lack their incomparable luster.

Characteristic of the gastropod is a broad, flat, muscular foot on top of which the animal's viscera are neatly coiled within the coiled shell. (The word "gastropod" appropriately means "stomach-footed.") The protection of its shell is, of course, less total than that of a bivalve, whose hinged double

The brightly colored sea slug below lives in the Red Sea. Like its relatives on land, this mollusk has dispensed with a protective shell; it has few natural enemies and grazes fearlessly on hydrozoan and anthozoan polyps, for it is immune to their stinging cells.

The mollusk family tree: though they look very different, all the mollusks are built on the same basic plan. The ancestral mollusk from which modern species have evolved was probably a marine snail with a shell secreted by a mantle that covered a soft body that had a head as well as a flat, creeping foot.

Below: A section through a piece of limestone showing mollusk borers in the burrow they have excavated. Under natural conditions, the siphon (red in the photograph) would be extended out of the burrow entrance.

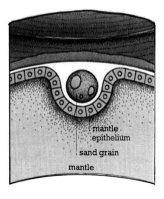

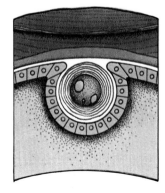

mantle
epithelium

sand grain

mantle

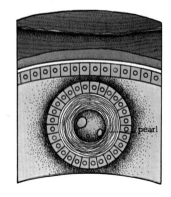

pearl

How pearls are formed: a tiny object, accidentally lodged within an oyster, is isolated from the mollusk's tissues by the secretion of many layers of a substance that eventually surrounds it. The mantle produces this secretion to protect the mollusk from mechanical damage and disease.

Man 1.80 meters

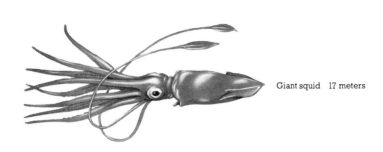

Giant squid   17 meters

Sperm whale   24 meters

Above: Relative sizes of a giant squid, its natural enemy the sperm whale, and man. Right: Three small squid maintaining station with one another in the water.

This drawing of a cuttlefish section shows the feature that chiefly distinguishes cuttlefish from squid – the internal shell, or *cuttlebone*, which consists of a series of chambers into which the animal secretes nitrogen to enable it to alter its buoyancy.

gas-filled chambers

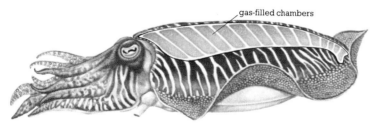

shell completely covers its soft body and is opened only when the animal needs to move, feed, and breathe. Though the bivalve cannot crawl about as snails do, its streamlined shell is well fitted for burrowing, which it accomplishes by pushing its tonguelike foot out into the sand and then expanding it at the tip to act as an anchor against which the shell is pulled downward as the foot is shortened. Simultaneously, the two valves close; this not only streamlines the animal's shape but also forces a jet of water from inside the shell to "fluidize" the sand and make it easier to penetrate – which is how the long, bladelike shells of the razor clam can burrow so quickly away from the probing beaks of hungry birds.

Bivalves use their large, flat gills, which hang within the spacious mantle cavity, for feeding as well as for breathing. Meshlike in structure, the gills are lined with cilia, whose beating draws in a powerful current of water, from which food particles are filtered. These living water strainers are so efficient that a bivalve can deal with as much as 30 to 60 times its own body volume of water in an hour, and it can filter out edible particles down to the size of bacteria.

But pride of place among the mollusks

(and perhaps among all marine invertebrates) must surely belong to the cephalopods – not simply because squids in particular swim so well, but because all the cephalopods have a highly developed brain and nervous system. Their name means "head-footed"; in these remarkable creatures, the typical snail's foot has become subdivided into eight or ten arms encircling the head. Behind this is the opening to the mantle cavity, the muscular walls of which surmount the lower part of the body. The shell has been reduced to a horny, quill-like vestige in squids and completely lost in octopuses. In short bursts, at least, the squid is the fastest of all salt-water swimmers; its body is streamlined for speed, and it propels itself forward by expelling a powerful jet of water through a conical tube from its mantle cavity by means of powerful muscular contractions. An octopus can also jet-propel itself forward over a few meters in order, say, to avoid a possible predator. But octopuses have generally adopted a more sedentary life style than the squids. Since they spend most of their time among rocks and thick clumps of seaweed, they have not even a trace of a shell; the muscular octopus body can insinuate itself

through narrow cracks and crevices. In the dark depths of the ocean there lurk many species of very large squid and octopus, which swim by pulsating a web of skin stretched out among their tentacles. The largest is the squidlike *Architeuthis*, which can be as much as 17 meters long, including its tentacles, and is a chief food of the sperm whale. In the cephalopod's death throes, its tentacles often grip the predator so forcefully that they leave permanent rows of circular scars on the whale's skin.

Later in this chapter we shall be discussing the remarkable intelligence of the cephalopods. Almost equally remarkable is a striking physical feature of these amazing invertebrates: their eyes. The eyes of a squid or octopus are much like the human eye in structure; yet they evolved independently of our own developmental history. Such an example of evolutionary convergence reminds us that elegant and highly efficient organs are by no means restricted to what we egotistically refer to as the "higher" forms of life. Many "lowly" creatures are capable of evolving organs as complex as some of ours if possessing such organs makes the species more successful in the competition for survival.

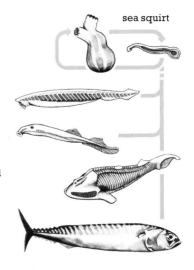

sea squirt

sea squirt larva

lancelet

hagfish

mudeating jawless fish

fish

It would be hard to guess from the flabby body of the adult sea squirt, or tunicate (left), that it is a direct descendant of a creature that stands at the foot of the vertebrates' family tree (right). Note, though, the trace of a backbone in its larval stage. This feature was further developed in lancelets, hagfish, and the now-extinct mud-eating jawless fish that was the probable ancestor of modern bony fishes – and us.

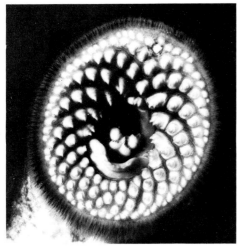

Left: Its suckerlike mouth attaches the lamprey, which feeds on fish, to the side of its prey, from which it sucks blood after puncturing the skin with its rasping tongue.

The notochord, a primitive form of backbone, is found at some stage in all chordate embryos. We see it clearly in the water-filtering lancelet amphioxus, right. The chord exists as a separate structure in the amphioxus, but in lampreys (cutaway drawing, below) it has become incorporated within a fibrous sheath, which also surrounds the spinal cord. .

Notochord

Vertebrate Beginnings

Where, among the marine invertebrates, do we find those whose forebears were probably the direct ancestors of fish and thus of all terrestrial backboned animals? Unlikely though it seems, one very plausible candidate is the lowly sea squirt. We believe that something like this flabby-bodied creature may stand at the foot of man's own family tree because its swimming larval form has a spinal cord and a concentration of nerves in the head that could pass for a brain. The resemblance to the vertebrates lasts only during the larval stage, for adult sea squirts are completely immobile and feed, like bivalves, on plankton by pumping water through their sievelike gills. But zoologists think that vertebrate life is likely to have evolved directly from the larval form, which, by a precocious development of its reproductive system, managed to cut the sedentary adult stage out of its life cycle.

A later step forward in man's evolutionary history may well have resembled the present-day fishlike lancelets. These little animals, only a few centimeters long, spend their days in shallow-water sand and mud, but ascend nightly into the plankton (where, in the waters off China, they are caught by the millions for food). The lance-

let has a body-stiffening rod – the notochord, which is the precursor of the backbone – and segmented muscle blocks like those of fishes. Thus, although it lacks a true head and feeds by filtering plankton through the gills, just as sea squirts do, its body anatomy is clearly related to that of the simplest vertebrates: fish without jaws. Among these are the eel-like hagfishes, which seem to spend most of their time buried in the mud, emerging only when their keen sense of smell tells them of the presence of decaying fish or other carrion. Such highly specialized fish represent the remnant of what was once an array of mud-eating jawless fishes. But these primitive creatures were so encumbered with heavy bony plates in their skin that they must have been very sluggish swimmers; they needed the armor to protect them from such voracious predators as the giant water scorpions, long-extinct relatives of modern spiders, which flourished in far-off times.

A prehistoric observer from outer space could justifiably have given the slow-moving jawless fishes only an outside chance of inheriting the earth as compared with the water scorpions or cephalopods. As it happened, however, fish won the day because their basic design was capable of great development. They evolved biting

jaws and paired fins, and eventually some of them left the bottom and became mid-water predators; along with changes of habitat came a change in body form to the streamlined shape that characterizes the fish of today. Jaws – which make possible big bites and the pre-processing of food before swallowing – probably developed from the hinged supporting skeleton of the gills. The paired fins evolved from skin folds and proved useful as control surfaces for checking the pitching, rolling, and yawing movements produced when a fish swims by the lateral undulations of its body and tail.

Like the bony fish, the sharks and rays evolved from freshwater jawed fish and gradually re-entered the sea. But unlike their bony relations, today's sharks and rays have probably changed little since they first began to take advantage of the sea's wealth of invertebrate food. Their bodies, which have cartilaginous rather than bony skeletons, are heavier than water, and their skin is covered in small toothlike scales – a relic of the continuous body armor of their ancestors – which are enlarged over the edge of the jaws to form rows of teeth. Terrifying though these may sometimes appear, their function is primarily to prevent the escape of prey rather than to bite, for sharks and rays usually swallow their food whole. As

the teeth wear away, they are continually being replaced by rows hidden under the skin of the inner jaw edge. The replacement of the milk teeth in mammals is an evolutionary relic of such serial replacement.

The rays (a group that includes the skates) generally inhabit the sea bed, where they feed on shellfish, which they crush with their wide expanse of small but powerful flattened teeth. But the biggest of all rays, the manta or devilfish, which can be five or six meters wide and a couple of meters thick, has abandoned the bottom way of life; instead, it "flies" through mid-water with its winglike fins, engulfing plankton and small fishes in its huge mouth. Sharks range in size from the harmless, plankton-eating whale shark and basking shark, which have been known to exceed 10 meters in length, down to the dogfishes that abound in many coastal waters. In between are such potential man-eaters as the various mackerel sharks.

Still, despite their deserved reputation as efficient predators, the sharks and rays are not an outstandingly successful group in evolutionary terms. For one thing, the 3,000 or so living species have not mastered a very wide range of habitats. Moreover, for sheer abundance and diversity of form they cannot hold a candle to the bony fishes, whose success story we shall consider next.

Above: A stingray searches among the sponges and soft corals of a Bahamian reef for shellfish, which it grubs out of the sand with its snout and crushes with rows of flattened teeth. The teeth of the 2.5-meter-long Australian gray nurse shark (left) are razor-sharp; instead of crushing prey, they either prevent the escape of fish that the shark has swallowed whole or tear massive chunks of flesh from larger animals. The woman diver here holds a 12-gauge shotgun for possible self-defense.

upturned tail

lift from swim bladder

swim bladder

hydrofoil fin

The Bony Fishes

It's both natural and right to think of bony fish as *the* water-dwelling animals. Their mastery of the ocean surpasses even that of the diverse and numerous crustaceans and cephalopods, for only the fishes appear to have successfully colonized every conceivable undersea habitat. The more than 20,000 existing species are specialized for all sorts of situations, and they have found countless ways of exploiting the ocean's varied stores of food; from the plankton to bottom-living invertebrates, from soft seaweeds to hard coral, there is little that is not utilized by one species of bony fish or another. The chief reason for their evolutionary success is their body structure, which is superior in several respects to that of the sharks and rays, as well as other marine creatures.

One important difference between many bony fish and the cartilaginous sharks is that, whereas a shark's body is heavier than water, the bony fish achieves buoyancy by means of its swim bladder, which is an internal organ not unlike the buoyancy tanks that keep modern submarine craft afloat in the depths. Into this thin-walled sac, gas – chiefly oxygen – is secreted by the blood as and when required. Since gases are compressible, the secreted amount increases with increasing depth, and some of it must be expelled as the fish ascends. Some fishes have retained the swim bladder's primitive connection with the gut, so that excess gas can escape through the mouth like a belch; in those that have lost this connection, the excess gas is reabsorbed into the bloodstream. Sometimes when a trawler hauls up fish very rapidly, the fish cannot cope

with the rapid change of pressure and the much-swollen swim bladder may force the foregut out of the mouth.

The great advantage of having a swim bladder is that, because bony fish do not need the flattened head, upturned tail, and hydrofoil-like pectoral fins that give the shark's body its ability to achieve lift, they have been able to evolve a side-to-side flattening of the body that facilitates the lateral movements that make for efficient swimming. In general, bony fish tend to be much better swimmers than sharks, for the body undulations have a larger amplitude, and the backward-moving push starts farther back on the body. As a result, the bony fish often seem only to wag their tails from side to side, with far less apparent wagging of the head end than in the sharks. Since the paired fins no longer have the hydrofoil (lifting) function, they can be used to control delicate maneuvers, or they can be oscillated rapidly to allow the fish to hover. In such speedy swimmers as mackerel, the two pairs of fins are close together, so that the rear pelvics can exert control over the slight lifting force produced by the forward pectorals when these are used for braking. In several other ways, the body of the mackerel, as of its close relative the tunny, has been evolved – like a man-made submarine hull – for high speed with a minimum expenditure of energy. To swim constantly and tirelessly, such fish ram water through their open mouths instead of pumping it through the gills as other fishes do. And their tails, typically sickle-shaped – a form that develops maximum thrust in relation to the surface area – vibrate at perhaps 10 times a second as their bodies cut through

Above: Some basic differences between the bony fish and cartilaginous sharks. One adaptation to aquatic life in many bony fish, the buoyant swim bladder, is lacking in all sharks, which must keep moving in order to remain afloat. Note, too, the shark's upturned tail and hydrofoil-shaped pectoral fins as compared with the symmetry of most bony fishes.

the water in search of prey. They have become so beautifully adapted for swimming, in fact, that they have dispensed with the normally important swim bladder.

But there are fishes of many shapes and ways of life. Even the wide range that we see in fish stores represents an extremely conservative selection. Some are so bizarre that they certainly wouldn't be offered for sale, and we might wonder whether they're fish at all. Consider, for instance, the snakelike pipe fish and their near relations the sea horses, whose stiff bodies have lost all swimming power; they lie concealed in seaweed, and their long snouts simply reach out and pick off little planktonic animals as they float by. Or consider the flying fishes, with their winglike pectoral fins. After a dash through the water at perhaps 30 kilometers an hour, they break surface and continue to generate lift by rapid beats of the enlarged lower lobe of the tail fin. Once a flying fish is airborne, it spreads its pelvic fins to provide additional lift, and it can manage glides of several hundred meters to avoid predators. Flying gurnards actually flap the pectoral fins up and down in the air (though rather feebly) like birds, and the pectorals of other gurnards are modified to act as legs for "walking" over the sea bed. Some strange fish, too, neither fly nor walk but sit down: in the deep ocean, observers

The rigid-bodied pipefish (far left) and the sea horse (left) are less mobile than even the lazy remora (below). They stay hidden in seaweed, where they feed on small planktonic animals that come drifting by. Pipefish and sea horses are unusual animals in that the males brood their young in a pouch at the front of the body after the eggs have been laid there by the female, which then takes no further part in the proceedings.

The remora hitches a ride on a larger fish such as a shark by means of a suckerlike modification of the forward part of its upper fin, but it sometimes takes temporary leave of its host to feed on smaller fish.

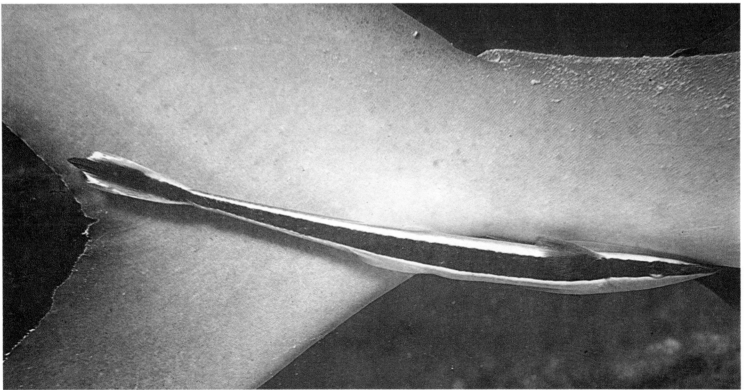

in underwater craft have seen the blind *Bathypterois* sitting motionless on its stiltlike pectorals and lower tail-fin lobe.

Many small fishes other than pipe fish and sea horses do no real swimming but rely entirely on vibratory fanlike fins for movement. Thus, unable to escape predators by speedily swimming away, the tropical trunkfish and globefish, for example, have found protection in heavily armored bodies. A different kind of protective device serves some fishes in good stead

against the batterings of the ocean itself: the pelvic fins of gobies and clingfish, which inhabit wave-beaten rocky coasts, are modified into sucker organs that permit the fish to cling to something solid when endangered by the surging sea. But suckers can have quite another purpose. The remora, for example, hitches a free ride on sharks and other large fish by means of a sucker plate developed from the dorsal fin.

So the bony fishes are wonderfully diverse in form and habits. What is perhaps of

most importance to our hungry world, though, is their sheer abundance. In the relatively shallow seas over the continental shelves, most of the familiar food fishes, such as herring and cod, are already being actively exploited, if not overexploited, and so there is probably little hope of significantly increasing the world catch from these areas. But, over the continental slope there seem to be huge stocks of large fishes, such as hake, grenadier, and sable fish, that we are barely beginning to fish for.

Returners to the Sea

Evolution has not always involved a one-way traffic from the sea to the land. The apparently retrograde step of returning to the ocean has been taken by such once-terrestrial vertebrates as the turtles, sea cows, seals, and whales. It may seem paradoxical that with nature apparently struggling for so long to evolve forms of life suitable for survival on dry land, some creatures should have abandoned the struggle, as it were, and gone into reverse; but there must have been good reasons for this – among them, perhaps, the pressures of increasing stocks of predators on land.

At any rate, once the latter-day marine creatures had re-entered the old environment, their land-developed brains and warm-blooded bodies probably gave them an edge over many stay-at-home inhabitants of the oceans in the competition for survival. On the debit side, however, their ancestral sojourn on land had resulted in fundamental body changes – particularly the development of air-breathing lungs – that are a liability in water. Body structures and mechanisms that have been lost in the evolutionary process cannot be easily regained.

Thus, although whales, for instance, are beautifully adapted to marine life, their loss of gills forces them to breathe air at the surface.

It was some of the terrestrial reptiles, which had once ruled the land and air, that first returned to the sea. Marine turtles and sea snakes constitute a fascinating remnant of the long-dead "Age of Reptiles." It is no wonder that the turtle's body seems somewhat clumsy and antiquated, for it has remained essentially unchanged longer than that of any other air-breathing vertebrate. Its legs form paddles with which it can fly through the water at considerable speeds (turtles have been known to cover about 100 meters in 10 seconds!), but it must come up for air; its hinged lower shell allows its chest to expand when it takes a big gulp at the beginning of a dive. We still know little about the habits of turtles, whose numbers are being badly depleted, for man not only prizes the adults because of their palatable flesh but also combs certain sea coasts for their eggs. Despite the turtles' millions of years of marine living, they are still bound to the land for reproduction, and the female arduously drags herself up above the high-water mark to lay her eggs

and bury them in the sand. During this difficult period she sheds copious tears – not from a broken heart, but because her normally submerged eyes need protection from the air. Some species migrate considerable distances across the sea. Every year, for example, green turtles swim as much as 2,000 kilometers from the coasts of Brazil in order to breed and lay their eggs on the beaches of Ascension Island in the middle of the South Atlantic. And their journey may be getting longer all the time. Professor Archie Carr of Gainesville, Florida, who is perhaps the world's greatest authority on turtles, thinks that sea-floor spreading has been slowly separating the green turtles' coastal pastures of turtle grass from the remote island breeding sites to which they have been instinctively returning for millions of years. But how do they find their way? No one knows.

The reptiles may have been the first ver-

Three hundred million years ago, a type of lungfish somehow left the water and managed to survive on land. From it, as shown in the chart below, evolved all terrestrial vertebrates as well as the reptiles and mammals that readapted to the marine environment even while retaining their air-breathing lungs.

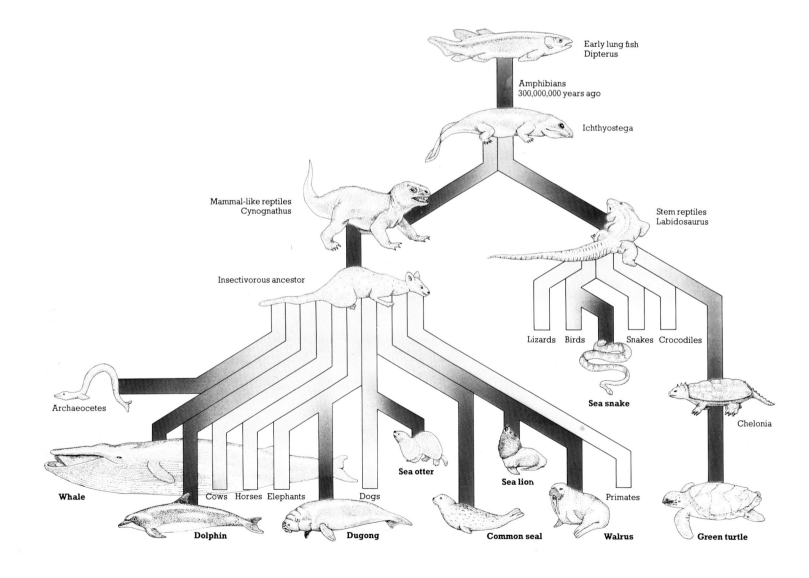

tebrates to re-enter the sea, but it is the mammals that adapted best to the marine environment. Indeed, the chief weakness of the four main groups of marine mammal – sea otters, seals and walruses, sea cows, and whales and dolphins – may be their susceptibility to human predators. Sea otters, for instance, have valuable fur and have therefore been hunted to the point of near extinction. They are now making a comeback – fortunately, for the extermination of these gentle little animals would be an ecological tragedy. One of their remarkable characteristics is the ability to grasp pebbles in their front paws and use them as tools for cracking open shellfish – surely an example of mammalian intelligence being put to good use in the sea.

Though superficially similar to sea otters, the seals are a distinct group. With their flipperlike limbs, indeed, they are rather better adapted to the marine environment than the otters, and they are certainly intelligent and swift in the water. Yet they all bear their young on land and may even drag their heavy bodies ashore to sleep. Most species of this fish- and shellfish-eating family, like their close relatives the walruses, inhabit northern waters; but the largest member of the group, the ugly-faced elephant seal, lives in the tropics. A bull elephant seal can grow to a length of nearly six meters and can weigh well over two tons. Almost as huge – and even less attractive to look at – are the bald, harelipped, mustached sea cows. It's hard to imagine even a hot-blooded sailor thinking of them as sea sirens (their zoological name is Sirenia); yet they are thought to have given rise to the fabled idea of mermaids – perhaps because, like mermaids, they lack hindlimbs, have an odd flattened tail and, in the female, prominent breasts, and are really quite gentle, retiring creatures.

We shall next take a brief look at the mighty whales. It is fitting to precede that look with a glance at the sea cows because they stand next to the whales in the degree of modification to marine life. For one thing, unlike most other returners to the sea, sea cows and whales give birth and suckle their young in the water. Like all these fascinating groups, however, they have been hunted to the edge of darkness by man, and only two of the three genera of sea cow – the manatee and the dugong – still remain. The single species of the third genus, Steller's sea cow, which was nearly seven meters long, was almost certainly exterminated more than two centuries ago. At last, though, we are beginning to realize how useful these herbivorous animals can be, both as a cheap source of food in the tropics and as efficient clearers of weed from the mouths of navigable rivers. We should learn to know all the returners to the sea better, not as easy prey, but as a carefully conserved resource.

The above map of the Brazilian coastline shows major feeding grounds of green turtles that lay their eggs on Ascension Island, up to 2,000 miles away. After struggling onto the beach well above the water line, the female buries her eggs in the sand (right). The newly hatched young must negotiate a dangerous route down the beach from the nest site, and vast numbers are killed by predators before they can start the long swim back to Brazilian waters.

Below: Floating on its back, with its belly as a chopping block, a sea otter can crack open shellfish by means of a stone – or, as here, another shellfish – held in the paws. This ability to use an object as a tool is rare in the marine world.

Whales, porpoises, and dolphins (collectively called the cetaceans) are the most highly specialized and successful of vertebrate returners to the ocean. The larger whales are the biggest living mammals, and the largest species, the blue whale (maximum length: over 30 meters; maximum weight: about 160 tons), is probably the biggest animal that has ever existed. Attainment of such enormous size is possible only in the marine environment, where an animal's bulk is supported by the density of water. When a large whale is stranded, it suffocates because it is crushed by its own weight in air.

The normal efficiency of whales, however, is phenomenal. Over long distances, for example, they are the swiftest of swimmers. The massive blue whale can move its huge bulk at 27 kilometers an hour for two hours or more, and can increase its speed to 37 kilometers an hour for short periods. The much smaller dolphins can move along at a steady 40 kilometers an hour with little apparent effort. This feat would require muscle power several times greater than that of any other mammal unless dolphins had some means of reducing the friction on their body surfaces; and it is indeed possible that ripples in their skin induce a non-turbulent "laminar" flow of water over their smooth, virtually hairless bodies. The insulating function of the hair of terrestrial mammals is accomplished in the cetaceans by layers of fat ("blubber") that may be up to 30 centimeters thick. The problems associated with insulating a warm-blooded marine animal explain why there are no cetaceans smaller than porpoises. A herring-sized whale, for example, would lose so much heat over its surface – which in a very small animal is large in proportion to its body volume – that it would be unable to maintain its necessarily high internal temperature. Even in a two-meter-long porpoise, the surface area is so great relative to the volume that 40% of the body weight must be composed of insulating blubber.

Whales must, of course, come to the surface to breathe, but their ability to dive and stay under without breathing makes man's diving achievements seem rather puny. The huge blue whale can remain submerged for 50 minutes, the bottle-nosed whale for two hours. During their dives, most whales probably go no deeper than a few tens of meters, but the sperm whale evidently dives much farther. The best evidence of this has come from several incidents in which cable ships have retrieved submarine telegraph cables with the corpses of sperm whales entangled in them, in one case from a depth of over 1,100 meters. What seems to happen is that in searching for the deep-water squids on which it feeds,

the sperm whale may swim close to the sea bed with its mouth open, so that the lower jaw gets caught in a loop of the cable. Then, in struggling to escape, the whale merely becomes further entangled.

Time spent breathing is short for the cetaceans; after the blowhole at the top of the head emerges, expiration and inhalation are completed within two to three seconds. How do they get enough oxygen from this quick lungful to make long dives to depths where the pressure may be as much as 100 times that of the inhaled air? Several mechanisms seem to be involved. For one thing, their body chemistry is such that their tissues can function without oxygen for much longer periods than ours can. For another, whale blood contains almost twice as much oxygen-carrying hemoglobin as man's, and their flesh is very dark red because of the quantity of oxygen-storing myoglobin in the muscles; thus they are able to utilize up to 90% of the breathed-in oxygen, as compared to only 20% for man.

The 80-odd species of cetacean fall into two groups: the toothed whales, including sperm whales, dolphins, and porpoises, all of which eat a variety of fishes and squids; and the baleen (whalebone) group, which have horny, fringed plates instead of teeth, for sieving plankton. The baleens feed largely on patches of krill, the shrimplike planktonic crustaceans that swarm in high-latitude oceans. It is the immense summer productivity of the surface waters of the cold Arctic and Antarctic seas that allows such mighty animals to get enough to eat, and most baleen whales migrate vast distances to these feeding grounds from their winter breeding areas in warmer latitudes. During the winter they eat little, living off stored reserves of fat.

Man has been a hunter of whales for a very long time, but intensive whale fishing dates from the mid-nineteenth century, with the coming of engine-powered catching vessels and harpoon guns. Since then, the stocks of several species have been badly depleted, especially those of the northern hemisphere. The Greenland right whale – so called because its slow speed and buoyancy after death made it the "right" sort of easy prey in the Moby Dick era – was almost exterminated in a little over a hundred years. Although it and other species are now protected by international agreement, it will be a long time before their populations recover – if they ever do. Whales have a very low reproduction rate: the blue whale, for instance, takes two full years to wean a single calf. And so, even if there is some hope for the threatened species, it may take 50 or more years for their former numbers to be restored. It is a sad fact that if the catching of the blue whale had been limited, some years ago, to a level where the maximum annual number of youngsters

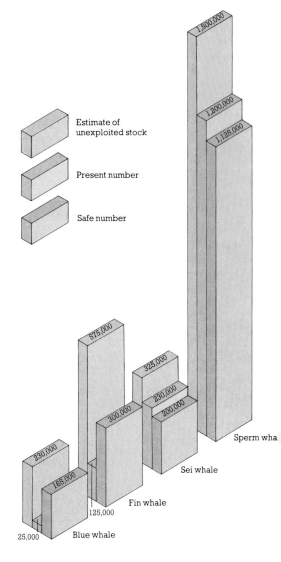

Above: Protected by international agreement but still threatened are these four species of whale. Present populations of two have fallen well below authoritative estimates of numbers needed for protection of the species, and the others are nearing the danger point.

Facing page, top: All cetaceans can put on extra bursts of speed in pursuit of prey or avoidance of danger. Here we see "sprinting" speeds attainable by various types. Dolphins and killer whales can easily keep pace with a small ship, and the blue whale can move almost as rapidly in spite of its bulk. Bottom: A group of right whales "playing" on the surface. Because they are comparatively slow swimmers, right whales are easy prey for hunters; thus, early whalers considered them the "right" species to chase.

was produced, this species alone might now be providing us with perhaps 6,000 whales a year – enough high-quality protein and oil for several million people. If stocks do not recover, we shall have lost more than a direct source of food. An improvement in our fragmentary understanding of the efficient physiology and communication mechanisms of these most fascinating creatures would surely help us to become more efficient in our own efforts to master the undersea environment.

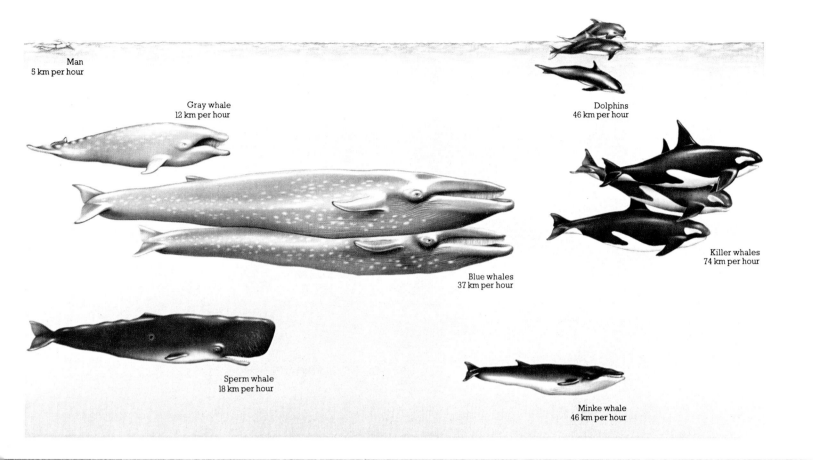

Man
5 km per hour

Gray whale
12 km per hour

Dolphins
46 km per hour

Blue whales
37 km per hour

Killer whales
74 km per hour

Sperm whale
18 km per hour

Minke whale
46 km per hour

Senses in the Sea

We know so little about how marine creatures communicate and react to external stimuli that it seems almost natural to assume that their senses are less well developed than those of land animals. In fact, though, marine biologists see a very different picture. Some ocean-dwelling species have senses of which we are not even aware, and they receive stimuli that terrestrial beings cannot experience.

Touch and smell, almost universal in marine animals, are perhaps the most elemental senses. This is probably because the ancestral environment of invertebrates was turbid and low in light, and it therefore favored the development of touch and smell rather than sight. The moving animal is usually most sensitive to touch or chemical stimuli at its head or, with such animals as starfish and jellyfish, on the tips of arms and tentacles. As for waterborne smell (technically known as "chemosense"), it is impossible to overestimate its importance to most creatures that live in water, where the chemical balance is affected by practically any substance as it dissolves.

Marine animals (apart from the mammals) are highly sensitive to even slight chemical changes. Some of the animals exude small quantities of substances ("pheromones") that act as chemical messengers to convey information to others, and that often elicit a specific response. For example, only a few molecules emitted by an approaching starfish may be sufficient to stimulate a scallop into violently closing its valves, thus producing a jet of water that propels it out of the predator's grasp. We now know of many other similar escape responses – mostly involving echinoderms and mollusks – that are triggered by pheromones, just as we know of many marine predators that smell out their prey.

Water masses are rarely still in the ocean, and smells carried by a current can enable an animal to direct its movement to or away from the source of the smell. Certain snails that have long inhalant siphons move the siphon from side to side in order to sample the water ahead so as to stay on course. But it is the fish that apparently have the keenest sense of smell. Eels and salmon are famous for their migrations to and from rivers and oceans in order to breed, and adult salmon and baby eels recognize the particular smell-spectrum of a river while still far out at sea. Female coral-reef fishes recognize their young, among crowds of others, from their smell; and some fishes can distinguish between individuals within their own species by smell, as dogs do.

The sense of sight is generally less important in the sea than on land (except, as we shall see, for shoaling fish). Many shallow-water invertebrates have simple light-sensing "eyes"; but, apart from the keen eyes of certain fishes and marine mammals, only the compound eyes of higher crustaceans and the wonderful vertebrate-like eyes of cephalopods are capable of distinguishing images. Hearing is another matter, however. The undersea is anything but the silent world it was once thought to be, as the American Navy discovered during World War II when it used sensitive hydrophones (underwater microphones) to detect enemy submarines: there was a deafening cacophony of undersea sounds made by snapping and crackling shrimps, groaning and grunting fish, and clicking, squealing, whistling cetaceans. At a stroke, a new research field opened up for marine biologists, and we are just beginning to learn a little about sound communication among the creatures of the sea. We now know, for instance, that drumfish can produce sounds to attract mates by vibrating certain muscles on the air sac; the pistol shrimp can actually stun potential enemies by clapping parts of its claws together so as to make a loud pistol-like noise; and by vibrating an apparatus at the base of its long antennae, a spiny lobster makes a sound that you can feel as well as hear if you pick it up.

In fact, marine creatures depend on a wide spectrum of low-frequency sounds that we might class as vibrations, not only for communication but even for navigation. A fish's lateral-line system, which we see as a row of dots running down the side and over the head, is used for receiving such vibrations from its fellows; by this means it maintains its position in a shoal. Sharks are particularly sensitive to the vibrations caused by the flapping of a wounded fish. Aided by its acute sense of smell for blood, a shark can detect prey from considerable distances; sight plays a part only in the closing stages, when it is ready to strike.

Utilization of underwater sounds and vibrations is most significant, however, in the marine mammals (which, incidentally, have no sense of smell). We still know little about whale communication beyond the fact that it is based on a complex pattern of frequency-modulated sounds. In the open ocean, whales may make use of a sound channel – a deep layer where sound, reflected back and forth from underlying layers of water, can travel for hundreds of kilometers. Even at the surface, a fin whale's call for a mate can carry through the water for 80 kilometers. Such toothed whales as the dolphins have extremely sophisticated pulse-modulated, echo-location mechanisms for finding prey and for navigation. They seem to scan the water with clicks of sound focused in pencil-thin lines from their sideways-swinging heads.

Among the more mystifying accomplishments of various undersea creatures is the apparent ability of sharks and rays to sense the electric field caused by the breathing movements of a buried fish. It is possible, too, that one type of electric ray uses its electric organs not only for defense but for navigation, by sensing irregularities in the electric field it sets up around itself. What is certain is that we should keep an open mind about the sensory abilities of marine animals. Our most advanced acoustic sounding equipment has been found to be primitive in comparison with what dolphins have – and we may make even more startling discoveries as we continue our studies.

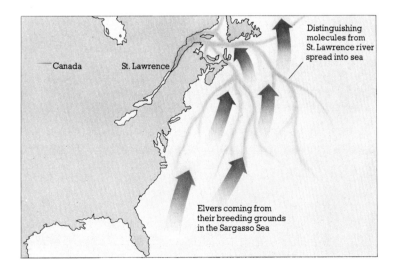

Distinguishing molecules from St. Lawrence river spread into sea

Canada   St. Lawrence

Elvers coming from their breeding grounds in the Sargasso Sea

Eels of the rivers of Europe and eastern North America breed in the Sargasso Sea. After drifting in prevailing currents for at least three years, the growing larvae swim to the "home" river, sensing the direction by means of characteristic molecules that flow from it (far left). At the river mouth they change into elvers (near left) before moving upstream, where mature eels remain until returning to the Sargasso Sea to spawn and die.

The eye of a squid (below) can distinguish images – an ability shared among marine animals only by the cephalopods, mammals, and certain fishes. Netted dogwhelks (left) cannot see the food they are after; they detect its presence chemically in water drawn in through their inhalant siphons.

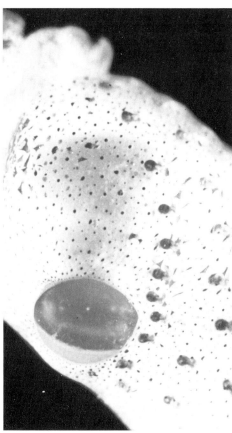

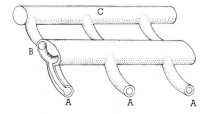

pressure waves

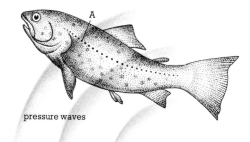

pressure waves

A Pores in lateral line
B Lateral line canal
C Nervous system

Two ways to "smell" danger. Above: The lateral-line system of the fish is sensitive to pressure waves in the water. Pressure on certain pores in the skin triggers a reaction that permits evasion of approaching predators, as well as maintenance of station in shoals. Left: These queen scallops have chemically sensed the foe – a starfish – and are trying to avoid it.

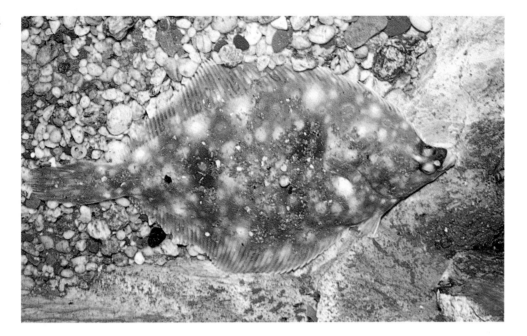

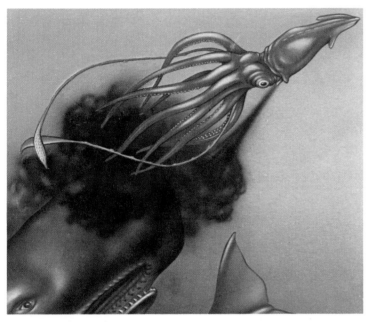

Sea-dwelling animals have evolved a variety of defenses against predators, from cryptic coloring to lethal stings. Like most bottom-dwelling flatfishes, the plaice (above) can camouflage itself by adjusting its skin pigmentation to match the sea bed. When threatened by its main predator, the sperm whale, the squid (left) emits a cloud of ink under cover of which it may manage an escape. There is some possibility that the ink also has a partly immobilizing narcotic effect on the attacker.

For protection, the sea mouse (left) has hollow pointed spines in among the iridescent bristles along its side. The spines probably contain a toxic substance that causes discomfort in the skin and mouth of any creature that may have accidentally disturbed, or that actually tries to attack, the sea mouse.

Survival

No creature in the sea is permanently safe from predation; even whales are preyed on by killer whales as well as man. As protection against their natural enemies, marine animals – including the simplest of invertebrates – have evolved special mechanisms of many kinds. One means of defense is just the state of being unappetizing. Since sponges seem to attract few hungry enemies, their prickly spicule skeleton and disagreeable odor are apparently successful deterrents. The coelenterates' stinging cells have a similar purpose – and, strangely enough, they help protect certain sea slugs that manage to eat them. The slugs somehow store the ingested stinging cells, undischarged, in brightly colored filaments on their backs. These advertise the slugs' unpleasantness to fish, which consequently seldom touch them.

It is quite common for animals with such weapons to show obvious warning signs to potential predators. Many fishes have dangerous, often poisonous spines, and the spine of the weever, for example, is tinted conspicuously black. Similarly, some of the electric rays, which can generate shocks up to 220 volts, carry an eyelike spot on the back as a warning.

Perhaps the most usual way to avoid being eaten is concealment. Flatfish can gradually camouflage themselves so as to blend with the color and texture of the bottom. And such bottom-living cephalopods as the cuttlefish and octopus are even more gifted, for they can change their color instantaneously under direct nervous control, whereas color changes in fish are under the slow control of hormones. What's more, cuttlefish can put on a bodily display (very confusing to the predator) of rapid waves of color, or they can startle an enemy by suddenly exhibiting a pattern of prominent black spots, with red around the fins. As a last resort, like all squids and octopuses, they will eject a cloud of ink, which is really a diversion rather than a smokescreen.

In the wide-open space of the open ocean lives a close-knit community of little animals associated with clumps of floating sargassum weed: the sargassum fish and some tiny crustaceans, all of which are not only colored exactly like the weed but even bear fleshy weedlike filaments on their bodies. It's easy for human observers to see how such creatures achieve concealment, but in many cases we cannot accurately judge the significance of specific colors and markings in salt-water animals. Our standards of vision are not those of marine predators. It may be, for instance, that the brilliant colors of tropical fish are only a means of producing a pattern of protective grays in the eyes of a color-blind attacker. And what we see as the conspicuously bright red color of the shrimplike crustaceans that live in the depths may well be *protective* coloration, for down there, where there is little or no light, and certainly none at the red end of the spectrum, the crustaceans must look black and inconspicuous.

Near the surface, however, most of the planktonic crustaceans are protectively transparent, as are arrow worms, medusae, and most young fish. To achieve something like transparency with increasing size, many marine animals have very flattened bodies. The filmy, elongated, wraithlike organism called Venus's girdle, which is akin to jellyfish, is surely not easy for predators to spot, and baby eels are so laterally flattened as to appear almost completely transparent. With much larger animals, this form of concealment is impossible, but it is not a coincidence that so many open-ocean species – the blue whale and blue shark, for example – are called "blue." The slate-blue color makes them hard to see in the water, particularly from above. Combined with its wavy, striped pattern, the blueness of a mackerel can make it well-nigh invisible to a gull winging overhead, and its silvery-white underside reflects the maximum amount of light, thus making it inconspicuous from below as well.

Less common than concealment are a number of other methods by which various species manage to remain alive in the face of constant danger. The outspread crown of the fan worm can retreat into its burrow at lightning speed because its highly sensitive eye-spots can detect the minutest change in light intensity caused by an approaching object, and the warning message is flashed to the muscles by fast-conducting giant nerves. Shellfish, too, have quick withdrawal reflexes for protecting their vulnerable fleshy parts; and many bivalves have evolved a two-compartment muscle: the fast part snaps the shell shut, while the other part can stay contracted long enough to frustrate the patience of most potential predators. Echinoderms, on the other hand, usually have batteries of spines and spicules and tend to fight back rather than retreat. When irritated, some sea cucumbers can swiftly eject from the anus a sticky mass of writhing threads, which swell and shoot about on their own, entangling and immobilizing an offender as large as a lobster. As a last resort, though, the sea cucumber retreats by completely eviscerating itself, in which case the escaping body eventually generates a new set of viscera.

This type of drastic defense against the catastrophe of being entirely eaten up may be more common than is generally realized among bottom-living animals. It is so common, in fact, that the inhalant tubes ejected in self-defense by little bivalves otherwise buried in the sands probably comprise a good part of the diet of some fishes. It is a truth worth emphasizing that the extensive animal-eating-animal series known as food chains do not necessarily involve the total destruction of all constituents. To eat just the bivalve's siphons is, for the predatory fish, a preferable alternative to eating the whole animal, because in this way the predator is guaranteed a better supply of food; it takes less time to regrow siphons than entire bivalves.

So nature is by no means all red blood and gore. Predators do not always catch their prey; even when they do, they do not always wholly destroy it. There is a delicate balance between their efforts to find and eat the prey and the success of survival mechanisms that the prey has evolved. That is the real balance of nature.

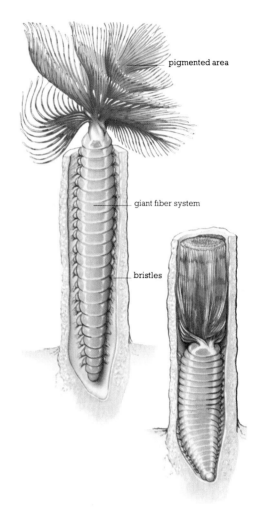

pigmented area

giant fiber system

bristles

When in its extended feeding position, the tube-dwelling fanworm is highly vulnerable. Light-sensitive spots on its delicate, waving "arms," however, react almost instantaneously to any passing shadow: impulses transmitted along giant nerve fibers cause swift retraction of the tentacles into the protective tube.

Below: When threatened or alarmed – as here by an underwater photographer – the sea cucumber shoots out long, sticky threads from the anal region. These swell up, lash about, and can enmesh and immobilize quite large predators. Under extreme provocation, a sea cucumber can expel its entire gut in order to make a fast escape. When this happens, the animal is able to regenerate a new intestine.

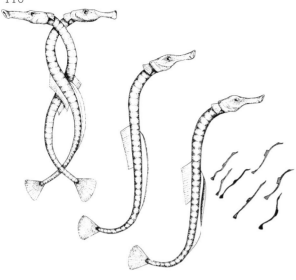

Pipefish (left) protect the young, whereas herring (below) abandon theirs. As mating pipefish entwine, the female lays a few eggs in the male pouch, where they rest secure till ready to hatch and swim away. Herring, on the other hand, lay thousands of eggs on the sea floor. Most are killed before the larval stage, and most larvae never reach adulthood – but because there are so many, a necessary few survive.

Reproduction and Regeneration

An unending cycle of reproduction in which the fusion of male sperm and female eggs subtly re-sorts the pack of cards that deals out the genetic makeup of every living species is at the heart of the evolutionary process. By natural selection of the fitter variants, each species may improve its chances for survival, and it is this pressure that has resulted in the immense radiation of different forms of undersea life. Methods of reproduction in the sea are almost as varied as the forms of life. Copulation is common, but so are such processes as vegetative budding or the free shedding of vast quantities of eggs and sperm into the water, where eventual fusion becomes a matter of fortunate accident.

Innumerable larvae do develop, however, from waterborne eggs and sperm, often because there is a synchronized breeding of the parents that seems to be controlled by the moon and tides. At certain times of the year, the larvae of many species, especially of those such as oysters and sea urchins that live in the sea bed, swarm so thickly in surface waters that they appear to make up the entire plankton. No wonder, since a single American oyster can spawn more than a hundred million eggs at a time. Yet such are the dangers to which the multitude of unprotected planktonic larvae are exposed that, despite such massive fecundity, most species just about hold their own.

The marine creatures that produce fewer offspring compensate by taking precautions to improve their chances of survival. In some species, the eggs are fertilized *before* the female sheds them; enclosed in a hard case to deter predators, they are surrounded with a nutritious yolk to provide for their early development. Other animals – lobsters and crabs, for instance – carry the young around with them, sometimes in special brood pouches on their undersides. Such larvae are deprived of the food-rich pastures of the plankton (which also acts as a means of population distribution), but they have a relatively secure early life. Among the fishes, this extra measure of security may be attained by the building of nests for the babies out of fragments of seaweed (as with the sticklebacks, gobies and wrasses) or even by retaining fertilized eggs within a parent's body. Male sea horses have kangaroo-like pouches in which to do this; the female merely puts the eggs there. The pouch then closes, and the eggs get oxygen from the father's blood. After a "pregnancy" of several weeks, the pouch opens, and muscular contractions

This swimming crab has been turned on her back to show how she carries her eggs. Attached by a mucous substance to the underside of the body, more eggs escape the notice of predators than if they were more exposed.

force the little sea horses to emerge.

Brooding, whether in nests or pouches, is often associated with special pre-breeding behavior to allow detection (sometimes at long range by means of waterborne "smells") and bringing together of the sexes for copulation. One common form of pre-breeding behavior involves elaborate courtship displays, which can be extremely complicated among certain fishes, and hardly less so in the cephalopods. Courting cuttlefish, for example, assume a striped body pattern and do a strange and wonderful dance with their tentacles. During copulation, the male and female take a head-to-head position, intertwining their tentacles, through one of which the male transfers his sperms in little packets (spermatophores) into a special pocket inside the female.

Many invertebrates, including a number of mollusk species, are actually hermaphrodites – that is, the same individual produces both eggs and sperm. One type of limpet is called *Crepidula fornicata* because it forms chains of "fornicating" individuals that lie in constantly mating heaps. Normally, a young limpet settles on the shell of an older individual at the top of a pile; within the chain, each limpet passes through male, hermaphroditic, and finally female phases, with the little male at the top fertilizing the hermaphrodite below it by means of a muscular penis. The oldest and largest individual, the female at the bottom of the heap, usually has a brood of babies protected in her shell; these eventually crawl off to seek out new chains.

Sexual reproduction is only one way to keep the species going. There are others. Reef-building corals, as well as the

A stack of individuals of the filter-feeding slipper limpet *(Crepidula fornicata)* attached to a stone on the shore. The oldest and largest animal at the bottom of the pile is always female, and the younger, smaller one at the top is always male. The hermaphroditic limpets in between are in process of changing sex from male to female – an invariable progression in this species.

little encrusting or mosslike animals that grow in profusion on the sea bed, form massive colonies by vegetatively budding off new individuals. Many different sorts of marine invertebrate multiply by dividing themselves up in one fashion or another. A flatworm can grow an entire new body from almost any piece it's cut up into; scientists have created many-headed flatworm monsters by repeatedly cutting into the head, as a result of which a new head grows from each little flap. This regenerative faculty is a useful way to keep a constantly threatened species from being exterminated by its numerous predators. Even creatures as complex as crustaceans and echinoderms can occasionally regenerate lost limbs. But, generally speaking, the power to regenerate becomes reduced with increased complexity, so that the marine mammals can barely replace even small areas of skin.

The extraordinary regenerative powers of the sea's invertebrates suggest that they probably retain the cellular flexibility that every life has during its earliest period of development from the fertilized egg, when all sorts of new cells are being formed – a flexibility that is lost in more highly specialized adult bodies. Many marine invertebrates even reverse the process; that is, they lose their specific form as their specialized cells revert back to a shapeless generalized blob. That is how sea squirts, for example, respond to unfavorable conditions, such as lack of oxygen. The blob of cells that was a sea squirt remains alive, and it can regain its previous body form when the environment improves. A rather similar phenomenon in human beings is the inexplicable growth of groups of generalized cells that we call "tumors." Unlike the sea squirts, we have lost the power to control the process. It is always possible, though, that our study of the processes of reproduction and regeneration in simple marine animals will bring us closer to the control of cancerous tumors in our own bodies.

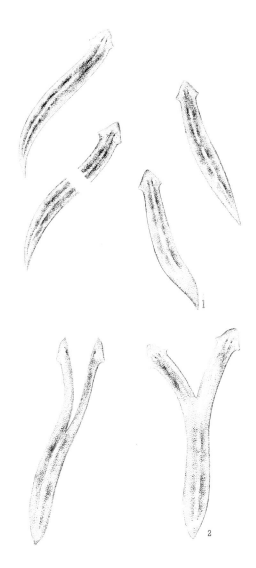

Some marine invertebrates can achieve a kind of immortality by regenerating damaged or destroyed body parts. The above diagram shows the regenerative powers of a flatworm, which can even grow a new head or two. Below: A hungry predator left this starfish only its central disk and one large arm. It is now growing four new arms. A question for metaphysicians: Will it still be the same starfish?

Brittle stars (far left) live in aggregations on the bottom. While feeding, as here, their arms are in a vertical position; but if the current becomes too strong, they avoid being washed away by lowering and linking arms to form a mat. King crabs group together, too. They link their legs to form pod-shaped structures that are often as much as five meters long. Note the lone crab in the foreground of this picture. It will doubtless soon join the rest, for such formations provide protection and facilitate reproduction.

Social Life in the Sea

Like terrestrial animals, most salt-water creatures live in groups, and the types of social organization range from the static colonies of reef corals or sea squirts to the almost mechanical shoaling of such open-water fishes as herring and mackerel, to the intelligent group activities of cetaceans. The fact that corals and sea squirts reproduce by vegetative budding of new individuals, which are not able to move away, underlies their formation of static colonies; and the resultant much-branched structures enable the polyps better to intercept floating food particles. But the value of a communal life for free-moving animals is less obvious. In looking at some representative communities, we shall therefore pay special attention to the apparent significance of aggregation to the animals involved.

Even the simple bottom-living invertebrates seem not to be just scattered about, but are frequently gathered together in clumps. These may sometimes be seen with the naked eye – for instance, the intertwined arms of colonies of brittle stars can form "carpets" that roll from one location to another in strong currents – or the grouping may be so much less obvious that only special sampling and mathematical techniques can detect it. In fact, such subtle grouping may be the outcome of opposing forces: on the one hand, a need for spacing out so that food supplies are never locally exhausted; on the other hand, a bunching together so that the sexes remain accessible to each other. The latter consideration is particularly important with static animals such as barnacles that must copulate to reproduce.

Thus, barnacles tend to live in clusters, and the mating with a neighbor, even as much as 8 or 10 centimeters away, is achieved by means of an elongated probing penis.

In the realm of the tiny organisms of the plankton, the vertical migration from normal daytime levels up to the surface at dusk may be a means of improving social aggregation, since it seems to permit organisms that have been scattered in three dimensions in the depths to become concentrated into a thin layer at night. At any rate, the different species do tend to form separate patches when they reach the surface, and so it seems possible that the night-time hours are the period during which sexual reproduction can best take place. Indeed, complicated prenuptial behavior has been observed even in little copepods; some species have gorgeous colored or iridescent

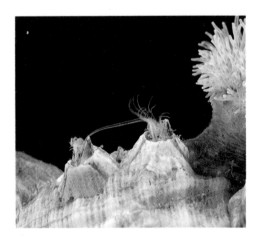

One reason for grouping is that it facilitates reproduction. Barnacles (above) must copulate to reproduce, and so they tend to live in clusters. It even seems probable that the nighttime rise of the plankton eases the breeding problem. As shown in the diagram (right), different planktonic species often form separate patches as they approach the surface; this may be their best time for copulation.

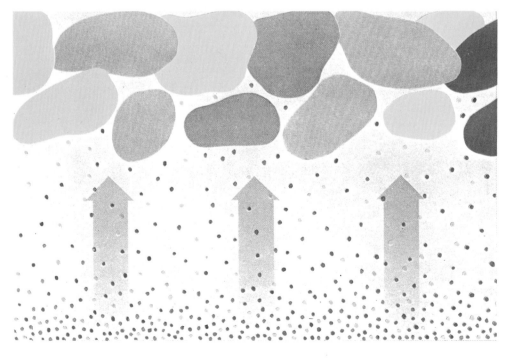

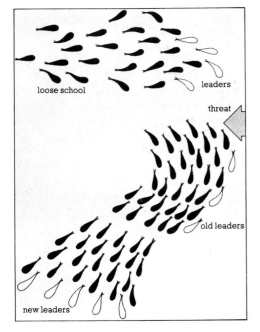

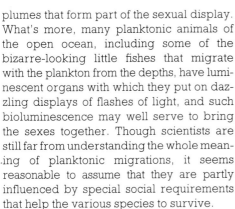

plumes that form part of the sexual display. What's more, many planktonic animals of the open ocean, including some of the bizarre-looking little fishes that migrate with the plankton from the depths, have luminescent organs with which they put on dazzling displays of flashes of light, and such bioluminescence may well serve to bring the sexes together. Though scientists are still far from understanding the whole meaning of planktonic migrations, it seems reasonable to assume that they are partly influenced by special social requirements that help the various species to survive.

The krill of Antarctic waters provide probably the thickest patches of plankton anywhere; each swarm contains such prodigious numbers that it may redden a very large area of surface water. Here we see that aggregation has dangers as well as benefits, for krill patches are literally lapped up by baleen whales. With most species, however, it is the benefits that count. For example, the close community of fishes that live in schools probably helps to reduce predation (except by man) in that enemies are confused by flashing color patterns and rapid movements as the school swims in unison, first this way, then that. About a fifth of all fish species have evolved the schooling habit, which is so deep-seated in some cases that such fish as herring cannot live singly but die of nervous tension.

Although fish schools are finely controlled, they are leaderless assemblages without pecking order or division of labor. It is apparently the visual sense that mainly keeps a school together, though such other receptors as scent and pressure waves picked up by the lateral-line system are important in the precise spacing out of individuals within the shoal. Schools break up at night when the fish can no longer see one another, and a blinded fish can never rejoin

The above photograph shows a school of scombrid fish (similar to mackerel) feeding in the Red Sea. From above, a large school can often look like an enormous swimming animal. As shown in the adjacent diagram, a threatened school does not scatter but, like a disciplined military column, protects its flank by turning.

its school. The fish seem to be irresistibly attracted to moving objects of their own size, and this probably explains why young and adult fish normally form separate schools. We rarely find herring of different ages swimming together, for example. The sometimes abruptly varying, but always synchronized, movements of a school resemble those of a super-individual; thus, when a horse mackerel attacks anchovies, the entire shoal dodges sideways, or it may swiftly split in two, to merge again later. It is only when fish are detached from the school that they become easy targets. Deep-sea fishermen have much to be thankful for in the shoaling instinct of pelagic fish, since the quarry are already concentrated before the trawl gets to them. A school even remains intact inside the opened-out net, and it is only when its individual members are exhausted that they fall back into the cod end.

We still know little about the social behavior of whales and dolphins beyond the fact that it is far more intricate than that of other marine creatures. With their alert mammal brains, the cetaceans are capable of flexible responses quite outside the ken of the rest, and their schools are integrated at least as much by acoustic signals and intelligence as by vision. By means of a complicated language of clicks and whistles, the members of a community communicate with one another to warn of dangers and to maintain the complex social structure of their group. The readiness of captive dolphins to cooperate with human trainers is partly due to the fact that, as intelligent social animals,

Lantern fish, which live in the black depths of the oceans, have light-producing organs on their bodies, which can be seen and recognized by other lantern fish. Some such ability to make contact is essential for the survival of the species.

they are by nature cooperative. A whale's cry of distress quickly brings fellow members of the school to its side – a community feeling that used to play into the hands of the men who managed the whaling fleets.

A substantial part of the cetaceans' social behavior, however, is not directly concerned with survival but seems to be merely boisterous and often imitative play. Under what heading other than play can we put the great body-spinning leaps of Pacific dolphins, which carry them out of the water, or the amusing, slightly awkward attempt of an Atlantic bottle-nosed dolphin to imitate such acrobatics when, in shared captivity, it gets its first sight of them?

Not only does the octopus have good senses of sight and touch, but it can also "remember" the past. The memory store sits in a central brain, from which radiate two optic lobes and a system of nerves reaching into each tentacle.

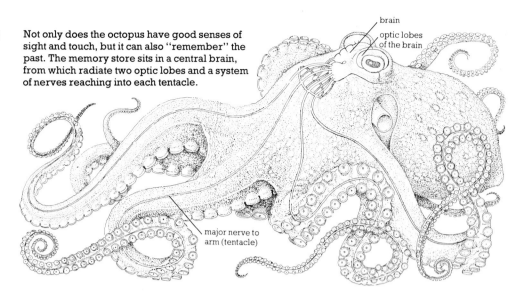

brain

optic lobes of the brain

major nerve to arm (tentacle)

Undersea "Brain" Power

Although scientists have barely begun to study the remarkable thought processes of the cetaceans, most laymen now know that whales and dolphins are highly intelligent. What may come as a surprise, however, is that many lesser marine animals have mental abilities involving quick learning and even "insight" in problem solving that are at least equal to those of our four-legged terrestrial friends. Take those glorified snails the cephalopods. One in particular, the octopus, has provided neurologists with a chance to study the structure and workings of a quite sophisticated "brain" without having to cope with the difficulties of experimental surgery on vertebrates.

An octopus senses its world largely through excellent eyes, and it also has a well-developed sense of touch in the ten-

tacles, but instinct seems to play only a small part in its actions and reactions. (The other main contenders for the title of salt-water invertebrates with the most complex behavior are such higher crustaceans as lobsters and crabs, whose behavior is chiefly instinctive.) Much of what the octopus does is an outgrowth of learning through experience. Ordinarily, for instance, the forever-hungry octopus will attack a crab and eat it with gusto; but experiments have shown that if an octopus is given a mildly unpleasant electric shock whenever it is offered a crab simultaneously with being shown a white disk on the end of a pole, the big cephalopod quickly learns to associate the two things and refuses to attack any crab offered along with a white disk. It can also learn to discriminate among different shapes and colors of disk, will react differently to each, and will retain the

memory of the distinctions for several weeks. Thus, its learning ability is certainly as good as that of smart dogs. (But if you have heard that an octopus can use stones as tools for opening oyster shells, don't believe it. Such complex manipulations are not possible. The tentacles are controlled strictly as reflexes by a lower brain center, and the animal cannot exercise the delicate feedback control that would be needed for handling a tool.)

Much lowlier animals than the octopus seem to behave in ways that suggest at least a modicum of intelligence. Even such an unlikely creature as the limpet sitting on the rocks of the shore acts as if it has a memory. Limpets are really quite active creatures: each individual goes off for feeding excursions at night, moving slowly on a broad, slimy foot – sometimes for distances of more than a meter. Amazingly, though, it returns to its original daytime position with unfailing accuracy time after time. It often does this even when moved a short distance away among a crowd of others. How does it "remember" where it belongs? Nobody knows.

In fact, animals as simple as polychaete worms have been shown to be able to learn by experience, although *what* they can learn is, of course, limited by what their simple nervous systems can sense in their

Below: Three steps in training an octopus. A crab is placed in front of a white card (1). If the octopus attacks the crab (2), an electric shock is transmitted through the water in the tank. Soon the octopus learns never to approach a crab – or any other prey – when a white card is present (3). Instead, it will shoot out a protective jet of water if the juxtaposed crab and card are moved toward it. But it will unhesitatingly seize and eat a crab that is unaccompanied by a white card.

surroundings, as well as by the small range of action that their bodies are capable of. Given the choice of two apparently identical holes to enter, the marine worm eventually learns to avoid the one where an unpleasant stimulus such as an electric shock awaits it. The polychaetes can even be taught to associate light changes with a reward of food, as octopuses do. What has so far baffled scientists, however, is that trained worms retain their memories even when their brains are surgically removed and can continue to be trained as easily as ever. We do not know, therefore, where their "memories" are sited.

Apart from the dolphins and whales, the most varied and complex behavior of saltwater creatures is found, naturally, among the fishes. A great deal, such as schooling and nuptial display, is basically inherited instinct, but there remain many actions that fish must learn during their lifetimes. Take the brightly colored little creatures that dart hither and thither on the coral reef, for instance. Most of them do not travel very far; they do not dare to, since to intrude on someone else's territory and to cross the invisible boundary line would provoke fierce aggression from a neighbor. Somehow, each individual has to learn to know its place, for it cannot know it instinctively, and most fishes do seem to recognize the intimate details of their surroundings. One type of goby, for instance, remembers its general area so well that it can jump from rock pool to rock pool at low tide with no risk of being stranded, even though it cannot possibly see the neighboring pools before leaping.

No doubt the most impressive memories are those of migrating fish such as salmon, with their unerring ability to swim from far out at sea to the portion of the river where they spent their first days as much as two to six years earlier. But the more we study fish in general, the more convinced we become that past generations of naturalists have tended to underestimate their intelligence. Goldfish, we now know, can learn the route through an artificial maze by trial and error; and, like the octopus, they can also be taught to associate specific shapes and objects with food. In particular, they seem to be adept at distinguishing subtle differences in the pattern of wriggly lines (perhaps because the lines resemble worms?). We have even discovered that some fish are fooled by the same optical illusions that trick the human eye. So, whether on land or at sea, the mammal is by no means the only creature with brain power.

The dolphin, one of the most intelligent of animals, can be trained to be useful as well as amusing. Here, during experiments by the U.S. Navy, a dolphin is learning how to carry a lifeline to a theoretically stranded diver whom the animal first locates, then guides back to safety.

Even ragworms can learn, though we don't know where their "memories" are physically sited. After a few tries, most worms discover that food always sits in the upper righthand corner of this T-maze. Thereafter they take the correct route without hesitation.

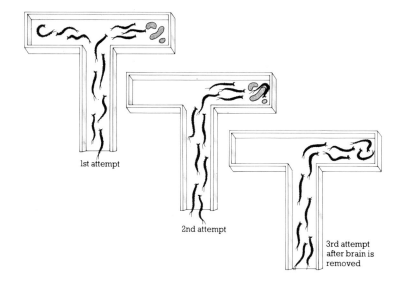

1st attempt

2nd attempt

3rd attempt after brain is removed

The Undersea Food Web

Much is said these days about the protein wealth of the sea, which we must learn to utilize more fully if twenty-first-century food needs are to be met. Without quarreling with that statement, it is nevertheless important to understand the realities of marine life – in particular of the sea's manifold food chains – in order to see the problems involved in human efforts to increase the ocean's productivity. So let's briefly review some familiar, if occasionally ignored, facts about the undersea food web:

The fish that makes a meal of a smaller fish will itself make part of a meal for a larger fish, and so on. At the top of this food chain are the largest predators, and man may be among them. The first link in virtually every marine chain, however, is the phytoplankton. The phytoplanktonic pastures, which comprise the bulk of marine plant life, are the all-important primary producers of food in the sea, and the herbivores that graze on them create a secondary production by digesting the plants into their tissues. It is the herbivorous members of the zooplankton that contribute most of the secondary production to the system – although, of course, many larger animals, including such fish as the mullet, are secondary producers because they eat

seaweed, and such bottom-dwelling filter-feeders as shellfish also feed mainly on the phytoplankton.

The third link consists of animals that prey on the secondary producers. Many of these predators are still small enough to be included in the zooplankton, but such smaller nonplanktonic fish as herring and anchovies enter the picture here, along with plaice, rays, and other shellfish-eating fish; and we can even include at the same level the krill-eating whales. From this point on, marine food chains vary enormously. Usually, though, they take on a number of further links as smaller animals are eaten by ever bigger, ever more specialized predators.

And just as it takes large numbers of grass-eating antelopes to support a few lions, so the total production in terms of bulk of marine animals must drop dramatically at each step, for the process of converting food into body tissues involves an immense loss of energy.

If we visualize the conversion process as a diminishing flow of energy that initially enters the life system from sunlight via photosynthesis, we can gain an idea of the loss incurred at each new link in a marine food chain by thinking of the bulk of herbivorous zooplanktonic production as a percentage of phytoplanktonic primary production. With young, actively growing animals, the figure may be as high as 40%, but the general average is closer to only 10%. That is a better conversion efficiency, actually, than we manage to get on land, because the ocean's secondary producers can use a great part of the simple marine plant as food, whereas only a fraction of most land plants is of much use to terrestrial herbivores. The land animal not only wastes much of the root, bark, and hard stem, but also needs to divert a lot of its energy from building its own tissues to reproduction and caring for the young. The average marine herbivore, on the other hand, just releases its eggs or larvae into the sea. So there may, indeed, be a staggering undersea total of more than 50 thousand million tons of secondary production every year. But since we tend to harvest only the higher links in the marine food chain, we utilize little more than a tenth of 1% of this total.

That may sound like rather poor "farming" on our part. The ocean itself, though, is far from being a well-managed farm, since by no means all the production at each link of the chain is eaten by the one above. A great many animals escape predation, and others, such as medusae and sea gooseberries (which can occur in enormous numbers), seem virtually unwanted by anything else. In fact, the conventional terrestrial notion of food chains is much too simple to apply unreservedly to the undersea. For example, scientists who have studied the feeding of commercially important fish such as the herring have found that the adult herring eats a number of smaller animals that themselves feed on baby herring. Thus, food chains can be, to say the least, complex and confused. And, though 50 thousand million tons a year of secondary production is a great deal, one reason why we utilize such a small percentage of the total is that we tend to harvest the sea's carnivores rather than the herbivores or the plants themselves; and so we permit the tremendous primary production of the marine plants to be dissipated into the bodies of a relatively small number of ultimate predators. We do this, of course, partly because, unlike our terrestrial crops and livestock, marine plants and most marine herbivores are just too small to be easily harvested.

And, in reality, the oceanic meadows are anything but rich. In terms of weight as compared with total weight of the animals that ultimately depend on them as the primary producers of the sea's food, the phytoplankton represents no more than about 5% of all marine life. This contrasts sharply with the picture on land, where the overall weight of plants is huge in comparison with that of animals. The phytoplankton is eaten nearly as fast as it is produced, and only the speed at which phytoplankton cells can reproduce enables them to keep up with the demand. If the same process operated on land, all green vegetation would be grazed as soon as it appeared; our trees would be bare and we'd have no grass. Instead, despite all the many sorts of animal that depend on terrestrial primary production, their numbers are never high enough under natural conditions to devour more than a small amount of the vegetation, and most plant production falls to the ground to rot uneaten. By contrast, only the availability of phytoplanktonic food limits the numbers of marine herbivores, most of which live just above starvation level. Truly, we do well to disbelieve any prediction about the impact on the marine food web of human efforts to wrest richer harvests from the sea if the predictions are based on analogies between the sea's ecology and that of dry land. The balance is utterly different!

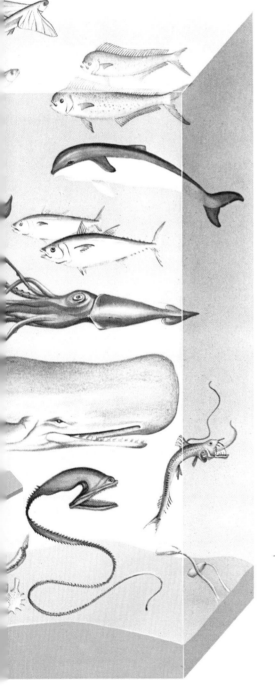

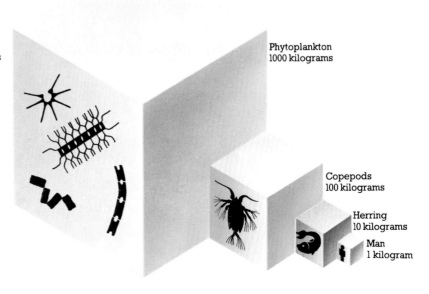

The flow of food energy in the sea can be viewed in two ways. Left: Here we see a cycle that starts with sunlight and chemicals and goes through phytoplankton, then herbivores, then various carnivores, to death, decay, and eventual recirculation of the chemicals through upwelling. But from man's standpoint, what matters is the amount of food available to *us*. Right: At each step from producer to consumer, roughly 90% is dissipated; from 1,000 kilograms of phytoplankton we get only one kilogram of human growth.

Phytoplankton
1000 kilograms

Copepods
100 kilograms

Herring
10 kilograms

Man
1 kilogram

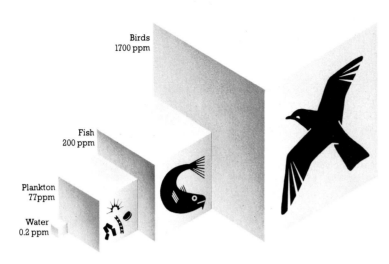

Chemical pesticides may improve a farmer's crops, but the run-off from his land could contribute to marine disaster. Left: How a tiny amount of DDT in the water – only 0.2 parts per million – becomes progressively concentrated in the body tissues of the animals in a typical food chain.

Birds
1700 ppm

Fish
200 ppm

Plankton
77 ppm

Water
0.2 ppm

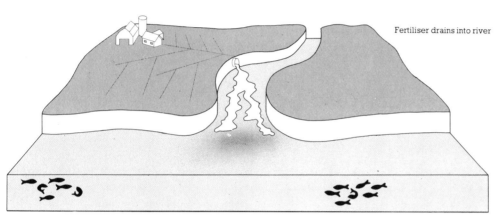

Fertiliser drains into river

Plankton growth is over stimultated and the oxygen content of the water is greatly reduced – fish begin to die

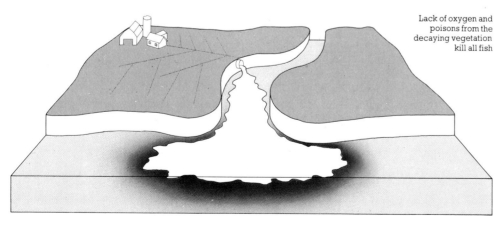

Lack of oxygen and poisons from the decaying vegetation kill all fish

Larder or Dump?

Until recently we have been complacent about the resilience of the marine ecosystem, taking it for granted that the oceans can survive any kind of physical catastrophe; and events have justified our complacency. Even our own increasingly severe and selective harvesting of salt-water animals has so far left the ecosystem apparently intact. We would no longer dare to kill vast numbers of predators in, for instance, the African savanna without being prepared for an explosive increase in the herbivore population, which would strip the grass and defoliate the trees. Yet we appear to be overfishing both the previously enormous stocks of copepod-eating herring in the North Atlantic and the krill-eating baleen whales in the southern ocean, and with no obvious effect on the phytoplankton. This may well be because the numbers of herbivorous copepods and krill, whose populations ought to be exploding (by terrestrial standards), are directly limited by the available quantities of phytoplankton. As was suggested in the preceding section of this chapter, the biggest single restraint on the salt-water food web seems to be, not the number of consumers, but the relatively small amount of phytoplankton existing at any one time.

The phytoplankton pastures *can* suddenly become lush. We know this from observing the occasional and apparently quite natural population explosion, or "bloom," of certain species of floating dinoflagellate that cause "red tides" – so called because the huge number of dinoflagellates imparts a red tinge to the water. The massive toxins in a red tide can poison everything, including fish and sea birds – but we do not yet know why the red tides happen when and where they do. And we cannot be sure of the eventual effect upon the phytoplankton of our increasing disposal of nitrogen- and phosphate-rich sewage into the sea. At best, it may stimulate the levels of primary production and, therefore, of fish production; and this perhaps partly explains why the fishing in coastal waters surrounded by dense population centers – the southern North Sea, say, or the Bay of Bengal – is especially good. But there certainly is potential danger in pouring increasing quantities of organic material into the oceans. Such loading can reach a point where the bacteria that break down organic matter have multiplied so greatly that they consume too much oxygen, and the water becomes

Even apparently harmless substances dumped into the sea in quantity can damage the environment. When biologically active wastes such as fertilizers (which environmentalists prefer to chemical compounds) drain into rivers, they may cause massive marine-phytoplankton bloom. The end result here:– death rather than life.

putrid. Because of the magnitude of the oceans, this could probably happen only in coastal waters. Even so, that is hardly cause for continuing complacency.

Nor is there room for complacency about man's detrimental impact even on the deep oceans, for we must face the problems caused by nonorganic pollutants in the marine ecosystem. The most highly publicized of these is crude oil from spillages at oil-tanker terminals and from tankers washing out their storage tanks at sea; we do not need further reminders of the harm that this can do to marine life. Far more insidious, however, are the effects of chemical pesticides, such as DDT and Dieldrin, that enter the sea via land run-off from farms. There is little evidence that crude oil has any other than a local, temporary effect, whereas DDT, ingested by the zooplankton (crustaceans, after all, are not a far cry from the insects that the pesticide is designed to hit) at low concentrations of a few parts per billion of sea water, is progressively concentrated in the bodies of successive links in the food chain. We are only just beginning to measure the levels of such poisons in the flesh of the fish that we catch and eat. So far, those levels are not high enough to cause great concern. But what of the future?

Still more disquieting are the amounts of industrial waste, and of by-products ranging from radioactive materials to heavy-metal-containing residues, that we have been dumping into the sea for many years without apparent ill effects. Take lead, for example; the quantity put into the ocean has risen enormously over the past couple of decades as a result of our use of lead-containing anti-knock additives in automobile fuels. The lead-rich exhaust fumes enter the atmosphere, and the lead is eventually rained into rivers or directly over the ocean. Like many other potentially poisonous heavy metals, including mercury, lead accumulates in animal tissues and is only very slowly excreted. Furthermore, it persists in the marine environment without being broken down, and so the creatures of the sea are bound to be ingesting more of it all the time.

As all such damaging pressures on the ocean grow, so does man's dependence on it as a larder. It would be tragic to sacrifice the long-term gain of an intelligent and increasing (though certainly not infinite) exploitation of the undersea food supply in favor of our present short-term – and destructive – gain from using the water as a convenient sewer.

A scene in a lab where research on the effects of both oil and detergents is being carried out. Tiny marine organisms are cultured in seawater and agar jellies, and their growth rates under exposure to oil and various cleansing agents are studied in an effort to determine which cleansers are least toxic.

Oil spillage from ships at sea is a much-publicized form of pollution largely because we can see the effects on land. Although an oil-damaged coast can be partially cleansed by special detergents – as in the spraying operation above – the detergents themselves may do harm; some sprays have been found to be as toxic as oil to marine organisms. Right: A modern bit of the coast (not quite appropriate for a picture postcard even though oil-free).

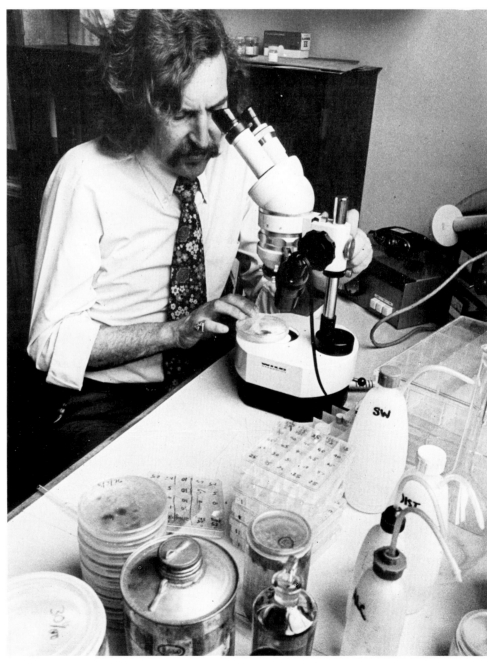

5
The Ocean's Resources

Robert Barton

World population is doubling every 30 years, and there is not enough food to go round. Demand for metals between now and the end of the century will probably exceed total needs of the last 2,000 years. Three times as much energy as has been used over the past hundred years may well be required during the next two decades. And the oceans contain enough protein, metals, and energy to meet most of those needs. If there were not a rapidly increasing demand for such commodities, little attention might be paid to harvesting the sea's resources, because working at sea is, in the main, far more costly and uncomfortable than working on land. But as land-based resources become depleted, it makes economic good sense to look to the sea to satisfy both the sophisticated appetites of Western technology and the protein hunger of the third world.

The world's fish catch, for instance, which now runs at around 65 million tons a year, could probably be nearly doubled by fishing unexploited areas of the sea and utilizing species that have traditionally been considered unpalatable; and this could be done without seriously depleting the sea's resources. At present, certain areas *are* being overfished, and the catch shows few signs of increasing beyond the 65- to 70-million-ton level. This chapter investigates some of the reasons for this, including questions of public taste, distribution of fish stocks, and the way in which overenthusiasm for fishing efficiency is causing the total catch of highly marketable species to diminish. But there are many areas, such as the Indian Ocean and the deep waters of the Atlantic west of Ireland, that still need to be explored if the sea is to yield its full protein benefit. There are also "new" species, where the problem is not so much one of catching the fish as of educating people to eat them. There is the possible further development of fish farming, so that man can have more control over the production of marine protein. And there are possibilities for exploiting the vast stocks of protein available one step down the food chain – the plankton. All these matters are examined in the following pages.

The living resources of the sea are not only valuable as a source of protein. Some marine organisms secrete toxins in a more concentrated form than any so far known to man, and further research into marine pharmacology could provide powerful weapons in the battle against disease. On a more mundane level, the shells of shellfish can be used for such diverse purposes as abrasives or to make cement. Indeed, the potential of the living resources of the sea is only just beginning to be realized.

A second great resource is, of course, oil. Undersea oil and gas exploration and production have a special glamour these days,

no doubt because the rewards are so high in an energy-starved world. In the section of the chapter dealing with this subject, we look at the origins of offshore hydrocarbons and at the equipment and the techniques necessary for finding them and bringing them to the consumer. Perhaps too much has been written about the offshore-oil industry's "here today, gone tomorrow" image – the way the industry is reputed to move into an area, find and produce oil, and move on, leaving a trail of pollution, wrecked social structures, and unsightly installations in its wake. What has not been sufficiently emphasized is that the offshore-oil industry is also responsible for some lasting benefits in terms of increased knowledge about the ocean environment. In the North Sea, for example, the industry desperately needs data on which to base calculations for the design and installation of platforms and pipelines. A crash program to obtain oceanographic and meteorological data has been mounted, as a result of which we shall soon know more about the workings of the North Sea than was learned throughout the past century – and this despite the fact that it has been one of the most closely studied patches of water in the world. Such information is of enormous value to all the fishermen, scientists, and ship operators who work in the sea. And similar investigations are under way in other parts of the world, from the Arctic to the Gulf of Thailand.

Less pressing than our growing need for food and fuels is the question of minerals, and present exploitation of the oceans' mineral wealth is less advanced than in the other fields. Such things as salt, magnesium, and bromine are already being extracted from seawater by shore-based factories, and dredgers working relatively close to the shore have for years been producing sand and gravel from the ocean floor. But there are likely to be more exciting days ahead. Present pilot-scale techiques are

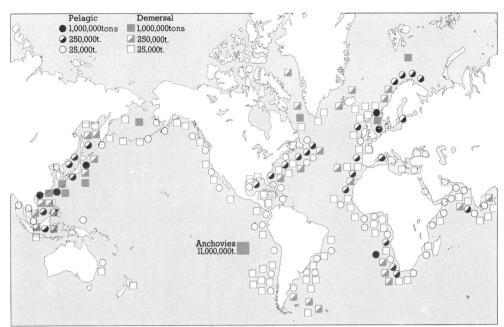

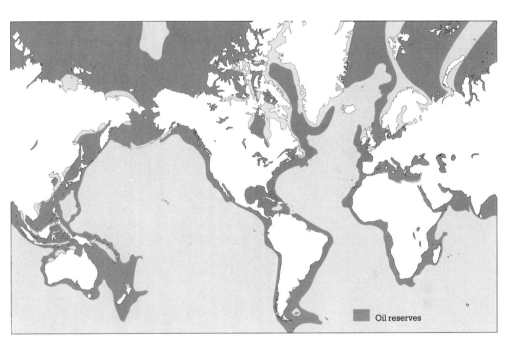

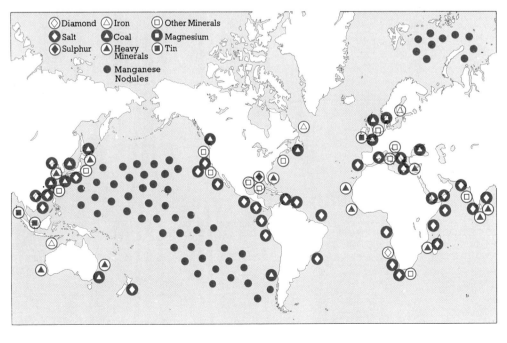

The most plentiful of all marine resources – fish – are also the most susceptible to tragic depletion. With modern detection and catching equipment, a group of fishing vessels such as the Norwegian purse-seiners in the photograph above can haul in as much as 400 tons at a time. These fish are herring, which are already in short supply in some areas of the North Atlantic.

How three major resources are distributed in the world's oceans: fish (top), oil (middle), and minerals (bottom). Partly because of their mobility, fish are the most difficult resource to map and assess, and the location and quantities of catches are bound to vary as conditions change. For instance, the relatively low figures for the Indian Ocean will probably rise sharply with the increasing use of modern fishing techniques. By contrast, the oil map is unlikely to be much altered in the foreseeable future. Present exploration and production areas, as well as those where there are known reserves, or where reserves are possible, may broaden. As for marine minerals, the important question is rather how to get them than where to find them. Although we have long been dredging for aggregates and extracting salt, bromine, and magnesium from seawater, we have not yet solved the technological problems involved in exploiting the enormous potential of the manganese nodules that lie on the deep-ocean floor.

aimed at exploiting hitherto unattainable mineral deposits from the deepest parts of the ocean; if they are successful, essential minerals such as manganese, nickel, copper, and cobalt will no longer be in such short supply as they are now.

So the useful riches of the oceans are vast and varied. The great areas at present untouched represent a golden opportunity for nations to work in harmony at exploiting those riches for everyone's benefit. At present it seems as if the slicing up of the huge cake will be subjected to the type of petty squabbling that has always attended the division of our comparatively meager land resources. But there is some hope in the very size of the ocean. It is so big that it gives statesmen and industrialists a chance to think big, too.

Fish and the Food Chain

With the constantly growing worldwide need for cheap sources of protein, the creatures of the sea would seem to provide an obvious target for rapid exploitation. Fish are catchable in many parts of the world's oceans; since fish stocks renew themselves, maintenance of the supply requires no investment in land, machinery, food, or fertilizers; fish are free to all comers for the taking; fishing is already carried on everywhere and needs merely to be improved by means of modern methods; the supply of fish seems almost limitless. But it is not as simple as that.

The world's fish catch rose from 29 million tons in 1955 to 53 million tons in 1965 and 70 million in 1970, but since then it has remained static, varying between 65 and 70 million. Yet the Food and Agriculture Organization of the United Nations estimates that we *could* harvest 100 million tons of fish every year without seriously depleting the sea's resources. Why are we not making the most of our opportunities? The first step toward finding an answer to that question is to try to reach an understanding of some of the truly agonizing complexities of the world fishing industry as it exists today.

Because fish are high up in the food chain, the total quantity produced in a year can be only a small proportion of the total amount of living matter in the oceans. A generally accepted estimate of the weight of carbon built into living salt-water plants by photosynthesis is about 15 thousand million tons a year. This consists mostly of phytoplankton, which are eaten by zooplankton, which are eaten by fish. Since the efficiency of conversion from any one link in the food chain to the body weight of the animal that eats it is only about 10%, it follows that total annual fish production can hardly exceed 10% of the zooplankton, or 1% of primary phytoplankton production – that is, 150 million tons. That places at least one fixed limit on the amount of fish that can be caught in a year. But neither the production nor catching effort, however, is evenly spread around the world.

Phytoplankton productivity is highest in areas where nutrient-rich water is brought up to the light, such as the upwelling water off Peru, California, and parts of Africa, as well as in the North Atlantic and North Pacific and in the shallow waters on all continental shelves, where tidal currents are strong. Abundant plant life leads naturally to abundant animal life, and so the places where dense fish populations are most likely to exist are limited; they generally coincide with those of high phytoplankton productivity. But the presence of fish stocks near the coast of a country does not necessarily mean that the country's people profit from them. For instance, the people of Norway, Japan, Peru and Chile (all mountainous countries with limited land for agriculture) take full advantage of nearby fishing grounds, whereas the United States and Canada have tended to neglect the protein riches on their own doorsteps because, until recently, North America has had more than adequate land-based resources; and many other countries – especially Russia, Britain, and Spain – are forced to meet their needs not only by exploiting their inshore stocks, but also by fishing intensively in seas far from home.

Thus, although the fish are present on a global scale to the tune of perhaps 150 million new tons every year, they are elusive. There is an endless hunt for more and richer fishing grounds – a hunt carried out partly by specialized research vessels, and partly by the initiative of fishermen themselves.

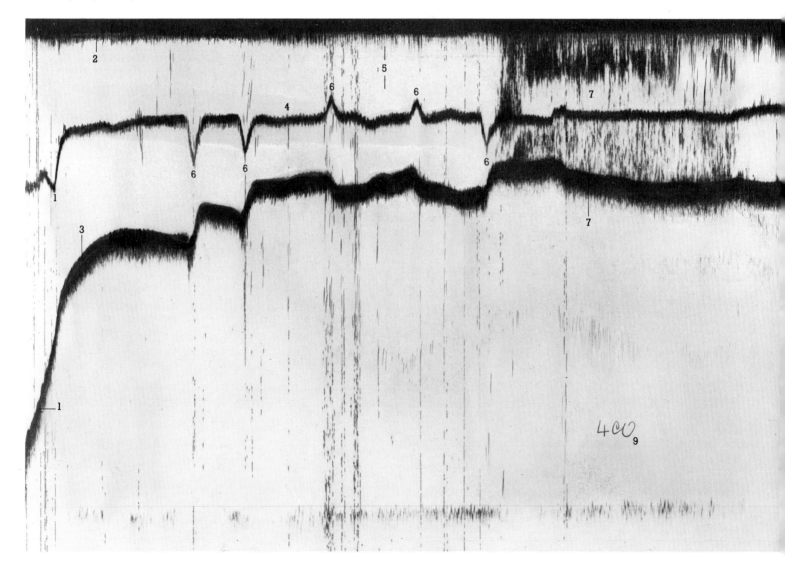

Navigation aids, radar, acoustic equipment that produces echoes off fish shoals, and meteorological and oceanographic forecasting all play a part in the hunt, as do studies of spawning areas and migration routes made by such major laboratories as the Fisheries University of Tokyo or the Lowestoft Fisheries Laboratory in Britain. Yet, in spite of our increasing knowledge of the life style of many species – for instance, the North Atlantic cod and herring, the North Pacific salmon and albacore, and the central Pacific tuna – so many unpredictable factors influence the presence or absence of a given species in a given place at a given time that technologists are still far from devising truly efficient ways of searching out much of the sea's vast supply of protein.

Where technology *has* made strides in this century is in the techniques of catching and utilizing such fish and other salt-water animals as are available. Indeed, we have become so proficient at taking certain highly desirable fish and crustaceans from the water (while virtually ignoring others that are equally nourishing) that we are endangering their survival as species. In the following pages we shall see how and why this is happening.

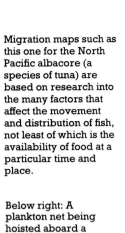

What the Trawler Captain Sees

This typical read-out from a net sounder shows how the net can be maneuvered as a result of the recorded information.

1 Start of trawl
2 Position of top of net mouth (a constant, since this is where the net sounder is fitted)
3 Sea bed
4 Varying positions of bottom of net mouth. Opening and closing movements appear as peaks.
5 Height of mouth opening
6 Maneuvers with trawl warps or ship's speed to alter size of opening
7 Fish shoal within and below net mouth
8 End of trawl
9 The catch: 400 baskets

Migration maps such as this one for the North Pacific albacore (a species of tuna) are based on research into the many factors that affect the movement and distribution of fish, not least of which is the availability of food at a particular time and place.

Below right: A plankton net being hoisted aboard a research vessel. The contents of such nets, which are equipped with current-flow meters and depth gauges, can tell scientists a good deal about the scarcity or abundance in a given area of the tiny floating creatures on which fish shoals feed. Moreover, a count of any fish eggs trapped in the net may provide a rough indication of future fish stocks.

Below: How sound waves help to catch fish. After a trawler has located a shoal by means of a forward-scan sonar (1), the depth at which the fish are swimming is checked by an echo sounder (2) as the ship passes above them. And a net sounder (3) attached to the trawl supplies information about the position of the shoal relative to the net, which is adjusted to make the best haul.

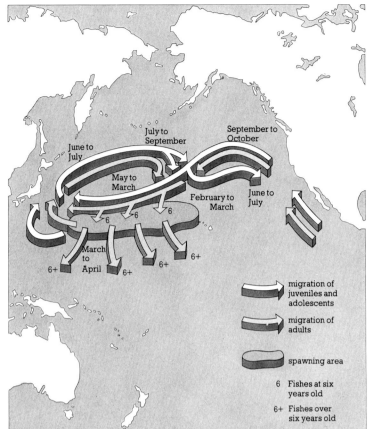

migration of juveniles and adolescents

migration of adults

spawning area

6 Fishes at six years old

6+ Fishes over six years old

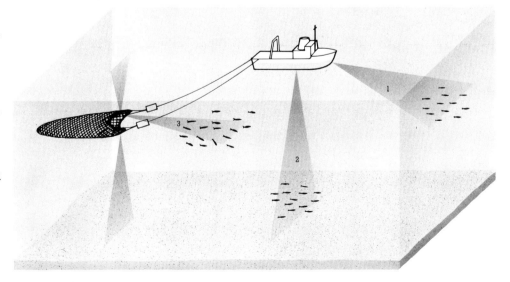

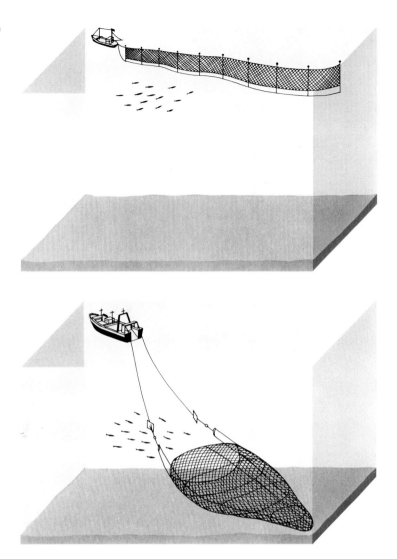

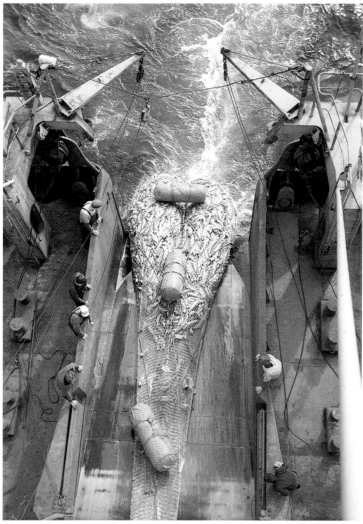

Nets and Lines

There are four basic ways to catch fish in quantity for profit: drift-netting with a net that floats near the sea surface; trawling, in which a net is towed on wires in mid-water or along the bottom; long-lining, in which the fishing vessel lays out a line with hundreds of hooks on it; and purse-seining – laying out a circle of net that can be closed around the fish like a bag. The most efficient technique for catching a particular fish species depends largely on the time of year, on the depth at which the fish are to be found, on whether they are in dense shoals or are scattered, and on their swimming speed. Of the four basic techniques, the one that can make the biggest catch per haul of pelagic (mid-water-swimming) fish is purse-seining; it is effective, however, only when the fish are shoaling densely and moving slowly within a hundred meters of the surface – and it may take a boat several days to find just one such shoal. Where the fish are either scattered or swimming swiftly in deeper water, one of the other techniques is called for.

Drift-netting, which was invented more than two centuries ago when cotton became available in bulk, works only in the top layer of water, where such pelagic fish as herring and salmon can become entangled in a drifting wall of netted fabric when they rise to the surface at nightfall in search of plankton. Modern drift-netting is highly mechanized. In the Japanese salmon industry, each of the mother ships that operate between Kamchatka and the Aleutian Islands supports 30-odd steel or wooden catching vessels, and each of these casts out a vertical net 6 meters tall and tremendously long – from 12 to 15 kilometers. The catchers operate up to 80 kilometers away from the mother ship, shooting their nets every afternoon and retrieving them soon after midnight. Each fleet also has several scout boats, which are constantly engaged in searching for new shoals to conquer.

Sailing vessels could easily deploy and recover drift-nets, but it took steam propulsion to make the technique of trawling entirely feasible, even though fishing smacks under sail had been towing nets (when the wind was right) ever since the seventeenth century. The principle of trawling, which is the only way to catch demersal (bottom-living) fish as far down in the water as 500 or more meters, is simple. The mouth of a cone-shaped net is kept open laterally, while the net is being dragged along the sea

Among various techniques for netting fish, two are drift-netting and trawling. Drift nets, sketched above (top left), are designed for fish in surface waters. Individual nets, which can be as much as 30 meters long and 14 meters tall, are suspended vertically from buoys, and up to a hundred of the nets can be joined to form a "fleet" stretching over several kilometers; great numbers of pelagic fish can become enmeshed in such drifting walls of fabric. As illustrated in the bottom drawing, trawling is designed for use in deeper water – in this case, for catching demersal (bottom-dwelling) fish. A bag-shaped net is held open by ropes, weights, and floats in conjunction with angled otter boards, which draw apart as the trawl is pulled along the sea bed. Above: A fair-sized catch fills the lower part of the trawl – known as its "cod" end – being winched onto the aft deck of a West German stern trawler. Cylindrical buoys keep the cod end from scraping the bottom when it is submerged.

bed, by "otter boards" – flat supports that shear outward when pulled through the water; steel balls attached to the bottom of the net allow it to follow the contours of the ocean floor, and metal or plastic floats at the top of the mouth give the net a vertical opening. The net with its contents is winched into the ship every few hours by means of cables ("warps") attached to the two otter boards. Today's bottom trawl, operated from a diesel-powered trawler, may be as much as

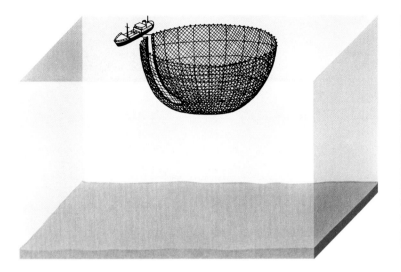

Purse-seining is the best way to catch great quantities of shoaling pelagic fish. As shown in the above diagram, a huge curtain of net is paid out around a shoal and closed – or pursed – under the catch by drawing up a wire that runs around three sides of the net. Once the net is closed, the slack is hauled in over a winch, or power block, aft of the deckhouse. The circle of red buoys in the photograph marks the top of the steadily decreasing area of net. The dense catch is forced to the surface, and the fish are lifted into the boat ("brailed") with a dip net or sucked up by a hydraulic pump.

Top: Long-lining, a method of catching pelagic fish not swimming in close-packed shoals, is chiefly practiced these days by the Japanese, but it is by no means entirely neglected in the West. Here a catch of spiny dogfish is being winched aboard a Norwegian vessel. Hundreds of meters of line upon which hang baited branch lines are suspended below the surface from buoys.

70 meters long, with an opening of up to 40 meters between the front "wings," which channel fish into the bag. The net is made of synthetic fiber, and the otter boards weigh over a ton each. Instruments attached to various parts of the trawl may measure such things as the degree of strain in the warps, movements and drag of the otter boards, the changing shape of the net itself, the flow of water through it, and even the behavior of fish in reaction to the net.

Long-lining involves the arduous task of baiting hooks and coiling and uncoiling very long lines to which the hooks are attached. In the nineteenth century, cod were generally fished for in this way. The line, usually paid out from schooners or small dories, was likely to be as much as 600 meters long, with a hook every couple of meters, and the wives and daughters of fishermen would often do the daily baiting job for their men. Today, long-lining has become mechanized. It is used chiefly in the Pacific for catching tuna that are not swimming in tightly packed shoals. A Japanese long line may have a total length of nearly 200 kilometers! Floats are attached at intervals, and branch lines, with hooks attached, hang down. Obviously, the process of recovering loaded lines is a delicate one, for which the Japanese have designed a special kind of winch.

In the late 1950s, in the American tuna-fishing industry centered around San Diego in California, the introduction of the hydraulic "power block" – a powered pulley through which a purse seine could be hauled – made it possible to handle nets as large as 900 meters long and 300 meters deep. The purse seine is streamed around a school of tuna – which swim in the ocean's upper layers – and its lower edge is then closed, or "pursed," by means of a steel purse wire. After the power block has taken in the bulk of the net, fish trapped in the purse are scooped out by a mechanical dip net, or "brailer." Two hundred tons of tuna can be gathered up in a single operation, or "set," of the net. Purse-seining, combined with sonar detection of fish shoals, brought an explosive expansion to the anchoveta fishing industry off Peru, where the annual harvest can reach 10 million tons. And in the northeastern Atlantic, purse-seiners from Norway and Iceland make significant catches of herring, capelin, and mackerel. Given the right conditions – dense shoals not too far down in the water – the purse-seining method is so efficient that many informed observers fear its effect on the already overfished oceans.

A number of futuristic fishing techniques have been talked about or brought to the prototype stage: the use of powerful lights to attract fish, or of electric currents in the water to direct fish into a net, or of various pumping devices to suck fish out of the water directly into a ship. But such ideas go on being futuristic; nothing has yet turned up that can completely displace the net and the line.

Number of vessels 1000 Tons and over

S. Korea
20

Norway
10

S. Korea
500

France
30

Peru
600

Britain
40

Norway
600

USA
40

France
580

Spain
80

Britain
580

Japan
110

USA
1700

Spain
1660

USSR
400

Japan
2980

USSR
2900

Number of vessels 100–999 Tons

This chart of the distribution of trawlers among major fishing nations does not take into account vessels of under 100 tons, since it is impossible to make a credible estimate of the number of smaller boats at work for a country's fisheries. Note the "weakness" of American and EEC trawler fleets in comparison with those of Russia, Japan, Spain, and Peru.

The Modern Trawler

Locating fish and catching them are only part of the commercial fishing business. The fish must be transported to harbor, unloaded, sold, and distributed to the consumer. Fish go bad quickly; and so when several tons hit the deck of a fishing boat, the first job is to prevent decay. In near-shore fisheries, the fish are gutted, packed loosely, often in crushed ice, and sold in quayside markets for distribution within a day or two. In much industrialized fishing, however, the fishing grounds may be three weeks' steaming from port, and so the catch is gutted, cleaned, and preserved in ice, salt, or oil, or else quick-frozen or canned. On conventional trawlers, the deck is slippery with scales and oil, and the fishermen spend hours bent over the fish and cleaning up afterward. The work is cold, wet, and dangerous.

Even today, the worldwide accident rate in deep-sea fishing is higher than that of any other organized industry – from two to four fatalities a year for every thousand men employed, compared with about 0.3 to 0.4 a thousand for mining. In other words, fishing is nearly ten times as dangerous as mining. It should be remembered, too, that a significant proportion of the world catch is still brought in by men working singly or in twos and threes, from small craft. The Food and Agriculture Organization of the United Nations has done splendid work in showing the fishermen of developing countries how to improve their boats or nets at a fairly elementary technical level, but the risks in fishing from small craft remain great.

Forty or fifty years ago, when a steam trawler headed up to the Arctic, it had to carry so much coal that some was stacked on deck, and its holds would contain tons of ice for preserving the catch. Ten days' fishing might produce as much as 100 tons of cod, haddock, ling, halibut, and plaice – then back to port. After three days there, it was off again for another 10-day stint. The trawler could not stay long at sea because ordinary ice does not keep fish fresh enough for several weeks at a time. Quick freezing, however, has removed these limitations. A freezer trawler's catch can be frozen to a temperature of −29°C, so that it can ultimately be sold as fresh food; or, without freezing, it can be canned or converted to fish meal or fish oil right on the spot.

The freezer trawler is the outcome of two essential developments of the 1950s: quick freezing at sea and stern-trawling. The conventional side-trawler tows and recovers its trawl over the dangerously low side of the vessel, and it tends to roll and take on a lot of water in bad weather. Trawling from the stern was a development from whaling factory ships, in which whales are towed up a ramp sloping down through the stern. The use of a ramp for launching and recovering trawls makes fishing safer, since the net-handling deck is much higher out of the water and better protected from the weather than the midship section of a side-trawler, while gutting, washing, and processing are carried out on a lower, completely sheltered "factory" deck. Such trawlers can often catch very-deep-water demersal fish at a fast rate: early in 1975, Russian stern-trawlers working off the eastern coast of the United States were known to be making catches at a constant rate of 17 to 20 tons per net haul.

For maximum efficiency, the skipper of a fishing vessel must know where he is, where the fish are, and where his net lies (which is not easy to find out if he is towing a net at a depth of 500 meters at the end of more than a kilometer of wire). Sonar and accurate position-fixing equipment have proved to be essential electronic devices for such needs. Sonar, for instance, was first used just to detect shoals; now it can determine the exact location of mid-water-swimming fish, estimate their size and den-

This cut-away diagram shows the equipment for processing and storing either quick-frozen whole fish or fillets on a freezer trawler. Thus, with no fear that its catch may decay, the ship can remain at sea until it has taken on a capacity load – sometimes as much as 700 tons.

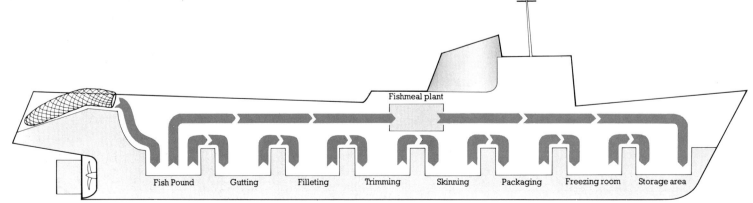

Fishmeal plant

Fish Pound — Gutting — Filleting — Trimming — Skinning — Packaging — Freezing room — Storage area

sity, and even monitor the fish and the net so precisely that skippers are able to steer the net onto the fish. As a result of such innovations and of navigational aids like gyrocompasses, radio, and radar, a modern trawler bridge looks rather like a power-station control room. In fact, the average freezer trawler has far more electronic equipment on its bridge than does a cruise liner such as the *Queen Elizabeth 2*.

Below: The bridge of a trawler is crowded with electronic aids to fish-finding and navigation. Items designed for finding and catching fish, identified in the key include, among others, sonar for searching in front of the vessel on a wide arc, depth sounders, instruments for distinguishing between the useless clutter of sea-floor objects and demersal fish, and water-temperature-gauging indicators. The ship's captain can operate the equipment and keep an eye on the various indicators from his centrally positioned chair.

Above left: The chief engineer of this British freezer trawler works in a soundproof control room overlooking the main engine room. He is responsible not only for the ship's power and propulsion but also for maintaining all nonelectronic machinery, including freezers, processing plant, and net equipment. Maintenance of the complete range of electronic equipment – sonar, echo sounder, radio, radar, etc. – is the responsibility of the wireless-telegraphy operator (right), who also handles communications and monitors weather reports.

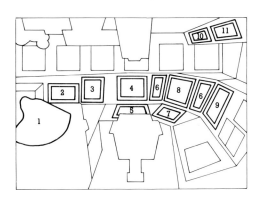

Key for trawler bridge

1 Ship's radar display
2 Forward-scan sonar read-out
3 Main forward-scan sonar display screen
4 Echo-sounder read-out
5 Echo-sounder display screen
6 Control panel
7 Net-sounder display screen
8 Net-sounder read-out
9 Secondary forward-scan sonar display screen
10 Mouth-of-net temperature indicator
11 Digital depth indicator

Fishing Fleets

Traditionally, the skipper of a fishing boat relied on a mixture of instinct and experience for knowing where the fish were, and individual skippers competed to bring home the largest catch. During the 1950s, Russia and other East European nations broke with this tradition and built up centralized industrial fisheries based on the fleet system, in which various stages of the fishing operation are divided among specialized vessels and the whole fleet is supported by meteorological and oceanographic data fed to it from shore stations. In this way, a number of catching vessels, usually trawlers, can be kept cooperatively employed on the best fishing grounds all the time.

The fleet system particularly attracts the Russians and Japanese because of their dependence on fish products, and so both governments invest heavily in fisheries research. The first step in creating efficient fleets involves the use of oceanographic research vessels for seeking out fishing areas, searching for fish, and studying the nutrients and planktonic productivity of the water. The countries that pioneered fleet fishing have also pioneered colleges for fishermen and fisheries scientists, such as the Fisheries University at Tokyo, where students can study such questions as the optimum balance of a fishing fleet, the design of its ships, strategies for fishing and marketing (including aspects of international law as it applies to these matters), methods of processing a catch, and so on. One of the best centers for such studies outside Japan is in Poland, which, despite the fact that it is not among the world's foremost fishing nations, maintains a group of research stations on the Baltic coast.

If preliminary research in a given area of the sea is encouraging, experimental vessels trawl through its waters to estimate the prospects for catches. These intensive investigations go far afield; they may well be carried out by a Japanese crew off the coast of West Africa or a Russian crew off Australia. And when a fleet finally sets out, it may include the following range of vessels: one or more mother ships of 10,000 to 40,000 tons; transport vessels of 2,000 to 15,000 tons, which convey preserved fish from fleet to shore; factory trawlers of 3,000 to 9,000 tons, which both catch and process fish; smaller catchers, which take their fish to the factory trawlers for processing; refrigeration trawlers (under 1,500 tons), which can store frozen fish for a time before offloading them into a mother ship; research vessels; and scout vessels.

Extra catchers of less than 70 or 80 tons are often carried on the deck of one of the larger ships. For example, mother ships of the 40,000-ton Vostok class, which the Russians began using in the early 1970s, carry 14 50-ton catchers on their decks. Fleets centered around Vostok-class mother ships can stay at sea for up to four months, during which they can produce 10,000 tons of fish products, 1,000 tons of fish meal, 10 million cans of fish, and 100 tons of fish oil. It is no wonder, then, that the fleet idea has appealed to other nations; the Spanish and South Africans, for instance, have been using the method since 1965.

In the area chosen for fleet operations, various types of trawler and small boat spread out to catch the fish as indicated by scout ships, and fish or fish products are transferred from one kind of vessel to another for further processing and transport. Since the major catching vessels are thus spared the necessity of making long voyages back to port between catches, they spend far more time fishing and waste far less traveling. A modern trawler in, say, a Polish fleet actually fishes for 60 per cent of its annual time at sea – a very high efficiency rate compared with that of a lone trawler.

So far, fleet fishing has not been adopted in most of the richer Western nations. One reason is that they are less dependent on enormous catches for their protein than are Japan and the countries of Eastern Europe. Another problem from the standpoint of the advanced industrial countries would be the difficulty of finding crews prepared to stay at sea for several months. That should not be an insurmountable problem, though; crew changes can be made in mid-voyage, with workers traveling back and forth on the fish-transport vessels or by air.

Some fishing fleets include one or more factory ships whose sole function is to process the fish caught by other vessels. Above: Russia's *Mikhail Tukhachevsky,* flanked by some of its 14 catchers, in the Sea of Okhotsk off the Siberian coast. The catchers, each of less than 70 tons, are hauled aboard for transporting whenever the fleet moves to new fishing grounds; when at work, they provide the big factory ship with a continual supply of fish for processing. Many of the highly automated processes are controlled and monitored at consoles like the one at the left. On the production line, the crew work at such jobs as salting and canning fish products (facing page, far right) or preparing fish for freezing (bottom left). Fleet factory ships remain at sea for months, periodically serviced by refrigeration ships – such as the one pictured in the middle of the facing page – which shuttle between the fleet and its home port, bringing out supplies and relieving the factory ships of their finished products.

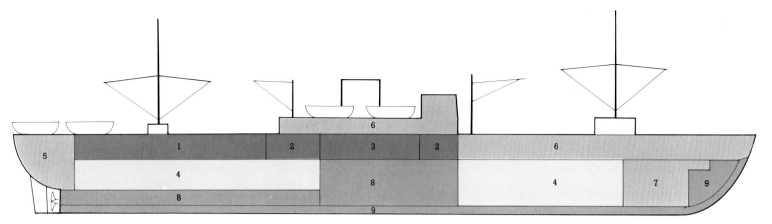

Fish processing areas

1 Cannery
2 Freezer factories
3 Freezer room

Refrigeration

4 Holds

Ships stores and crew areas

5 Stores
6 Crew
7 Salted Cargo

Ship-support machinery

8 Engine room
9 Fuel

Above: This is the layout of one of Japan's fleet factory ships, which processes salmon and trout in the North Pacific. Thirty catcher boats bring their daily catch to it, and it has a productive capacity of roughly 1,800 tons of canned, 500 tons of salted, and 1,000 tons of frozen fish and fish products.

Making the Most of the Catch

Three-quarters of the world's fish catch – some 45 of an annual 65 million tons – is landed by 14 countries, listed here in order of quantity of takings in the early 1970s: Peru, Japan, Russia, Norway, the United States, Thailand, Spain, South Africa, the Philippines, Great Britain, India, Iceland, North Korea, and France. Of these countries, only Peru, Iceland, and the United States fish almost exclusively off their own coasts; the others roam widely, with Japan and Russia especially active on a global basis. Since 90% of the world's sea food is caught over the continental shelves – the main exceptions being tuna and whales – this means that most coastal states are in a position where foreigners take more fish within a hundred miles of their coast than they themselves do. That accounts for the rather tense atmosphere in political negotiations about conservation of fish stocks – a subject to be more fully discussed in a later chapter of this volume.

According to the Food and Agriculture Organization of the UN, nearly 80% of the present catch comes from only five of the 15 major sea areas: the northwest and southeast Pacific and the north, southeast, and east-central Atlantic. All these areas are apparently being fished to about 90% of their potential yield. "Overfished!" some would argue, although there remain vast stocks of octopus, squid, and other animals not eaten by many Western people.

Obviously, an immediate target for any attempt to raise the world fish catch is to exploit both the relatively unexploited sea areas and the sea's relatively unexploited stocks of edible creatures. About 500 species of fish are now being hunted and eaten by somebody somewhere, but in most places the actual range of acceptable sea foods is very limited. In the Western world, every country or region tends to have a preference for a handful of fish species – herring for the Scandinavians and Germans, cod or plaice for the British, tunafish sandwiches for the Americans, and so on. Such ingrained tastes (and mirror-image prejudices, like the typical Anglo-Saxon revulsion from France's beloved sea urchins) make the task of increasing the food yield all the more difficult. Nevertheless, there are many ways in which fisheries scientists and technologists are managing to make use of previously unwanted fish stocks.

A prolific sea creature is the Antarctic krill, a shrimplike plankton-dweller that constitutes the basic food of baleen whales. Both the Russians and Japanese have been exploring possibilities for human consumption of these protein-rich animals; clearly, there must be possibilities in any creature that whales can live on simply by swimming along and filtering them through their "teeth." Large hauls of krill made by Russian research ships have been processed into a nutritious and tasty paste. Other countries are considering ways to exploit species that till now have been used chiefly for conversion to fish meal or fertilizer. Capelin, for example, can be found in abundance off northern American coasts, and the Canadians have begun to process them for the supper table. Why not squid and octopus, too? In most sea areas, says the Food and Agriculture Organization, a serious effort to catch them would add about 10% to our supply of sea food. Even when an edible animal is extremely unappetizing, it can generally be processed – and disguised – into some kind of fishcake or other nourishing product.

There is a new technology, too, for getting more meat from present catches: a flesh-bone separator that consists of a perforated stainless-steel drum against which fish are forced by a rubber roller. Some idea of its efficiency can be gained from the fact that a cod skeleton left after conventional machine filleting can be made to yield another 3% to 5% of flesh when passed through the flesh-bone separator. This device makes it possible to consider using "new" species such as the extremely bony deep-sea grenadier. It also improves our chances of making perfectly good fishsticks, cakes, and spreads out of odd species that are now caught and discarded.

Even today, getting the most out of the sea does not stop at exploiting familiar fish. Apart from the shellfish industry – lobsters, crayfish, mussels, shrimps, oysters, and so on – there are specialist markets for such delicacies as sea-urchin roes and sea cucumbers. There is also a market for a host of unusual chemicals and drugs extracted from ocean-dwelling organisms. Kelp is cropped off California to produce alginate, a food additive, and the Japanese eat numerous seaweeds. An anti-tumor compound has been extracted from sea cucumbers by scientists at Queensland University in Australia. And scientists at the University of Washington in Seattle have extracted a chemical from a luminescent jellyfish that enables doctors to detect changes of calcium concentration in the human body.

As if all the searching for expansion were not enough, the pattern of fishing activity may soon be radically altered for political reasons. If, as seems likely, international law accepts the contention that coastal states should be permitted to claim fishing rights out to a distance of 200 miles from their shores, many of the world's most far-reaching fleets may be forced either to work in deep waters outside the 200-mile limit, or else to fish more intensively within the 200-mile limit of a few friendly states. This would surely intensify the already feverish quest for exploitable sources of protein and would directly affect the design of fishing gear and ships, thus changing the pattern of the industry.

These three fish – the grenadier (top), director (center), and black scabbard – may not be beautiful, but they are perfectly edible. Yet they are among the many living resources of the sea that remain largely untapped. If enough potential consumers were "educated" to accept such fish as food, we could doubtless harvest them in commercial quantities.

With her extreme shortage of food-producing land, Japan relies more than most countries on the protein resources of the ocean. These men are harvesting some of the iron-rich seaweed called *wakame* (or "sea lettuce") that the Japanese often use as an ingredient in soup. *Wakame* and other edible seaweeds grow in coastal waters either wild or, as here, in cultivated "fields."

The demarcation lines on this map divide the world into 15 main fishing areas as defined by the UN Food and Agriculture Organization. A few areas, comprising the northwest and southeast Pacific and almost all the North Atlantic, are now being fished to the limit and produce 80% of total world catch, while stocks in other waters remain underexploited.

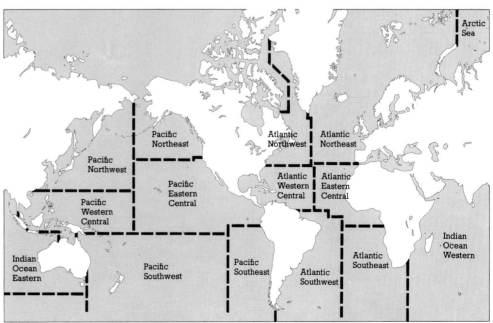

138

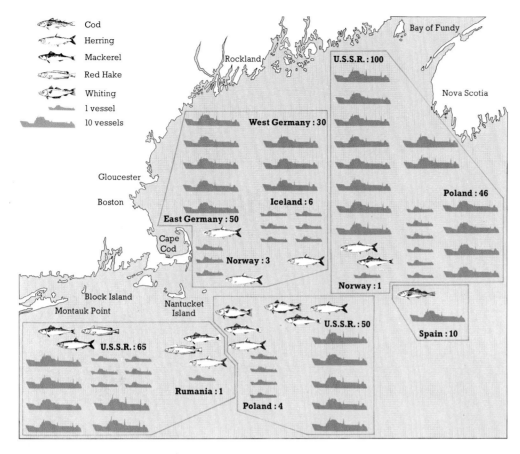

Cod
Herring
Mackerel
Red Hake
Whiting
1 vessel
10 vessels

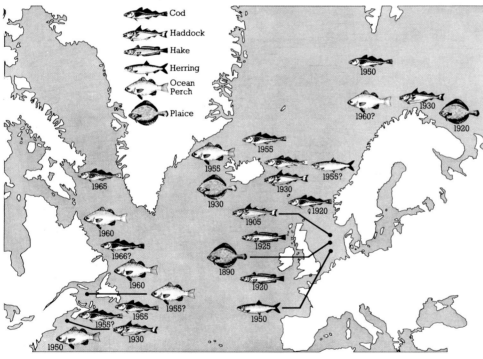

Cod
Haddock
Hake
Herring
Ocean Perch
Plaice

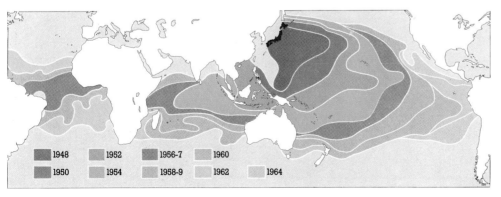

1948 1950 1952 1954 1956-7 1958-9 1960 1962 1964

The Need to Conserve

We used to think that the living resources of the sea were inexhaustible, but we were wrong. Twentieth-century fishing methods have been depleting the most marketable fish stocks at an alarming rate. By the early 1970s, unmistakable signs of scarcity had begun to appear. Icelandic and Norwegian purse-seiners appeared to have fished the Atlanto-Scandian herring almost out of existence; in the northwest Atlantic, vessels of 30 nations were fishing the Newfoundland cod to a point where stocks were dangerously low; and fish prices in markets everywhere were reflecting the increasing difficulties of finding enough of the "right" kinds of sea food to satisfy consumer demands.

Conservation has therefore become a matter of some urgency. Not, of course, that there has been no earlier concern about the risks of overfishing; there has been a great deal – but to little effect. For many years, scientists have been pointing out that the simplest way to reduce the pressure on a fish stock is to take only larger fish, which have already spawned and replaced themselves. So the size of a net's meshes has been prescribed by international agreement, in order to permit younger fish to slip through. But although fishermen of some nations have respected regulations forbidding the landing of undersized fish, those of many countries have bent the rules, so that the average size of the most popular species – cod, haddock, flounder, halibut, and so on – is becoming smaller, a sure sign that fewer full-grown fish remain to be caught. Limiting mesh size, then, is not a satisfying conservation measure unless it can be rigidly enforced.

Part of the fishery scientists' job is to monitor catch rates, total catches, and the composition of catches according to size for

Three maps indicating how the problem of overfishing has snowballed in our century. The map at the top gives a bird's-eye view of the international competition for protein that brings the fleets of far-off nations into a single small patch of the sea – in this case, the rich fishing waters off Cape Cod, Massachusetts, where Spanish vessels search for cod while the Scandinavians and Eastern Europeans gather up vast quantities of herring, mackerel, whiting, and red hake. The second map shows the overexploitation of certain popular species in the North Atlantic. The date with each type of fish pinpoints the year when it ceased to be possible to catch greater quantities of the species by means of increased or more efficient fishing; after that date there were no longer enough to go round. And the need for worldwide conservation is perhaps most dramatically illustrated by the bottom map, which shows how Japan pushed her tuna-fishing activities farther and farther afield during the 1950s and '60s. By now her fleets are taking nearly 400,000 tons of tuna a year; in virtually no waters are the fish safe from pursuit.

commercially fished species. As a result of their calculations it is possible to ascertain the maximum weight of fish that can be taken from a given area without depleting the stock so far that the total catch would not increase if more vessels arrived to take them. Obviously, if more and more boats with more and more efficient hunting and trapping methods chase declining stocks of fish, the cost of catching each fish soon reaches an uneconomic level. Everyone realizes this, and yet it is very difficult to legislate international cooperation to limit catching efforts. The problem is intensified because different groups of countries fish different areas for different fish stocks, and no country can be forced to join a self-regulating group of nations. For example, there is a UN-sponsored Committee for East-Central Atlantic Fisheries, set up in 1967, which includes 20 nations; but nearly 24% of the fish caught in the east-central Atlantic is taken by Russia – which is not a signatory to the agreement. In 1958, a number of countries ratified the Geneva Convention on Conservation of Biological Resources of the High Seas, but among them they catch only 25% of the world total of fish; Peru, Japan, the Soviet Union, and Norway, all of them among the world leaders, have never signed up!

Nevertheless, over the years, dozens of regional agreements *have* come into force, and on the whole they have done a good job. But the economic drive to find supplies for the marketplace tends to outpace unenforceable agreements. Perhaps the best way to take the strain off the fish population is for each nation to establish its own catch quota system or to make rigidly defined bilateral agreements to catch limited tonnage of popular fish species. Thus, to avoid unpleasant "cod wars" off Iceland's shores, Iceland and West Germany signed an agreement in 1975 on catch quotas for their competitive fishing fleets; and Iceland, Norway, and the U.S.S.R. have reacted to the slaughter of Atlanto-Scandian herring by agreeing among themselves that they would no longer fish for the species. Such patterns must continue. The international fishing grounds of the world, once an outlet for rugged individualists, must be subjected to new rules if stocks of marine protein are to endure.

A complicating factor is the current confusion over territorial limits. For years, a number of South American nations have claimed exclusive rights to all fish within 200 miles of their coasts, contrary to the internationally accepted limit of three miles. As they see it, their claim received indirect support in 1959 with the ratification of the United Nations Convention on the Continental Shelf, which declared that coastal states own all minerals and living resources, *excluding swimming fish*, on and under the

shelf. While countries such as Iceland, with lots of fish and no known undersea mineral resources, felt cheated, Peru and Chile, because they have no shelves at all, felt confirmed in their "possession" of a wide expanse of water unrelated to the continental-shelf principle. Iceland then declared a 12-mile fishing limit for her own waters; she raised this to 50 miles in 1973, and then, in 1975, to 200. Other nations, including Australia, Denmark, Haiti, Taiwan, and South Africa, have raised their declared limits to 12 miles or more. Until recently, the United States accepted the 12-mile limit, but the frequent sight of 40,000-ton Russian mother ships surrounded by their catching vessels working around the clock in coastal waters just outside that limit is probably the reason why she too has changed her position and has now adopted the idea of a 200-mile zone.

The ultimate move in the conservation game is thus a political carve-up of the ocean's offshore resources, with each state taking the responsibility – or trying to – for fish stocks in its own area. The full political implications of this development will be discussed in a later chapter, but as new fishing frontiers are established, there are bound to be dramatic repercussions within the fishing industry.

Two determined adversaries collide during the 1975–76 "cod war" sparked by harassment of British trawlers by armed Icelandic coastguard vessels in unilaterally declared Icelandic waters. Though both countries agreed that it is essential to conserve fish stocks, they were the victims of conflict between this conviction and the demands of their respective fishing industries.

Man proposes, but nature often disposes: when, years ago, Peru decided to protect her vital anchoveta industry by claiming exclusive fishing rights within a 200-mile coastal zone, scenes of extreme abundance like this one seemed likely to persist. But the catch fell by 80% in the 1970s after shifting currents took already overfished stocks far out to sea.

These two-year-old oysters, reared from embryos caught in a New South Wales (Australia) estuary, were kept in nursery beds until old enough to be moved upriver. Once laid out on racks, they will be left to mature in the brackish water from three to five years, when they will have reached marketable size. Such careful cultivation produces much better crops than would the hazards of the wild.

Fish Farming

Although freshwater fish have been farmed for centuries, the captive breeding of sea fish is a relatively new concept. There seemed little reason for breeding fish when the supply appeared inexhaustible, but the sea is no longer as full as it was, and the cost of "hunting" the kind of fish most valued by the consumer is rising so steeply that economic reality may eventually justify the production of a number of species under "farm" conditions. Sea fish have already been bred to market size in captivity, but feeding costs – it takes about 20 kilos of food to bring a fish up to a weight of only 2 kilos – plus other expenses have so far made large-scale farming impractical.

It is easier with freshwater fish, some of which, such as trout and catfish, grow quickly on little food, or feed happily on weeds or algae, as carp do. All can be grown in ponds and require little supervision, since the farming of them merely exploits a basically natural situation. In contrast, sea-fish farming necessitates either the creation of a miniature replica of open-sea conditions, with a sea-like environment maintained in defiance of rainfall, wind, heat, and pollution, or else an elaborate system of cages placed in the ocean. Either method costs money. As a result, farming is practicable – at least for the present – only for very high-priced fish or shellfish. It is certainly not the

answer to the world's need for cheap food. Research is intensive, however, and techniques may be so improved in a few years that costs will come down drastically.

Let us clarify what we mean by farming. If a sheep farmer tried to make money by catching wild sheep, fattening them in a pen for a few months, and then selling them, we would not rate his chances very high. Wild sheep do not fatten up much even if you feed them well. Or suppose the farmer caught wild sheep when they were pregnant, looked after the lambs for a few weeks to make sure they survived, then let them loose again, to be rounded up for slaughter when fully grown. Again, a pretty inefficient method. Yet these processes, by analogy, are the nearest to farming that has been achieved with most fish species. For some years, the Americans, Canadians, and Japanese have captured fertilized female salmon in rivers, protected them to ensure that the maximum number of eggs hatch safely, and then released them into the sea for catching later. The Japanese go a step further with yellowtail mullet: when the young hatched from eggs of captured fish are over five centimeters long, they are put into the sea in cages, where they grow to a weight of a kilogram in a few months and are ready for sale. The weak link in all such operations is that there is no control over the whole life cycle of the fish. In particular, the actual breeding is not controlled (marine species

do not generally seem capable of breeding in captivity), and there is no selection of individuals that grow fastest, eat least, resist disease, and taste best. So fish farming is a long way from achieving the efficiency of, say, chicken farming.

At the moment, sea-fish farming, with partial control of the life cycle, is carried out commercially – but on a comparatively small scale – for grey mullet in Israel; yellowtail mullet and puffers in Japan; salmon and rainbow trout in the United States, Canada, Japan, Norway, and several other European countries; and milkfish in the Philippines and Taiwan. All these species are relatively expensive delicacies, much prized in the countries where they are cultivated. In Britain, fish-farming experiments have been concentrated since 1957 on the somewhat less exalted plaice, sole, and turbot. Originally, after plaice eggs had been hatched, the larvae were reared for nine weeks until they had metamorphosed into little flatfish, and were then released to boost the sea's natural stocks. Over the years, though, it became apparent that the baby fish died very quickly in the sea, possibly because they had not developed proper reflexes for escaping from predators. And so, since 1965, young fish have been kept in captivity up to salable size by means of an adaptation of the Japanese system of floating cages, within which the fish are stocked at high density. The cages have

141

The farming of salt-water fish involves protecting stock from disease and predation throughout their lifetime, from the egg stage on. Thus, they neither migrate naturally nor feed in the open sea, but must – like any domestic herd – be kept under constant supervision. Left: Each floating salt-water pen houses a thousand Scottish salmon for their last two years before being harvested. Some of the mature fish are taken to a freshwater-hatchery building, where females are "stripped" of their eggs for fertilization. The fertilized eggs are inspected regularly and the dead ones removed with a pipette (below, top). Live eggs hatch in about 18 weeks; then the alevins (newly formed salmon) are transferred to their own tanks (middle), where they stay for two months before being transferred to outdoor tanks (bottom). Following nine months in this – the parr – stage, they are ready for "migration" to salt-water pens.

proved quite successful, with the sea bringing in some food and a supply of clean, oxygenated water. Additional food – fishmeal and chopped mussel – is provided daily, and the fish reach marketable size rapidly.

Most sea-fish farms have an interconnected system of hatching and rearing tanks, followed by larger salt-water ponds to which young fish are transferred. The fish require different water conditions and different food at each stage, and so they must be segregated into age or size groups. This also prevents cannibalism, which otherwise develops in crowded conditions. Antibiotics have to be added to the water, too, in order to prevent diseases from spreading in the close quarters. All such controls of environment and water purity cost money, and that is why the cage-in-the-sea idea, where the ocean itself keeps the water right, has a special appeal for some breeders. Still, the closed-circuit tank system has an advantage in that it permits a faster feeding rate and can be located conveniently near urban markets. Moreover, the cage-in-the-sea system is only possible on unpolluted, protected coasts.

Fish are not the only marine creatures that can be bred in captivity. In fact, other sea foods are easier to handle. In Japan, where many of the modern farming techniques have been pioneered, a large species of prawn is farmed in long concrete tanks, where the newly hatched young are fed on brine-shrimp eggs, cockles, and other small shellfish reared in nearby tanks. The final stage of growth is in pools 200 meters square, from which a million edible prawns can be culled each year. Oysters are also hatched and planted out as "spat" in many countries. But only Japan has gone so far as to cultivate seaweed: certain species of large red seaweed are considered a luxury food by the Japanese, who collect the spores of the seaweed on special nets, which are then strung out in the sea on buoys. The variety of fish, shellfish, octopus, and seaweed farmed in Japan is astonishing. Many miles of the Inland Sea are festooned with buoys, poles, fences, catwalks, bamboo screens, boxes, cages, plastic nets, and fish-farmer pontoons, huts, and paraphernalia.

To sum up our brief review of the living resources of the sea, it can be concluded that fishing as now practiced might, with much further investment not only of money and human energy but of self-discipline, come close to doubling our annual yield of around 65 million tons of fish in another 10 to 20 years – but this will certainly not fill the gap in the world food supply. Some form of farming is the only way in which the many species of sea animal can ultimately be made to yield more than the natural productivity of the sea. But it will take years of research and experiment before the quantities produced in this artificial fashion can make a significant contribution.

Undersea Oil

Although oil and natural gas are found under rocks on land as well as under the sea floor, such substances were formed, several hundred million years ago, in the sea. Just as organic materials rotted in the brackish swamps of past eras to form peat and coal, so decaying plant and animal life sank to the bottom of the sea, was covered in a rain of fine sand and silt, and was eventually enfolded between layers of rock in movements of the earth's crust. Even today, similar deposits appear to be accumulating in deep basins on the margins of continents – places like the Gulf of Mexico and the Pacific waters off southern California. Pressure, heat, and – above all – time converted the ancient deposits into liquid and gaseous hydrocarbons. Some of the liquid was squeezed out of the rock layers and lost. But great quantities contained in porous limestone and sandstone rocks were trapped by nonporous rock above and water below. It is these vast reservoirs of oil that constitute the oil fields so eagerly sought after today. Hydrocarbons in gaseous form – natural gas – were similarly trapped and are often found associated with petroleum deposits.

Wherever there is oil, then – even in the arid deserts of the Middle East – there was once deep water over a continental margin. It is continental drift which, by pushing Africa and Arabia northeast against Asia, trapped oil-bearing rocks between two land masses, creating the easily accessible oil riches of the Middle East. Wherever there is oil, moreover, the environment was probably tropical in the distant past, even if (as in the North Sea and Alaska) the area is now far from the tropics. Oil-bearing rocks of geologically recent origin (up to 30 million years) lie exclusively in tropical latitudes; when oil fields are discovered in far northern and southern regions, it is certain that the rocks are more than 30 million years old and that they *were* in the tropics when the oil was laid down. But there would be little point in wandering around the tropics looking for vast undiscovered oil fields these days. Virtually all the world's major fields on what is now land have probably been found and are already being worked.

It is the immense reserves that still lie underwater that interest us. Why are politicians and engineers willing to pay so much and work so hard to exploit offshore oil, considering the strikingly difficult conditions under which the work must be done? How are the undersea wells found? And how do we tap them and get the oil ashore once it has been found?

Most of the presently accounted-for reserves in the non-Communist world lie under the sands of the Arabian deserts. To achieve independence from Middle East supplies, the world's non-Arab states are naturally eager to develop their own oil potential; and for most of them this means going into the water. The dream of developing undersea oil wells is far from new, but it was only in the 1960s and early '70s, as oil prices began to rise sharply, that the search for alternative supplies became a national priority in country after country. But one fact seems incontrovertible: the massive deposits of the Middle East are unlikely to be matched anywhere else. There are rich hydrocarbon reserves in such areas as the Arctic Ocean, the north Aegean and South China seas, and – perhaps especially – on the huge plateau off the coast of Siberia, which is yet to be explored. Although these fields are extensive, their proven wealth is small, however, in comparison with that of the Middle East. The continental slopes, from about 200 meters down to 3,000 meters, may hold further stores of petroleum, but it will be hard to get at them because of the great depths as well as the sea-bed configuration. As for the deep-ocean floors, they are geologically young and largely composed of basalt, which is not the kind of nonporous rock beneath which oil is likely to be trapped. So it seems improbable that there will be many more spectacular finds.

The search for new fields continues, of course. It is being carried on more unremittingly than ever, for the product will remain indispensable for years to come. That is why self-sufficiency in oil has become the active dream of all industrialized nations that border on the sea, despite the difficulties and dangers involved in exploring for hydrocarbons and retrieving them from beneath the sea bed. When appreciable reservoirs *are* discovered, the effects tend to be enormous. Apart from the obvious commercial benefits for the country in whose waters the oil is found, there are inevitably outside pressures – economic, political, and legal.

In the North Sea, for instance, Britain has been relying on her offshore oil wealth to bring her a measure of economic salvation; her European partners in the Common Market, or at least some of them, feel they ought to have a share in this new-found treasure; a Scottish faction insists that, in view of its location, much of the oil should belong to Scotland alone; Norway, in her novel role as an oil exporter, is worried about resultant changes in her way of life; and Japan and America, with their insatiable thirst for the commodity, are showing in various ways their eagerness to become commercially involved in North Sea as well as offshore Indonesian and Australian operations. Meanwhile, there remains the constant, though never clearly defined, threat to the environment posed by the procuring and transporting of oil.

In the following pages, emphasis is placed on the techniques of exploration, drilling, production, and transmission. It should hardly be necessary to emphasize the vital importance of the need to exploit undersea oil with due respect for the environment. No other undersea resource has more power than petroleum to give mankind either pleasure or pain, depending on how intelligently he makes use of it.

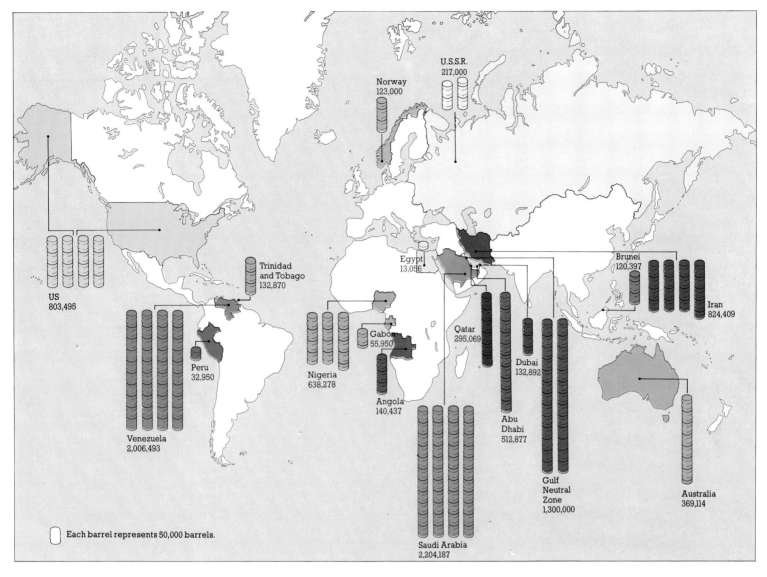

US
803,495

Trinidad
and Tobago
132,870

Venezuela
2,006,493

Peru
32,950

Nigeria
638,278

Gabon
55,950

Angola
140,437

Egypt
13,056

Norway
123,000

U.S.S.R.
217,000

Qatar
295,069

Dubai
132,892

Abu
Dhabi
512,877

Gulf
Neutral
Zone
1,300,000

Saudi Arabia
2,204,187

Brunei
120,397

Iran
824,409

Australia
369,114

Each barrel represents 50,000 barrels.

The map above shows the major producers of offshore oil and their daily output in barrels (1 barrel = 159 liters) for 1975 – when, apart from a few wells in Norwegian waters, the North Sea had not yet begun to yield appreciable quantities. Even today, as then, about half the total amount of undersea oil still comes from the Middle East.

Far left: Typically rough conditions prevail in the North Sea as the American rig Transworld 61 is towed to an exploratory drilling site, to work in conditions where waves can be 30 meters high and winds can gust to 200 kilometers an hour. A sister rig to this one was lost, a few years ago, during a North Sea storm.
Left: Storms are not the only hazard encountered in the quest for undersea oil. As exploration moves to the far north, extreme climatic conditions in all seasons make drilling particularly hard and production especially costly. These barges loaded with equipment for a rig in Prudhoe Bay, Alaska, are being pre-cariously towed through Arctic Ocean pack ice.

The only way to find undersea oil is to drill for it. But there are technological pointers to areas where drilling should pay off. A geologist begins by suggesting broad areas based on his knowledge of how oil-bearing structures were formed. When he has indicated a likely sedimentary basin covered by nonporous rocks of a suitable configuration, geophysicists gain an idea of the structure of the rocks by measuring changes in the earth's magnetic field from the air, or else by working out local variations in gravity from a ship. Next comes a thorough study of specific areas by means of a seismic survey that involves firing sound waves into underwater rock strata and picking up the echoes on hydrophones. Different kinds of rock transmit sound at different speeds, and a recording of the echoes gives a cross-sectional picture of subsea geological strata. If the picture looks promising, the obvious next step is to start drilling. This is the stage at which the costs of undersea exploration begin to hurt, and they rise sharply with increasing depth, distance from shore, and rough weather – all the factors that make a barrel of oil from the North Sea or offshore Alaska ten times as expensive to bring up as a barrel from the Middle East.

Exploratory drilling in an untried area is called "wildcatting." In this stage, oilmen are intent on identifying the position and depth of oil-bearing rocks, and on seeing how fast oil flows from the well. When they know this, they can decide how many more wells to sink into the field in order to get oil at an economic rate. Many wildcat wells turn out to be dry, and the cost of exploratory failures must be added on to that of successful wells when the selling price of oil is finally determined.

At the turn of the century, prospectors drilled off the ends of piers in the Caspian Sea and the Gulf of Mexico. Then, as they went deeper, mobile drilling barges were developed. The jack-up barge is an example of a mobile rig. It has a floating hull with legs; after it has been towed out to sea, the legs are jacked to the bottom so that the hull stands above sea level. Such fixed barges can be used in depths no greater than 30 or 40 meters, but they are still a standard exploration tool. As recently as 1975, 137 of them were at work in the world's waters, with another 63 under construction. In deeper areas, though, the wildcat driller uses one of two types of floating barge – a semi-submersible or a drillship.

The basic semi-submersible unit consists of a platform mounted on cylindrical legs attached to pontoons. Despite its over-all height of up to 100 meters, this vast barge generally has some self-propulsion and can move to its site with the assistance of a tug. Once in place, its ballast tanks are flooded, and it sinks until the deck is about 15 meters above the water. Up to 10 heavy anchors are spread out around the rig to keep it from drifting sideways and bending the drill; in very heavy seas, however, the whole "drillstring" has to be protected by bringing it up to the deck of the rig. The rig itself gets most of its buoyancy from the pontoon hulls deep in the water below the influence of waves, and so up-and-down movement is minimized. The biggest, toughest semi-submersibles, capable of drilling down to 9,000 meters below the sea bed in 300 meters of water, are designed to operate in the North Sea, where they can remain stable in 100-knot winds and 30-meter-high waves. As of early 1975, no fewer than 40 of them (out of a world fleet of 74) were working in the North Sea.

Drillships, which have a hole (known as the "moon-pool") amidships through which the drill pipe is lowered, lack the semi-submersibles' stability in heavy seas but have compensating advantages. Totally self-contained, a 150-meter-long ship can stay at sea for months without support; and it is, too, much cheaper than the semi-submersible to move from one site to another. Today's drillships can also remain on station in water too deep for anchoring – over about 400 meters – by means of dynamic positioning, a tech-

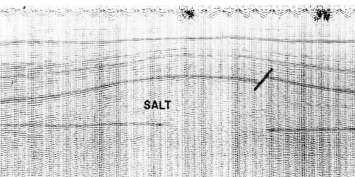

SALT

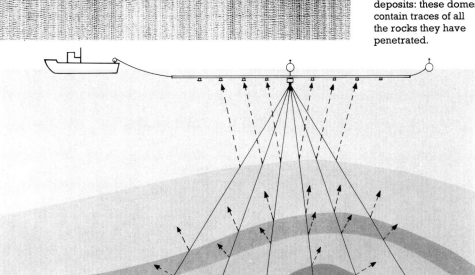

Compressed-air guns attached to a metal frame are being retrieved here from waters where scientists are making a seismic survey – the most common method of surveying undersea rock to determine whether it might contain oil. When in action, the gun carriage is towed by the survey ship and kept from sinking by being suspended from a float (seen beyond the carriage). As illustrated in the diagram, below, the guns fire a pattern of sound waves into subsea strata, and echoes picked up by hydrophones positioned along the towing cable give a cross-sectional view of the area's geology. Above the diagram is a read-out from a survey indicating a "salt dome" – an upthrust from a buried layer of close-packed salt – which may have oil in it or suggest nearby deposits: these domes contain traces of all the rocks they have penetrated.

nique involving a shipborne computer that automatically restrains the ship's movements by activating a number of propellers around the hull. Drillships have been used effectively for wildcatting in such quiet waters as those in the Mediterranean and Southeast Asia areas, as well as, during the summer months, the North Sea.

Semi-submersibles and drillships alike are being increasingly geared for work in the deep ocean. Already, there are designs for units capable of drilling in 1,000 meters of water. The offshore-oil industry cannot afford to stand still on the continental shelves, or even in the lower latitudes. Exploration is now spreading into Arctic regions, even though drilling in Arctic waters holds two major terrors for the prospector: the possibility of great weights of ice forming on the superstructures of rigs and ships, and the threat of icebergs. For facilitating further exploration in deep, stormy, and icy waters, a better system of gathering information about the marine environment is badly needed. Until lately, such data as exist were collected mainly for oceanographic purposes; now the oil companies are mounting their own programs for studying winds, waves, and currents.

When an exploratory jack-up, semi-submersible, or drill-ship strikes oil, the well is temporarily sealed off until a full program of production drilling can get under way. The rig now moves away and starts drilling wells elsewhere. Since so many of the problems of exploratory drilling arise from weather conditions at the surface, however, the era of the mobile rig may be approaching its end. Some far-thinking engineering companies are trying to design sea-floor drilling rigs – large spheres or domes within which men could live at sea-level pressure (a so-called shirtsleeve environment) and operate a complete drilling system independent of the surface. This is not yet a reality, but it may well come if industrial states continue to strive for self-sufficiency in oil.

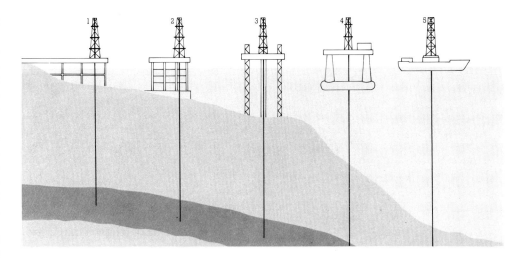

Above: Five kinds of exploratory drilling. The earliest type of offshore rig (1) was merely a platform built on stilts and connected to land by a boardwalk. The submerged drilling barge (2) came next; floated out to slightly deeper water, it is sunk to provide a base for the drilling outfit. Jack-up barges (3), which are floating hulls with long legs, are used in up to 100 meters of water. Semi-submersibles (4) can drill in much deeper waters because kept afloat by pontoons. And the mobility of drill-ships (5) makes drilling feasible in the greatest water depths. Older types of rig have not been entirely super-seded but remain in use in shallow-water areas.

Left: Much drilling in the Caspian Sea is still being done off piers such as this 150-kilometer-long one. Workers' houses stand on piles in artificial islands.

Right: The American drillship *Glomar Coral Sea* at work in about 400 meters of water in the Nile Delta. Although drillships are ideal for deep-water drilling and are, obviously, much more mobile than other exploratory rigs, they are less stable in rough seas, and therefore less suitable for winter use in such areas as the North Sea.

Far right: Crew members on a semi-submersible rig prepare to screw the top joint of the drillstring into the "kelly" – a 10-meter-long square (or sometimes hexagonal) pipe that transmits the turning motion of the rotary table to the string.

60 meters

300 meters

900 meters

2000 meters

3000 meters

1
2
3
5
4
10
9
8
7
6

1. Mud pump.
2. Telescopic joint.
3. Marine riser.
4. Well head
5. BOP (Blow-out preventer) with hydraulic rams.
6. "Mud".
7. Bit.
8. Drill pipe.
9. Casing.
10. Cement.

How to Drill an Exploration Well

Imagine trying to drill a hole in the wall of a room using a hand drill with a six-meter-long bit. Imagine, too, that as the drill bit penetrates the surface, the underlying formations are of different consistency, so that sometimes it is hard work to gain an inch of penetration while at others the drill lurches forward through almost jellylike substances. And there are still further problems. The formations through which the drill bit makes its steady progress can have trapped pockets of gas or liquid under high pressure; hit one of those and it will spurt out dangerously. The drill bit will become blunt from time to time, and so the whole drillstring will need to be removed from the hole and the cutting teeth replaced. These problems, on a vastly magnified scale, face the oil-well driller. They are severe enough on land; at sea, with perhaps 300 meters of water separating him from the point at which the drill bit first begins to do its work, they become even more complex.

Geologists decide the best site at which a rig should drill. It arrives at a predetermined location, often using satellite navigation systems, and positions itself within a few meters of a spot where its chances of hitting oil or gas have been determined, perhaps up to two years before. Tugs place up to ten anchors to hold the rig steady.

The first step is "spudding-in." This involves drilling, jetting, or driving a hole 0.75 meters in diameter to a depth of 100 meters or more. Conductor pipe (casing) is run into this hole. At the bottom of the pipe is a concrete plug and a valve. Cement is pumped down the hole, through the valve, and out between the casing and the surrounding formation until it forms a bond. A foundation plate – the temporary guide base – is lowered to the sea bed to "mate" with the top of the first run of casing; this guide base is linked by wires to the rig. To contain any pressure rises in the well which may bubble

dangerously to the surface, a unit consisting of a series of valves, cutters, and rams is lowered down the guide wires to sit on the sea bed. The valves are controlled hydraulically from the surface and operate a series of rams that can shut across the hole to prevent gas or liquid under pressure from flying up to the surface. Some rams are so powerful that they are able to sever the drill pipe in the hole to make an effective seal. This pressure-leak preventive unit is known as the blow-out preventer stack (BOP).

When the first run of casing has been successfully cemented into place, a smaller drilling bit is run into the hole to drill deeper into the formation. More casing, of smaller diameter, is then cemented into place. This process continues until the depth at which hydrocarbons have been estimated to be present is reached.

The drill bit is secured to the end of the drillstring, which is composed of several-meter-long sections of hollow pipe, each of which is hauled upright under the derrick on the rig and lowered onto the previous section as the hole becomes deeper. The drillstring is turned by a turntable (the rotary table) in the rig floor. As the bit makes its hole, and before the casing is cemented into place, the sides of the hole have to be kept from caving in. This is achieved by pumping a liquid known as "mud" down the hole. "Mud" is a drastically oversimplified term for a complex mixture of chemicals, the composition of which has to be carefully formulated not only for each well but for various stages of individual wells. In addition to keeping the hole open, the mud also cools the drill bit, displaces cuttings, and returns them to the surface via a pipe, known as the marine riser, installed between the BOP and the rig. When it reaches the surface, the mud passes through a degasser and sieve (shale shaker) and into storage tanks (mud pits), ready to begin its downhole journey again.

From time to time – the intervals depend-

Left: Major components of a typical exploratory drilling project. The hole is drilled through the blow-out preventer in sections that narrow down as strata density increases; each section is lined with a steel tube encased in concrete. Chemical "mud" is pumped down the well to keep the bottom of the shaft open and the drill bit cool. It returns under pressure to the rig via the marine riser and is treated for recycling.

Left: Two sections of conductor pipe are fitted together during the exploratory-drilling process known as "spudding in" – the preliminary job of drilling a deep hole and lining it securely with a protective casing of pipe and cement so as to prevent collapse.

ing on the consistency of the formation through which drilling is taking place – the drill bit becomes blunt and has to be replaced. The entire drillstring has to be hauled up from the hole, broken apart, and, when the new bit is in position, rejoined.

When the bit hits oil, what happens is – nothing. There is a saying in the industry that "if you see oil on the rig, you're in trouble." The complex systems of pressure-control devices on and beneath the rig ensure that oil will not gush uncontrollably up the pipe, and the announcement of a discovery is usually made by a geologist after hours of poring over records obtained from instruments run into the hole and from studying debris returned by the mud.

Just discovering oil proves nothing. What counts in a potential field is the amount and accessibility, not the mere presence, of oil. From a range of tests that engineers have devised, they can tell a great deal about whether a field is worth commercial exploration, and only if it looks promising will further wells be drilled after the first strike. In such an event, so as to know the size and shape of the field, the company will drill several more wells stepped out from the first one and will measure the depth at which oil is found in each one. So the superstructure – BOP, riser, etc – is removed from the first well, which is temporarily abandoned, and the mobile rig moves to a new site. To make sure that oil pressure cannot blow out the well, it is blocked with cement and its top is covered with a steel "trash" cap.

After several step-out wells have been drilled and sealed, oil-company experts decide whether to go into full-scale production. If the answer is yes, they must make preparations for several years of costly drilling from fixed platforms, decide on a suitable way of transporting the product ashore, and prepare for the construction of refineries to process the oil or gas. Meanwhile, the mobile rig, its task completed, sails away.

When the drill approaches target depth for an exploration well, samples of drilling "mud" and cores of rock strata taken by a core barrel are examined for traces of hydrocarbon. Left: A core is released from the barrel, to be placed in correct sequence in a sample box. Below: A "mud engineer" tests a returned sample. Only one out of 25 exploratory holes is apt to prove promising.

The drill bit is changed when it becomes blunt or rock strata alter in character. This involves raising and dismantling the drillstring. Bottom of page: The drilling crew on this semi-submersible rig are using "pipe tongs" to uncouple the drillstring's nine-meter-long sections, some of which can be seen already stacked up to the left of the photograph.

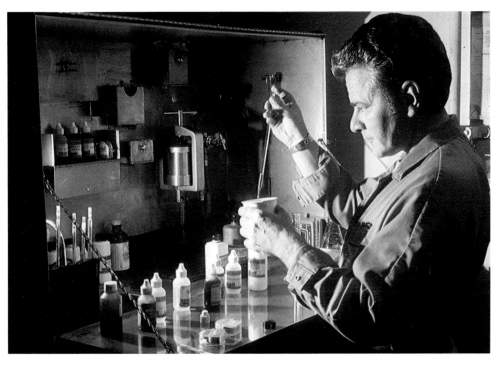

Left: Once gas or oil has been discovered, the well must be tested to confirm its commercial viability. Here crude gas and oil are being flared from the semi-submersible rig Sedco 135-F off New Zealand, to determine the size of the field by monitoring flow pressure over a period of several days.

Above: These fish seem to be exploring the intricacies of a typical "Christmas tree" (so called because of its many branches). All wellheads are topped with similar complex arrangements of valves designed to control the flow of oil. This tree is part of a Persian Gulf subsea system, which pipes a mixture of gas and oil directly ashore from the wellhead, thus avoiding the need for an expensive platform. The many valves and gauges are linked by radio to a shore base from which the well can be shut down electronically in case of emergency.

Setting Up a Fixed Platform

Above: The gigantic (32,000-ton) jacket section of the British platform Graythorp II on its way to the Forties oil field in the North Sea.

At the site, the vast flotation tanks on which the Graythorp II jacket has rested are gradually flooded, and it settles onto the sea bed.

Before the flotation tanks are removed, an attendant crane barge secures the jacket to the sea floor with four enormous steel piles.

Below, right: Four examples of production platforms. Note their depth as indicated in the accompanying scale. A straight-sided steel tower (1), the most common and easiest to construct, is peculiarly vulnerable to weather conditions until its long piles are secured to the sea bed. The broader base of a template structure (2) provides more stability, but the prefab sections are more costly to build. In very deep, rough water, a ballasted reinforced-concrete platform (3) can do double duty, since the many cylindrical cells of its base can also serve as storage tanks; drilling is done through the columns. An alternative design for this type (4) has only one large column, whose wall perforations reduce the force of the waves.

Now firmly pinned down with the addition of 16 more piles, the jacket is ready to be topped by 1,750-ton deck modules.

Drilling for Production

Exploratory drilling is expensive (and economically risky, too, because the area may prove to be inadequately productive). Drilling for production is even more expensive; it involves the building of fixed platforms, which must initially support multi-well drilling equipment, then oil- and gas-processing equipment, for perhaps 30 years.

The earliest fixed platforms were wooden trestles. Seventy years ago they were used in the Gulf of Mexico and the Caspian Sea in no more than 10 to 20 meters of water. By the 1960s they had evolved into steel structures developed for water depths of up to 100 meters in the Gulf of Mexico, the Persian Gulf, and off the coast of Nigeria. By the 1970s the traditional design had to be much modified for use in the stormy weather of the North Sea, where protection against corrosion and cracking presents greater problems than anywhere else in the world.

The legs and cross-braces of a platform, without its top deck, are called the "jacket." A jacket is normally secured to the sea floor by steel piles driven down sockets on the side of the legs. When the jacket is secure, the platform deck is floated out on a barge, lifted into position, and furnished with a drilling derrick, accommodation quarters for the workers, a power generator, and equipment to carry out the initial stages of oil processing before the oil is transported ashore. In the North Sea, jackets have been installed in 140 meters of water. To survive the battering of 30-meter-high waves, the decks of these platforms must stand more than 160 meters above the sea bed. Each of the largest North Sea steel platforms (built in Britain) weighs over 30,000 tons, contains enough steel to make four Eiffel Towers, and is fixed to the sea bed by piles driven 80 meters into the sand and clay. The packages installed on the jackets weigh up to 2,000 tons, and floating cranes – the biggest in the world – have had to be designed to lift them.

Regardless of size, the principle of production drilling is always the same. To get a steady flow of oil from all over the field, wells must be drilled across the whole area of the subsea reservoir. It is unnecessary to build a separate platform for each one since up to 60 wells may be drilled outward at an angle from each platform. If they are slanted outward at, say, 30° angles into an oil field 3,000 meters below the surface, they will tap an area about 3 kilometers across. (The original exploration wells may or may not be included, depending on their location; it is not difficult to redrill them.) The biggest modern platforms may have two derricks drilling simultaneously, with double crews working round the clock. A fleet of work boats and helicopters keeps bringing in a stream of pipe, casing, cement, mud, men, fuel, and food.

When all the wells have been drilled, the production equipment is put into place; this includes a complex stack of control valves – the "Christmas tree" – on each isolated wellhead. From now on the operation is largely automatic. Each well has an automatic valve at the sea bed, which shuts instantly in the event of damage to the well; and on the deck of the platform the various components of the crude-oil or gas mixture are separated and the oil and gas transferred to a tanker or pumped through a pipeline to the shore. Once all such equipment is working efficiently, the complete platform can be automated and controlled from a computerized shore base.

Beyond depths of around 200 meters it becomes increasingly difficult to install steel jackets and their associated packages. This has led to the development of gravity

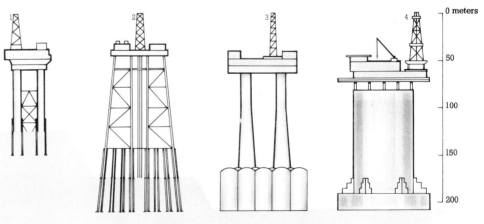

0 meters

50

100

150

200

structures – so called because they rely entirely on their own weight to keep them in place on the sea bed. They are built of pre-stressed, reinforced concrete; only the decks are steel. Within their bulk they can store up to a million barrels of oil. They can be built quickly (in two years) and are easy to install: they are towed out to sea and ballasted into position. There is still some debate, though, as to how concrete will perform under the hammering of year after year of storm waves. Over a dozen of these mighty structures will be in position in the North Sea by 1980, and their platforms will be closely monitored.

An alternative to platforms is subsea production, in which all items of equipment are mounted on the sea floor. Sea-bed wellheads are encapsulated by a chamber in which there is a one-atmosphere environment. A small diving bell transports operatives from a surface ship and mates with the sea-bed chamber, into which the men can pass to work in dry, warm, sea-level-pressure conditions. Eventually, it is envisaged, a series of these sea-bed wellheads dotted over an oil field will feed a centralized sea-bed gathering station, from where crude oil will be piped to a platform in shallower water, or to tankers, for processing and transportation. Systems of this type represent the most likely answer to future oil production from water depths greater than 600 meters.

Oil and gas, though usually taken from separate wells, can be produced simultaneously in fields where both abound. This complex in the Gulf of Mexico has some twenty wells sunk beneath the drilling platform (foreground). The other big platform contains equipment for separating oil from gas before piping them ashore. The third structure contains crew quarters.

Two diagrams of a large oil-production platform currently operating in the Forties Field of the North Sea. The jacket, pipelines, and stripped-down platform are shown in the inset (right), while major features of the three storied 11,600-ton deck structure are illustrated below.

1. The jacket (in this case a tapered steel tower)
2. Incoming oil lines from satellite platforms
3. Hydraulic control for blow-out preventers
4. Cement storage tanks and pump
5. Radio mast
6. Living area for about 140 men
7. "Mud" storage tanks and pump
8. Production skimmer tanks, where "separated" water is cleansed before dumping
9. Conductor pipes for bringing up the oil/gas/water mixture
10. Drilling derrick
11. Flare stack for burning excess gas, or oil in an emergency
12. Oil-line pump exhausts
13. Control room for pumps and wells
14. Production separators for removing water and gas from the oil
15. "Pig" launcher (the pig is a device for probing and cleansing the subsea pipeline)
16. Control Panel for separators
17. Air intakes for electricity generators
18. Main oil-line pumps for pumping oil ashore
19. Compressed-air storage tank
20. Oil pipeline to shore
21. Workshops
22. Life boats
23. Aviation-fuel tank
24. Helicopter pad
25. Cranes for lifting pipes to the derrick or moving heavy machinery
26. Pipe rack
27. Drawworks – the winch that raises and lowers the drillpipe and casing loads
28. Generator exhausts
29. Electricity power station
30. Offices
31. Drinking-water tank

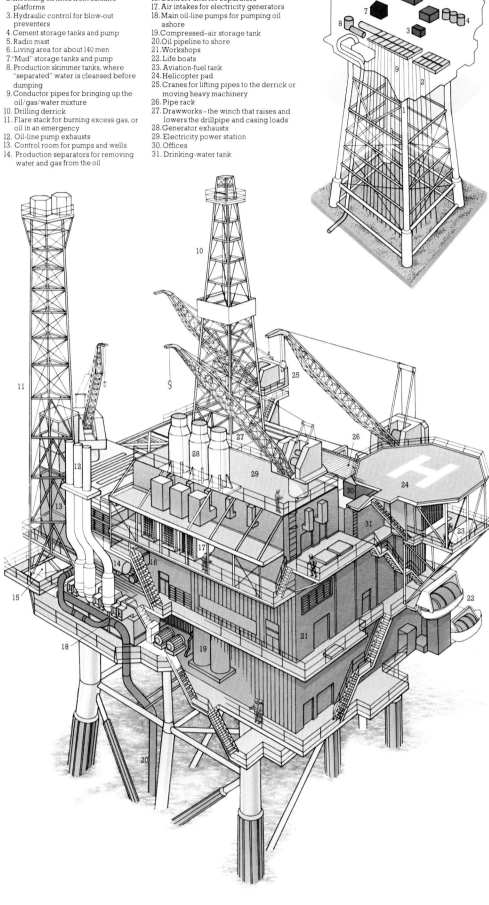

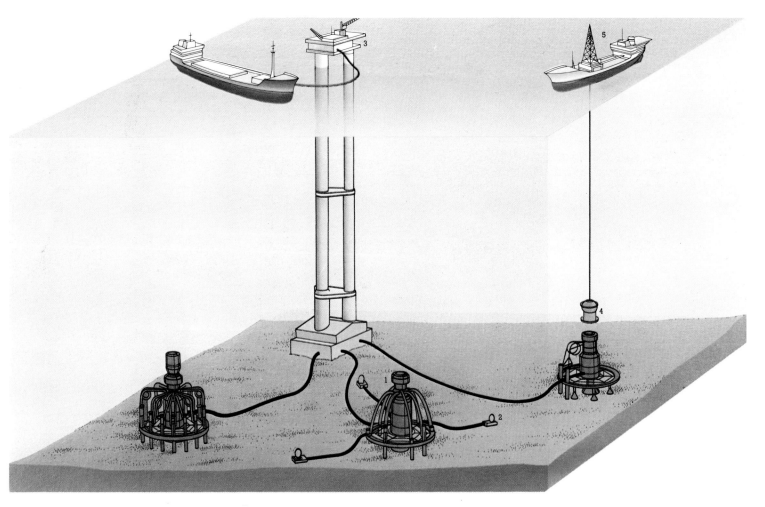

Left: In July, 1975, five enormous tugs towed the first production platform of the new Condeep (concrete deep-water structure) type out into the North Sea. The 350,000-ton platform had been built up on its 145-meter-tall substructure in sheltered waters, thus reducing the difficulties of final installation in rough seas. Platforms will be used to depths of around 200 meters. For much greater depths, subsea systems are now being developed. Above: The French SEAL (Subsea Equipment Associates, Ltd.) system consists of a number of sea-floor production stations (1), each of which taps several wells (2) to feed a central storage platform (3). All equipment is sealed in a controlled atmosphere and serviced by technicians lowered in a diving bell (4) from a surface ship (5). The U.S. Exxon system (below) is even more highly automated. A large template (1) supports equipment for a cluster of up to 40 wells. Servicing is by remote control from a workboat (2), which lowers a usually unmanned "maintenance manipulator" (3). Gas or oil is pumped to a surface buoy (4) via a self-standing pipe (5) and articulated riser (6).

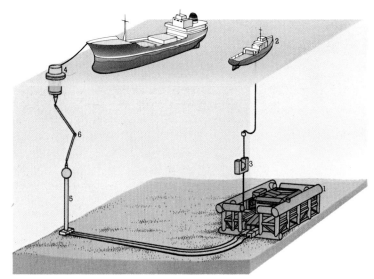

The most efficient way to bring oil or gas ashore is by means of pipelines installed along the ocean bed. Left: A load of 12 meter-long steel pipes, coated with protective concrete and tar except on their ends (which are soon to be welded together), has just been delivered by tugboat to this flat lay-barge in the Gulf of Mexico. As the pipes are lifted aboard, they are fed into the production shed located at the upper right end of the barge. In the foreground is the sloping ramp – the "stinger" – down which the joined lengths of pipe are lowered into the sea. Right, top: Inside the production shed, the sections of pipe, each of which is about a meter in diameter, are joined at one of several welding stations. Right, bottom: After the exposed joins have been X-rayed to make sure there are no defects, they are protectively coated like the rest of the pipe. Finally, the continuous pipeline is paid out down the stinger as the barge creeps slowly forward, pulling itself along on its anchors.

Getting the Oil Ashore

Offshore oil- or gas-producing areas such as the Gulf of Mexico, the Niger Delta, or the North Sea usually consist of numerous fields, each a few kilometers across, scattered over several hundred kilometers of the continental shelf and slope. As we have seen, many wells are drilled from each production platform, and the oil and gas flowing to the platforms (or to subsea gathering points) must then be brought ashore. It would cost too much to lay pipelines from every far-off platform directly to the shore; and so, within each field, small-diameter pipes are usually laid from all the platforms to a central point, from which one large pipe can carry the product ashore. If the fields in a given area are rather small, the product from several fields may be handled in this way. To protect the pipes from damage by ships' anchors or fishermen's trawls, they are often buried in the soft sediment, and detailed maps of the locations of both

platforms and pipelines are published.

Installing pipelines in areas like the Persian Gulf, where a well or platform can be less than 10 kilometers offshore, is relatively simple. For such short distances, on-shore welders join lengths of steel pipe into a single pipe about a kilometer long. Several such lengths are lined up alongside one another, pointing straight out to sea, and – for a distance of no more than about five kilometers – a tug simply drags each pre-welded pipe along the bottom. When one section has been pulled out till its end rests on the shore, the tug halts until the welders join on the next section.

In deeper water, a lay-barge – a flat vessel of 3,000 to 5,000 tons and as much as 120 meters long – is held in position over the route of the pipeline by anchors. The barge carries a stock of pipe lengths and several welding stations where the pipes are welded together before being paid out continuously over the stern. A sloping ramp supports the pipe on its way to the sea

floor, and this ramp – called the "stinger" because it resembles the sting in a scorpion's tail – has buoyancy tanks controlled from the deck of the barge, so that the slope of the steel pipe can be adjusted to keep it from bending too much as it leaves the barge or settles on the sea floor. Such barges can lay pipes with diameters of nearly a meter and a half in water depths as great as 130 meters.

A steel pipe full of oil or gas would tend to float off the sea bed, since both products are lighter than water. For this reason, as well as to prevent corrosion and damage from anchors, undersea pipe sections are coated with tar, fiber glass, and concrete before supply vessels carry them out to the lay-barge, with their steel ends left free for welding. Before the welded joins pass over the stinger, they are inspected by X-rays to make sure they are sound; then tar and quick-setting concrete are wrapped around the join before it hits the water. Once the pipe is on the bottom, another barge directs

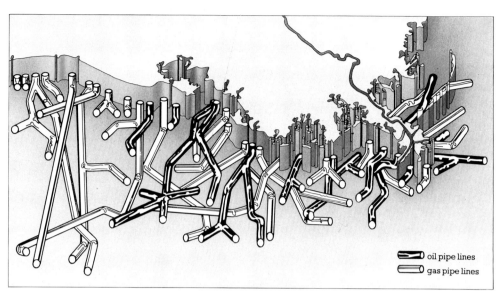

oil pipe lines
gas pipe lines

For an idea of the immense number of pipes already laid out in some of the older offshore fields, consider the above map of gas and oil pipelines that criss-cross the floor of the Gulf of Mexico off Louisiana. The position of such pipelines must be accurately charted so that clear shipping lanes can be established, and in shallow water the pipes are usually buried to protect them from damage by ships' anchors or fishermen's trawls.

In some places, pipelines are not the most feasible oil-transport system. One alternative is to load the oil onto tankers right at the field. Left: Oil flows from the production platform to this single-buoy mooring, from which it can be transferred to moored tankers via the floating hose. This type of operation is difficult in rough weather, but it saves a lot of money; in 1976 the cost of laying one kilometer of pipe in the North Sea was approaching $2,000,000! Below: The first major oil pipeline from a North Sea field being winched ashore in the Shetland Islands. Estimated cost of the 150-odd kilometers: $200,000,000.

a system of water jets over it, so that it sinks into a trench and is covered with sand.

Not all oil is transported to shore by pipeline. The flow of oil from the field must be at a sufficient rate to justify the installation of a line, and not all fields can produce at such a rate. In this event, the oil is loaded at the oil field directly into tankers after initial processing on the platform; it is pumped away from the platform by pipeline to a mooring a short distance away. These moorings, known as SBMs (single-buoy moorings), are anchored to the sea bed, and a hose carries the oil from the sea-bed pipe through the buoy to an outlet above the sea surface. Tankers moor to a large swivel at the top of the buoy so that they are able to rotate, if necessary, through 360 degrees, to obey the dictates of wind, waves, currents, and tides. The oil flows from the buoy to the tanker via a floating hose. There are drawbacks to this method of transporting oil, since it is so vulnerable to bad weather, but it is nevertheless used in most major offshore areas.

The Hazards

There can be no technological progress without occasional setbacks. It is a credit to the scientists who explore the basic techniques of subsea development, and to the engineers who put them into practice, that accidents are, in fact, so rare. One kind of technical mistake has been much publicized in recent years: the advance that brings financial rewards at serious risk to the environment. There have indeed been a few unfortunate occasions when oil has been accidentally spilled into the sea and spewed onto beaches. There have also been fatal accidents to divers working on commercial undersea projects, and an occasional tanker or trawler has foundered. But all such incidents should be viewed against a background of constant efforts to maintain and improve safety, not only for the benefit of personnel and equipment but for the health of the sea itself.

Some errors and consequent accidents are inevitable when the offshore industry is working continuously at the very limits of technology. But what is impressive is the infrequency rather than the frequency of accidents brought on by stepped-up efforts to tap the ocean's resources. From 1955 through 1964, for instance, the number of total write-off accidents to the world's mobile drilling rigs averaged fewer than two a year. In 1965 there were eight, but the rate has dropped considerably since then, despite the greater number of rigs in oper-

ation. Most of the accidents have occurred to jack-up rigs, during a period when working depths were increased beyond 30 meters, particularly – and understandably – at the start of drilling the stormy North Sea. The worst such incident was the disastrous collapse of the jack-up rig Sea Gem, with a loss of 13 lives, in December, 1965.

Semi-submersible platforms have an excellent safety record, even though they run the constant risk of dragging anchors or excessive stress in their support legs and cross-members in stormy weather. Perhaps the most serious loss of such a structure occurred in January, 1974, when the newly built semi-submersible rig Transocean 3 sank – but without loss of life. Transocean 3 had legs that were retractable for ease of towing between drilling sites. In the last weeks of 1973, her legs were extended and she was towed out into the North Sea to do her first job; but as the wind strengthened to up to 60 knots, the legs began to work loose. On the afternoon of New Year's Day, 1974, a helicopter lifted 38 men off to safety, leaving 19 behind, but by evening one of the legs had bent outward, and since it was clear that the rig was doomed, the rest of the crew were taken off. Later that night she capsized, and she sank a few days afterward. The economic costs of such a disaster are, of course, gigantic. Still, there have always been shipwrecks of one kind or another, and there always will be.

More worrying from the standpoint of society is the type of accident that results in

severe pollution. One obvious danger spot is the Gulf of Mexico, where there is a recurring risk of hurricanes' striking drilling platforms so suddenly that engineers do not have time enough to stop drilling, retrieve drillstrings, or close down producing wells. Even if a production well is hit by a hurricane, the so-called storm choke – an automatic valve at the bottom of the well – should close down and prevent leakage, but storm chokes are sometimes blocked by sand. In the notorious Chevron Oil Company fire off the coast of Louisiana in February, 1970, storm chokes failed to work. It was not a hurricane but a drilling accident that started the fire, which raged for 50 days as well after well on the platform was ignited. The fire could not be extinguished directly, since this would have resulted in vast quantities of unburned oil pouring into the sea. As it was, floating barriers were arranged to help to minimize pollution, while fire hoses played on the platform itself to keep it from melting. Meanwhile, fresh holes were drilled from nearby platforms to intersect the wells and block them with cement. As a result of the failure of storm chokes, another safety device, the downhole safety valve, has been developed and is now standard equipment on mobile rigs.

The technique of intersecting an existing well had been used previously to stop a catastrophic oil leak in a platform off the coast of Santa Barbara, California. In January, 1969, just as the drill was being removed from the fourth of several wells

155

An ever-present hazard of exploratory drilling is the risk of hitting a pocket of gas before protective concrete casing or blow-out valves can be installed in the hole. The possibly disastrous effect is graphically illustrated in these pictures of what happened, not long ago, to a jack-up rig off Southeast Asia.
Left: Erupting gas engulfs the rig, which bursts into flame as the gas ignites (below). Weakened by the blow-out, the sea floor beneath the rig's legs caves in and the rig collapses (bottom).

Above: A view of the coast in 1967 after the *Torrey Canyon* had run aground and spilled much of her 117,000-ton cargo. Thousands of animals and birds perished, victims of oil pollution and of the detergents used to combat it.

Facing page: The end of a nightmare. After raging for seven months, the fire caused in 1974 by a blow-out from a wellhead on the Wicked Witch platform in the Gulf of Mexico was conquered in early '75 by the noted American firefighter "Red" Adair, who used the crane on the barge in the foreground to position an oxygen-consuming explosive charge that snuffed out the flame.

being drilled, the well started to blow back. The crew abandoned 700 meters of drill pipe in the well and closed down the blow-out preventers, but great quantities of oil leaked through the concrete wall of the well. In the next ten days a total of 56,000 barrels of oil had leaked out before the well could be intersected and blocked far below ground. The floating oil devastated much marine life as well as many miles of beautiful shore front, and the effect on public and political opinion was enormous. As one good result of the disaster, oil companies everywhere redoubled their efforts to prevent oil spills from offshore wells, and from tankers, too.

Tanker accidents are, of course, bound to happen, since there are thousands of tankers in operation around the world. Here, too, however, surprisingly few bad collisions, groundings, or fires occur. The most spectacular tanker incident so far was the grounding of the *Torrey Canyon* off the Cornish coast of England in 1967, when most of the ship's cargo of 117,000 tons of crude oil spilled into the sea and eventually fouled some of the rocky beauty spots for which western Cornwall is renowned. It should be emphasized that the wreck of the *Torrey Canyon* is remembered not because it was typical of what happens when tankers get in trouble, but because it was exceptional.

Engineers, designers, governments, and industry learn something from every such incident, as well as from the constant run of minor setbacks and near-accidents that do not attract global attention. Recent losses of trawlers in the ice and gales of the northern Atlantic have led to greatly increased research into the causes of icing on rigging and superstructure and on the ways of preventing ice from building up. No diver or submarine craft is lost, no machine is wrecked, no oil spilled without an official investigation to determine precisely what happened and why. And each new answer to these questions brings at least an attempt to forestall similar calamities by means of changes in design technique, and operational procedure. Democratic governments and industrial interests cannot afford to be complacent.

The Mines of the Sea

There was a marine mining boom in the 1960s. Within the space of two or three years, the world shortage of essential materials, from sand to titanium, was not merely eased but ended. The rich nations became richer; the starving millions of the third world became fat and prosperous. We all wallowed in precious metals recovered from the oceans.

At any rate, that's how it would have been if the predictions and promises on a million sheets of paper and in the warm air that puffed from a hundred conference platforms could have been fractionally fulfilled. But the marine-minerals boom of the '60s was only a paper boom. There were people who took hold of the paper ores and actually went to sea to grab some of the promised wealth. But paper quickly goes soggy in seawater. Most of the would-be miners returned from this latter-day gold rush sadder, wiser, and poorer.

How did it start, this rush to the mines of the oceans? As with most crazes that burst upon the public, there were many factors that led to the flowering of the minerals mania: for instance, a growing awareness of the potential of ocean resources, an increasing shortage of essential minerals, and the political instability of many of the land areas in which great quantities of essential minerals were locked. But if one event brought all the boom-contributing factors into sharp focus, it was the publication in 1964 of a book called *The Mineral Resources of the Sea* by Dr. John Mero, an American scientist. Over half the book's description of undersea mineral resources was a scholarly and soberly argued discussion of deep-ocean manganese-nodule deposits. The rest dealt with minerals in solution in seawater, with deposits on the continental shelf, and with deep-ocean deposits other than the nodules. To what degree Dr. Mero was responsible for what followed is hard to say, but shortly after the publication of his book the rush began – led, in the United States, by some of the giant aerospace and electronics corporations that were eager to diversify from their declining space activities. In Europe, meanwhile, advanced-technology companies with little marine experience, backed up by a handful of academic-glory-seeking scientists, propounded sophisticated solutions to knotty marine problems.

It is not difficult to understand why businessmen and scientists were dazzled by the riches they thought could be plucked from the oceans. The mineral potential is huge.

Diamonds from the sea floor? The chances for bringing them up in commercial quantities in the near future look slim. This sampling equipment is still being used off the southwest coast of Africa, but large-scale efforts to mine these waters have been abandoned.

Seawater itself is a complex solution containing 50 million million tons of dissolved salts. That works out at 40 million tons for each cubic kilometer; and the last time anyone went through the rather pointless exercise of calculating how much that little lot was worth, he came up with $6,000 million. There's gold in those waves – and silver, uranium, tin, titanium, zinc, copper, and hundreds of other precious elements. Gold lies, too (along with tin and diamonds), in the sands of the continental shelf, in which concentrations of heavy minerals have formed where streams, waves, and currents have carried away lighter materials. Lying on the continental shelf there are phosphorite nodules. And beneath the sea bed, in addition to oil and gas, lie vast deposits of coal and sulphur.

Above all, though, it was the fantastic wealth lying in lumps on the deep-ocean floor – the manganese nodules to which Dr. Mero devoted so many pages of his book – that fired the imaginations of the hopeful marine miners of the 1960s. To the entrepreneur it must have seemed almost too good to be true. Down there, as Mero pointed out, lay enough titanium to last two

million years, nickel for 150,000 years, cobalt for 200,000 years, manganese for 400,000 years, aluminum for 20,000 years, and so on. All this was there in handy-sized lumps just a few kilometers under the Pacific, in a decade when the seemingly invincible technology of the aerospace and electronics companies was proving capable of putting a man on the moon. How easy it seemed!

And there was more. Mero discussed the commercial potential of the deep-ocean sediments – the red clay and calcareous and siliceous oozes that contain materials suitable for use in making concrete, or to use as insulation, abrasives, absorbents, or filters. And more! At the bottom of the Red Sea, in narrow troughs over 2,000 meters below the surface, occurs a hot, salty brine rich in iron, zinc, copper, and manganese, and under it lie sediments scores of meters thick that contain up to 5% zinc and 1% copper. In the late 1960s, when copper was in very short supply, the thought of these sediments was enough to induce one marine miner to place a firm order for 3,000 meters of piano wire with a British company as the first stage of his Red Sea dredging project.

Above: A bucket-dredging barge (with its attendant tug) digging for tin in 32 meters of water off the west coast of Thailand. It can dredge nearly 8,000 cubic meters of ocean floor every day. Tin crystals are separated out and graded in a treatment plant on board, then sent ashore for smelting.

There it sat: a potential salt-water Klondike. Plans for technological expeditions to recover this wealth envisioned submarines zooming about the oceans; remote-controlled underwater dredgers gobbling up sea-bed deposits; manned sea-bed-crawling vehicles trundling happily across the sea bed to dig in the sand; underwater habitats with men living down below for weeks at a time. So what happened? Not much. Few of the plans got beyond the feasibility-study stage. As life returned to normal, experienced marine miners, who had watched with bewilderment (and an occasional giggle) the sudden influx of corporate giants and trendy scientists into their fairly mundane world, shrugged and settled down to normal work again.

What is "normal" undersea mining work? There *is* such an industry, and it makes money. We'll take a look at it now.

The simplest way to go marine mining is not to go to sea but to sit on the shore and extract what you need from the salt water. The obvious substances to extract are salt and water, but there are others as well. In fact, almost all of the 92 chemical elements that occur naturally on earth occur in reasonable concentrations in seawater – concentrations ranging down from over 1% for chlorine and sodium (which together make salt) through about 1 part per million for boron and silicon and 1 part in 100 million for tin and arsenic, to less than 1 part in 10,000 million for gold. The lower the concentration, of course, the more difficult and expensive the extraction process becomes.

There are few problems in extracting salt in places where sunshine and wind abound. Seawater is simply put into evaporation ponds and allowed to evaporate. Calcium sulphate and calcium carbonate precipitate first; the remaining brine is then removed to crystallizing ponds, where salt (sodium chloride) begins to precipitate; after removal of the remaining liquid, fresh concentrated brine is pumped in, and the process is repeated until a layer of salt is available for harvesting. Where there is not enough sunshine, the extraction process is much more costly. Salt can be obtained by freezing sea water or by boiling it, but both methods use expensive energy.

The lack of cheap sources of energy is an even greater obstacle to the extraction of fresh water from the sea. Large-scale research projects have been aimed at developing techniques for the desalination of seawater, but simple evaporation still seems to be the most practical method; and sunshine does not provide the kind of energy required for the job. A system built in Greece a few years ago, for example, passed the sun's rays through a glass roof onto a pool of seawater covering a black, heat-absorbing plastic floor. Cooled by winds, condensation formed on the glass and ran into a collecting trough. On the sunniest days, the fresh-water production of this system amounted to rather less than 5 liters a day per 5 square meters of glass!

It is possible to evaporate seawater by a number of methods using "artificial" heat that is not in itself especially costly – steam, for instance. Every such technique has a serious drawback, though: large quantities of expensive metals, such as titanium and copper alloyed with nickel or zinc, are required for building the necessary apparatus so that it can function without corroding. As for the production of fresh water by freezing salt water, although the machinery does permit use of such cheaper materials as steel and plastics, the process is very slow, and the end result – the desalinated water itself – is not entirely pure.

In evaporation ponds like this one in sunny California, a five-year-long course of evaporation from seawater to crystallized salt culminates in a rich harvest as a special machine cuts a swath of salt and loads it on miniature railway cars. The tracks are shifted along the solid layer of salt as the harvester moves across the pond.

And of several other desalination methods that have been explored (reverse osmosis, for example), all have so far proved prohibitively expensive for large-scale operations.

It can be seen, therefore, that the most obvious of all substances to be extracted from seawater – the water itself, without mineral content – is one of the hardest to come by in significant quantities. Two of the elements in the ocean's waters, on the other hand, are now being harvested with relative ease. One such substance is bromine, which occurs in concentrations of 65 parts per million of seawater. Some 70% of the world's total output of this important chemical – it is used in the manufacture of dyes and in both medicine and metallurgy – comes from extraction plants in the United States. Another element that can be mined without going to sea is magnesium, which is used for production of light alloys, as an anti-corrosion protection agent, and for various phar-maceutical products. Well over half the world output of metallic magnesium is extracted from seawater in Norwegian and American extraction plants.

There are silver, tin, titanium, and uranium in seawater, too. But because 96% of the content of seawater is pure water, it is the complications of removing this pure water that makes the extraction of such valuable elements too expensive to be worth the effort. Unlike bromine and magnesium, which occur in relatively high concentrations, these metallic elements are just too dilute to be capable of easy extraction. Our optimistic marine miners of the 1960s considered the possibility of building multipurpose extraction facilities, where fresh water and a variety of other elements could all be extracted in a single plant. But even on paper it did not work out. In order to produce a few ounces of, say, tin, hundreds of tons of salt, water, bromine and magnesium compounds, and other substances would also have to be separated out – so much, in fact, that the world market for all such commodities would become glutted. Prices would fall, extraction plants would become uneconomic to run, and economic failure would follow hard on the heels of technological success.

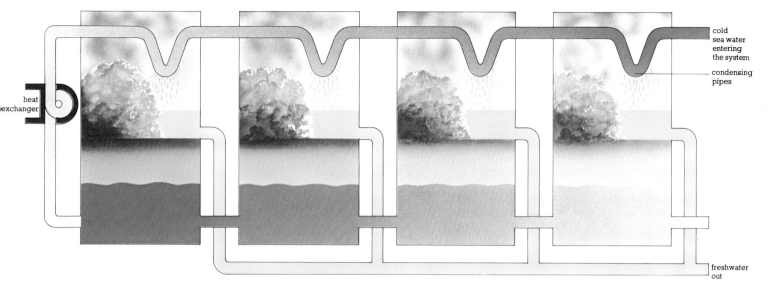

cold sea water entering the system

condensing pipes

heat exchanger

freshwater out

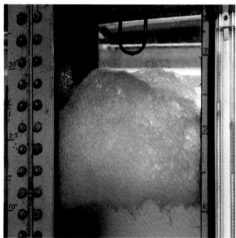

Despite the cost of extracting fresh water from seawater, desalination plants provide much of the water supply in a number of places – for instance, Mexico (roughly 30 million liters daily) and Abu Dhabi (some 10 million). The system most commonly used is multistage flash distillation (illustrated above). It works on the principle that when heated seawater passes through a low-pressure chamber a portion of the water instantaneously evaporates ("flashes") to steam. This is then condensed in contact with cool pipes, and the resulting fresh water is collected in trays and pumped away for use. The multistage system utilizes a series of such chambers, each kept at a lower pressure than the one before, a "flash" being produced in each chamber. At each stage the heat from the condensing steam warms up the incoming seawater so that less heat has to be put in at the main brine heater. Left: an actual flash.

Below: These immense settling tanks are a prominent feature of the Kaiser Corporation's California plant for extracting magnesium compounds from salt water. Here magnesium hydroxide, precipitated from seawater by a chemical process, is allowed to settle. Later it will be washed, filtered to reduce its water content, and eventually used in the manufacture of paper, insulating materials, etc.

Sand and Gravel

When you leave the safety and comfort of land and go to sea in search of mineral wealth, you learn some hard truths that no amount of paperwork can teach you. The rewards, to be sure, are there; phosphorite, gold, and diamonds do lie a few miles offshore in relatively shallow waters (anything from 30 to 130 meters deep). Yet the most important minerals being harvested from the continental shelf in the late 1970s are sand and gravel, just as they were before the "boom" of the '60s. All efforts to recover more exotic minerals are proceeding at the same slow pace as in the pre-"boom" days, or have been abandoned.

What the 1960s miners didn't appreciate was just how difficult it is to do *anything* at sea. If gold nuggets are strewn across the hills of California, you need your hands and a wheelbarrow to pick them up and cart them off. Identify a similar gold mine lying a mere three meters below the waves, and you need at least a rowboat and crude dredge to get at it. Calculate the difference in costs, and you have a rough indication of only one difference between land and sea mining. Another big difference is that you can always leave your nugget-strewn hillside and find it again easily by means of landmarks. At sea, out of sight of land, you can quickly be lost without expensive navigation equipment.

It gets worse. With land-based ore deposits, there are reasonable surface clues to the underlying geological formations, and many nations also produce geological maps. There are no surface clues to marine deposits, and only the major maritime nations have even begun to provide geological information for the continental shelves. Furthermore, once a land-based deposit has been discovered, it can be harvested by the use of a bulldozer seven days a week, whereas several million dollars' worth of dredging machinery may be able to work only one day in seven if there's an east wind blowing. Finally, there are the problems of processing and transporting the minerals once they are dredged up. If, for example, tin is being mined at an offshore site, it would be unwise under normal circumstances to transport the ore ashore without some initial processing, for the expensive barges would be perhaps 90% full of rubbish. But conventional mineral-processing equipment – sluices and spirals that rely on gravity and weight differences for separating the valuable mineral from the dross – often fails to work on the pitching, rolling deck of a ship.

Those were among the problems that, in the 1960s, defeated every attempt to harvest tin off Cornwall or diamonds off South Africa. Such ventures failed, not because the basic plans or equipment were wrong, but

because of a multitude of minor irritants: the vagaries of position-fixing; the difficulties of sampling ore; turbid water that made direct observation of sea-bed conditions by divers virtually impossible; and the logistic problems of changing crews and coping with mechanical failures in remote areas. And weather – always the weather! When trying to assess the profitability of a mineral-dredging operation, how can you allow for the fact that your key personnel may sometimes be so seasick that they can't lift their heads from their bunks, let alone work?

So we are left, with a few minor exceptions, with sand and gravel as by far the greatest mineral harvest from the world's continental shelves. Why has this industry succeeded where others have failed? First, sand and gravel (technically known as aggregates) are easily found, and the extent and depth of a deposit can be quickly and accurately gauged by taking core samples. Then there are market forces. Aggregate is needed for making cement for building. Since building activity flourishes mainly in places where land values are high, there is little sense in ruining valuable land by digging in it for aggregate; and so the sand and gravel must be conveyed to building sites, often from fairly far away. Thus, it frequently becomes cheaper to dredge aggregate from sea-bed deposits lying close to the center of demand than to pay the heavy costs of transporting land-based deposits over hundreds of kilometers of road.

Two modern methods of offloading aggregate. Above: The Dutch suction dredger *Ahoy* pumps saturated sand and gravel into a transporter barge. The *Ahoy* can lift aggregate from 40-meter depths and discharge it into flanking barges simultaneously. As the barge fills, excess water is forced out over the top of its open hold. Right: Water brought up with sand and gravel by the British suction dredger *Deepstone* has been pumped overboard before mechanical grabs suspended from a movable gantry deposit the aggregate onto the central conveyor belt for offloading. Some 1,500 tons an hour are handled in this manner.

It is on this economic basis that the European and American marine-aggregates industry has been built. By now the equipment for the industry is quite sophisticated, with dredging ships capable of carrying up to 4,000 tons of material, which can be loaded at a rate of 1,800 tons an hour by a diesel-powered dredge pump. A pipe trailed from the vessel sucks up the aggregate from a maximum depth of about 45 meters, and the dredged material passes through vibrating screens that can separate the material into all-sand, all-gravel, or mixed cargoes. To unload, the most modern type of vessel has scraper buckets that bite into the cargo, drag it to the top of the hold, and discharge it into a hopper, from which conveyors carry the aggregate ashore at an hourly rate, again, of up to 1,800 tons.

In some places – around Britain and off the east coast of the United States, for instance – the industry has a slightly uneasy re-

lationship with fishermen, many of whom complain that continual dredging can ruin spawning and nursery grounds. For the time being, however, there is no threat of open warfare. After all, aggregate dredging is a well-established industry by now, with a tradition that even fishermen can respect. It is a tradition from which newcomers to the exciting, but risky, business of marine mining could learn a lot. One lesson it teaches is that, where the sea's mineral wealth is concerned, slow growth involving technological evolution may be more profitable than an attempt to make a quick killing through a massive injection of revolutionary technology.

Below: Two types of aggregate dredger, one for use inshore, the other for deeper water. The cutter-suction dredger to the right, held in position by a "spud" driven into the sea bed, has a rotating cutting head for loosening the aggregates on the sea floor; these are sucked up onto the barge through a pipe in the "ladder." The aggregate is then pumped to waiting barges or directly ashore via a floating pipeline. In the second system – operated in deeper waters by a trailing-suction-hopper dredger – aggregates disturbed by a dredge head are pumped from the sea bed by a trailing suction arm connected to the slowly moving ship. The vessel can hold large quantities of graded sand and gravel in its hoppers until ready for offloading.

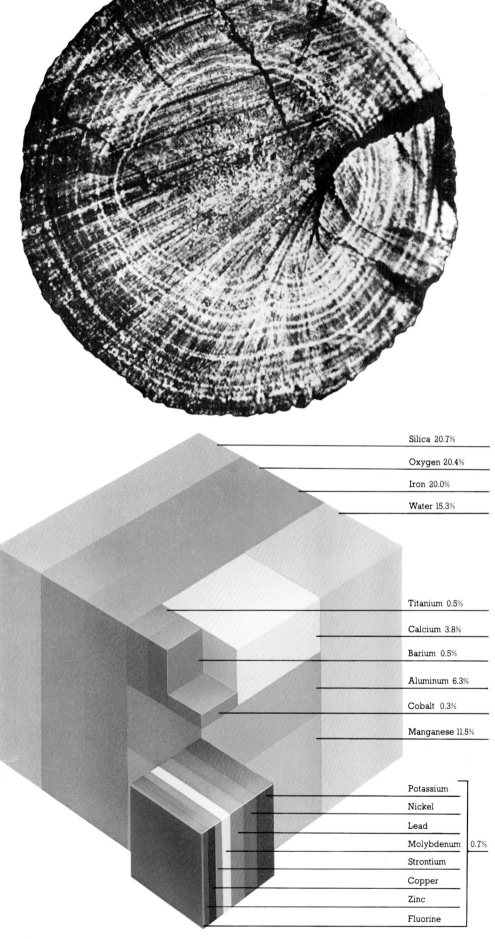

Silica	20.7%
Oxygen	20.4%
Iron	20.0%
Water	15.3%
Titanium	0.5%
Calcium	3.8%
Barium	0.5%
Aluminum	6.3%
Cobalt	0.3%
Manganese	11.5%
Potassium	
Nickel	
Lead	
Molybdenum	0.7%
Strontium	
Copper	
Zinc	
Fluorine	

Manganese Nodules

The problems inherent in all types of marine mining become especially acute as the prospector moves away from the shore and the continental shelf to the deep ocean. But the potential rewards are greater, too. There are enough manganese-nodule deposits in the Pacific alone to keep industrialized society in such essential metals as nickel, copper, and cobalt for the foreseeable future.

Manganese nodules are not a recent discovery. The British research ship *Challenger* dredged them up from the Pacific and Indian oceans on its epic voyage a century ago, and we have known since the late '50s that vast quantities of them are spread over millions of square kilometers of the great oceans. How these roundish objects – ranging in size from a small pebble to a rock weighing a ton – are formed remains something of a mystery; it is generally thought that substances precipitated from the water gather around some nucleus such as a fish tooth or fish bone, or else around a piece of pumice or other volcanic material that has found its way to the ocean floor. The rate of nodule growth is probably about 0.01 millimeter per 1,000 years. That may not seem a lot, but multiplied over thousands of millions of individual nodules it adds up to a much faster accumulation of metals than the present rate of world consumption. So the nodules represent a metal farm in the deep ocean – a self-sowing farm that needs no tending.

In the 1960s, however, when the first serious attempts to survey and begin pilot-scale harvesting of the nodules were made, the difficulties soon became evident. For one thing, the mineral content of individual nodules over quite small areas is not constant, and the resultant complications could make nonsense of the economics of recovery and processing, since inordinately complex processing methods are needed in order to extract the metals from the nodules. Moreover, if such processes did become economically feasible, the world market for

The above cross section of a nodule shows the irregular growth rings that have formed around the nucleus – in this case a bit of volcanic rock – as successive layers of mineral deposits have accrued. Components of nodules vary considerably; the diagram, based on average contents of samples recovered from the Atlantic, illustrates the large number of substances that can be contained in a single specimen. Note, as indicated in the accompanying map, that although the Pacific has so far proved the most productive field for the nodules, samples have been dredged from all the great oceans, and underwater TV systems indicate that potential mining sites are remarkably widespread.

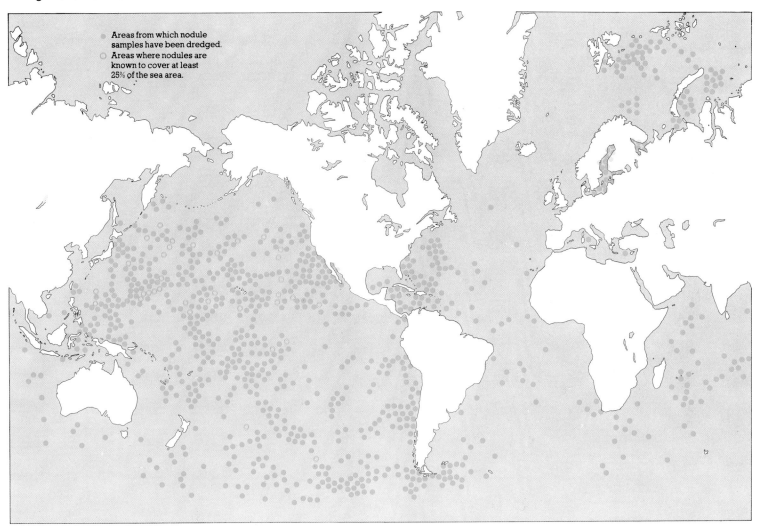

Areas from which nodule
samples have been dredged.
Areas where nodules are
known to cover at least
25% of the sea area.

certain metals could be swamped by an overenthusiastic harvesting of them, thus bringing down prices and ruining the economics of the whole operation.

More discouraging than any other factor was the depth at which the nodules lie. To recover objects from the sea floor at a depth of 4,000 to 5,000 meters may not seem a technologically insuperable problem in an age when men have walked on the moon. But we have to consider this depth in relation to present capabilities in marine mineral recovery. Recovery of large quantities of unconsolidated deposits from the deep-ocean floor would require some form of suction dredge that could bring the stuff up from 4,000 meters or more; the most sophisticated commercial suction dredge now working can suck up sand and gravel from no more than 45 meters!

The possible rewards, though, are so great that several heavily financed organizations are still moving ahead with plans to exploit the nodules – four American consortia and one each from Canada, Japan, France, and West Germany. None of them is close to the mass-production mining stage as yet, but they appear to be making steady progress in working out recovery and pro-

cessing techniques. Their first task, of course, has been to survey potential mining sites and bring up samples over the area of a given "mine" (which can cover as much as 10,000 square kilometers). To survey the nodules, the consortia use underwater TV systems, usually mounted on sleds that also carry lights and still cameras; and some recently developed forms of acoustic instrumentation make it possible to record depth to an accuracy of a few centimeters.

Improved TV systems and precision depth recorders are actually by-products of the manganese-nodule boom. Even, therefore, if the mining program is brought to a halt tomorrow, it will have been of lasting benefit to the oceanographic community. Another of its valuable by-products is the free-fall sampler that has been developed for the industry. This device has aluminum jaws, which are held open as it drops to the deep-ocean floor; when it hits the bottom, the impact activates a trigger that snaps the jaws shut. Ballast weights fall off, and the buoyant sampler rises to the surface with a supply of nodules trapped in its jaws. A number of these devices can be deployed over the area being surveyed.

Work that has already been carried out

has revealed a few somewhat disappointing facts about manganese nodules. There may perhaps be only about a twentieth as many as enthusiastic estimates of ten or fifteen years ago led us to expect (though that still leaves a very great number). Furthermore, as already pointed out, the nodules vary in quantity and quality over short distances. And their average ore content is not as high as was once believed. Even so, the total amount is considerable. A typical nodule contains from 27 to 30% manganese; from 1.1 to 1.4% nickel; from 1 to 1.3% copper; and from 0.2 to 0.4% cobalt.

What has emerged, then, is the to-be-expected conclusion that the miner cannot sail happily out into the Pacific, put down a dredge wherever it suits him, and raise mineral wealth to the surface. As in any other mining venture, he must begin by choosing a mining site, carefully delineating its area, and taking extensive samples – an initial job that is far harder to accomplish at sea than on land. The major groups investigating nodule deposits have just about finished that job by now. Ahead of them looms the gigantic task of bringing the nodules to the surface and making the operation pay.

Hopes and Fears

After a dozen years of looking for manganese nodules and analyzing what has been found, the people who believe in a rosy future for deep-ocean mining are at last facing up to the difficult job of proving their point. To do this, they must find economically viable ways of bringing up, not samples, but tons upon tons of the nodules and extracting valuable metals from the conglomerate mass.

So far, two main recovery methods have been developed and tested. One, favored by America's Deepsea Ventures Inc., involves placing TV cameras near the sea bed. These send pictures to a monitor on the mining ship, which then directs a giant suction tube along the bottom to suck nodules up into the hold. The other scheme, developed by the Japan Ocean Resources Association, is a massive extension of the traditional line-bucket dredging method, which has been for many years a common procedure for dredging shipping channels: an endless chain of buckets is lowered from the bow of the ship; they scrape along the ocean floor, collecting nodules as they go, and are then hauled up to the stern, where they unload their contents before moving on for another dip. Tests indicate that both the American and Japanese systems are promising, but there is a long way to go before either can come into effective operation. An investment of hundreds of millions of dollars will be needed in order to get full-scale production under way, and refinement of the ores requires the building of highly specialized factories.

Exploration of techniques for processing the nodules is still extremely tentative. Technologists of Deepsea Ventures Inc. have investigated possible ways of extracting the four principal metals – manganese, nickel, copper, and cobalt – and have found themselves beset by problems. Already, in fact, they have been forced to try out more than a hundred different processes. One big problem is the fact that, in addition to the four principal ingredients, each nodule contains 30 or more other metals in sufficient concentrations to interfere with many extraction processes. In what seems to be the most nearly feasible method, a form of chemical hydrometallurgy, the nodules must be ground, dried, chemically reacted, leached, filtered, and separated before the final extraction is possible. It will be a long time before any such process can be carried on profitably as part of a full-scale mining operation.

Huge though the technical problems, they *are* being slowly solved. But there remains a nontechnical problem that overshadows all the rest. It is a question that applies not only to the manganese nodules but to all the vast resources of the deep ocean: who owns them?

The resources on and under continental shelves are owned by the country adjacent to them; where two countries have a common shelf, you draw a line down the middle. There is no similarly easy, logical answer for the deep ocean, for it is "stateless." Its nodules, like its fish, would appear to belong to nobody and everybody. But no such situation can be allowed to persist – not now any longer, not with the citizens of several rival nations poised to start making a lot of money from manganese nodules. Until our time, arguments about jurisdiction over the deep ocean were rather academic, but they have become increasingly urgent. Can we have a legal regime for the deep-ocean floor, so that it is divided among nations? Or do we accept a technological free-for-all in which those who have the technology to exploit the sea's riches reap the benefits? If we have a legal regime, how do we work it out? What about landlocked states? Where is the deep-ocean floor deemed to begin? If we have a free-for-all, what happens if mining ships from different nations arrive at the same site at the same time?

These and many related questions must be answered if nodule mining is to become a peaceful reality. As is shown in the chapter of this book that deals with marine law and politics, progress in resolving ownership problems is excruciatingly slow, with every nation in the world – even the landlocked ones – using the opportunity of the nodule

Only years of research and testing can tell us whether or not large-scale mining of the deep oceans is economically feasible. Above: The *Prospector*, a research vessel of America's Deepsea Ventures Inc., carries TV and still cameras that can operate down to nearly 6,000 meters, as well as many other kinds of specialist equipment for surveying and sampling manganese-nodule deposits. The cameras are mounted on a huge tripod – shown here being lowered into the sea – which is also equipped with high-intensity lights and a suspended dredge. The tripod, which the *Prospector* tows slowly at an accurately measured depth just above the ocean floor, records deposits of nodules and scoops up samples for analysis.

debate to air political views that often seem to have little to do with minerals, fish, oil, or any other of the sea's resources. Some conclusions must be reached soon, however, if the world is to benefit from an intelligent exploitation of the oceans.

Meanwhile, there appears to be an encouragingly widespread trend toward realizing that exploitation of the oceans is not synonymous with, so to speak, squeezing them dry. Our industrialized societies, at least, are becoming more conscious of the degree to which industrial processes and their waste products have been allowed to interfere with natural biochemical processes. Because of its vastness, the sea used to be looked upon as having a limitless capacity for absorbing rubbish as well as for providing fishmeat and whalebone. Most of us know better today. Aware of the dangers of pollution and of overfishing, we are being more careful than we would have been a generation ago about where and how we drill for oil or where and how we try to lift minerals from the sea floor. And this is certainly a step in the right direction.

The sea is our key to survival. It not only represents our last storehouse of water, protein, and mineral resources, but is the earth's lung, the vital organ of exchange of oxygen, carbon dioxide, and chemicals with the atmosphere. As we prepare for an unprecedented exploitation of its treasures, it is right for us to be as concerned for the health and welfare of the earth's lung as we are for our own.

Below: Aboard the *Prospector*, pictures transmitted from the towed tripod are viewed on this TV monitor, which shows a nodule deposit on the sea floor at a depth of about 5,500 meters. Such pictures are recorded on tape and, together with nodules brought up in the sample basket (left), undergo initial analysis on the ship. Further samples go for processing to an experimental plant in Virginia, where metals are separated out by chemical means. Bottom: The 7,500-ton *Deepsea Miner* used for mining trials, has a 23-meter-high derrick mounted over its "moon pool" for deploying and recovering a prototype production dredge.

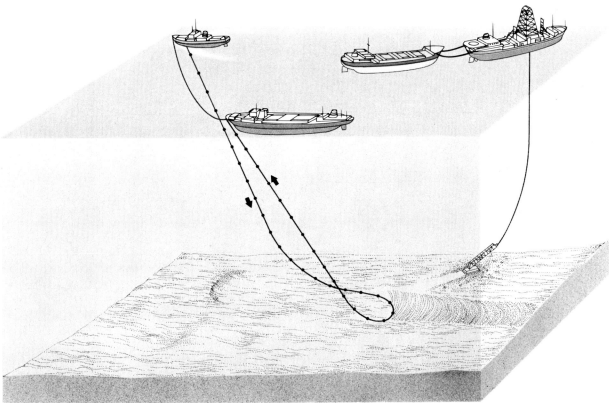

Left: The two deep-ocean-mining methods most likely to prove practicable. The suction dredge is a hydraulic "vacuum cleaner" successfully tested by America's Deepsea Ventures – so far, though, not much below 1,000 meters. The other system is a French modification of a Japanese technique that has managed to reach down to 3,600 meters: two ships work in tandem to operate an immensely long bucket chain.

6
Using Ocean Space

Robin Clarke

The ocean, which has been called "our last resource," is like most natural resources, finite. Its volume of water – 1,350 million cubic kilometers – is limited; its surface area – 381 million square kilometers – is limited; the length of its coastline – 504 thousand kilometers, ignoring small bays and inlets – is limited; its tonnages of salt-water fish and minerals are limited. As human industrial activities increase in scale, and as the world's population proliferates, so the total stress on the ocean grows, not just in terms of overfishing or pollution but in the sheer crowding of conflicting activities into a finite space. Cables, pipes, fishing, shipping lanes, sailing, water skiing, waste disposal, bridge building, fish farming, dredging, tunneling, tourist parks, nature reserves – there seems no end to the things people want to do in, on, or under the sea.

Ultimately, the confusion must be controlled by national and international laws and regulations, which are discussed in chapter 10, but lawyers and politicians are involved only where conflicts are immediately apparent, or where military or economic advantages are at risk. Most coastal, offshore, and oceanic problems never reach a point of evident crisis; yet thoughtless, careless, or stupid decisions frequently lead to disputes, damage to the environment, or under- or overutilization of a valuable resource or amenity. A new vision is needed to see how offshore activities interact with one another and with onshore populations; how they compete; and how we can assess the beneficial and harmful effects of multiple use of ocean space. All of us need to try, at least in a general way, to identify guidelines for safe and socially sound development. And so, as we proceed to look at a wide range of uses of oceanic and coastal space, one of our aims will be to see what strains these uses place on the environment and on the people who take advantage of that crowded space in so many different ways.

In the preceding chapter we saw how the extraction of fish, oil, and minerals from the sea may have adverse effects upon the sea itself. In this chapter we'll consider the question of balancing other uses of the ocean – for transportation, communications, recreation, as a military base, etc. – against the risk of damage to both the water and the coast. Obviously, I cannot discuss all the complexities of regulating and planning procedures, but I shall mention the general problems and draw attention to special conditions and activities that result in uncommon planning problems and require uncommon solutions.

Sea uses of the kind to be considered here have important social effects, and so the planning authority must keep its eye steadily on land as well as sea. It must take into account the structure of coastal populations in

The vast expanses of the oceans constitute a usable resource in their own right. Splashdowns such as the one pictured here present virtually no risk to man or his environment, whereas dumping the domestic and industrial wastes produced by urban societies can do great harm. Far right: As protection from the results of indiscriminate dumping, Baltimore, on the densely populated East Coast of the United States, has erected a boom in an effort to keep waterborne garbage out of its harbor.

terms of age, sex distribution, standards of living, employment patterns, and migration of people into and out of the planning area. For instance, the introduction of a new industry into a coastal zone may involve construction of new power stations and expansion of such services as roads, drainage, housing, schools, and public transport. The effect of all these activities on the environment must be predicted, then monitored, so that pollution can be controlled and overcrowding avoided. And all this must be done within the economic, political, and legal framework of the country and region. Such constraints are common to all planning that isn't dictatorial, but they are especially applicable to sea-use and coastal planning.

Our most intensive use of the sea takes place in offshore waters, particularly in the neighborhood of the world's great deltas and estuaries. As soon as we talk about planning in the coastal zone – Coastal Zone Management (CZM), as it is often called – we have to try to define its landward limits. Clearly, integrated CZM implies control over a huge range of factors that are mainly terrestrial. To build roads or dams in a coastal zone may be no different from building them far inland, but the environmental and social effects are bound to be different. A dam can alter the entire water circulation in an estuary; a road can create a tourist influx that will ruin miles of beaches. So full treatment of the subject of CZM must include the terrestrial factors over quite a wide zone. The landward extent may be determined by local laws, or by the extent of such natural marine phenomena as salt marshes, tidal estuaries, and alluvial plains, or of such man-made phenomena as a continuous

coastal conurbation.

Depending on the case, therefore, the coastal zone may be regarded as extending from one to 50 kilometers inland. We must assume that any marine and coastal planning involves a consideration of the terrestrial area, whatever its width. In these pages, though, I shall restrict explicit discussion to activities that actually take place in the water or that directly affect the water as well as the land.

To assess the full impact on the marine environment of various uses, we must examine each of them singly in terms of its scale and nature before judging their combined effect upon the ocean and coast. For example, waste disposal and the use of sea water for industrial cooling combine to make a heavy demand on the volume of clean water available; shipping and tourism together place a stress on the available surface area of both water and coastline, and so on. Such factors as water volume, sea-surface area, and coastline length are, to a certain extent, independent of each other. Thus different configurations of coast and sea floor define areas that are suitable for different uses.

With this in mind, let us proceed to our examination of the major nonextractive industries and uses of the sea. In noting the benefits, requirements, and risks attached to each activity, we shall see whether it principally affects the coastal zone or the entire ocean. Finally, in looking at all the activities as a complex whole, we can begin to understand the true aim of sea-use planning: how to make the best possible use of a given area with the least possible disturbance of its essential well-being.

Monitoring the Ocean

Virtually no decision at sea can be safely made without accurate information about what is happening, and what is likely to happen next, under, on, and above the surface. Fishermen need information to help them estimate productivity in a given area and to assess future fish stocks, fish migrations, and maximum fishing levels. Even more at sea than on land, workers and managers are guided in planning their daily routines by the meteorologists' studies of air masses and their weather forecasts. Planners of waste disposal and deep-ocean dumping must know the essential environmental factors. Technologists working out plans for extracting energy from the sea can form no opinions until they have mastered the science of oceanography. And although government authorities, both civil and military, do not necessarily publish all their finds, they never stop studying such matters as wind strength, direction, and humidity; the size spectrum and directional spectrum of waves; temperature layers in the water; the speed and direction of currents at the surface and various depths; the tides; the speed of sound at different water levels; and the concentrations of oxygen, nutrients, plankton, and pollutants.

Scientific research, as outlined in earlier chapters, reveals the basic facts about all such variables, the laws that more or less govern them, and even the mean values to be expected in a given area at a given time of year. But the mean value is never an adequate basis on which to build practical programs of fishing, engineering, waste disposal, or military maneuvers. It is the variability that counts. A ship or an oil rig must be designed to cope with extreme storms. Pollutants must be disposed of in such a way as to disperse and contaminate neither the shore nor fishing grounds even in the most unusual current conditions. And to get the kind of information that can provide such security under stress requires closely spaced observations of oceanic conditions at frequent intervals over months or years, followed by statistical analysis on which to base predictions and trends. Continuous monitoring programs of this type are already being funded by oil companies and governments in coastal waters of the Gulf of Mexico, the North Sea, and the Arabian Gulf; but coastal waters comprise only a small portion of the oceans, and the possibilities of more far-flung monitoring have been widely considered in recent years. It will not be easy, however, to transform such possibilities into actuality.

As an example of the difficulties, we can look at weather forecasting. This is supported at present by a worldwide system of thousands of meteorological stations, large and small. With few exceptions, they are land-based, and each of them measures several variables many times a day and transmits them to national and international computing centers. Predictions are then worked out on the basis of sophisticated models employing some of the largest computers in the world. All this must be completed and turned into maps and simple reports for broadcasting within hours; if it took longer, it would not be worth doing. As everyone knows, the results are good – but far from infallible. Forecasting would improve enormously if comparable continuous monitoring could be done at sea.

But to monitor the ocean in this way would entail staggering problems. A small meteorological station on land can be operated part-time by someone on an airfield, or even by an enthusiastic schoolmaster. At sea, weather reports from ships in transit, though useful, are erratic and widely spaced, and they generally come only from established shipping routes. There are special weather ships – four in the eastern Atlantic, for instance – but such vessels are expensive to maintain, and they produce data only from very small areas. Of the 500 oceanographic research ships in the world, about 50 are large enough to be capable of doing continuous work in mid-ocean, but they are dedicated to obtaining original oceanographic data and testing new theories; a monitoring ship must routinely measure the same factors over and over again. The solution to this problem could only be to use vast numbers of unmanned buoys, each carrying a chain of submerged instruments to measure oceanographic factors, each topped by meteorological instruments, each with a radio aerial to transmit the data to land, either directly or via satellite. Well – why not?

One reason why not is that the costs and technical difficulties of building and maintaining such buoys have turned out to be enormous. In 1965, the General Dynamics Company of America experimentally moored what it called its "Monster" – a buoy weighing over 40 tons – in the water off the east coast of Florida, and some government planners were so impressed that they expected the entire ocean to be covered with hundreds of accurate monitoring buoys within a decade. But it has not worked out that way. Now, a dozen years later, there are only 45 small Nomad buoys, each weighing 12 tons, that have been emplaced around the American coast by the National Oceanic and Atmospheric Administration. And a very limited number of experimental oceanographic monitoring and weather buoys are also being operated by government research agencies of Japan, Britain, France, and Germany, usually in their coastal waters.

Among the other problems that have arisen in connection with these experimental buoys are the difficulties of mooring a buoy securely for years at a time in deep water, of providing a long-term power supply, of solving the logistics involved in setting up periodic inspection and maintenance programs, and of ensuring the reliability of components in all kinds of weather. Even if everything else goes smoothly, it has been found that the movements of a buoy and its mooring cables as it is buffeted by the wind and waves introduce errors into the instrumental readings, and that such errors cannot be easily or swiftly corrected. This points up a distinction that should be made between scientific research and monitoring of all kinds (not merely meteorological) at sea.

When a marine scientist designs instruments himself, uses them, and works out the results carefully in the laboratory, whether at sea or on land, he can continually note every time when the instrument was not performing properly or when, for any reason, errors crept in. He can then interpret the results accordingly. But a monitoring system, which by definition must process a very great quantity of data against a constant succession of deadlines, has to detect its own errors and adjust for them automatically – or, better still, be utterly error-free – and there is not time or opportunity for the

This Russian research ship is doing a job that no great nation can neglect: gathering information about every possible aspect of marine environmental conditions. Unlike many other such floating laboratories, however, the *Cosmonaut Vladimir Komarov* also doubles as a satellite tracker, using radar housed in the white domes.

experienced scientist to intervene. Thus, the design of monitoring instruments, their moorings, and the techniques for analyzing their data pose totally different problems from those faced by scientists and their human research assistants. What is needed, of course, is a prior effort by scientists themselves to establish basic principles upon which automated marine-monitoring systems could operate more efficiently than they now do.

Some progress is already being made. Space satellites are an answer strongly advocated by many experts, and satellites are being used to photograph cloud cover over the oceans and to provide information on sea-surface temperatures by measuring them with infrared devices. Aircraft, too, may prove useful. But what is really needed for monitoring the deep ocean is a system located on and in the water itself. An effective, economical system would be a boon to humanity, but we are not yet close to having one. Because the full and efficient monitoring of coastal waters must obviously be achieved first, it will probably be another decade or more before we see the beginning of a true mid-ocean program. When it comes, it will almost certainly be carried on by means of moored buoys equipped with oceanographic instruments.

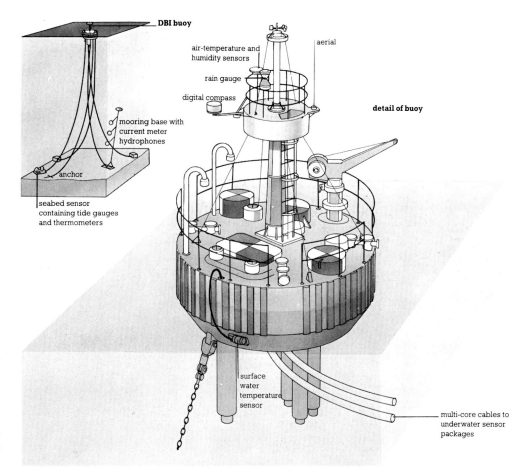

In the late 1960s some observers believed that the ocean would soon be dotted with hundreds of weather-monitoring buoys. They were over-optimistic; such buoys are enormously difficult to build and maintain, and as yet only a few are in operation. The diagram above shows the variety of equipment and sensors fitted to Britain's 40-ton DB-1, which radios information to shore from its North Sea mooring. A comparable system is at work in the U.S. Navy's huge (60-ton) Alpha buoy, left, now anchored off the California coast.

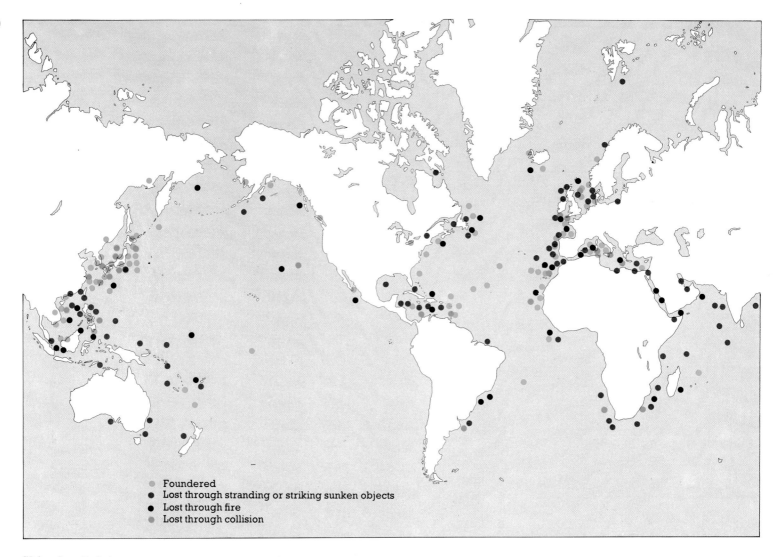

- Foundered
- Lost through stranding or striking sunken objects
- Lost through fire
- Lost through collision

Shipping Safety

The safety of a ship at sea depends not only on a solid understanding of water and weather conditions, but on such factors as the engineering design of the ship itself; adequate safety devices in the form of fire prevention, lifeboats, etc.; universally accepted standard systems of radio frequencies, lights, distress signals, and rules of the road; certified qualifications for officers and skilled crew members; and relatively high standards for ordinary crewing and working conditions. Considering the amount of traffic that the sea has had to bear in this century, bad accidents are wonderfully rare. Yet the dangers are manifold, and everyone can think of a dreadful example of one or another kind of disaster: collision on the high seas (Italy's passenger liner *Andrea Doria*, which went down near Nantucket Island in 1956 after being hit by the Swedish ship *Stockholm*); collision with an iceberg (the *Titanic* in 1912, with the loss of 1,500 lives); fire at sea (the *Morro Castle*, which sank off the New Jersey coast in 1934); explosions in empty oil tanks (there have been several notorious examples in the 1970s); the gradual destruction of vessels in severe storms (the Athens-

Crete ferry in the 1960s); and so forth.

To a certain extent, the number of accidents that cause loss of life or property, or damage through pollution, is bound to increase as the number of operating vessels increases and their tonnage also grows. Constant monitoring of international shipping by the International Maritime Consultative Organization (IMCO), which was established some years ago by the United Nations, is designed to prevent trouble, and most seafaring nations abide by all IMCO regulations and agreements. Unfortunately, though, a few countries – notably Liberia and Panama – provide flags of registration for ships without requiring full compliance with international standards. The problem of such types of noncooperation is discussed at length in the concluding chapter of this book. What I want to set forth here is a broad picture of our present use of the oceans for shipping, with some of the problems that this involves, along with some of our methods of solving the problems.

Total tonnage of world shipping is increasing at 8% a year, which is enough to double the tonnage in under ten years. Because of competition from airlines, the increase over the last two decades has been accompanied by a continuous decline in

This map shows positions of some of the 363 ships over 100 tons gross that were lost in one recent year (1974). Most went down in the crowded waters around Europe and the Far East. Losses nearly doubled between 1961 and '71, but dropped slightly in the early '70s.

passenger traffic, but this has been more than offset by the spectacular growth in size and number of oil tankers. The boom in tanker activities results, of course, from the fact that the industrial economies of Europe, America, and Japan require more and more oil from the Middle East, North Africa, and Venezuela, in spite of an occasional slowdown such as the general recession in the mid-1970s. Tankers began to be built on a larger and larger scale after the Suez Canal was closed in 1967 and it became necessary to transport vast quantities of oil for the West all the way around the Cape of Good Hope. Many modern supertankers are too large to go through the canal even though it is now open, and so the round-the-Cape route continues to be used and big tankers to be built. By now tankers account for almost half the world's gross tonnage of seagoing ships!

A world map of shipping routes and density of traffic provides a quick view of where the concentrations are thickest: in the Atlantic as a whole; throughout the Carib-

The development of such supertankers as Sweden's 356,000-ton *Sea Scape* (left) has brought additional navigational hazards into busy shipping lanes, for huge vessels react sluggishly when changing direction or speed. Above: With exploration and production of oil moving into arctic areas, icebergs pose an increasing threat to safety. Here, in Labradorean waters, an iceberg is being towed out of the route used by supply ships.

bean and Gulf of Mexico; around the coast of Africa; across the Mediterranean; through the English Channel; and from the Arabian Gulf, around India and Sri Lanka, through the Strait of Malacca to Japan. Compared with these areas, the rest of the seas are relatively empty. In a later section I'll be discussing the question of traffic separation in narrow straits. But it must be obvious to the reader that even on the open seas some agreement on routing is not only desirable but increasingly necessary. Such agreements are in the economic interest of every ship operator, for they help him avoid fog, ice, and storms as well as collisions.

The idea of prescribed shipping lanes is not new. It was proposed, in fact, back in the mid-nineteenth century by an officer of the American Navy, Matthew Fontaine Maury, and put into practice first for sailing ships, then for steamers. In 1898, all major passenger lines adopted a formal North Atlantic Track Agreement, which has been in force in the Atlantic, with periodical modifications, ever since. The prescribed routes for east-, west-, north-, and southbound surface craft take into account the weather, danger from ice and fog, and the presence of fishing fleets in certain areas. And, of course, similar agreements now apply in the Pacific and other oceans. Today, moreover, special routing attention is being given to tankers, whose cargoes present a potentially lethal threat to nearby vessels. A number of recognized sea lanes have been earmarked for them on all the main oil-trade routes from the Arabian Gulf to Europe, from the Caribbean to Europe and North Africa, and from the Middle East to Japan and America.

In spite of much-publicized incidents of the grounding of big tankers and of explosions aboard them, there is no inherent reason why a single ship should be more likely to cause an accident than ten small ones. Still, there are problems that have arisen because of innovations in design and structural techniques necessitated by the unprecedented size of such craft. For instance, it has taken years for tanker operators to learn how to control the danger of explosions in large empty tanks full of oil vapor. Nowadays such tanks are washed out with the utmost care so as to prevent the build-up of static electricity, and the highly charged atmosphere in them is replaced with a safe mixture of inert gases from the ship's engine exhausts. Technically, this sort of danger is entirely controllable. It should no longer *be* a danger unless eco-

nomic pressures force the employment of inadequately trained crews.

One drawback, however, to very large ships – those over 100,000 tons – is a worrying one, and we have not yet been able to work out a technique for overcoming it: when a ship of this size moves through the sea, it entrains with it a huge mass of water, which adds to its inertia; and so any attempt to turn quickly or slow down suddenly meets with little success, since the rudder or reversed propeller has to operate within the moving mass of water. Even if the ship manages to turn its bows off the original direction of movement, it tends to continue sliding sideways in the direction of travel. Five to ten kilometres are needed as adequate room for such ships to slow down in. They must be navigated with particular caution in coastal seas and straits, but they require skilled seamanship in any water.

Safety at sea for all vessels demands eternal vigilance. Fortunately, though, this is one aspect of sea-use planning that gets more than mere lip service from almost every organization and individual concerned with marine matters. After all, most of the world's trade cargoes are carried by sea – and the cheapness of sea transport will keep things that way.

Canals and Straits

Canals and straits are channels not only for ships but also for the water itself, the plankton, fish and other swimming creatures, and the sediment suspended in the water. Most canals and straits also serve as busy cross-channel communications routes involving cables and pipelines as well as passenger and freight traffic. There are almost always special problems of congestion and conflict of interest in the region of canals and straits, and modern technology aggravates some of the problems while providing the means of solving others.

Three big sea canals have been in existence for many years. The Suez Canal from the Mediterranean to the Red Sea was opened to the strains of Verdi's *Aida* in 1869, Greece's Corinth Canal in 1893, and the Panama Canal from the Atlantic to the Pacific in 1915. The first two are sea-level canals through which water, plankton, and fish can travel freely, but this is not the case with the Panama Canal. The distinction is an important one from the standpoint of ecology. For example, some Red Sea species of fish and sharks that are known to have migrated through the Suez Canal are threatening to replace Mediterranean species. Ecologists consider it a matter of accidental good fortune that the Panama Canal poses no such threat to the oceans it joins together by cutting through the Isthmus of Panama.

The Panama Canal was a much bigger construction problem than the others, partly, of course, because of the tropical climate, but largely because of the hilly terrain. It takes six sets of 300-meter-long locks to raise and lower ships across the hills of the isthmus; and those locks are neither wide nor deep enough to accommodate today's big ships. Because of their limitations (not to mention the delays and

costs of the lock system), there has been a good deal of serious talk about digging a more ambitious sea-level canal at a point where the isthmus is lower, though wider. To accomplish such a task rapidly and economically, something other than conventional explosives, bulldozers, and excavators would be needed, and, a decade ago, American engineers were considering the feasibility of a controlled series of underground nuclear explosions to dig the initial ditch. In spite of assurances from experts that there would be no radioactive fallout, and in spite of evidence that the Russians had used nuclear explosions for excavations in Siberia, the American people wisely concluded that an experiment on this scale was unacceptable. One argument against building a wide sea-level canal by whatever method came from the ecologists: there is no telling what might happen if it became easy for species to migrate between the tropical Atlantic and the tropical Pacific. Such migrations would have an incalculable effect on fisheries, algal growth, fouling of ships and harbors, and the growth of coral reefs. The plan for a sea-level canal has now been shelved and seems likely to stay on the shelf.

It is a striking fact that no major sea-to-sea canals have been built since 1915. A few years ago, work was begun on a project that was to join the Gulf of Mexico to the Atlantic across northern Florida, but it was quickly stopped as a result of pressure from conservation groups that convincingly stressed the probable damaging effects of the canal on the swamps and marshes that give Florida such a rich natural wildlife. But the ecological argument, strong as it is, would not deter canal builders if they had much faith in the profitability of new canals. In fact, the economics and technology of shipping are changing so fast that it is questionable

Above, left: The 6-kilometer-long Corinth Canal facilitates passage between the Ionian and Aegean seas not only for ships but for marine animals and plants. Such short-cut migration of salt-water life from one sea to another is not possible via the Panama Canal (above), since its freshwater locks prevent the movement of marine creatures through the 82-kilometer-long waterway. That's all to the good, for species newly introduced into strange waters might displace indigenous ones.

whether a canal built today will be either needed or usable in ten years' time. The cost-saving to shipping lines of large, highly automated ships with small crews is so great that it is cheaper, for example, to send one 300,000-ton tanker around the Cape of Good Hope than to sail six smaller ones through the Suez Canal. In this context, too, there may be political resistance to canal construction simply because canals attract ships, and thus compound the problem of marine pollution. Taking the Suez Canal as an example again, if current proposals to widen and deepen it to accommodate larger tankers were implemented, oil pollution in the· Mediterranean would undoubtedly increase. So the enlargement might be fiercely opposed by many Mediterranean states.

The chief difficulty presented by natural straits, such as the Florida, Dover, Gibraltar, and Malacca straits, is that they are becoming increasingly unsafe because of traffic congestion. In the open ocean a ship is considered dangerously close to another if the separation is less than several kilometers. In a crowded strait, with ferries and fishing vessels zipping in and out of the dense through traffic, there are appalling risks of collision or grounding. The imposition of one-way traffic rules is only a partial answer to this problem. By 1976, some 60 traffic-separation schemes were in operation around the world's oceans, but most of

Completed in 1973, this 1,500-meter-long suspension bridge across the Bosporus at Istanbul, in Turkey, where Europe meets Asia, provides a splendid new intercontinental land link, thus easing traffic on the busy waters of the narrow strait.

them were voluntary, not mandatory. In the few that *are* mandatory, as in the Strait of Dover, enforcement is extremely difficult; ships' captains are reluctant to surrender any of their traditional total authority at sea.

The best hope for minimizing shipping risks in the busy straits may lie in the building of more high bridges and undersea tunnels, which can carry much of the traffic now being transported by cross-channel ships. One such tunnel, by far the longest in the world, is now under construction from the main Japanese island of Honshu to the northern island of Hokkaido. When completed – the target date is 1979 – this remarkable tunnel will be 54 kilometers long, 23 kilometers of it under the sea. At its deepest point it goes 100 meters below the sea floor, which itself lies 140 meters below the surface.

A famous recent achievement in bridging sea straits was the completion in 1973 of the Bosporus Bridge, a single suspension span that links Europe and Asia. A much more complex project is again Japanese: a series of bridges connecting Honshu with the central island of Shikoku across several smaller islands of the Inland Sea. This project is already under way, but will take many years to complete. Even more complex are the engineering problems involved in a proposal to join Africa and Europe by a bridge across the Strait of Gibraltar. Since the strait is 20 kilometers wide and becomes 200 meters deep at the point where it could best be spanned, a conventional bridge is out of the question. Engineers have proposed that the spans of the bridge be suspended from pylons supported on anchored, floating islands rather like the semi-submersible oil rigs described in chapter 5 – only much bigger.

It is hard to predict whether or not some such bridge will ever be built. An additional complexity for any group that attempts to plan how to make the best possible use of straits is that they are often bordered – as is the Strait of Gibraltar – by different countries, with perhaps opposing interests.

The Japanese are at work on an unprecedentedly ambitious maritime bridge-and tunnel-building project. As indicated in the maps, some 20 bridges are planned, and these will eventually hop between many small islands to join the major islands of Honshu and Skikoku; and the world's longest railway tunnel – about 54 kilometres – will link the town of Aomori in Honshu with Hakodate in the northern island of Hokkaido. The photograph shows an already completed segment of the island-bridging project.

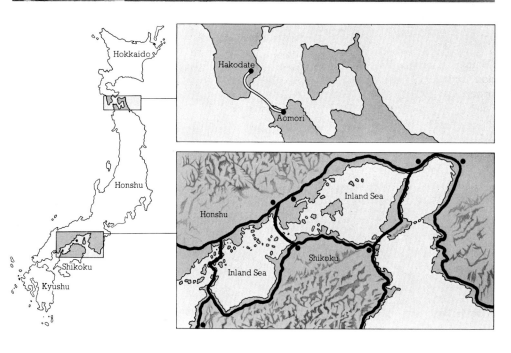

Three of the world's ports – Rotterdam, New York, and New Orleans – handle annual totals of more than 100 million tons of cargo each, and thirty others handle more than 30 million tons each. The enormous quantities of goods that converge on these and many smaller city ports bring a number of problems in their wake – among them, congested shipping lanes, high concentrations of industrial and domestic pollution, and the often frustrated recreational demands of large local populations that expect to find attractive areas of coast within reasonable reach. Because of such conflicting interests, there has been a recent trend toward the development of new harbors far away from centers of population, where such specialist materials as oil or iron ore can be handled in virtual isolation. This trend, in turn, is leading to problems of its own. To see why, let's begin by looking at the general requirements for any good harbor.

A harbor is the natural coastal configuration that provides shelter for ships, and a port is the assemblage of docks, quays, cranes, warehouses, railways, roads, and offices that provide the necessary servicing for the ships and their cargoes. Several small harbors can be conjoined into a single port (Piraeus in Greece is an example), or a single large harbor, such as San Francisco Bay, may contain several different ports. But it is very expensive to build a port on a coast where there is not good natural harborage. The task is made easier in a semi-enclosed sea with diminished wave heights, where ships can be anchored or moored to jetties with relatively little protection, as in the Arabian Gulf or the Gulf of Paria (between Trinidad and Venezuela). If a country has a straight coast, it can build a port only by constructing exceptionally big breakwaters to protect ships, as Israel has done at Ashdod. The best natural harbors are found in fjords, as at Stavanger in Norway; in large river mouths, such as those of Rotterdam or London; or where several rivers flow into a broad indented bay, as in New York City, San Francisco, Chesapeake Bay, Sydney, or the Welsh port of Milford Haven. It is this last type of configuration, known as a "ria," that is most favored by modern engineers for building new ports to cope with modern ships that draw 20 or more meters of water. And the overriding necessity of finding and exploiting new rias accessible to land transport raises an immediate and obvious problem: an outcry from conservationists, who object to the defacing of regions formerly valuable chiefly for their unspoiled beauty.

Yet the search for new harbors continues. Not only does the traditional combination of a large port with a large city involve inherent social and technical conflict, but

technological developments of recent decades have meant that ports built before the 1950s are very difficult to adapt for modern usage. Apart from the increased draft and size of ships, especially tankers, which require much deeper water and much more maneuvering space than they used to, modern shipping methods call for bigger docks and dry docks, larger oil-pumping and oil-storage "farms," and enormous flat areas for stacking containers. The container is a very big standardized metal box that can carry great amounts of many kinds of cargo. Specialized cranes can unload or load a container ship in a matter of hours, thus cutting the turnaround time enormously. But the containers cannot be as swiftly whisked onto or off their trucks and trains. The price of turning ships around fast is that many acres of land near the dock have to be available for storage of containers that are awaiting either rapid loading onto a ship or slow loading onto vehicles with a number of different terrestrial destinations.

Add up the requirements – deeper water, wider basins, more acres of empty flat land – and you can see why a once-booming port in the heart of a city – London or Liverpool, for instance – simply cannot be adapted to late-twentieth-century traffic. Even in more suitable sites such as Rotterdam and New York, where the natural estuary and harbor are large enough to accommodate the necessary expansion and modernization, this can only be done by the expenditure of enormous sums of money. And is the expansion actually worth the expense in view of the fact that so many cargoes such as oil, gas,

Rotterdam, the world's largest and fastest-growing port, services over 32,000 ships a year. Situated at the mouth of the converging Rhine and Maas rivers, Rotterdam provides dredged deep-water moorings that enable even the biggest ships to load and unload close to Europe's main industrial markets. This Dutch port handles all types of general and container cargo, as well as oil, ores, and grain.

and possibly iron ore and coal are often going to a single "consumer" – a pipeline, refinery, power station, or giant steel mill – which is unlikely to be located in the port city itself? Indeed, there is no good reason in the modern world why the port for many a commodity such as those I have mentioned should be built anywhere near a big city. This reasoning has led to the increasing exploitation of natural harbors as terminals for specific kinds of cargo. Thus, Bantry Bay in southwest Ireland has become a busy focus for supertankers, Fos-sur-Mer in southern France (near Marseille) for iron ore and oil.

Because transport by sea is so much cheaper than by land, planners who are establishing new industrial complexes designed to exploit the economies of large-scale production and distribution tend automatically to think of siting these industries on the coast and fully integrating them with a new harbor. It is easy to anticipate the end result of such huge complexes: in spite of the fact that automation means that modern refineries and steel plants employ fewer people than they did twenty years ago, the location of an industry joined to a port in a new site inevitably draws an increasing population to the area. What we are seeing, therefore, is the birth of new

Specific kinds of cargo require special facilities. Above: This enormous crane is used for loading containers up to 12 meters long at London's docks. Above, right: Submerged pipelines in Ireland's Bantry Bay carry oil from tankers to onshore storage tanks. The tankers may well have loaded their oil from a single-buoy mooring (right); fed from adjacent oil fields, such moorings obviate the need for very long pipe lines and harbor installations.

cities in what used to be sparsely populated places. In other words, it's back to square one again.

Ocean-going ships used to spend half their life tied to the land. Now that a supertanker of nearly half a million tons or a container ship of 20,000 tons can be turned around in hours – or, at most, in a couple of days – it can spend most of its life really earning its keep. The relocation of ports to adapt to such ships is bound to cause further movements of industry and consequent further movements of population. Very far-sighted planning is needed in order to assess in advance the impact of the new ports on the human and natural environment both onshore and off. In most of the world such planning is not being done. The machinery of local government is poorly geared for coping with integrated planning of developments ranging from roads to docks to the sea floor out to ten or twenty kilometers from the coast. Yet, since only that kind of planning will do the U.S. Federal Government is now offering grants to U.S. coastal states to support coastal-management planning, and similar government pressures are being applied in several European countries.

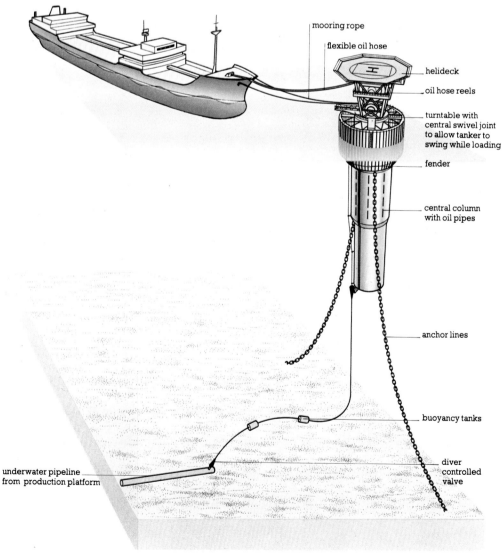

mooring rope

flexible oil hose

helideck

oil hose reels

turntable with
central swivel joint
to allow tanker to
swing while loading

fender

central column
with oil pipes

anchor lines

buoyancy tanks

diver
controlled
valve

underwater pipeline
from production platform

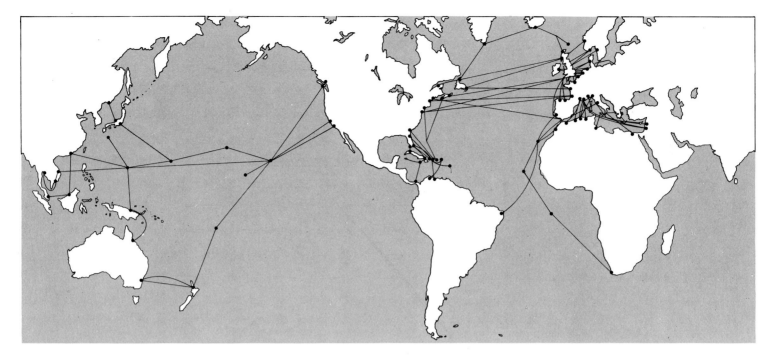

Cables on the Sea Bed

Submarine cables have three major functions: through them are transmitted large quantities of electric power, or of pulsed telegraph signals, or of telephonic voice communications. They have a number of other uses as well, ranging from the control of oceanographic instruments to the furnishing of power to listening posts for tracking submarines. But the vast majority of the world's great network of marine cables are devoted to the single task of providing mankind with swift, easy transoceanic means of communication. Each type of activity requires its own special design of cable. Whatever the design, though, all marine cables are alike in needing extraordinarily efficient insulation, for salt water conducts electricity, and the slightest leak in a cable would instantly destroy its ability to function.

When cables are laid, they sink to the bottom under their own weight. And, particularly in continental-shelf depths, their position much be charted accurately, so that ships can be advised not to anchor in the vicinity and trawlers and dredgers warned against picking up or cutting the wire. Damage of the latter kind has become so common in the last decade that such cable-laying authorities as the British Post Office or the American Telephone and Telegraph Company have resorted to chartering submersibles with water-jet pumps to bury their cables in trenches up to half a meter deep. This isn't necessary in the deep ocean, of course; cable burying is usually carried out only from the shore to the edge of the continental shelf, in depths of up to 200 meters. The problem is steadily increasing, however, in some coastal seas as pipelines for gas and oil join with the cables in a maze of sensitive zones on the sea floor.

Submarine telephone cables cross almost all oceans to link up with land lines – though, as the map shows, some areas are better served than others. The first transatlantic phone cable was not laid till 1956, 90 years after the establishment of transatlantic telegraphy.

Subsea power cables designed to carry megawatts of electricity between far-flung areas of a large grid system are the least common of the three main types. Examples are the power cables that connect Britain with France or that run between the north and south islands of New Zealand. Such cables allow peak power loads to be shared over a wide range of power stations; and when peak loads occur ar different times, power can be transmitted either way to even out the load on each side of the water. But the power cables do not, obviously, need to be laid across very wide expanses. It is for telegraphic and telephonic communications that the technology of marine-cable laying has been largely developed.

A subsea cable is not just a long, strong, thick piece of insulated wire. It is a complex bundle of conductors, insulating materials, strengthening sheaths of steel and plastic, and waterproofing. The electrical signals that make up a voice message are bound to be distorted and weakened as they pass along the cable and are mixed with random electrical noises. And so, to improve the signal, small amplifier stations (called repeaters) are built into telephonic cable at intervals of 50 to 100 kilometers, and power must be provided for these stations. In modern telephone cables, power is generated within each repeater by heat from a radioactive isotope. This is quite different from what goes on in a nuclear reactor, where enormous temperatures are reached. All that happens in the repeater is that as the isotope slowly disintegrates

over a period of 20 years or so, the sub-atomic particles pass through a surrounding thick screen of material, warming it up in the process; and the heat generates electricity by direct thermoelectric processes without any moving parts.

The present world network of undersea communication cables consists of over a million kilometers of telegraph and 200,000 kilometers of telephone cables – a vast system that has been built up during the past 100-odd years by slow and laborious stages. The first subsea telegraph cable was laid in the mid-nineteenth century, when Morse code was used for transmitting messages. A transatlantic link between the New World and Europe was established in 1866 by a Newfoundland-to-Ireland cable, which was laid from the converted hull of a very famous ship designed by the English engineer I. K. Brunel – the unprecedentedly large 22,000-ton passenger steamship *Great Eastern*. By the 1860s it was already possible to send Morse code over thousands of kilometers of submerged wire because the abrupt on-off pulses of the code suffer much less deterioration in intelligibility than do the variable signals needed for transmitting vocal sounds. Later improvements in pulsed coding enabled much more complex information to be carried than the Morse system was geared for, and the improved telegraphic systems are still in constant use for sending transoceanic telegrams, stock-market reports, and routine data of all kinds.

It may surprise some readers to learn that nearly a full century passed between the *Great Eastern*'s exploit of 1866 and the laying of the first transatlantic telephone cable, but it is a fact that this latter achievement was not realized until 1956. As long ago as 1891, telephone conversations between France

Left: While in port, men of the British cable-layer *Mercury* wind in part of the 1,650 kilometers of 37-millimeter cable that will fill the ship's three storage vats; the work must be done entirely by hand. Below: To improve vocal clarity, amplifiers, or "repeaters," are built into subsea telephone cables. This one, along with 500 others placed at 100-kilometer intervals, was incorporated in a cable laid between England and Canada. Repeaters allow nearly 2,000 voice circuits to be carried without distortion across 5,000 kilometers of sea bed.

and Britain were made possible by the laying of a short cross-Channel cable. There were, however, two obstacles to deeper and longer cables: the difficulty of protecting the fragile multi-strand conductors and insulators from water pressure, and the inability to cope with vocal distortions. The pressure problem was solved early in the twentieth century, and in 1924 a 160-kilometer-long cable was strung along the sea floor, at a maximum depth of about 1,000 meters, between Florida and Havana. By 1927, radio telephony had made across-the-water vocal communication relatively easy without the need for underwater wires, but the interference and distortion of long-distance radio kept research engineers on their toes until eventually the subsea repeater was perfected. In 1943 repeaters were successfully installed in a subsea cable across the Irish Sea from the northern coast of Wales to the Isle of Man – a distance of around 100 kilometers – and then, in 1950, in another link between Florida and Havana. The cable laid across the Atlantic in 1956 went from Scotland to Newfoundland. It consisted of twin links, with repeaters every 64 kilometers, and with each of the two links capable of carrying 36 simultaneous circuits. Since then the whole world has been criss-crossed by cables that carry many hundreds of telephone circuits, and there are few remaining salt-water barriers to long-distance gossiping.

Subsea cable is now laid by specially constructed vessels, usually with two propellers aft and one forward to enable them to hold a steady position even in very heavy seas. The hold of a cable-laying vessel consists of large vats for storing cable, and as much as 4,000 kilometers of continuous cable is coiled down into the vats by hand before the ship sails. Because of the obvious need for careful coiling, it generally takes more than two weeks to load the ship. At sea the cable is winched out at a maximum speed of some 8 knots (12 kilometers an hour). At that rate a single cable-laying job may last a month or more.

The same vessels are used for servicing the cables. If an undersea cable breaks or deteriorates internally, electrical tests on land determine the approximate location of the trouble. A ship proceeds to that spot, grapples for the cable, and then, by means of grapnels designed for the purpose, cuts it and brings both ends to the surface. After making tests to find out in which half of the cable the fault lies, the engineers seal off and buoy the good end. Thereafter, the vessel continues to haul up the faulty end until tests show that she has passed the point where the fault lies. The rest is easy: the faulty section is cut out and a new length of cable spliced on. The repaired cable is then once again dropped to the ocean floor to carry on its service to our global communications network.

Since 1960, when the first Polaris submarine was launched, the balance of terror between the great powers has depended largely on the possession of apparently undetectable submarine weaponry. Every year more money is spent on designing submarines that will go deeper, quieter, faster, and farther, and from which deadlier missiles can be launched over greater ranges. On the defensive side, meanwhile, more money is also being spent on huge arrays of listening devices, heat sensors, wake detectors, and satellite surveillance of harbors.

This ghastly game of hide-and-seek depends on the fundamental properties of sound and water. As we saw in an earlier chapter, shortwave radio does not penetrate water, and so radar systems cannot track enemy submarines. The only reliable way to do so is by either picking up the sound of the submerged craft or bouncing a projected sound wave off its hull. But sound does not travel underwater in straight lines; it bends and curves, bounces off the surface and the floor, and leaves great gaps where, from a given listening or projecting point, no sound at all can be heard. The paths of sound, and hence the dead points, vary constantly, for they are determined by the density and temperature structure of the ocean at any moment. Thus, a submarine crew that knows the oceanic structure in its neighborhood can often exploit it to stay hidden. Conversely, an antisubmarine force with up-to-the-minute information on subsea weather can dunk its listening devices at carefully chosen spots to get maximum coverage of the volume of ocean nearby, with the fewest possible dead spaces.

Four nations – America, Russia, Britain, and France – now operate submarine fleets capable of launching nuclear missiles, and the same four have corresponding programs of antisubmarine warfare (ASW for short). Many nuclear missiles, such as America's Polaris and Poseidon, can be launched underwater and guided with an accuracy of better than two kilometers to land deep within a continent. This threat against population centers is the ultimate strategic weapon in the maintenance of our current brand of heavily armed peace. At present the U.S. Navy has about 120 submarines of all types, including 40-odd nuclear-powered missile launchers, while the Russian underwater fleet consists of close to 350 vessels, of which roughly a hundred are nuclear missile launchers – a fact that gives bad dreams to ASW experts in the Western alliance. The imbalance is not as startling as it may seem, however, for a number of the Russian missile launchers cannot launch their rockets while submerged, and many of the U.S. missiles carry

from 3 to 14 independently targetable warheads. In fact, a single American sub can launch as many as 224 individual nuclear warheads within five minutes while remaining concealed beneath the surface.

On paper, that looks like the perfect nuclear deterrent: no country could risk a nuclear attack on another for fear of massive submarine retaliation. Unfortunately, the nuclear submarine also represents a potential first-strike weapon of inestimable power, and no nation enjoys being aimed at by lethal machines whose undersea positions they do not know. The science of antisubmarine warfare therefore assumes increasingly colossal importance, with ever-more-sophisticated computers being brought to bear on predicting the subsea environment, calculating sound paths, and distinguishing the scores of possible targets from random noises and freak echoes. Enormous strings of hydrophones, or listening posts, are now deployed at strategic points both along the ocean floor and at certain depths in which sound is known to travel better than in others. And, of course, surface ships and aircraft of NATO, Warsaw Pact, and other nations constantly patrol strategic areas using all the latest devices for detecting underwater activity.

One result of the elaborate cat-and-mouse games is that the arms race keeps pushing deeper below the ocean surface. For every extra hundred meters that a submarine can dive, there are an extra 10 million cubic kilometers of water in which to hide. So it is not surprising that of the many presumably "nonmilitary" modern submersibles known to be able to dive below 500 meters,

A crewman in Britain's Polaris submarine *Resolution* monitors the preparedness of its 16 nuclear-tipped missiles. More and more sophisticated devices are being developed to track down undersea launch platforms, which are themselves equipped to minimize detection.

all but four have been, are being, or will be used for military research. One aim is to develop submarines capable of reaching much greater depths than is now possible. Another, less immediate aim is to supplement the nuclear submarine at some future date with stationary undersea nuclear arsenals that could be serviced and manned by military personnel living in underwater "habitats" protected from the water pressure.

Seamounts, which are natural features of the sea bed – flat-topped mountains standing at least 1,000 meters above the general level of the floor – are therefore of potential interest to military planners. In 1964, the American Navy built its first experimental habitat on a seamount off Bermuda. This was known as the Sealab 1 experiment, and the inconclusive outcome of a series of such experiments with undersea living quarters is discussed in chapter 8 of this book. The Americans once hoped that by the early 1980s they would have an operational habitat, within which normal atmospheric pressure would be maintained at a depth of 2,000 meters, somewhere on the Mid-Atlantic Ridge, but that now seems unlikely. Still, many projects of this nature continue to be investigated. The U.S. Navy's Project Rocksite, the most ambitious of them, envisages the drilling of tunnels from land to an established base in the deep ocean. One

The game of underwater hide-and-seek is highly complex. On the ocean floor, chains of hydrophones (1) search for elusive missile-firing (2) and attack submarines (3), which are themselves trying to find each other. The mid-water is scanned by long-range sonar arrays (4) fixed to the edge of the continental shelf. On the surface, submarine chasers (5) and static buoys (6) help the search; shipborne helicopters (7) dunk sonar buoys (8) for local detection. Above the sea fly medium-range anti-submarine planes (9); long-range shore-based aircraft (10 and 11) drop sonar buoys (12) far out at sea. Control of the hunt is assisted by man-made satellites (13).

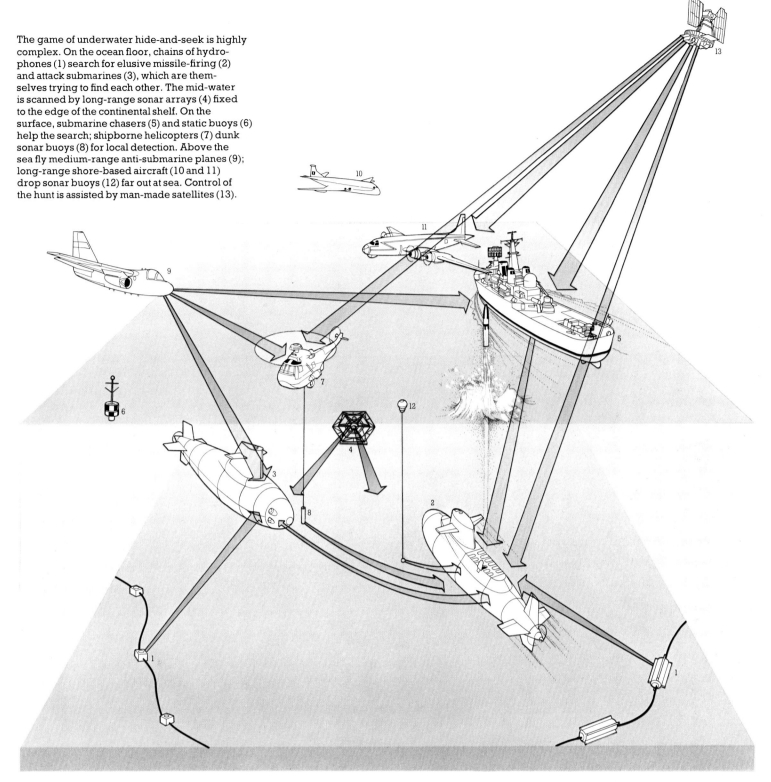

Navy expert has outlined how such a base could support a population of 50 men, who would get oxygen from electrolysis of seawater and power from a nuclear generator.

Even if underwater bases did not replace submarines, they could be of great strategic significance, for subs could be serviced underwater and need never come to the surface. That would remove what is still the easiest means of checking the whereabouts of undersea enemy craft. The only trace of such a system would be a series of underwater hatches on the deep-ocean floor.

We know much less about the Russian undersea-planning program than about the American. The Soviet Union certainly has a massive program of physical and chemical oceanography to support her fishing and military activities, and most of the data could doubtless be used for both. We know too that she constantly conducts gravity surveys and magnetic and topographic surveys on a global basis, all directed at improving the navigation and control of missiles and missile launchers. But we can only guess that her long-range planning is likely to be no less thorough than that of her rivals. All in all, the military aspect of sea-use planning is by its very nature too secret and too vast to be anything but a source of

nightmares all round. The possible occupation by rival powers of certain strategic areas of the ocean floor and of strategic seamounts is sure to complicate an already difficult problem of international legislation, in which attempts are constantly being made to preserve the world's oceans for the well-being of mankind. And the current atmosphere of mistrust, suspicion, and undersea menace casts a shadow of doubt on research ships of all kinds, no matter how innocent their stated objectives. No coastal state, alas, can be sure that the so-called "oceanographic" research of another state does not have some arcane military motive.

Planning for Disarmament?

In 1967 the tiny state of Malta asked the UN General Assembly to debate a proposal that the sea bed should be reserved exclusively for peaceful purposes. This proposal led through slow degrees to the Sea-bed Arms Limitation Treaty of 1971, which has been ratified by all the great powers except France and China. That treaty is, however, a very weak restraint on the undersea arms race, and the reasons are not far to seek. Apart from the obscurity that veils much of the vast battlefield upon which "cold-war" undersea warfare is being waged, there is the never-ending diplomatic maneuvering that thwarts all who genuinely yearn for disarmament.

To start with, we need to clarify what is meant by disarmament. In theory, disarmament simply involves a general reduction or abolition of existing forces or weapons. Under the sea this should mean, at the very least, restrictions on the numbers of submarines, missiles, and antisubmarine-warfare installations, but the term "disarmament" is usually applied only to offensive weapons. The phrase "arms control" is a more general expression. It implies some form of agreed limits to numbers of offensive weapons, together with a system of inspection or verification; and to suggest that the sea floor be used only for "peaceful purposes" further implies a banning even of acoustic listening devices. Since many countries other than the superpowers are now capable of installing undersea listening devices, and since most of them believe that it is only prudent to do so, the Maltese proposal of 1967 may have led to a widely ratified treaty, but the treaty has come nowhere near fulfilling the original aim.

Any agreement to control arms under the sea has to overcome the following obstacles: the old-established principle of freedom of the high seas; the technical problem of monitoring what is happening in the deep ocean; the economic rivalry among states competing for offshore resources; and the lack of any existing form of authoritative control by government policing or civil inspectorate outside territorial waters. The legal problems are discussed in a later chapter of this book. Here let us examine some of the technical aspects of undersea arms control.

During the early phases of discussion of the issue, it was common to compare the question of sea-bed disarmament with such

The aim of SALT (Strategic Arms Limitation Talks) is to prevent either Russia or the U.S. from amassing overwhelming numbers of long-range weapons. But agreement depends on defining the word "strategic," and the talks are made more difficult by the development of such new arms as America's Tomahawk sea-launched cruise missile – here being test-fired.

existent treaties as the nonmilitarization of Antarctica (1959), the partial test-ban treaty (1963), the banning of orbiting nuclear weapons in outer space (1967), and the nuclear nonproliferation treaty (1968). In each of those cases, however, it is not at all difficult – if a bit cynical – to show that the agreements merely prevent the superpowers from doing what they do not want to do anyway. The Antarctic is an easily defined area, and not strategically valuable; by the 1960s the testing of nuclear weapons both in the atmosphere and underwater had become manifestly dangerous, and underground tests have continued at such a rate that there has been no reduction in total testing; any orbiting nuclear weapons circling the earth could rapidly be identified by the enemy and shot down by rockets; and nonproliferation of nuclear weapons – which means merely not selling them to other nations – preserves the dominance of the big powers.

In spite of the enormous sums that the two superpowers are now spending on antisubmarine warfare, they cannot reliably track each other's nuclear submarines, and so it is obvious that no international agreement requiring supervision of arms control under the sea would have any chance of success unless both America and Russia wanted it to succeed. If some miraculous improvement of tracking techniques made it suddenly possible to scrutinize all movements of submarines, there might be a change in attitudes. At the moment, though, Britain and France as well as the superpowers are deploying nuclear missile-launching submarines that cannot be tracked, and, as a corollary, they all reserve the right to deploy tracking systems and are cool to the idea of any international sea-bed agreement with teeth in it. As a consequence, the arms-limitation treaty of 1971 binds the nations only ''not to emplant or emplace on the sea bed and the ocean floor and in the subsoil thereof . . . any nuclear weapons or other types of weapons of mass destruction as well as structures, launching installations, or any other facilities specifically designed for storing, testing, or using such weapons.'' Thus, the prohibition applies solely to offensive weapons; and since – as of now, at least – fixed installations or launching sites for offensive purposes would be of much less strategic value than mobile ones, the agreement places no great restraint on the undersea arms race.

The Strategic Arms Limitation Talks

(SALT), of which the first phase was agreed between the U.S.A. and U.S.S.R. at Vladivostok in 1974, places an upper limit on the number of missiles with nuclear warheads that the two powers may hold. This does mean that there is some sort of limit to the quantity of nuclear weapons likely to be deployed under the sea, for both countries must undoubtedly retain many missiles for launching from land bases. But such limitations do not preclude improvements in accuracy to make the missiles more deadly, nor improvements in their navigational control to make defense more difficult. In this context, America's development of the

cruise missile, a long-range missile that can be launched either underwater or on land, and that can be navigated remotely throughout its flight, must be an unnerving event for the Russians, who are unlikely to agree to further controls until *they* have a better missile.

As if all this were not pessimistic enough, the major undersea nuclear powers are no doubt sensitive to the probability that if they agree to slow down the lethal competition, they may merely be making it easier for China to catch up. And so, at the moment, there is nothing really optimistic to be said on this subject.

Detection techniques of U.S. ships and sonar-dunking helicopters (top right) are also available to Russia, and America doubtless uses listening devices similar to this Russian one found by Icelanders in their waters. But because no detection system is wholly efficient, there can be no certainty anywhere about the deployment of subsea weapons.

Pollution by Chemicals

Among the unpleasant by-products of industrial and technological progress, none is more universally and rightly condemned than the generating and dispersing, whether on purpose or otherwise, of substances that do damage to the air we breathe, the land we live in, and the water that we drink, sail upon, bathe in, or get food from. In planning how to make the best possible use of the sea we cannot overlook the need to guard against harming it and ourselves. That is why I shall devote this and the next two sections to the question of pollution.

A primary concern of scientists all over the world, of course, is marine pollution, its causes and potential cures. One of the most informed scientific bodies is known as GESAMP (for the Group of Experts on the Scientific Aspects of Marine Pollution), which consists of experts drawn from various United Nations agencies and national marine laboratories. Since 1968, when it was established, GESAMP has been responsible for sifting vast quantities of data on marine

The oil on this beach after a tanker wreck off northern Spain in 1976 is only a tiny drop of the six million tons spilled annually into the sea. Such pollution not only harms plant and animal life but can mean economic disaster for fishing and tourist industries.

pollution and sorting out the facts that will eventually, it is hoped, become the basis for both national and international laws affecting the dumping of chemicals, oil, nuclear waste, sewage, and waste heat into the world's oceans. In this section let's concentrate on the first three of these pollutants.

The GESAMP scientists have prepared a list of 500 substances frequently carried at sea, ranging literally from acetaldehyde to xylene and zircon, and have classified them according to their tendency to accumulate in plants and animal tissues, their toxicity to plants, animals, and people, and their potentially damaging effects on beaches and tourist resorts. Several different tests have to be carried out to assess the toxicity of a chemical, but the most obvious is to measure the amount of concentration in seawater that will kill such susceptible animals as fish. To ensure statistical reliability, many fishes are simultaneously exposed to a given chemical, and equal numbers of new fishes are exposed to differing concentrations of the chemical. The definitive index of toxicity is taken as the amount of concentration that kills half the fishes in three days – that is, 96 hours. This concentration is called the 96-hour toxicity limit (TLm96); and the higher the TLm96, obviously, the lower the toxicity of the substance, since it takes more of it to kill the fishes. If a substance can be tolerated

at a concentration of more than one part per thousand for three days without killing fish, it is considered a negligible toxic hazard. But if the TLm96 is less than one part per million, it is extremely toxic indeed.

A final factor to take into account is whether or not cargoes of the substance are likely to be large. For example, ammonia is classified as only moderately toxic, while cyanide is highly toxic; but ships are likely to carry ammonia in hundreds or thousands of tons, cyanide in much smaller quantities. Thus the two may in reality pose risks to the marine environment out of proportion with the risk presented by equal amounts of each.

Many of the classified chemicals are used in bulk in industry, and so they find their way into the ocean through rivers, estuaries, and waste pipes, as well as from damaged ships or by deliberate dumping at sea. The dumping of industrial waste chemicals from ships is the easiest of these sources to control, as is evidenced by the relative speed with which a number of European countries have signed recent agreements on common standards for the northeast Atlantic. But the piping of industrial waste into estuaries or the open sea is a greater problem. To begin with, control of such practices is usually the responsibility of national or local authorities who have no

special interest in the sea, and recognized marine groups like the Intergovernmental Maritime Consultative Organization (IMCO) have no right to intervene. The discharges are generally of very large quantities of chemicals in extremely dilute form and there may be many compounds mixed in the waste, with the proportions varying from day to day. This makes it difficult to check the contents of effluents or to enforce any regulations that just might be in effect.

There are two possible approaches to limiting chemical-waste disposal. One is based on the maximum permitted concentration in any effluent pipe, the other on the assumed maximum safe concentration of the pollutant in neighboring sea areas. Taken at their simplest, these criteria differ in that the first discriminates equally against all factories, but sets no upper limit on total pollution if new factories are built; the second appears to favor already established factories, since it limits the total amount of waste that can go into the water. A subtle combination of the two approaches has to be devised before workable regulations can be adopted, and there must be regular and vigilant monitoring of pollutant concentrations. A classic instance of what can happen when industrial effluent is not controlled was the tragic poisoning of fishermen and their families during the 1950s in the region of Japan's Minamata Bay, when industrial mercury waste discharged into the bay was gradually absorbed by fish, and mercury in the diet of the fish-eating population caused blindness, muscular weakness, brain damage, and even death.

The most widespread dispersal of potentially harmful chemicals, however, results from large-scale agricultural use of artificial fertilizers and pesticides, which are spread into the sea by the winds, rivers, and rain. Everybody knows about the menace of DDT and polychlorinated biphenyls (PCBs), but even the complete banning of them would not end our worries. The assimilation of DDT into marine life is so gradual that plants and animals will go on absorbing the DDT already in the sea for many years to come. A computer model of its spread shows that if a worldwide program to reduce its use had been started in 1970 with a view to eliminating it completely by the year 2000, the concentration in fish would continue to rise until 1980, and would only drop back to the 1970 level by 1985. The use of DDT is now, in fact, restricted in many countries, but world usage is still about 100,000 tons per year. And – again this is something that most people do not realize – the same sort of time-scale applies not only to PCBs but to almost *every* marine pollutant. Whatever you do to control it now may have no effect on marine life for a further 30 years; if you wait 20 years before instituting controls, the problem may well get radically out of hand.

Every year, more than six million tons of oil are spilled into the oceans, and this figure is bound to increase. Dramatic accidents such as the *Torrey Canyon* wreck or the Santa Barbara oil leak actually contribute no more than about a tenth of the full amount of such pollution. The rest comes from leakage from ships, deliberate dumping from tankers, and waste from land-based oil users. In chapter 10 you will find a discussion of some of the steps that the international community is taking (and rather successfully, too) to deal with the deliberate discharge of oil from tankers. But very little has been done so far about many other aspects of the problem. And as yet there appears to be almost no way to cope with the highly emotive issue of radioactive fallout from bomb tests and nuclear waste from power stations and experimental reactors. People can ingest only very limited quantities of radioactive material without damage, and the limits have been accurately measured for every known radioactive source. Yet, with nuclear power stations being built in coastal areas all over the world, the International Atomic Energy Agency has its work cut out to monitor and restrict pollution of the water. In spite of vigorous research, we are still largely ignorant of what nuclear wastes may be doing to various forms of marine life.

Substances	Bioaccumulation	Damage to living resources	Hazard to human health		Reduction of amenities
			Oral intake	Skin contact + inhalation (solution)	
	A	B	C	D	E
Acetaldehyde	0	2	1	0	x
Acetic acid	-	2	0	0	0
Acetic anhydride	0	2	0	0	0
Acetone	0	1	0	0	0
Acetone cyanohydrin	0	4	3	11	xx
Bordeaux arsenites	+	2	3	0	xx
Bran pellets	0	0/BOD	0	0	0
Brazil nuts	0	0	0	0	0
Bricks	0	0	0	0	0
Bromine	0	3	2	11	xx
Bromoacetone	(z)	2	(2)	1	xx
Bromobenzyl cyanide	(z)	2	(3)	11	xx
Ethyl acetate	0	1/BOD	0	0	0
Ethyl acrylate	0	1	1	0	x
Ethyl alcohol	0	0/BOD	0	0	0
Ethyl amyl ketone	0	(2)	1	0	x
Ethylbenzene	0	2	0	0	x
Ethyl bromoacetate	0	1	3	1	xxx

The above extract from a list of 500 commodities that an international group of scientists considers potentially harmful to marine ecology suggests the heavy burden that modern trading places on the ocean. All these things – solid articles as well as chemicals – are being carried in great quantities across the seas.

Some of the 20 million tons of taconite effluent that flows into Lake Superior from one U.S. mining company's treatment plant every year. Uncontrolled dumping into freshwater can result in ecological damage to the sea and to sea-coast areas as well.

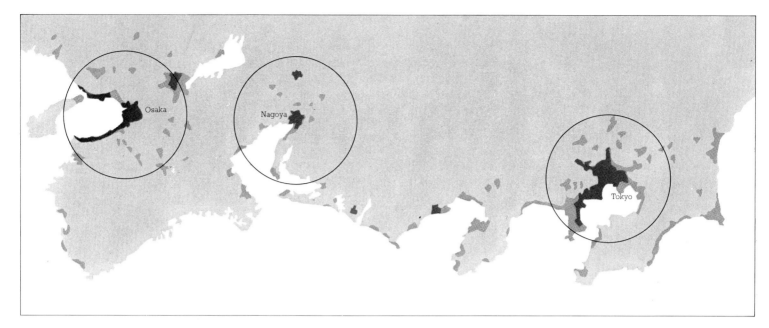

Pollution by People

The five largest conurbations in the world – Tokyo/Yokohama, New York, London, Shanghai, and Osaka/Kobe – are all on estuaries or inlets. Of the 65 cities with populations well over a million, 60% are on estuaries, deltas, or sea coasts, and 6% are on large lakes. Over half the population of the United States lives within 200 kilometers, nearly a third within 100 kilometers, of the coast. In Japan, if you draw circles with 50-kilometer radii around the three coastal cities of Tokyo, Osaka, and Nagoya, you include within the three small circles some 33 million people – that is, 32% of the total population of the country. And similar, if less spectacular, figures apply all over the world. Moreover, if present trends continue, the concentration of populations in limited areas is likely to increase; demographers predict that the percentage of the world's people who live in cities with populations over a million will have risen from the 12 or 13% of the early 1970s to 16% by 1985 and 25% by the year 2000. Since these percentages apply to a population that is expected to double during the next 25 years, the impact of population pressure on coastal areas is horrifying to contemplate.

Large cities mean large amounts of sewage, far more than can be disposed of in the time-honored fashion of dumping it into the nearest river, or directly into the sea. Both fresh and salt water contain organisms capable of breaking down organic waste and turning it into inoffensive and nontoxic materials, but this natural sanitation system begins to fail when overloaded; and as the level of harmful bacteria and other organisms rises, the water rapidly becomes a health hazard. There are efficient sewage-treatment-and-disposal systems in which organic products are broken down in sewage plants before waste is disposed of,

and dangerous microorganisms are killed off by simple chemical treatment. But such plants are expensive to build and run, and many communities – even in the 'enlightened' West – prefer to take the less costly though risky step of dumping untreated sewage straight into the sea via a pipe of moderate length.

In the developing world numerous major rivers, notably in India, have already become gigantic sewers, dangerous to bathe in let alone drink from. With the swift growth of urbanization in the industrialized countries, the trouble is spreading to formerly unexpected places in spite of the more frequent use of efficient sewage-disposal machinery. Rivers and estuaries have become polluted, and sewage is beginning to contaminate coastal and marginal seas. A recent survey of water conditions at a number of European beaches indicates that tourists in Italy, France, Spain, and Belgium are twice as likely to acquire an infection if they go swimming as they are if they stay dry.

Coastal tourist resorts have a special problem because of the crowds that swell their numbers during the holiday season (usually in summer). Even if a resort town has an efficient sewage-disposal plant for its year-round population, no such plant can cope with double or even triple the amount of normal waste that piles up in the busy weeks. Thus, just at the time when most tourists are likely to be swimming, raw sewage has to be dumped into the sea. In theory, there need be no such abuse of coastal waters; every tourist center should have a sewage-treatment plant big enough to handle the effluent from its maximum seasonal population. But again we are faced with the problem of high costs.

Quite apart from the fact that it endangers human health, human sewage can damage life in the sea. The best fertilizer, as every-

Some 33 million Japanese (almost one-third of the country's population) are concentrated in areas that lie within a 50-kilometer radius of each of three coastal cities: Tokyo, Osaka, and Nagoya. Harbors, factories, sewage, transport, and recreation all make demands on the surrounding sea.

one knows, is manure. Human fecal matter is no exception to this rule, and it will fertilize an ocean as well as a field. I once fished in Lake Neuchâtel in Switzerland and for the only time in my life caught fish as fast as I could haul them in. Inspection revealed that my line was only a few meters from an open sewage pipe, around which the fish gathered to feed off the other organisms that were thriving there. (We did not, I hasten to add, eat the fish I'd caught.) In the sea, the first effect of human sewage, even when it has been treated, is to promote a bloom of algae. If, as often happens in thickly populated coastal areas, there is also a power station nearby, effluents from the station are sure to raise the temperature of the water, and this increases the speed and size of the bloom. More algae might be expected to mean more zooplankton and consequently more fish – but, as we saw in an earlier chapter, this does not always happen. Instead, the constantly burgeoning plant life may lead to an overheavy demand on the supply of oxygen dissolved in the salt water. As a result, oxygen is denied to the animals in the neighborhood, and their numbers are drastically reduced. This is now happening in many rivers, lakes, estuaries, and seas that have become polluted with organic matter.

The problem in itself is not new, but the scale of the problem keeps increasing. Extreme cases of sewage pollution have led to eutrophication – a condition already described in the chapter on plant life – in a few limited areas of coastal water where the biological oxygen demand has become so

great that all the oxygen is now gone from the lower levels of the water column. There are signs of such man-made eutrophication in parts of the Baltic Sea, for instance. But the most notorious example of eutrophication as a result of human activity is a freshwater one: America's Lake Erie, which had become a nearly dead body of water by the mid-1960s. So far, the open sea remains very much alive, but we have little reason to be complacent about it.

National and coastal authorities almost everywhere now realize that large estuaries, and perhaps even the sea itself in time, could suffer the same fate as Lake Erie if drastic measures are not taken to control the dumping of human as well as industrial wastes. In recent years, regional groupings of countries have combined to measure and predict safe levels of treated sewage disposal in enclosed seas such as the Baltic and the Mediterranean. But action is almost always taken very late, after the problem has become too serious to be speedily and effectively solved. Its ultimate solution involves two components: first, a better way to dispose of sewage; secondly, a new way of planning coastal cities so as to avoid creating vast estuarine megalopolises. Why not put sewage on the land, where it does good, and not in the sea, where it does harm? The Chinese have been doing this for thousands of years, and small agricultural communities in Europe traditionally mix animal and human waste for fertilizing fields. The difficulty under urban conditions is to make absolutely sure that human diseases and parasites will not be transferred to food for

human consumption – but that should be possible with improved sewage-treatment methods.

Though redesigning coastal cities is an ambitious notion, it is not as crazy as it may sound. People come to live by the sea for reasons of trade and jobs, for cheap transport, and for a pleasant atmosphere. Since there is only a limited amount of habitable coastline, why not increase the effective supply of the coast? This could be done by avoiding the building of ribbon cities along

the shore or of massive coastal highways, and by encouraging the establishment of small cities just inland or offshore. Cities on artificial and natural islands, or on floating islands such as Aquapolis – a "water city" constructed for Japan's Ocean Exposition of 1975 – would spread large populations over a much wider coastal zone. Thus, with industrial and recreational uses of the sea applied to a bigger volume of water, less damage would be done to the undersea environment.

The pleasure and beauty of surfing draw holidaymakers and tourists from all over the world to the beaches with the best surf. But marine sports run the risk of attracting too many people for the good of the environment.

Might cities built over the water relieve some of the congestion in today's overcrowded coastal areas? A possible preview of the shape of things to come was the floating "city" named Aquapolis. Erected in 1975 for Japan's Ocean Exposition in Okinawa, the semi-submersible platform had a 10,000-square-meter upper deck and life-support facilities that included accommodations for 40 inhabitants.

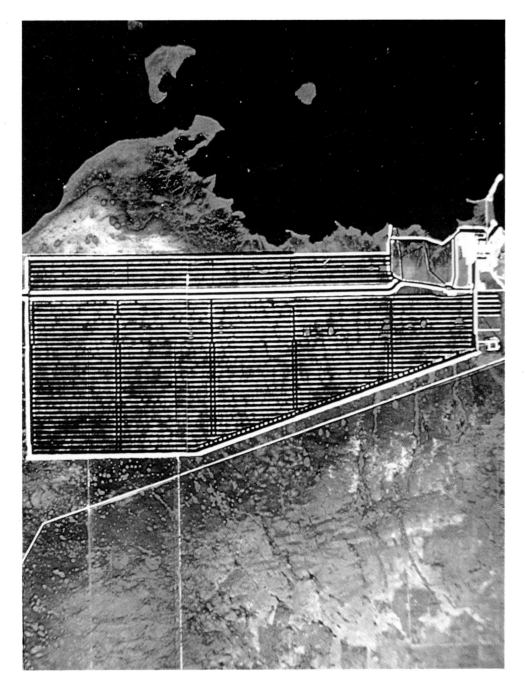

Left: An aerial view of the 270 kilometers of cooling canals through which hot waste water from four nuclear power stations at Turkey Point, Florida, must be run before being discharged into coastal waters. Because such stations need vast amounts of cooling water, many – such as the one at San Clemente, California (above) – are sited at the ocean's edge, preferably in places where waters are constantly mixed by strong currents, thus dispersing the heat.

Pollution by Heat

Man's lust for energy has produced a remarkable list of side-effects, almost none of which could have been foreseen when artificial methods of producing energy were first discovered. One of the most important of the side effects stems from the fact that all power stations work with much less than 100-per-cent efficiency. In other words, in the process of converting the chemical energy of conventional fuels or the nuclear energy of uranium into electricity, a good deal of waste heat is generated by high-temperature steam turbines. The turbines drop the steam temperature from over 500°C to a bit over 100°C, which allows the extraction of a lot of energy; but the final step of extracting energy from the 100°C steam is extremely uneconomic and thermodynamically inefficient. This extra

heat must somehow be dissipated, and the easiest way to get rid of it is to throw it into the nearest body of water – often a river, sometimes an estuary, more and more commonly the ocean itself.

The world's oceans being so vast, it might be thought that this could never amount to a substantial problem. It does. The world total of waste heat from power stations today amounts to about 7.5 million megawatts and is likely to reach 30 million megawatts before the end of the century. Part of this waste goes up the chimney as smoke and fumes from the furnace and is lost in the atmosphere. But most of it is carried away by cooling water. To put what happens in concrete terms: a nuclear power station that generates, say, 1,000 megawatts of power must dispose of roughly 700 kilowatt-hours of heat every second, and when this much heat is dumped into a nearby river with a flow

rate of about 30 cubic meters a second, the entire river temperature rises by nearly 20°C.

That amount of wasted heat could provide comfortable central heating for perhaps 40,000 two-story homes in a cold climate. Why isn't it doing so? Mainly because few people like to live close to power stations, particularly nuclear ones. As a result, environmental pressures are causing new stations to be sited farther and farther away from population centers, and the stations have no alternative to disposing of their waste heat than by drowning it. The seriousness of the problem can be demonstrated with some simple figures. By the end of the century the world's power stations will probably be using about 18 million million liters of water a day. The world's total daily fresh-water runoff from the continents is only in the region of 45 million million liters – and so

Waste heat need not necessarily become a pollutant; it can be put to good use in fish farming, as here in the vicinity of a Welsh power station, where trout are being restocked in warmed water. The warmth hastens growth, and the mature fish are eventually used for replenishing nearby lakes.

if we wanted to use fresh water for cooling all power stations, we'd soon need to channel and control nearly half of all the fresh-water flow on earth. Even today in some of the industrialized regions of the world, among them the United States, *all* fresh water is heated by passing through an industrial cooling system somewhere on its route to the sea. That is why engineers are being forced to look more and more toward the ocean as a giant heat sink.

Many nuclear power stations are already sited on the ocean's edge, where they can use the ocean as a sump both for deposition of controlled amounts of radioactive waste and for the elimination of surplus heat. If the heat from such stations and all others could be evenly dissipated in the great expanses of the sea, there would be no problem. As it is, however, power stations vent their heated waste water only as far out to sea as engineers find feasible from the standpoint of good economics. For the most part, they do try to choose sites where there are enough currents or a sufficiently strong tide to cause rapid mixing of the natural water with the heated effluent. But, even so, the coastal waters are affected by their activities. Surveys have shown that the water in the immediate vicinity of a power station is always several degrees hotter than the surrounding waters. And such temperature

changes inevitably have substantial effects on the local ecology. For example, as any tropical-fish enthusiast knows, a sudden change of even one or two degrees may well cause the death of a species. Temperature changes can also alter predation rates and susceptibility to poisons, can disrupt biological rhythms and migration patterns, and can increase rates of plant growth to the detriment of all marine life in the area.

Every type of environmental disruption need not necessarily be counted a disaster. By and large, ocean productivity increases with temperature, and there is reason to believe that in some instances catches of fish and shellfish have risen when new power stations have begun to warm up the surrounding waters. But, so far, too few attempts have been made either to predict the levels at which artificial heating of a given portion of the ocean becomes harmful or to find ways of putting the heat to clearly good use. In other words, waste heat has been sadly mismanaged – or perhaps I should say "unmanaged" – in too many places.

In others, fortunately, things are different. Experimental fish farms are already being supplied here and there with waste hot water from power stations, and there are good grounds for assuming that more such farms will be established in the vicinity of stations in the coming years. Then, instead

of being piped out to sea, waste heat will help maintain artificially heated tanks of sea-water fish, which can either be bred to edible size in captivity or loosed into the open ocean when past a certain critical size. Indeed, there is no reason why such techniques should be limited to salt-water fish. Power-station effluents could equally well be used for heating fresh-water tanks containing such somewhat-easier-to-grow – and often more temperature-dependent – fish as tilapia and carp.

What is needed *now* is much more research into ways of integrating the power stations of the future into the ocean environment. Proper management will come eventually – but why wait until the pressures of increasing population and dwindling food supplies force us into a desperate examination of every possible untried method of food production? The wastes that create pollution and constitute a major ecological hazard need not be either wasted or hazardous if we can learn how to turn them to good account.

No coast is static. The ceaseless pounding of winds and waves erodes cliffs and beaches, washing material out to sea where tidal currents pick it up. Swept along, the debris is deposited somewhere else, where conditions and exposure are different, and the land builds forward into the sea. Thus erosion and deposition cause horizontal retreat and advance while, quite independently, various processes in the earth's crust can cause the land to rise and sink, with the obvious consequence of either drying out the shore or drowning the land. Because in all cases – advance or retreat, rise or sink – the result can be economically unpleasant, engineers work hard, whenever possible, to combat or counteract such processes.

Advances of the shoreline are most spectacular near large deltas such as the Mississippi or the Rhône, where unhindered deposition of sand and mud can rapidly block a harbor entrance. There are two traditional methods of solving this problem: either constant dredging or the building of artificial breakwaters or training moles. In the slow course of time, though, accretion of the land almost always wins out – or, at least, has in the past – as we can see from the fact that the ancient harbors of Rome, Utica in Tunisia, and Ephesus in Turkey now stand some kilometers away from the sea. Accretion can also take place in coastal areas where wind and waves from different directions converge, as at Dungeness on the southeast coast of England. Although this is an exposed area, sand and gravel have built a pointed foreland 10 kilometers out to sea into a depth of 20 meters of water. There are now two nuclear power stations on the promontory, and they have to be protected from the incessant drift of gravel along the beach; the only way of doing this is to maintain a constant beach profile by continually carrying the gravel back in trucks from the tip of the point to the mainland.

Erosion is an even more serious problem. If King Cnut actually wished he could hold back the advancing tide, many a modern engineer has felt real sympathy for him. Where the shoreline is of unconsolidated material such as sandy cliffs, clay, or soft chalk, the waves may devour it at the rate of one to two meters a year. Where fallen sand or rocks protect the base of a cliff, erosion is slower, but waves and currents often combine to remove the fallen debris, and then the destructive process speeds up again. The simplest way to combat erosion is standard the world over: groins of wood or concrete are built like small breakwaters across the beach and may well extend into the sea as far as low-tide level. Such structures prevent longshore drift – that is, the transport of sand and gravel along the beach in the surf zone – and when transport is slowed, cliff falls remain in the area longer, as does any material reaching the beach from offshore or "upstream." Thus the beach becomes wider and higher, and its expanse protects the soft cliff. At least that's the theory. Too often, in practice, when longshore drift is stopped in a given place, the area "downstream" is deprived of protection. So it is a constant battle of trial, error, and compromise to keep a long stretch of shoreline from being washed away here if not there.

But there is a more subtle form of prevention that actually encourages the accumulation of new sand from offshore. The technique was used successfully on part of Russia's Black Sea coast in the late 1960s, and it is now being tried in other places with a low tidal range. A straight breakwater a few hundred meters long is built parallel to the beach at a distance of 100 meters or so offshore. The breakwater only just rises above the sea, and waves swing around its end and converge behind it, creating convergent longshore drift. With sand being swept into the zone behind the breakwater over a broad area, a wide spit like the natural feature known as a "tombolo" is formed. A series of such offshore breakwaters produces a sequence of tombolos, and these work as natural groins, slowing longshore drift but allowing equilibrium to be established.

Similarly imaginative, but less successful, have been recent experiments with plastic "seaweed." Theory predicted, and experiment confirmed, that mats of plastic strands anchored in the shallow water off a beach would encourage the deposition of sand. Unfortunately, it has proved impractically expensive to anchor strong enough mats in the surf zone, and so this method of defeating erosion is now used only to protect vulnerable areas near such valuable deep-water structures as oil rigs.

A moral to be drawn from all such experience is that the beach itself is the best protector of the shore, and that the stability of the beach is in fragile balance. It follows that a common practice, especially in de-

To protect London from severe flooding, a barrage across a 520-meter-wide downstream stretch of the Thames has been projected. When it is built, ships will normally pass between a series of concrete piers. On a flood tide, however, crescent-shaped gates resting on the bottom will swivel upward to close gaps.

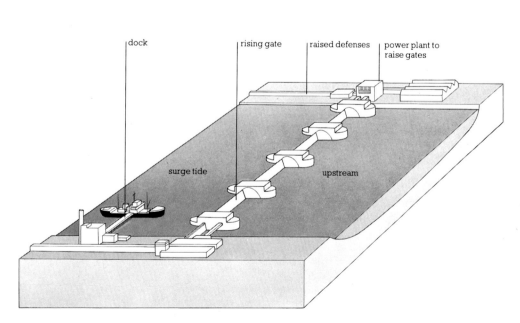

dock | rising gate | raised defenses | power plant to raise gates

surge tide   upstream

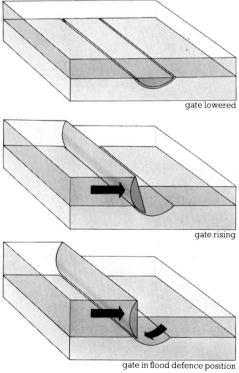

gate lowered

gate rising

gate in flood defence position

veloping countries, of mining beaches for sand to make cement and concrete is short-sighted, since it will inevitably lead to the loss of good land.

One essential way to protect sensitive low-lying coastal areas is through improving our techniques for predicting exceptional storms and high tides. In the tropics, particularly in the Caribbean and Gulf of Mexico, this means hurricane prediction. But in some other places, less ferocious winds can combine with a moving atmospheric low-pressure system to create a two-to four-meter surge of sea level. If this coincides with high tide, as in the North Sea a quarter of a century ago, the resulting floods can be devastating; the flood of 1953 caused thousands of deaths in Britain and Holland. A more restricted surge in the Elbe estuary killed 300 people in Hamburg in 1963, and it was a somewhat broader atmospheric effect moving up the Bay of Bengal that resulted in the tragic Bangladesh floods in 1970. Mathematical models of coastal sea areas can now give quite accurate pictures of atmospheric conditions that cause surges, and by combining these models with normal meteorological forecasting it is increasingly possible to know when to strengthen flood defenses or move the populations of low-lying areas.

Where vertical shifts are the problem – in other words, if the coast is actually rising or sinking – there is almost nothing engineers can do about it unless the problem itself is man-made. For instance, the subsidence of Venice, which is probably due to water pumping and industrial processes, might well be arrested by careful management. But such things as uplift of the coast of the Gulf of Bothnia in the northern Baltic and subsidence of the New Jersey coast are facts of life that cannot be balked. (For their natural causes see the opening chapters of this book.) Still, there *are* ways in which we can, as it were, turn the tide against nature. Land reclamation, for example, carries the anti-erosion battle onto the offensive. As most people know, about two-fifths of the Netherlands is below sea level and would be subject to flooding at high tide if it were not for the extensive system of dikes and drainage that not only protects the western part of the country but creates good agricultural soil where only the salt-sea bed existed before – a remarkable engineering project that started in the eighth and ninth centuries and has never stopped.

On a smaller scale, land is reclaimed from the sea in many other countries, wherever the price of land is high and the supply is limited. Garbage, building rubble, and other kinds of refuse are being used for filling in marshes, lagoons, and creeks so as to make valuable land near such big cities as Tokyo and New York among others.

Left: One way to keep the sea at bay is to construct a protective barrier of "tetrapods" – four-legged blocks of concrete that break up the incoming waves. Here some 7,000 interlocking 20-ton tetrapods are being installed at Tripoli, in Libya, to protect jetties.

Below: This model of the coast near Rotterdam helps engineers to study the effectiveness of present and projected flood defenses. Currents, tides, and even earth's rotation are simulated, and sensors record water pressures at various places, thus indicating when and where new dikes may be needed.

Tidal and Current Power

The great tides and currents of the open oceans represent a continuously self-renewing source of energy, but it is not an easy source to tap. In mid-ocean the amplitude of the tide is only about 50 centimeters, and ocean currents, though strong, are slow and tend to meander. In engineering terms the energy is diffuse and low-grade – that is, it is thinly spread over such a wide area that it would take enormous collection devices to obtain large amounts of power from the general expanse of the sea. That is why we haven't yet managed to exploit the power of currents, and why present-day efforts to harness tidal power are concentrated on limited coastal areas where the natural force of the tides themselves becomes concentrated. This happens at many places on the continental shelves and in the marginal seas, where tides driven by the low oceanic tides may reach amplitudes of several meters and the speed and amplitude are further increased as the water is funneled into narrow channels, bays, or estuaries.

The earliest recorded use of tidal power is a reference to a Dover Harbor tidal mill in William the Conqueror's Domesday Book of 1086, and such mills – quite apart from those that run in conjunction with freshwater streams – have functioned in various places ever since. But the first serious consideration of large-scale efforts to use tidal energy did not occur till 1920, when similar schemes happened to be almost simultaneously proposed for Passamaquoddy Bay between Canada and the state of Maine, for Aber-Wrach near Brest in France, and for the Severn estuary in England. All these sites offered good prospects. The tidal range in each was large; they were estuaries where a lot of water could be dammed up during the ebb tide and then released slowly through turbines to generate power; and they were not too far away from centers of heavy energy demand.

These same factors still determine the placement of tidal-power projects. An average tidal range of some five meters is required for profitable production of energy, and appropriate sites with such a range and within, say, 150 kilometers of a center of heavy demand for electricity are rare outside northern Europe. There are perhaps four in Canada, two in the United States, and six each in South America and Asia. In addition, vast amounts of tidal power could be generated along the northwest coast of Australia, but that area is too sparsely populated – at least at present – to warrant the effort and expense. In northern Europe, though, there are several places – particularly in Russia's White Sea region and on the north coast of France and the west coast of Britain – where the situation seems right for exploitation of the tides.

The Russians intend to build a 50-kilometers-long dam to generate tidal power at Mezenskaya Guba in the northern White Sea. As a preparatory measure, this prefabricated experimental section of a barrage complete with turbines and generators, has been installed at Kislaya Guba in the Barents Sea, where the workings of tidal power in northern waters can be observed and studied.

Nothing came of the forward-looking schemes that were proposed back in the 1920s. Since then, however, an occasional project has come to fruition – notably the remarkable French tidal-power station on the Rance estuary in Brittany, which has been in operation since 1966. This was an unusually ambitious project in that the engineers, not content with generating electricity only on the ebb tide, designed their turbines to run when the water flowed into, as well as out of, the estuary. Furthermore, the turbines double as pumps, and this overcomes one of the main problems of tidal power, which is that unless tidal flow happens to coincide with peak energy demand, much of the produced electricity cannot be usefully used. The French system, which has been officially reckoned an economic success, enables off-peak power to be used to pump water between the sea and the basin reservoir, one of which is lower than the other, depending on the tide, and to release the stored energy later on when power is required.

The main physical feature of the station is a long artificial bar (a "barrage") across the estuary. In it are housed 24 turbines, each capable of generating 10,000 kilowatts; and some 180,000 cubic meters of water per second flows through these turbines at both ebb and flow tides. As a result, the French get about 544 million units of "free" electrical energy a year. But it would be a big mistake to regard the power as *actually* free, for, despite the fact that the costs of running the station are minimal, the initial capital costs were enormous. All the truly big expenditure involved in building a tidal-power system such as the one at Rance comes at the beginning, when huge sums of money, all of which must be repaid with interest, have to be borrowed. Payments to creditors over the course of many years must then come out of the income from selling the power. Thus, only countries, or groups of countries, with highly industrialized economies can seriously consider investing in tidal-power undertakings.

The next most likely place for a station is probably at the giant Severn estuary in England. Calculations there have shown that the energy gain could be very much bigger than at Rance. A Severn tidal-power station could generate as much as 14,000 million units of electricity a year – nearly 8% of total British demand. The construction work needed will be gigantic, involving a barrage about 12 kilometers long and the artificial enclosure of some 400 square kilometers of sea area. Because such outsize engineering schemes are certain to alter the ecology of the region concerned, a number of studies have been made of the total effect of tidal-power control on the Severn estuary. Preliminary findings are reassuring; they suggest that there would probably be rather few damaging results and several beneficial ones. So it appears likely that not only the Severn but virtually all possible tidal-power stations in the world will eventually be constructed. In the past, when fossil fuels were relatively plentiful and cheap, tidal power seemed only marginally competitive. From now on, this nonpolluting, inexhaustible energy source is sure to become more and more competitive.

The future for harnessing some of the energy of the sea's currents looks less promising. The most obviously utilizable ocean current in the world is the Gulf Stream in the Straits of Florida. Futuristic schemes have been proposed for anchoring huge turbines in currents such as the Cromwell or the Kuroshio in the Pacific, but it will be a long time before such ideas become realistic rather than visionary. On the other hand, modern technology would make it possible to build floating power stations with submerged turbines for tapping the energy of the Gulf Stream at its maximum speed between Florida and the Bahamas. It is unlikely, however, that we shall do this while more feasible projects remain to be tackled.

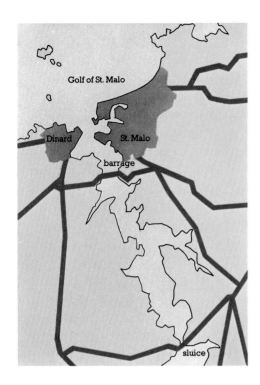

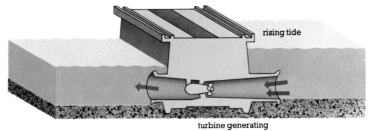

rising tide

turbine generating

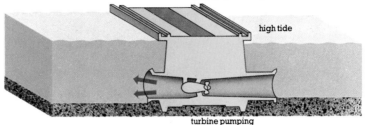

high tide

turbine pumping

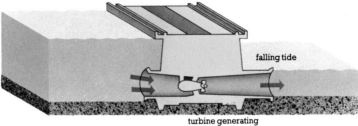

falling tide

turbine generating

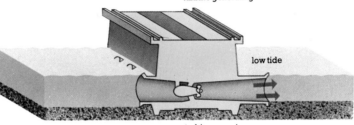

low tide

turbine pumping

As the tide rises, the great difference in level between it and the basin reservoir forces water through turbines to generate electricity.

Just before high tide, electric power (5% of the total generated) pumps more water into the basin to raise the level for the start of the next phase.

As the tide falls, there is a maximum height difference between the basin and the open sea, and most electricity is generated by the outflow during this period.

At low tide, another 5% of the power is used for lowering the level of the basin so as to get the greatest possible height difference at the start of the whole cycle.

At Europe's best-known tidal-power station, on the Rance estuary in northwestern France, engineers have taken advantage of the 13.5-meter tidal range to form a huge basin reservoir by building a barrage across the river mouth (aerial photo below). Seawater enters and leaves the basin through ducts containing turbines that power generators; a sluice at the river end prevents flooding upstream. When full, the basin holds 184 million cubic meters of water, and the plant produces 544 million kilowatt-hours annually.

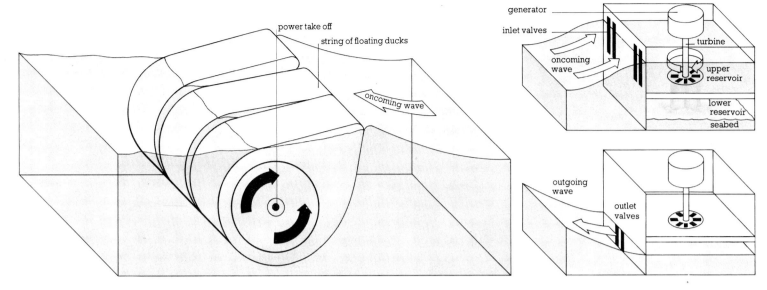

Energy From the Waves

The average Atlantic wave carries power equivalent to about 70 kilowatts a meter. This means that a one-meter length of the wave has enough power to run a large automobile at 80 kilometers an hour or to heat seven medium-sized houses in a temperate climate. Small wonder that the power of the oceans' waves has now become a prime target for an energy-hungry world! The waves accumulate within themselves energy from the winds that blow over all the thousands of square kilometers of sea surface; and in this sense they are rather like an extensive system of windmills, which store widespread energy that can be extracted at a single point, or along one line.

A particularly powerful storm wave may contain 20 times more energy than the average wave in a given area. That is why such coastal waters as those off Morocco, Brazil, California, or western Australia seem especially attractive as locales for the possible placement of wave-energy power stations: even when they are not being buffeted by large storm waves – even, in fact, when there is no local wind blowing – there is a steady inrush of long swells on this type of unsheltered coast. Storm waves in sheltered areas like the Irish Sea or English Channel might occasionally provide sizable bursts of energy, but wave sizes there are highly dependent on day-to-day weather changes, and an erratic source of power is of dubious value. Thus, the experts agree that a successful wave-power station must be designed so as to operate on oceanic waves in average conditions and still be able to cope with storms. If stations were built with due regard to placement in favorable parts of the oceans, and if wave energy could be extracted with 25% efficiency, a string of stations 1,000 kilometers long could supply about half the total electrical-power demand of a country such as West Germany, Japan, or France. And one of the attractive features of energy drawn from the waves is

Above: This wave-power converter (designed by a British hydraulics lab) has two reservoirs; water flows down to the lower one through a turbine generator. For power in quantity, long chains of strong concrete structures would need to be set up in coastal waters.

that the amount available varies seasonally in a fairly dependable, and very convenient, way – more in the winter when it is most needed, less in the summer when demand drops.

So far, it must be stressed, not a single commercial wave-power station has been built. The idea is still confined to the drawing boards and to model situations where prototypes can be tested on a small scale in specially designed tanks that incorporate miniature wave generators. However, now that other forms of energy have become so scarce and expensive, it seems likely that wave power will eventually take its rightful place alongside other, more traditional

means of energy production. When it does – and the first prototypes are sure to be operating within a decade – it will pose stiff problems for ocean management.

What will a wave-power station look like? First of all, it will have to be built to withstand the worst storms known at sea. For greatest efficiency, though, it will be designed not to extract maximum energy from the biggest waves – which occur only a few times a year – but from the average-sized waves that run most of the time. One particularly promising design now on the drawing board is known

Wave power as an energy source is still only a technological dream. No device pictured here has progressed beyond the experimental stage; someday, though, one of them may well prove viable. Below: A model of a "duck" being subjected to a lab test. A long series of heavy ducks positioned so as to be rhythmically rocked by incoming waves would drive a generator within the axle linking them.

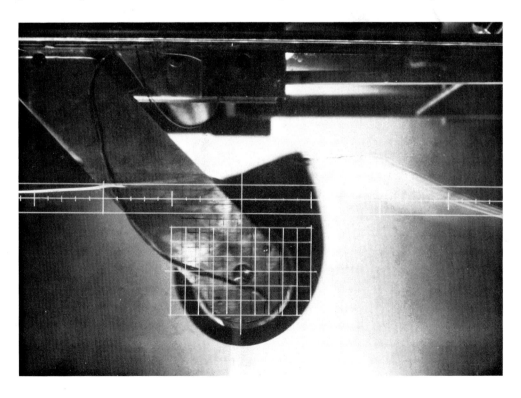

as the rocking boom, or "duck." This egg-shaped device floats on the waves with its more pointed end facing them and its rounded rear end lying below the water surface. As a wave approaches, the duck rides it, rising up in front and then settling down, with its heavy rear end remaining below the surface. The mechanical energy of the rocking front end is converted by devices inside the duck to electrical energy, which is carried ashore. Such ducks would be very big – probably over 20 meters from end to end – and they would be strung together in groups. These groups, each of which would be as bulky as a supertanker, would have to be sited many kilometers out to sea in order to avoid fouling coastal shipping lanes, and also so as to get the full impact of waves coming in from the open ocean. Farther inshore, wave patterns are disturbed by reflections back from the shoreline, and areas of disturbed water are much less suitable than the open sea for wave-power production: waves coming from different directions tend to cancel one another out.

Among other possible ideas being investigated is a float hinged in several places so that it would ripple up and down as it traveled over a wave, and pumps positioned at each hinge would be designed to extract the energy that was causing the alternating up-and-down motion. One of the keys to commercial success in this search for the best way to tap wave energy is to design a system that will extract the greatest possible percentage of total energy. And engineers at Wavepower Ltd., the British firm that has devised the float system, believe it might achieve as much as 50% efficiency – might, in other words, extract up to half of every wave's energy. In comparison, a conventional power station cannot get better than about 40% efficiency in terms of the amount of electricity it generates from such fossil fuels as coal or oil. In production, each of the Wavepower Ltd. floats could be roughly 10 meters long and from 20 to 40 meters wide. As with the ducks, several of them joined together might form even larger arrays, and they'd be built entirely of parts that could be mass produced even today without the need for any new technology.

In Japan, a firm called Masuda is experimenting with a different design: electric generators mounted in ring-shaped buoys; the doughnut-like form permits a single buoy to span several waves and, for the most part, to avoid being lifted by any one wave. Its underside is divided into a number of chambers, each of which is open at the bottom. As a wave passes under the buoy, it rhythmically displaces air in each chamber, and the changes in air pressure are used for driving a low-pressure air turbine to produce electricity. In the open-ocean conditions of the Pacific or Atlantic,

such a buoy would need to have about a 300-meter diameter in order to generate industrial quantities of electric power.

No matter how the wave-power station of the future is designed, some means has to be found to bring the energy ashore and hook it up to existing systems. The most obvious method would be by means of undersea electrical cables carrying high-voltage direct current. But there are other possibilities that need not involve housing a generator out at sea. For example, if the power devices in the wave-power station operated pumps, the energy might simply be piped ashore in the form of compressed air, which would then be transformed into electricity. Or – an even more advanced idea – electricity could be generated on the power station itself, but could be used at the site to electrolyze seawater into its two constituents of hydrogen and oxygen. Hydrogen, of course, is an excellent, if explosive, fuel, and it too could be brought ashore by pipelines. Moreover, it has one great advantage over electricity: until the fuel is wanted, it can be stored in gas holders for as long as need be. And hydrogen is also used in the manufacture of inorganic fertilizer – which has led Wavepower Ltd.

Below: An alternative to wave power is thermal energy, which exploits temperature differences between surface and deep water. This platform – the result of a feasibility study by Lockheed for the U.S. National Science Foundation – would measure 500 meters from top to bottom.

to the idea that we might in future see the establishment of wave-powered plants to produce both storable energy and synthetic fertilizer. Provided the products were used responsibly, such factories would be of great value, particularly to developing countries, which are frequently short of both energy *and* fertilizer, yet have long coastlines exposed to the wave energy of the Pacific, Atlantic, and Indian Oceans.

The speed with which we shall move toward the practical use of wave power depends on a number of factors. As I pointed out in an earlier section of this chapter, devices that utilize "free" energy – the tidal-power station at Rance, for instance – tend to be very capital-expensive. That is, although they can *operate* on a low budget, they cost a lot of money to build. Then, too, wave-power stations would undoubtedly add to the hazards of already overcrowded coastal waters. In addition to being sited well clear of shipping lanes, they would have to be equipped with superb early-warning signals easily detectable by radar, by eye, and by ear to avoid the chance of costly collisions. Finally, like many offshore oil rigs that must withstand the buffeting of heavy seas, they would require constant and difficult maintenance. But if such problems can be successfully solved, we can expect at least some of our electricity to come in time from the free energy of the waves – a tremendous and wild force that has never before been tamed.

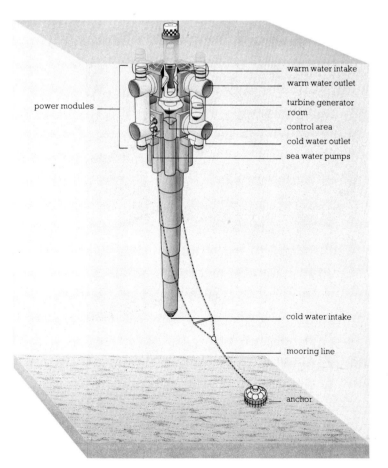

warm water intake
warm water outlet
turbine generator room
control area
cold water outlet
sea water pumps
power modules
cold water intake
mooring line
anchor

Warm surface water pumped in initially by an external power vaporizes a permanent supply of ammonia (which has a very low boiling point). The vapor generates power by passing through turbines, while near-freezing deep water, pumped up through the column, condenses the ammonia and returns it to the cycle. Four separate generation plants are fitted to each platform.

Power From Nuclear Fusion?

All projections of energy demand for the next three decades show that such fossil fuels as coal and oil will prove inadequate for total world consumption. Thus, there is one idea about future exploitation of the world's oceans that far exceeds all others in its appeal to the late-twentieth-century mind: the probability that we shall someday be able to transform salt water into energy by a process known as controlled thermonuclear fusion.

Thermonuclear fusion is not an entirely unknown form of energy. It is, in fact, the energy of the dreaded hydrogen bomb, which was first released on an unsuspecting Pacific atoll that, as a result, no longer forms part of the world's geography. The H-bomb is an example of *uncontrolled* thermonuclear fusion – a terrible weapon forged by a process quite unsuited to the peaceful manufacture of energy. *Controlled* energy is a very different matter.

Nuclear fusion is the opposite of nuclear fission, the process that currently powers our nuclear reactors. In the fission process, a heavy element such as uranium is bombarded by tiny particles known as neutrons, and thus is forced to split into smaller fragments, giving off a vast amount of energy as it does so. Scientists have learned how to use the reaction of splitting, or fission, in two ways. The first is brought to mind instantly by the name Hiroshima, where the first atomic bomb was dropped in 1945. The second, a later and more constructive use, is in the nuclear reactor, where controlled chain reactions release energy for peaceful consumption.

Whereas fission splits an atom apart, fusion joins atoms together. To produce nuclear fusion, two lightweight atoms are forced to unite into a single heavier atom. To date, that very difficult result can be achieved only by using an atomic bomb to trigger off the implosion of materials that produces nuclear fusion. But it's worth the trouble; the energy given off in the fusion process is spectacularly greater than that attained by fission. Hence the much more devastating power of the hydrogen bomb compared with the original atomic bomb.

For over 25 years, scientists of a number of countries – notably the Soviet Union (which apparently leads the world in this field), the U.S.A., and Britain – have been trying to tame nuclear fusion in the laboratory. The problems they face are immense. Atomic nuclei can be made to fuse, for instance, only when heated to startlingly high temperatures – in the region of hundreds of millions of degrees – at which point the materials to be fused take the form of a highly ionized gas (called a "plasma"), in which the atoms are stripped of their outer electrons. The sun is composed of just such a plasma, and its power – like the power of all other stars – is obtained from nuclear fusion. The main scientific problem, of course, is to control the reaction – in other words, to contain this very hot plasma in one

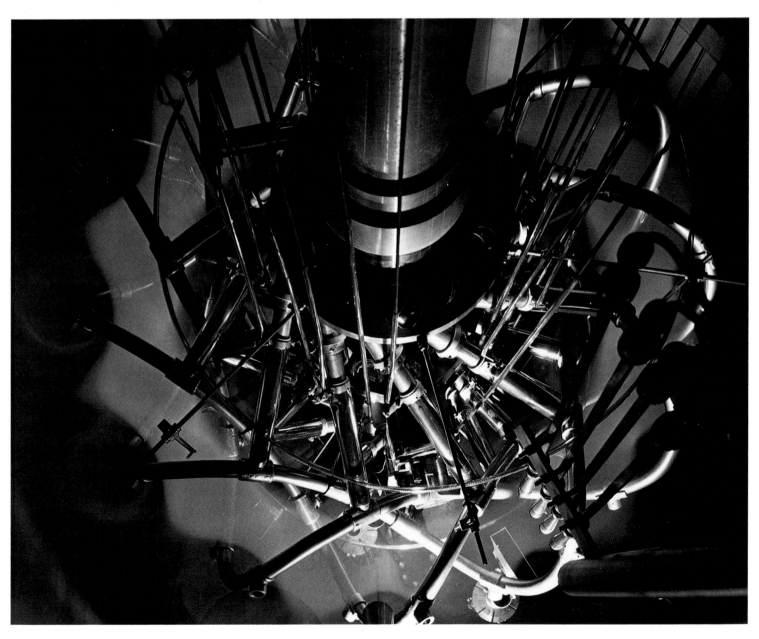

place long enough to increase its temperature to a point where the atoms will fuse into new elements, thus releasing an excess of energy that can be harnessed. Obviously, this cannot be done in any conventional way, for once the plasma touched anything material – metal or glass or brick walls, for instance – the result would simply be vaporization. Plasma, however, is electrically charged (i.e., the electrons have split off from the atomic nuclei, and the negatively charged electrons and positively charged nuclei are moving separately in the gas); and the movements of charged particles can be influenced and even "trapped" by strong magnetic fields. To contain the plasma in this way is, as one scientist has put it, like trying to hold a fistful of water together with an elastic band. So far nobody has been able to design just the right electromagnetic arrangement to do the job for more than a fraction of a second. But the problem is sure to be solved in time.

What has all this to do with the oceans?

The answer is that the oceans contain a superabundance of the very element that is best used for the fusion process. And it has enough of that element to provide mankind with sufficient energy to last until the sun itself dies. The ideal material for nuclear fusion is deuterium, which is simply a heavier form of hydrogen. Atoms of heavy hydrogen will fuse to form the heavier element helium – and water, as everyone knows, is composed of two parts of hydrogen to every part of oxygen. Moreover, for every 3,000 normal atoms of hydrogen bound up in seawater there is one that occurs naturally in the form of deuterium. This means that the world's oceans contain something like 40 million million tons of nuclear-fusion fuel.

If we could succeed in the controlled fusing together of a one-kilogram mass of deuterium, the reaction would produce an amount of energy equivalent to the energy in a staggering 3,000 tons of coal. In practice, to be sure, there would be a good deal

of wasted heat and many other inefficiencies in the system, and so we might end by getting less than a third that much useful power. Even so, it is easy to see that this energy source would enable the world to go on burning energy at its present rate of some 6,000 million tons of fossil fuel a year for another few thousand million years – and that is longer than most of us would care to think about.

So once the laboratory problems are solved – and although they're still enormous, most authorities believe they'll be solved by the turn of the century – we can expect to find fusion reactors springing up along our crowded coasts. In many respects these would be a great improvement upon fission reactors. For one thing, the helium produced by the fusion of deuterium is not radioactive, and so there would be none of the disposal difficulties with radioactive waste that accompany nuclear fission. For another, the system is bound to be "fail-safe"; in other words, should anything go wrong in the fusion factory, the fusion process would, because of its very difficulty, inevitably break down and stop. By contrast, it would be possible (though improbable) for the chain reaction in an atomic-fission reactor to continue and to release radioactive material into the atmosphere if the automatic control systems were to break down.

In one sense, though, fusion is worse than fission: one of its unavoidable side effects could do great damage to the sea itself. Because of the high temperatures involved, widespread use of nuclear fusion would undoubtedly produce great quantities of waste heat. And the only way this could be dissipated under present circumstances is into the ocean – as, indeed, is much of the waste heat from both conventional and fission power stations. I have already discussed the dangers of thermal pollution at the present time. Those dangers could be multiplied many times over once we start building nuclear-fusion reactors, unless we start planning *now* to overcome the menace of waste heat.

Sometime in the next century, then, the first nuclear-fusion reactors will probably begin to be built on our coasts. Pipes will draw in ocean water, which will be chemically treated to isolate the deuterium. This element will be passed into the adjoining reactor, and electric power will be generated from the reaction, most likely by using the fusion-produced energy to run conventional electrical turbines. Let's hope that by then we shall have found good ways to collect waste heat and put it to useful work instead of pumping it back into the ocean. One possibility is that we may have perfected fish farming to a point where the heat can be used to increase inshore protein productivity of ocean waters.

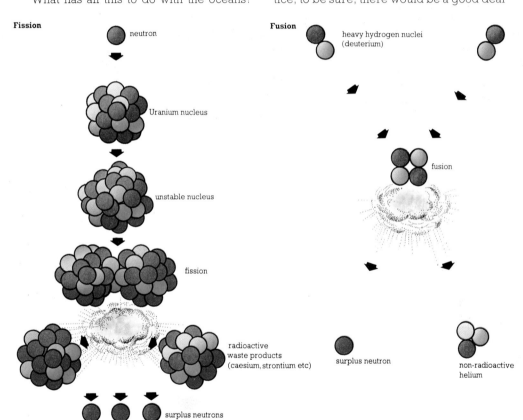

Fission

neutron

Uranium nucleus

unstable nucleus

fission

radioactive waste products (caesium, strontium etc)

surplus neutrons

Fusion

heavy hydrogen nuclei (deuterium)

fusion

surplus neutron

non-radioactive helium

Existing nuclear power stations depend on nuclear *fission* (as in the above diagram), which requires scarce uranium fuel and produces radioactive wastes. By contrast, *fusion* can use heavy hydrogen, which is abundant in seawater, and its end products are not radioactive. But we have not yet fully mastered the problem of controlling fusion reactions, which can occur only in a gaseous plasma at many millions of degrees centigrade (shown as a purplish glow in the photograph on the facing page); and the plasma can be enclosed and contained only by a complex array of magnets. Left: An experimental electromagnetic apparatus being assembled at a laboratory in Culham, England.

Marine Parks

In the last few pages I have discussed some ways to make the sea work for us. Now let's consider a few ways in which we can work for the sea, apart from the obvious task of keeping the water as pollution-free as possible. There are two approaches to such work – first, the preservation or protection of threatened marine species; secondly, the designation of whole areas as nature reserves or parks, where the water and all the life in it are protected from damage. Much has been said in this book about the need to preserve such creatures as the whale, dugong, manatee, and certain species of edible fish. I want to say something now about the second approach.

There is a difficult question about parks and reserves: whom should they really be designed for – people or animals and plants? A marine area dedicated wholly to

A new and fascinating experience for many tourists is this underwater observation tower, which forms part of a hotel complex on Japan's east coast. Access is by gantry from the shore to above-water viewing platforms (left), from which as many as 15 people at a time can proceed to the subsea viewing deck (below). There, at a depth of 10 meters, local marine plants and animals, as well as cavorting divers, can be observed through portholes.

preserving wildlife in as natural a condition, and keeping the water as pure, as possible would exclude human beings, apart from an occasional visit by scientists. A few such severely restricted reserves are needed in certain parts of the world, in fact, if we are to ensure the survival of all coastal and shallow-water species in a pristine state, undisturbed by industry, pollution, or underwater noise. But this sort of nature reserve has very little appeal for the general public, who never see it; and, to speak bluntly, it generates no income for the organization that must spend money managing and policing it. These are obvious drawbacks.

At the other extreme are parks designed to encourage tourism, with underwater viewing towers, glass-bottomed boats, subsea signposts for amateur divers interested in coral reefs and wrecks, etc. In between the extremes are various combinations of zoned coastal areas and offshore zones where industry, fishing, building, or dumping may be banned, or where roads, domestic building, and tourism may be restricted. A proper official policy for conservation might combine a sensible mixture of all types of reserve or park, since no one type can perform all the desirable functions. Perhaps the closed reserve is best confined to very remote areas, while tourist parks belong in places where there are nearby urban populations.

Almost all the protected areas established so far are in coastal waters; many, indeed, are merely terrestrial parks that have been extended a little way into the sea. The purpose of a marine reserve of any kind must be to guard a suitably extensive body of seawater against interference, so that, at least in large parts of it, plant and marine-animal life can continue normally. Because pollution travels so easily through the aqueous environment, and because many species themselves either move long distances or have a planktonic stage in their life cycle when eggs or young are carried long distances by currents, the park must cover many kilometers both alongshore and offshore if it is to achieve its purpose. At the very least, it should stretch out to the 20-meter depth contour in order to include a reasonable body of water within which natural conditions are maintained.

There must also be strictly enforced rules protecting the area from disturbance. This is a complex matter, for laws have to be passed defining the boundaries of the park, both on land, across the shore, and in the sea, and there must be further laws defining what is and is not permitted, what the penalties are, and how they shall be enforced. Once the park is set up, staff to patrol the entire area must be employed – among them, in some places, divers who can make sure that souvenir-hunters do not pick coral or loot wrecks. Obviously, then, the planning and control of a park take time and money.

There are now more than 175 marine parks and reserves in the world, but nearly a third of them are essentially coastal land parks with "wet" areas. The first entirely oceanic park to be established anywhere was the John Pennekamp Coral Reef State Park at Key Largo in Florida, which was opened to the public in 1960. Unfortunately, within a dozen years scientists found that the chosen site was too close to excessive concentrations of sewage and turbidity caused by dredging and that the corals were beginning to die. This highlights the general problems of protection of coral reefs, which, along with their associated ecological systems of reef-dwelling plants and animals, are extremely susceptible to damage from industrialization, construction work, or water-temperature changes brought on, for example, by waste-heat discharges from power stations. Because of their natural beauty and ecological fragility, coral reefs are, therefore, obvious choices as locales for special protection, and many of the world's marine reserves are centered on such reefs and islands. So far, Australia's Great Barrier Reef – the largest coral reef in the world, containing thousands of coral islands – has not been declared a park, but it deserves to become one.

Australia is not yet among the countries that have made serious attempts to set up marine reserves at regular intervals along their coasts. This *is* being done in Canada, Japan, Kenya, the Philippines, South Africa, and the United States. But most European countries have been slow to recognize the importance of marine conservation by means of reserves. Denmark is a notable exception; a large part of its North Sea coast is now officially protected from industrialization. And Holland apparently has plans for designating the Waddenzee, a huge area of shallow tidal flats between the Frisian Islands and the mainland, as a marine park where tourism, pollution, and drilling for oil and gas will be discouraged.

In 1975, representatives of such organizations as the International Union for the Conservation of Nature and Natural Resources, UNESCO, the World Wildlife Fund, and the UN Environment Program met in Teheran to discuss various aspects of marine conservation in and around the Middle East. One outcome of the meeting was a long list of recommendations for marine parks – 24 in the Persian Gulf, 6 in the Red Sea and Gulf of Aden, and 22 along the Indian, Pakistan, and African coastlines. There are plans for similar conferences to be held elsewhere in order to examine the world's oceans one by one. This is very much a step in the right direction. Particularly satisfying is its indication that interest in conservation is by no means confined to Western enthusiasts. Indeed, we already have a World Marine Park, which was presented to the international community "for the benefit of mankind" in May, 1975, by the government of the Cook Islands in the central Pacific. The park centers around Manuae, the first of the atolls discovered by Captain Cook. Other nations would have a hard time matching the generosity of the Cook Islands, but we can at least hope that some will try.

The world's oldest oceanic park – it was opened in 1960 – is John Pennekamp Coral Reef State Park in Florida, where snorkeling divers can swim past such unexpected sights as this statue, which adds a spiritual touch to one of the park's submerged nature trails.

The nature reserve around the remote island of Lundy in England's Bristol Channel is not for careless pleasure-seekers. Since 1973, its waters and steeply shelving sea floor, out to a kilometer from the cliffs, have been preserved for biological studies, and diving is closely supervised.

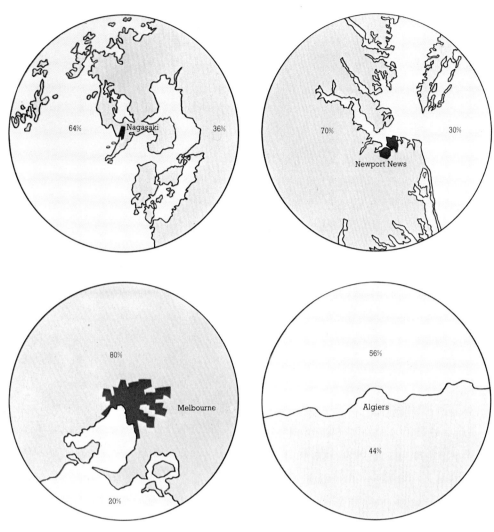

Some coastal activities use only the coastline, others the total area or volume of water. And good planning takes into consideration the quantities of coastal land and water available in a given locale. Compare, for example, the potential facilities within a 100-kilometer radius of each of these four major ports: Nagasaki, with its long coastline and large water area; Newport News (Virginia), with a comparable coastline but less water; Melbourne, with a shorter coastline and little water; Algiers, where the straight coast means nearly equal amounts of land and water.

Coastal-Zone Management

The coast is where land, sea, and air meet. The air and sea interact to produce waves and currents; the land and air interact to produce the environment for terrestrial life; the sea and land interact, as we have seen, to produce erosion, deposition, and the complex environment of the undersea world. Within a 200-kilometer distance of the coast we have not only the coastal zone of the great oceans but also most of the area of such marginal and semi-enclosed seas as the Gulf of Mexico, the Sea of Okhotsk, the Mediterranean, and the North Sea. And it is in this limited area and limited depth that most of the uses of the sea take place.

The coastline of the whole world is only 504 thousand kilometers long, apart from indentations that are less than 18 kilometers wide at the mouth. That is not, relatively, a very great distance. Yet, fringing this limited amount of coastal land lies virtually the entire realm of intensive sea use – no more than about 7% of the total ocean. Within the 7% of marine space is an even smaller percentage of ocean volume – less than 1% – and within these waters lie more than 50% of our fisheries, nearly 70% of the world's recoverable oil and gas, and virtually all our marine waste disposal apart from deep-

ocean dumping of highly toxic and radioactive materials. Even most shipping lanes are in coastal waters, as are, of course, practically all recreational activities. When, on top of this, you consider the concentration of population and industry around estuaries and harbors, it becomes quite clear that the coastal zone represents the worst example of what modern civilization can do to the sea.

The great international questions of pollution in the deep oceans, the global undersea arms race, and the preservation of threatened marine species must not be overlooked; but if the attentions of scientists, engineers, and politicians were spread evenly over the volume of all the oceans, we could never apply enough effort to cope with the problems of any one location. We have limited resources in terms of money and skilled manpower, and these must be focused primarily on improving our management of the coastal zone. If, as seems likely, the nations of the world eventually agree that coastal states should have economic control over 200-mile zones, many key management decisions will remain in the hands of national and local governments. International agreements, treaties, and laws will require coastal states to administer their sea areas in certain ways, but the ultimate responsibility will be national.

Unfortunately, most countries do not have expert teams of scientists experienced in dealing with the intricacies of oceanography, biology, ecology, technological forecasting, etc. Governments in general are geared to deal with land-based functions, not with offshore undersea areas. The optimum use, or mixture of uses, of a coastal region depends on two broad groups of factors: first, the natural characteristics of the coast and fringing waters; and, secondly, the economic and social requirements of the local population. Yet, although excellent maps of geology, soil types, vegetation, climate, and so on usually exist for land areas, equivalent data for the nearby sea floor and water are likely to be sparse. A government official or company executive faced with a decision about routing a road or siting a factory can quickly get necessary background facts. Faced with offshore decision-making, the same individual is literally at sea.

This is not, happily, true everywhere. As the intensity of coastal-zone use grows, forward-looking authorities in some countries are placing more and more instruments on ships, buoys, and platforms so as to obtain a continuous flow of information. They have begun to realize that we need to know with extreme precision the size and direction of waves at every time of year and at numerous points all over the coastal seas, for such things affect just about everything: the design of all structures from rigs to harbors; the rates of coastal erosion and deposition; the kinds of ships, pipe-laying barge, and dredger that can operate in a given stretch of water; and the amount of energy available for, say, wave-power generation. We need to know the speed and direction of currents at various points, too, since these affect not only the design of structures again, but also the mixing of polluted waters and the movement of nutrients – hence, primary productivity. Finally, we need accurate studies of marine weather patterns, since wind speed and atmospheric pressure cause waves and changes of sea level, while air temperature, precipitation, and evaporation affect offshore working conditions as well as the temperature and salinity of the sea surface.

There is nothing new about the gathering of such data. Special ships have been assisting weather forecasters for decades, with

Port Grimaud on the south coast of France is a good example of how ingenious planning and engineering can lengthen and enlarge a coastal strip. This artificial lagoon provides every inhabitant of the village with an individual dock and bit of sea front.

oceanographers measuring the variables at widely spaced points all over the oceans. But what is happening now is that progressive governments and industrial organizations are combining to install instruments permanently in coastal seas; millions of digits of data are being annually recorded on magnetic tape at each measuring station, and computerized data banks are processing the information into official maps and charts. Furthermore, the oceanographic and meteorological researches are now supplemented in many places by geological data on the sediments of the continental shelf and other aspects of subsea geology and by the logs of subsea oil wells and expert observations on fisheries and plankton.

Offshore data require onshore supplementation as well, since coastal-planning decisions must relate to the shape of the coast itself and the distribution of towns, industry, and road and rail links. Plans for the English Channel Tunnel were abandoned a few years ago, for example, not because of the complexities of undersea work but because land-based factors militated against the project, whereas an equivalent set of onshore statistics operated in favor of Japan's Hokkaido Tunnel. Coastal land forms and industrial requirements also, of course, influence the location of new ports. And coastal-zone management must increasingly presuppose an understanding of the complex movements of salt and fresh waters in estuaries, as well as the ecological importance of salt marshes, swamps, lagoons, and wetlands in general.

No dictatorial system of planning is likely to work well, and I do not mean to imply that coastal-zone management can produce a "right" way to plan the development of any or all coastal seas. But one thing is certain: a sure recipe for disaster is to go on intensifying the utilization of coastal zones without attempting to understand better the effects on both land and water. In particular, the interactions between industry and oceanographic factors need to be studied by specialists, who can then make decisions in the knowledge of predictable consequences. As I have suggested, the evolutionary process of developing new ideas and techniques has begun. But the health of the world's coastal zones depends on continuing and accelerating the evolution.

This timber pulp mill in Georgia is discharging polluting chemicals into nearby salt marshes. Any attempt at coastal-zone management must take into account the possible effects of inland industry on the coastal and offshore environment.

A Summing Up

How to manage and preserve the world ocean that stretches far beyond the coastal and semienclosed seas may seem to be a much less urgent problem than that of coastal-zone control. It lacks the immediacy, the demand for a concentrated effort of analysis and expert planning *now*. Even here, however, there is no room for complacency. It will take many years to reverse the damage already done by DDT, and similar long-term dangers may be building up, only to emerge fully in the years ahead. Control of overfishing is obviously urgent and there are other pressing issues.

International agreements imply consistent national maritime policies for each of the signatories, but national policies relevant to modern marine technology have only been evolved laboriously over the last two decades. Since 1965 the governments of the major industrial states have been reorganizing, shuffling, and rearranging the administrative structures of departments concerned with national marine affairs and marine policy. The aim in every case has been to solve the problem of how traditionally oriented departments of transport, defense, local government, agriculture, etc, can be coordinated so as to develop coherent policies of marine exploitation, even though marine affairs are only a small part of their concern. Lobbyists for a strong national marine policy tend to recommend the creation of a new central government department with some such title as "Department of Ocean Affairs," but this idea has been rejected by most countries because it would entail an immense effort to reorganize the marine and terrestrial segments of agriculture, transport, mining, and so on. The debate has been carried on at the very highest levels of policy making, notably in the U.S.A., France, and Japan.

By 1970 several countries had centralized marine research and development agencies in such bodies as America's National Oceanic and Atmospheric Administration (NOAA) and France's Centre National pour l'Exploitation des Océans (CNEXO). These agencies, backed by programs to stimulate investment in marine technology and the training of specialist engineers, have been designed to increase the national economic benefit from the sea. In all cases, however, the original targets established for the agencies also anticipated the importance of environmental conservation, and government funds have been channeled into their research projects. Nevertheless, so much has changed so quickly, and the machinery for planning at an international level is so slow, that we are still in danger of being overtaken by events.

It is hard to separate planning problems from political and legal ones. In theory, we ought to be able to use an ocean area sensibly regardless of who owns it; in practice, though, if ownership is disputed, governments are unlikely to agree on technical details. That is why the UN's international Law

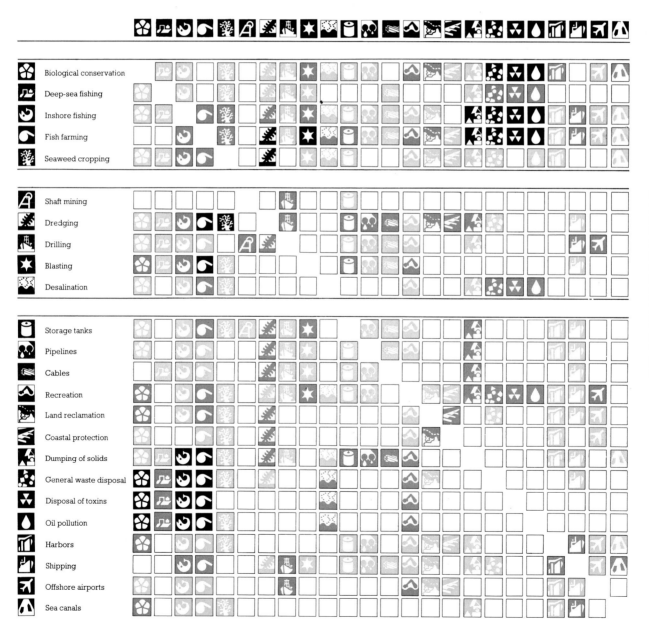

Human uses of the sea can be broadly divided into 24 classes, which often interact within a particular sea area to the detriment of one another – or, as in competitive fishing interests, to the detriment of another aspect of the same activity. In this diagram (to be read like an intercity mileage chart) the potential impact of such interactions is indicated as negligible (empty square), slight (light gray), moderate (dark gray), or severe (black). Right: A problem posed for undersea communications by the fishing industry. This trawl gear has hopelessly fouled a submarine cable.

of the Sea Conference is so important. As a result of it, the world will sooner or later have a new legal and political framework for the sea and sea floor beyond national limits. Within that framework, agreements on pollution, safety, dumping, overfishing, navigation, etc., can be implemented.

Meanwhile, various international and regional agencies are coordinating their studies of the worldwide effects of industrial use of the sea, and are promoting special and limited agreements or regulations concerning already identified harmful factors. Examples of their achievements are the international standards of safety at sea produced by the Intergovernmental Maritime Consultative Organization (IMCO); regulations on tanker design and tank washing at sea, also arranged by IMCO; various regional fisheries agreements promoted by the UN Food and Agriculture Organization (FAO); large-scale studies of the energy exchange between the ocean and atmosphere coordinated by the World Meteorological Organization (WMO) and the Intergovernmental Oceanographic Commission (IOC); and the applied oceanographic research conducted under a broad program known as the International Decade of Ocean Exploration, which began in 1969 and is being coordinated by IOC. In an earlier section I mentioned the work of GESAMP on marine pollution, and the United Nations Environmental Program (UNEP) also concerns itself with pollution monitoring and prevention at sea. Finally, there is a small department of the UN secretariat, the Office of Ocean Economics and Technology, that produces reports bringing together the findings of all the other agencies, so that politicians and diplomats can get up-to-date information in a readily accessible form for use in negotiations.

It is only too easy to lampoon these activities as rampant bureaucracy, acronymics, empire-building, or Parkinson's Law run riot. No doubt, too, there is wasteful duplication of effort; vitally important issues that are politically touchy get left till last; and reams of unreadable – and largely unread – reports are produced and seldom acted upon. But in the world of international politics, every delegate or technical expert is subjected to countless conflicting pressures and motives. More than 130 nations are represented at the UN, so the accumulation of specialized agencies, each with its limited terms of reference, is not surprising.

Any plan for constructive ocean management has to take into account the multiplicity of activities in and under the oceans, the multiplicity of potential conflicts among them, and the multiplicity of interested nations, international agencies, and commercial organizations. A simple way to analyze the potential interaction-conflict pattern is to consider the current range of marine interests in broad classes – from, say, biological conservation through mineral extraction to nuclear-waste disposal – and work out the network of their relationships. Briefly, an analysis of this kind produces over 20 general classes of activity, each of which can interact with several others and with different groups within its own class. Fishermen, for instance, can clash with other fishermen, whether for different kinds of fish or from rival countries or organizations, as well as with dredgers or drillers. So there are altogether about 300 types of potential interaction of varying degrees of severity in the coastal seas and deep ocean. These must again be analyzed for the probable scale of each kind of activity in a given sea area, and for the various ways in which they are likely or unlikely to conflict. Thus, for instance, toxic-waste disposal is incompatible with local fish farming but not with drilling for oil.

International planning agreements are therefore needed to help us make the best possible mixture of uses of sea areas, in much the same way as we need local planning measures for coastal zones.

The two short-term problems of management are: (1) how to prevent the worst kinds of harmful excess (usually in the coastal zone), and (2) how to avoid serious political or military conflict in the squabble over who owns what. But even if we solve these two problems, the end result might still be nothing better than an agreement as to how, and under whose auspices, the sea is to be gradually depleted. We *must* press for the sort of long-term management that will try to achieve permanently sustainable yields of all marine benefits, from food and other material goods to the production of energy or the creation of salt-water playgrounds. To do this means to make intelligent choices between apparently baffling alternatives. Thus, for example, it might be much more sensible to encourage the increase of the krill-eating whale population, and then catch restricted numbers of the whales, rather than permit them to become extinct, and then try to find ways of catching and eating krill ourselves.

We must not exhaust one resource or one area and move on to the next, as has so often happened on land. This only produces a series of crises, cheapness and expense, plenty and scarcity. In the oceans we must do better. And there is still – just – time enough to learn how.

7
Underwater Archaeologists

Colin Martin
Nicholas Flemming

Man's relationship with the sea is inextricably linked with his progress on this planet. When early representatives of *Homo sapiens* wandered away from the kindly environments of warmth and plenty in which they had evolved, and to which their highly specialized but vulnerable bodies were adapted, they showed a remarkable aptitude for survival through applied technology. The process of worldwide distribution was, of course, extremely gradual; we can only speculate about how and when significant numbers of human beings moved to previously uninhabited parts of the earth. But we know that by roughly 8000 BC man had penetrated to most of the regions that he now inhabits – and that he must have been able to cross stretches of salt water in order to do so.

Many places that are now part of the oceans were, to be sure, dry land at various times in the distant past. During the Ice Age, starting a million or more years ago, the volume of ice fluctuated continuously, causing the sea level to rise and fall through a vertical range of over 100 meters during periods lasting hundreds of thousands of years. And so, as early people followed the great animal herds that provided their food, they pushed forward against a background of strongly varying sea levels, and they usually managed to migrate by land rather than water. Whenever the sea level was as high as it is now, or even higher, they were no doubt largely restricted to the separate land masses that have been traditionally thought of as the natural geography of the earth. But whenever the sea dropped far below its present level, as it did about half a million years ago, all sorts of new land routes opened up.

Thus, the first Americans were small tribes of Arctic mongoloids, who trekked across the ice and tundra of what is now the floor of the Bering Sea between Siberia and Alaska. There was a prolonged period from around 20,000 to 12,000 years ago when such migrations could have taken place. And there are many places elsewhere in the world which, though now isolated by the sea, could have been reached by foot at various times. Others, though, were never easily accessible. Recent discoveries, for example, suggest that human beings crossed from the Celebes Islands of Indonesia into the empty continent of Australia sometime during the late Pleistocene epoch; and these people could have come only by sea.

Man's development of boats as a form of transport was as notable an evolutionary advance as was his control over fire. It is not a coincidence that the rise of civilization runs parallel with the technical development of seafaring. Even little boats gave people a freedom of movement and an ability to carry loads, to gather food, and to trade with others on a scale and over distances previously unimagined. In many different places throughout the eastern Mediterranean, we have found tools made of obsidian (a hard, glassy stone) that date back to 7000 BC, and obsidian in that large area could only have come from the small island of Milos in the Aegean Sea. So man was already an accomplished seafarer by the Neolithic period of the Stone Age.

The sea has always been a perilous element for those who venture upon it. Even today, in spite of technological advances in propulsion, navigation, and design, we lose a tragic number of vessels and people every year in both coastal and mid-ocean waters. Until the quite recent past, the risks and losses were enormous, for survival or shipwreck on any voyage depended largely on luck, pluck, and the weather. During thousands of years, vast numbers of ships have gone down, sometimes to remain intact or partly so in mud or silt, more often to be smashed to bits and scattered around the reef or cliff base on which they foundered. But material deposited in water seldom disintegrates completely. Normally, after a short period of destruction and decay in which much may be lost, the remainder settles and interacts slowly with environmental processes until a state of balance is reached and physical and chemical disintegration slows down to a virtual halt. The resultant "wreck formation" (as archaeologists term it) now includes natural as well as man-made elements, and it may be preserved underwater almost indefinitely.

Thus, the remains and contents of countless large and small craft – fishing boats, coasters, traders, and warships of every type and period in the history of maritime endeavor – lie around the world's coasts, and to a lesser extent (but potentially better preserved) in the deep oceans. Marine archaeologists consider each such wreck to be precious and irreplaceable – but not for the material treasure it may hold. Far more valuable than any gold bullion is the "time capsule" of information unique to its particular function and date that lies buried within every submerged wreck or man-made object. No two are exactly alike. All are speaking records of man's past. Nor is it only the remains of early vessels that lie beneath the sea. The key evidence for man's early migrations can be found in the form of camp sites or once-inhabited caves now under 50 to 100 meters of water. And earth movements and changes in sea level have inundated dwelling places, harbor works, and even whole cities, often preserving them to an extent that would have been impossible had they remained on dry land.

The almost limitless possibilities that such material opens up to archaeologists have only recently been appreciated. Perhaps because of the physically demanding nature of the undersea environment, and perhaps because of the glamorous notion of diving for "buried treasure," undersea antiquities attracted adventurers before they attracted scholars. But archaeology is a science – the science of deducing, from the exact position of fallen buildings, bones, religious objects, and discarded tools and pottery, as much as can be deduced about past cultures. Archaeologists are real-life mystery-story detectives; in an imaginative, rational search for clues, whether on land or underwater, they try by means of careful logic to re-create the lives of long-gone people. The avaricious treasure hunter, on the other hand, may destroy the clues in his haste. Only, therefore, when scholars themselves became adventurers and went below the surface to examine the sea's treasure trove was the full richness of our submerged heritage established. But it is a vulnerable heritage, subject to ruthless exploitation by greedy men or nations. In a later section of this chapter we shall look at some important questions concerning responsibility for, and control of, individual findings.

Above right: During the Ice Age, when the sea level was much lower than it is now, early man undoubtedly migrated between land masses by walking across the "land bridges" indicated on this map. Travel by sea came later.

Many caves and stone huts of late-Ice Age man are now deep under water. **Right:** Divers at Gibraltar prepare to explore partly submerged caves that were occupied from Neolithic to Roman times. **Below:** These loose stones of a Neolithic wall were uncovered in 1968, when a storm stripped sand away from an Israeli beach.

The starting point for surveying an undersea archaeological site is likely to be on shore. Left: This photograph of the submerged city of Halieis in the Peloponnesus, taken from a tethered balloon, shows room layouts at a depth of 1.5 meters. Above: Also at Halieis, archaeologists work from a line on the beach to fix offshore positions with a theodolite and surveyor's staff. The scaffolding hut gives divers a base directly above the ruins.

It isn't easy for divers to find an undersea wreck of which the presence is merely suspected, and the best way is to follow methodical swim patterns. Below, left: This is an actual map of the random paths taken by searchers each day around a Mediterranean island some years ago. Though they swam for many days (the numbers show the day of each swim), and though several men swam in formation on some paths (the wider ones), the search produced only a few amphorae. Modern "swimline" searches (below, right) are much more efficient. Up to 10 divers, swimming abreast and holding a line, follow another line strung between buoys. They signal finds by tugging on the connecting line and release marker floats. On completion of a sweep the buoys are moved slightly, so that successive sweeps overlap.

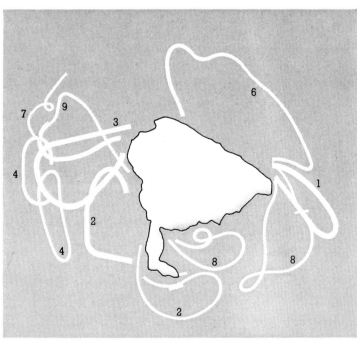

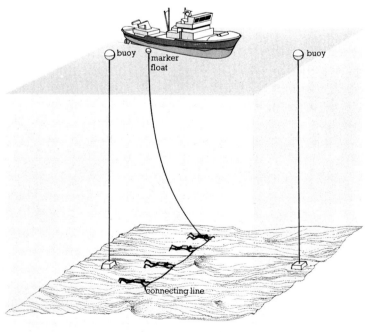

Tools and Techniques

The location of an ancient wreck or submerged city is generally known through accidental discovery by fishermen, sponge or sports divers, or enterprising travelers. Often, though, the existence of something interesting somewhere in a given area of the sea is only suspected, and a full-scale search must be made in order to track it down – and such searches often fail. In any case, the archaeological process does not start until the exact position of the ship or drowned settlement is pinpointed and the marine historian is in command of secure information about local currents, depth, underwater visibility, bottom cover of sand or weed, and weather conditions in the work area. All such matters affect the speed – and hence the cost – of excavation.

Until quite recently, the only way to search for wrecks was with your eyes; and where some historical evidence exists to narrow the possibly productive area down to a few miles of coast or open sea, a simple visual search with teams of divers swimming along bottom lines can have very good results. For instance, one of the Spanish Armada's ships known to have sunk off a certain section of the Irish coast in the sixteenth century was the *Santa Maria de la Rosa*, which was found in 1968 after coordinated teams of divers had examined every square meter of several square kilometers over the course of many months. Visual search methods are hampered, however, by low underwater visibility – no more than 20 meters under the best conditions – and the possibility that the target may be hidden under mounds of sand or weed. Such problems can be surmounted today by nonvisual electronic search equipment, which first began to be used for marine archaeology about a decade ago. And these tools work wonders.

Side-scanning sonar, for example, can be used for building up an oblique picture of the sea floor so that objects that project even slightly out of the floor can be detected. It was by this means that a brick cooking stove from the American Revolutionary brigantine *Defense* was found, a few years ago, off the coast of Maine, although it projected only 30 centimeters above the bottom – a small find that led to the discovery of the ship itself. High-power sonar pulses can actually penetrate several meters of sand and mud when directed vertically downward; several buried wrecks have been spotted in this fashion.

Still another technological aid for persistent searchers is the magnetometer, a standard geological tool for measuring the local strength of the magnetic field. Because the earth's magnetic field is altered slightly by the presence of iron, other metals, or even stone and pottery in a background of sand

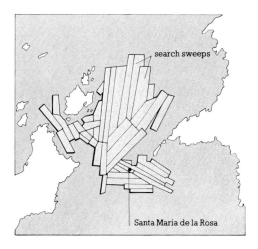

search sweeps

Santa Maria de la Rosa

In 1968 this pattern of swimline-search sweeps led to the finding of the sunken Armada ship *Santa Maria de la Rosa* off the coast of Ireland. Once such remains are discovered, the archaeologist faces a more delicate problem: how to locate artifacts within the wreck so that they can be recovered with the least possible damage. In the photograph below, a magnetometer probe is being used on the *Amsterdam*, an eighteenth-century Dutch ship that lies in English Channel quicksand, its hull timbers sometimes exposed at low tide. The probe can detect metal objects or even pottery as it is played over a marked-off area.

or mud, a magnetometer adapted for underwater use becomes a valuable guide to the size and shape of a buried vessel. Lastly, there are numerous devices for achieving accurate surface navigation in a search boat – accuracy being essential, of course, in a situation where hundreds of likely targets are identified for closer examination later on. Not only sextants and shore-based theodolites, but also radio navigation systems, radar, and acoustic beacons dropped to the sea floor are used for fixing the position of suspected wreck sites.

When the archaeologist finally starts work at an underwater site, he wants as far as possible to work with the same precision as his colleagues on dry land. Archaeological precision requires that during the long, slow process of mapping, photographing, excavating, identifying, and protecting all structures and artifacts (such as ship's timbers, pottery, walls, tombs, coins, etc.), there must be an on-the-spot recording of

the exact position in which every article is found. A jumbled heap of objects dragged from the water and piled up on the deck of a boat is merely salvage, not archaeology. Apart from trying to identify and date each object with accuracy, the marine archaeologist must apply his mind and skill to interpreting the juxtaposition of finds (much as accident-investigation experts study the position of all the scattered wreckage of a plane in order to figure out what happened). It is obviously much harder to do this with buried artifacts underwater than on land, where a trench or pit dug into the ground provides a permanent vertical record of stratigraphy. On land, the finds that are lifted out can be related to the layers of soil (known as "horizons") that are exposed in the vertical walls of the pit. Measuring of distances can be carried out at leisure; photography, drawing, sketching, and note-taking are easy; when in doubt, you can ask a colleague to sit down with you in the excavation so as to look at the problem while discussing it. Underwater, there's little leisure or ease; and no helpfully permanent trenches can be dug in the sea bed.

But the marine archaeologist is by no means without helpful tools. For instance, there *are* ways for divers to keep track of the stages of a survey or excavation by making underwater sketches or writing underwater notes. And instead of starting by digging a trench or pit, the undersea archaeologist lays out reference points – either a series of carefully placed markers or a whole grid of tapes – which, along with the measurements needed to locate each item relative to the reference, comprise the essential components of an underwater survey. The choice of reference system depends on the size of the site and its distance from the shore. The total area within which the hull and cargo of an ancient ship lie is seldom more than about 30 by 60 meters, and the water depth is unlikely to exceed 50 or 60 meters (at any rate, no wrecks in deeper water have been excavated so far). A submerged city or harbor will cover a much greater area, naturally, but in water that is usually shallower than 10 meters. (Stone Age caves may lie as deep as 100 meters – which is one reason why they have not yet been studied seriously.)

On a small site, a grid of plastic tapes, with a spacing of 1 to 5 meters, is commonly laid over the whole area. Some archaeologists, though, prefer a rigid grid of steel bars, for the rigidity not only improves linear accuracy but also constitutes a fixed base to which the divers can attach cameras and various kinds of measuring equipment. On a city site up to a kilometer square, a complete grid is not practicable, but the intersections of hypothetical grid lines are generally surveyed in and marked with either numbered stones or plastic labels. Tapes can then be laid between any four markers to define a particular small area where the archaeologists want to work.

How measurements are made within the designated site depends on the slope of the sea floor as well as on the size and depth of the spot. Working on a small, flat site, you can fix the position of each item by stretching short tapes between it and two reference points or grid lines (a method called "trilateration"). Quicker, because it does not involve moving tapes around, is the old navigational technique of triangulation – i.e., using an underwater theodolite to measure the angles from two reference points to each target in turn. This assumes that you can see a reasonable distance – up to 15 meters, say; the tape-trilateration system is more practical over distances where you cannot see from end to end of the tape, and obviously has to be used in dirty water. If the site is steeply sloping, or is obstructed by rock outcrops, it's also essential to measure the difference in altitude between points. You can do this again by means of a theodolite, or else you can either measure the pressure difference between two loca-

tions or – if the site is small enough – fix some additional rigid grid bars in position, so as to take direct measurements of vertical distances.

None of this is done, of course, until a preliminary survey of the undisturbed site tells you that excavation of a new underwater discovery is likely to prove fruitful. Archaeologists know of many hundreds of wreck sites, chiefly in the Mediterranean, western and northern European waters, and the Caribbean, and the choice of which ones to excavate is difficult. If an examination of a sampling of material indicates that the results of a full study would justify its costs, the site is marked out and divers must prepare to lift off accumulations of sand, coral, or rocks layer by layer in such a way as to get at archaeological material without damaging or displacing it. And throughout the lengthy process, surveying must continue, since each removal of a layer of debris is likely to reveal new deposits, and these must be plotted *in situ* for the sake of archaeological precision. Thus a three-dimensional map of the whole deposit will be slowly built up as the excavation progresses.

For detailed excavation work and the freeing of fragile objects from sand or mud, there is no substitute for the diver's experienced hand; a gentle manual winnowing and picking away of clay or concretions ensures that no damage is done. Power-driven suction tools are usually used only for clearing already discarded waste

Once the position of a wreck has been pinpointed, excavation can start. Below: A rigid steel or plastic grid (1) is laid over the remains to facilitate excavation and recording in three dimensions. A camera mounted on a tower over the grid (2) takes stereoscopic pictures for the purpose of precise reconstruction. An airlift pump (3) powered by a compressor on the support boat (4) removes sand and mud and catches artifacts in a net at the barrel end (5). Labeled small finds are lifted to the surface by an inflated bag (6). Right, top: Underwater notes are written by an ordinary hard pencil on a roughened plastic surface. Bottom: An underwater theodolite for measuring angles rests on a drawing table.

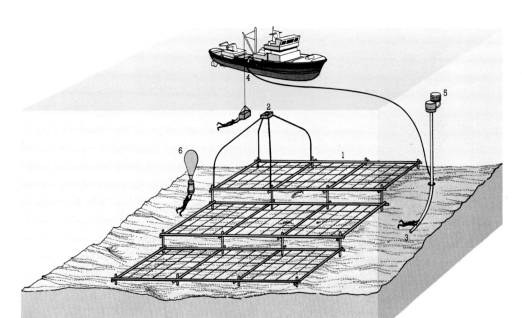

material away from the excavation area. Particularly delicate artifacts require very special treatment, naturally. Such things as baskets, wooden bowls, fragile pottery, or wooden wheels may have to be specially reinforced or packed in frames before being lifted out of the water. Larger wooden structures, such as a ship's hull, may have to be cut up or taken to pieces on the sea floor, and this reinforces the importance of precise prior surveying. Consider the problems involved in rebuilding an ancient ship from timbers which have been taken completely apart, and which, after being raised, must be chemically treated to prevent decay during the course of perhaps a year or more before they can be re-assembled!

The marine archaeologist's work goes on long after the team has finished the job of excavation. Details and photographs of preliminary and work-in-progress surveys have to be interpreted and drawn up accurately for publication. All finds have to be preserved, identified, and dated. And the mass of accumulated data has to be synthesized into a large-scale report that analyzes every possible piece of information gleaned from the excavated material about the people and society that created it. Only if we do all these things well and thoroughly can we justify interfering with the sea's archaeological treasures, which, if undisturbed, would preserve their secrets intact for future – and perhaps wiser – generations to unlock.

As indicated in the diagram on the facing page, a good way to remove objects worth procuring from the excavation site is to lift them to the surface by means of air-filled bags. Left: Here such bags are being used for lifting cargo from the Dutch merchant ship *Batavia*, which sank off the coast of Australia in 1629. Below: On the same site, a diver holds a tube that sucks up sand and mud, which is then deposited on the sea floor some distance away – but not before being filtered through a net to recover possibly valuable material.

Because much can be learned from the relative position of artifacts within a wreck or ruin, accurate measurements are essential. The diver below is checking the position of an object in relation to a grid line.

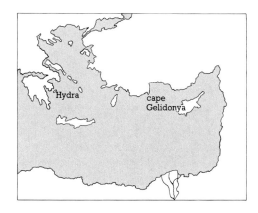

The 1960 excavation of a Bronze Age wreck off Cape Gelidonya in Turkey is considered to have been the first truly disciplined underwater "dig." This diver is systematically prying away brushwood dunnage – still intact after more than 3,000 years – that was used for wedging cargo in the hold.

Setting the Standards

Until 1960 the academic world largely ignored undersea archaeology. Conventional archaeologists often expressed interest in antiquities raised by divers, but they regarded such finds as little more than accidental discoveries picked up at random from the sea floor – which, indeed, most of them were. When archaeologists were involved in salvage operations (as were, for instance, Professors Fernand Benoît and Nino Zamboglia, each of whom helped to supervise the raising of cargoes from ancient Roman wrecks during the early 1950s), the underwater work was carried out by engineering rather than archaeological methods, while the archaeologist observed from the surface or from his university chair. The idea that the context in which submerged objects were found might be as significant as the objects themselves was ignored, even though archaeology on land had evolved into a highly sophisticated investigative discipline. Then, in 1960, this appalling situation was ended, mainly through the efforts of a virtually unknown American archaeologist with a taste for adventure: George Bass, who has been called the father of underwater archaeology.

George Bass had just completed his training as a classical archaeologist at the University of Pennsylvania when he was chosen to lead an expedition to investigate a Bronze Age wreck lying off Cape Gelidonya in southern Turkey. The wreck had been discovered in 1959 by Peter Throckmorton, an American journalist with an interest in archaeology, and Mustafa Kapkin, a Turkish underwater photographer. In ancient times, Cape Gelidonya, a rugged headland inaccessible from the land itself, was a critical turning point in coastal voyages from Egypt, Cyprus, and the Levant to Crete and Greece, and many ships must have gone down within sight of the towering cliffs that descend into deepening water. Having heard from a sponge diver of a submerged

wreck that apparently held a cargo of bronze, Throckmorton and Kapkin suspected it might be a rare relic of the Bronze Age, and so they set out to explore the Mediterranean waters near the cape.

After a long, grueling search, they did indeed come upon a strange undersea mound that proved to contain copper ingots in the shape of oxhides. A sampling of the ingots were identified as products of the thirteenth century BC by Professor Rodney Young of the University of Pennsylvania Archaeological Museum, and the university decided to sponsor a further exploration of the wreck under the leadership of George Bass.

Deeply interested in the project, he prepared for it by – of all things! – learning to dive (a much easier task, as he pointed out later, than learning to be an archaeologist). His aim was simple: to survey and excavate the wreck *at first hand* according to the principles he had been taught to apply when working on "normal" sites. In organizing operations in this way, he dispelled at a stroke the mystique surrounding ancient wrecks that divers themselves had done much to propagate. And in rejecting the hit-or-miss methods of the past, he shattered the complacency of landlocked academics and launched the era of the "wet" archaeologist.

The Cape Gelidonya wreck lay on a rocky bottom, where the lack of a protective cover of mud or sand had caused the wooden hull to disintegrate almost completely. Only the cargo – a ton or so of copper and tin ingots and baskets of scrap bronze – remained relatively intact. In the course of thousands of years, the metal objects had become encased in a rock-hard matrix of concretion. First, Bass and his team plotted the position of the concretion by triangulation, the classical trigonometric method of surveying the earth's surface; then they pried it from the sea bed for reassembly and cleaning on the narrow beach where they were encamped. As they freed each object from the con-

cretion, they recorded its precise location on a master plan of the site.

Within the hard mass of material, Bass found fragments of the ship itself, including the dowel-fastened planks of its outer hull and the brushwood dunnage that had been used for wedging in the cargo. By studying in minute detail all such recovered material, in addition to having recorded the sea-bed features during excavation, the archaeological group was able to piece together a great deal of information about the ship, its purpose, and its origin. The vessel, Bass deduced, had been a trader about 11 meters long belonging to an itinerant bronze-smith, and it was carrying a cargo of tools, scrap metal, and ingots from Cyprus when it went down. The copper ingots bore Cypro-Minoan inscriptions, and their hide-shaped form suggested a date of about 1200 BC – a date later confirmed by radiocarbon analysis of the brushwood dunnage. A Syrian cylinder seal and five scarabs made on the Syro-Palestinian coast identified the ship's origin; and sets of balance-pan weights conforming to standards then in use in Syria, Cyprus, Egypt, Crete, and Asia Minor indicated the zone over which it had traded.

Though American-led, the team of divers who worked on the wreck – a historical operation in so many ways – included archaeologists and competent amateurs from France, Germany, and Britain. They had to do their underwater work on a steeply sloping rockface at a depth of about 30 meters, and so there was always the risk of accidents or of decompression sickness when ascending from a dive. In their determination to do the job safely as well as thoroughly, George Bass and his principal assistants established techniques for systematic control, both of diving safety and archaeological accuracy, that have since become standard in marine excavations. Their dedication and hard work yielded the most important accumulation of pre-Classical copper and bronze artifacts ever found in the Aegean area – an unparalleled contribution to archaeology.

More than that, the 1960 Cape Gelidonya expedition proved that systematic undersea archaeology is a highly valuable scientific discipline.

The finds from the Gelidonya wreck are now displayed in the Turkish Aegean seaport of Bodrum, where a special museum has been created for them. Until a couple of years ago, no other Bronze Age wreck had been found. Then, in the summer of 1975, Peter Throckmorton did it again: diving off the island of Hydra with a team from the Greek Institute of Marine Archaeology, he found a mound of pottery in 20 meters of water. The pottery, which has been identified as of the pre-Mycenaean Cycladic type, probably indicates the site of a wreck that must date to about 2500 BC! If the ship's timbers have survived, excavation will push our knowledge of shipping back more than a thousand years to the days of the early navigators who first colonized the Mediterranean islands.

Below: How a wooden hull is likely to disintegrate in salt water after settling on its side at the bottom (1). Within ten years the sides will have collapsed and the decks caved in, sandwiching all the ship's contents between them (2). After another couple of decades, marine borers and currents will have broken up most of the hull, while the lower sections will have begun to sink into the mud (3). A full century after sinking, almost nothing remains unburied; but the buried timbers and other relics are now protected and will last indefinitely (4).

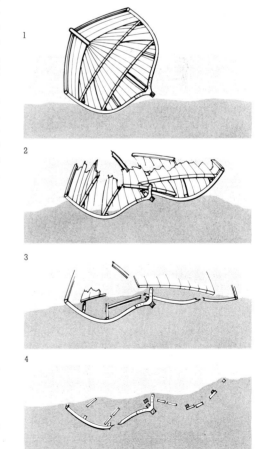

Left: Some of the 34 hide-shaped copper ingots – cleaned and rearranged as they were found on the sea bed – that comprised the main cargo of the Cape Gelidonya wreck. The ship quite probably belonged to a traveling copper-smith.

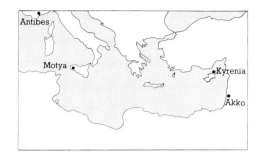

Wrecks of the Ancient World

From the eighth through the third centuries BC the whole Mediterranean area was colonized by two competing maritime civilizations, the Greeks and the Phoenicians. These powers carved up the Mediterranean world between themselves, expanding their cultures almost entirely by sea, and there was a good deal of piracy and marine warfare, as much among rival cities within each culture as between the two. So there must be hundreds of ships, warships as well as merchantmen, lying on the bottom in those busy waters. From written records, sculptures, and painted vases, we know how varied the ships and their contents must have been: sleek military vessels rowed by two, three, or even five rows of oarsmen working in various combinations on several banks of oars; slow-moving heavily armed ships carrying hundreds of troops and catapults that could hurl half-ton rocks at city defenses; fat-bellied merchantmen trading in oil, fish, nuts, grain, spices, wool, tar, silver, gold, salt, wine, pottery, lead, copper, and bronze; and all sorts of smaller craft for fishing and working in harbors or as ferries.

Our knowledge of daily life in ancient times is broadened considerably whenever we find and excavate one such vessel. Yet, numerous as they must be, we have so far fully excavated only three wrecks (apart from the Cape Gelidonya one) that antedate the end of the third century BC, when Rome began to become the dominant Mediterranean power.

Divers have discovered hundreds of ancient craft scattered here and there on the Mediterranean floor and have salvaged bits and pieces of many of them, along with parts of the cargo; but truly significant excavation from the archaeological standpoint involves removal of the cargo down to the ship's timbers and reconstruction of the ship itself. The only boats of the great Greek-Phoenician period that have been excavated to that extent are three Greek merchantmen. One fifth-century Phoenician wreck off the coast of Israel was explored in 1971–72 by Israeli divers, who brought up several statuettes of the Carthaginian goddess Tanit and masses of Phoenician pottery, but no timbers of the wreck itself could be found. In 1971, a British diving archaeologist came upon another – and possibly more exciting – relic of Phoenician power in the shallow water off the coast of western Sicily; partial salvage of its timbers indicates that it was probably a third-century warship, but only the stern part of the ship has been found (and very few artifacts). Of course, one would not expect a large cargo on a warship, but the absence of any special military structures or weapons is puzzling. Work continues on this wreck, and it may yet prove to be the first known example of an ancient fighting ship. In short, the total sample of pre-Roman ships is both limited and unrepresentative; we have only the tiniest evidence for the ship-construction techniques of the Phoenicians, who were certainly the most accomplished sailors of the first half of the first millennium BC, and we have found no Greek warships.

Of the excavated Greek wrecks, one – a boat of the sixth century BC found off Antibes in France in the 1950s – was not as carefully worked on as it might have been in more modern times; and the second – a fifth-century wreck discovered near the northeastern tip of Sicily – was, unfortunately, not very well preserved in the water. The most completely reconstructed of the finds is a fourth-century boat that a Cypriot diver discovered in 30 meters of water northeast of the city of Kyrenia and pointed out to Michael Katzev, a University of Pennsylvania colleague of George Bass, in 1967. A preliminary survey of the site by means of metal probes and a proton magnetometer revealed a cargo area of about 10 by 20 meters, within a small patch of which Katzev and his team of divers could spot a number of fourth-century amphorae. During the next two years a rigid plastic grid was mounted over the wreck, and 404 amphorae – most of them from Rhodes, for transporting wine – were brought up. The cargo also included 10,000 almonds – their cloth containers rotted away but their shells intact – and 29 heavy grinding stones, which, since they were not matched pairs, were probably being used as ballast.

Twice a day, a diver swam over the grid and photographed the progress of the excavation with pairs of cameras, so as to

In 1967 a Cypriot diver found the hull of a fourth-century-BC merchant ship in 30 meters of water off the northern coast of Cyprus. After a visual search that spotted a number of amphorae on the sea bed (right), followed by a proton magnometer scan, excavation work began toward the end of the 1960s. Left: Here the buried timbers, now exposed and straddled by a plastic grid, are being further uncovered by divers. In 1973 the fully reconstructed ship was put on public display in nearby Kyrenia; it is our best example so far of an ancient Greek vessel.

prepare a stereoscopic reconstruction. By 1973, after half of the hull had been uncovered intact and each piece had been subjected to a preservation process, the entire hull was finally reconstructed and mounted for public viewing in an air-conditioned room of the castle at Kyrenia. Built almost entirely of pinewood, the 18-meter-long ship was sheathed in lead fastened with copper tacks, to prevent marine borers from attacking the wood. It was undoubtedly constructed according to the "shell-first" technique: the ancient ship-wrights first laid the keel, then built up the sides of the vessel over a temporary frame, fitting the planks together with mortise-and-tenon joints, secured by wooden dowels. When the shell was complete, the interior frames were fitted and fastened by copper nails. The step for the mast was only about a third of the length back from the bow – a placement that Katzev thinks may mean that the sail was rigged fore-and-aft rather than square. From the number of domestic pots and plates aboard, we can deduce that the crew consisted of only four men, including the captain. And the vessel was far from new; carbon 14 dating of the timber gives 389 BC as the date of her construction, while a coin found in the wreckage was certainly minted no earlier than 306 – and so the ship

must have seen at least 80 years of service before she went down.

Lively as the sea was during the centuries of Phoenician and Greek maritime glory, the traffic in Roman times must have been much heavier – which is at least a partial reason why we have been able to excavate more Roman than Greek vessels. By the beginning of the Christian era, Rome's imports of grain alone had reached about 150,000 tons a year, most of them seaborne. Cargoes of marble and granite for the façades of public buildings, theaters, and temples were transported all over the Mediterranean world, as were sculptures and statues of bronze and marble. Enormous quantities of other goods must also have been carried by sea, and there were in addition the ceaseless military patrols of fleets based at Misenum near Naples, Caesarea in Algeria, and Seleucia Pieria near the modern boundary between Turkey and Syria. Surprisingly, as with the earlier period, no warships have been excavated so far. But we been able to study several cargo ships.

Divers working off the coasts of Spain, France, Italy, and Tunisia have excavated a sufficient number of Roman wrecks to reveal not only the general nature of the timbers but the minutest details of construction.

Jacques-Yves Cousteau, commenting on the quality of workmanship, has said that it is more like cabinet-making than ship's carpentry. All the ships studied so far probably had square-rigged sails, carried cargoes weighing from 120 to 350 tons, and appear to have been better equipped for long voyages than were their Greek predecessors. For instance, in contrast with the earlier vessels, where there was no provision for cooking on board, almost all the Roman merchantmen have a kitchen area of hearth tiles and roof tiles furnished with terracotta braziers and cooking utensils. Clearly, food was prepared on board during long voyages, with due precautions taken against the risk of fire.

There remains a need for further research on the existing information revealed by our successful excavations of the Greek and Roman wrecks. But there are many more ancient ships still to be found. From the analysis of their cargoes we will someday know a great deal more than we now do about trade routes in past centuries, and reconstruction of more and more hulls will help to trace the evolution of shipping techniques. One day soon, too, we shall no doubt discover the remains of a warship that will shed light on naval warfare as practiced by the ancients.

In excavating the wreck of a very large Roman merchantman found off Torre Sgarrata in the Gulf of Taranto, divers were able to recover her entire cargo of 150 tons of marble, which the second-century-AD ship was carrying from Asia Minor to the Italian mainland. Right: Air-filled bags are being used to lift a stone sarcophagus to the surface.

Above: An aerial view of the cofferdam-surrounded Roskilde site in Denmark within which lay the drowned remains of five Viking ships, here seen high and dry after the water had been pumped out. Numbers 2 and 4 in the insert are fragments of the same fighting longship. Top right: Wreck 1 close up, showing some of the scattered stones that were evidently loaded aboard the ships so as to sink them and blockade the harbor. Bottom: After years of archaeological work on this wreck (a deep-sea trader), its timbers, strengthened by special frames, are lifted into position on the reconstructed hull.

Byzantine Intricacies

Nine wrecks – five in Danish waters, one in Germany, one off Sicily, and two off the west coast of Turkey – are our only marine witnesses to seagoing trade and war in the thousand years between the fall of Rome in AD 476 and the fall of Constantinople. The boundaries of states and empires fluctuated dramatically during this period as northern European tribes expanded southward, Turks moved westward, and Arabs pushed westward through Africa and northward into Spain. Meanwhile, the magnificent Byzantine Empire, founded by Constantine in AD 330, grew and then shrank until it was crushed by the Turks, who captured Constantinople in 1453. Marine warfare played a vital role in both the rising and falling fortunes of Byzantium.

At the beginning of the period, existing seagoing vessels ranged from the heavy cargo ships and oared fighting galleys of the Mediterranean to fragile Nordic boats built of wood stitched together with thongs (rather like an Eskimo kayak). The stitched boats were not strong enough to support a mast, but nails came into use in the fifth century, and later Viking ships had a strong keel, mast, and sail while retaining much of the speed and lightness of the earlier craft. By 800 the Vikings were raiding all through western Europe, and Charlemagne, the Frankish emperor who died in 814, is said to have wept in his Mediterranean castle as he watched the Viking ships heading eastward on the horizon.

There are many medieval paintings, fres-

coes, and tapestries in which ships are portrayed; perhaps the most famous is a representation in the Bayeux tapestry of William of Normandy preparing to invade England. There are also descriptions of sea battles in the Norse sagas and other narratives. As always, however, artistic and literary descriptions of ships are suspect from a technical standpoint. We can really understand how the ships of a long-gone age were built and used only if we study their remains. The Vikings, to be sure, often buried complete hulls along with seafaring chieftains, but such ships are so highly decorated that it seems likely they were mainly used for ceremonial purposes. Once more, then, it is actual wrecks upon which we must depend for facts rather than fancy.

In 1956, amateur divers found the wrecks of five ships blocking Roskilde Fjord in Denmark, and from 1957 to '59 a team of archaeological divers charted the wrecks and discovered that they dated from about AD 1000 and had been deliberately loaded with heavy stones to sink them and blockade the fjord during a battle. Since the boats were in only three meters of water, the archaeologists were able to build a steel coffer dam around the site and pump out the water so as to survey and salvage every fragment and artifact. The work took years, and it was not until 1969 that the ships were moved to the Viking Ship Museum in Roskilde. Fortunately for students, the five wrecks are all different. There is a deep-sea trader 17 meters long with a 5-meter beam, a coasting trader 14 meters by 3, a fighting longship 27 meters long but of uncertain width, a small

warship, and a little ferry or fishing boat. The big trading ship is of such heavy design that Dr. Ole Crumlin-Pedersen, who directed the entire operation, thinks it was built for North Atlantic voyages and is probably similar to the ships used by Eric the Red and Leif the Lucky; if Pedersen is right, this is the only known example of such ships.

Moving southward, we come to the Hanseatic League towns of Lübeck, Hamburg, and Bremen – the northern German ports that dominated European trade in the thirteenth and fourteenth centuries, chiefly by means of a type of ship known as the "cog." Cogs were broad, high-sided, single-masted vessels; they were built for carrying heavy cargoes through rough seas, but could be converted into troopships with castles fore and aft, from which bowmen could fire down on the enemy. Until the discovery of a cog at a depth of eight meters in the estuary of the Weser at Bremen in 1962, no modern man had ever seen one. Most of the hull was recovered in late '62, and nearly all the timbers had been lifted out of the mud by '65. Reconstruction shows that the 24-meter-long hull had a very straight stem and stern to support the high castles, and that its cargo capacity was some 130 tons – a rather large amount for a late-medieval ship plying the stormy tidal waters of the north. Actually, the Bremen cog marks the end of a line, for European shipwrights began to develop three-masted ships toward the early 1400s.

Continuing our medieval journey southeastward we come to Sicily, off whose rocky coast a large ship carrying nearly 300 tons

Top left: When the harbor at Bremen on the Weser estuary in Germany was dredged in 1962, engineers uncovered the remains of a fourteenth-century "cog" preserved in the mud. The overlapping planks and heavy rib structure of the clinker-built ship can be clearly seen here; they show through where the planks are broken. Bottom left: A barge positioned near the wreck, now freed from the mud and protected by tall wooden posts, acts as a support ship for divers engaged in raising the hull.

Below: Wooden-hull construction methods. Ancient Mediterranean ships had planks joined by wooden tongues (tenons) fitted into mortise slots (1). In northern European clinker construction, overlapping planks were pegged together (2). Early vessels of both types were built shell first, with strengthening ribs inserted later (3). Eventually this process was reversed, with planks meeting flush (carvel-jointed) and secured directly to a prebuilt rib structure (4).

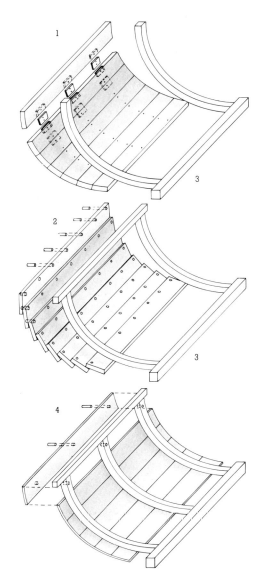

of marble from Byzantium was wrecked in about AD 540. In a series of expeditions between 1960 and 1967, this fascinating cargo was excavated by divers led by the German archaeologist Gerhard Kapitän. It turned out to be prefabricated stonework for a basilica. The divers raised 28 marble columns with separate bases and capitals, as well as panels and sections carved ready for a large altar with flanking stairways and a stone canopy, all made of white marble and decorated with green porphyry. Though there are very few preserved timbers of the ship itself, the cargo alone indicates the scale of Byzantine trade in architectural stone – and prefab stone at that!

Our last wrecks were discovered within 500 kilometers of Byzantium itself (later called Constantinople, now Istanbul). One of these dates back to the fourth century, the other to the seventh, and both lay in deep water off the rocky island of Yassi Ada near the Aegean coast of Turkey. They were mapped and excavated during many working seasons from 1961 to 1969 by teams of divers from the University of Pennsylvania led by George Bass, and Bass's meticulous reconstruction of the original curves of the hull has given us an almost perfect "shipwright's blueprint" of them. Both were light Byzantine merchant ships about 19 meters long, and both were carrying nearly a thousand amphorae of wine when they went down. It may seem a remarkable coincidence that both sank in 30 meters of water off the same tiny, virtually inaccessible island, but they are only two among many casualties of the Yassi Ada rocks. The

organization required to get all the equipment and accommodations for 20-odd divers onto the deserted island was extraordinary. But Bass had plenty of experience. Now he was able to improve and streamline the undersea-archaeology techniques he had pioneered at Cape Gelidonya and apply them to the complex hulls and cargoes of the Byzantine vessels.

The two ships, separated in age by three centuries, make an interesting pair, since they show the deterioration of Rome's shipbuilding tradition of edge-jointed planks held together with close-spaced wooden mortise-and-tenon joints. The spacing between tenons in ships at the height of the Roman Empire was usually about 10 centimeters; in the fourth-century ship it was 17 centimeters; in the seventh-century ship it varied from 40 to 90 centimeters. As the quality of "cabinet work" declined, the Byzantine shipwrights used an increasing number of bolts and copper or iron nails, and they fitted planks to a skeleton of ribs rather than building the shell first and then fitting the ribs. Thus their boats, while less craftsmanlike than earlier ones, are closer to modern wooden vessels in construction.

The Arabs first besieged Byzantium in 673, but their ships were beaten off by the use of Greek fire, a kind of early napalm. The episode shows, though, how early the Arabs started to be active at sea. As yet we have found no Arab or Turkish ships from the Byzantine period, but one day perhaps we shall. Only then will we begin to get a truly balanced picture of seafaring in the turbulent Byzantine millennium.

— — — Route of the *Amsterdam*

Left: The hull of the Swedish warship *Wasa*, sunk soon after her launching at Stockholm in 1628, breaks surface after over three centuries on the bottom. Below: How the *Wasa* was lifted and moved in easy stages up the sloping floor of the channel. Buoyant pontoons were attached to the ship, then flooded so as to begin sinking. Next the cables were tightened, and the pontoons – now filled with air to lift the hull – were tugged forward till the suspended ship touched the upward-sloping bottom. This process had to be repeated many times.

Raising Whole Ships

Though most wrecked ships are at least partly broken up, there have been rare instances in which they have remained virtually intact. The best-known example is that of the *Wasa*, a handsome Swedish warship that was struck by a sudden squall and sank in 33 meters of water at the very start of her maiden voyage from Stockholm harbor in 1628. Divers salvaged most of her guns shortly afterward, but nothing further was done about the wreck until 1956, when an amateur historian discovered that she still lay in the muddy bottom of the channel where she sank, partly buried but almost whole. In 1958, a massive recovery operation, powerfully backed by the Swedish Government, got under way. Although George Bass had not yet shown that marine archaeologists could be active divers and could themselves do underwater excavating, shore-based archaeologists worked with engineers to solve the formidable technical problems of raising the bulky yet fragile structure without harming it.

With great courage and skill, Swedish navy divers drove six tunnels beneath the hull and passed heavy lifting cables through them. Thus cradled, the *Wasa* was securely fastened to two large pontoons, which gently lifted her free of the clinging mud and brought her by easy stages into shallow water, where she remained submerged for two years while divers strengthened and consolidated the 300-year-old hull for the final lift into the air. After the ship broke surface, she was pumped out, floated into dry dock, and placed in position on a specially built concrete pontoon. This now serves as the base of an enormous enclosed capsule in which the *Wasa* is still being restored and treated for conservation.

The task of restoration is immeasurably eased by the fact that the ship was so extraordinarily well preserved during her centuries at the bottom in Stockholm harbor. One reason for this is that the water of the

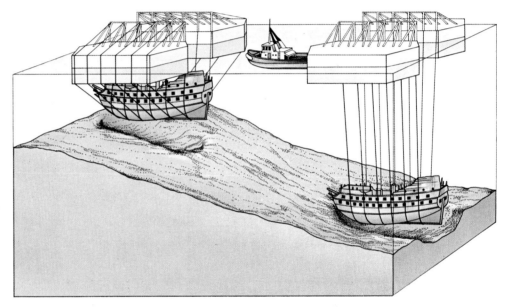

inner portion of the Baltic, where the *Wasa* sank, is almost completely fresh. Thus, the salt-water teredo worm, which is known as the shipworm because of its hearty appetite for timbers not protected by sand and mud, cannot survive there. Seventeenth-century wrecks in the Atlantic, Mediterranean, or any other salt sea would have been eaten away down to the sand line, except for possible fragments encased in coralline growth. Even the elaborate wooden carvings that decorated the *Wasa's* exterior were salvaged practically uninjured; divers found most of them in the soft sea bed, where they dropped when their iron fastenings had rusted away. And when the mud-filled hull reached the surface, archaeologists who excavated it deck by deck found most of the gun carriages in position, along with a few bronze guns the seventeenth-century divers failed to salvage.

Although the *Wasa* had not yet been fully fitted out when she sank, the raised hull nevertheless contained many things of tremendous archaeological interest, including not only a carefully folded set of sails in the sail locker and sea chests belonging to crew members, but also the bodies of several of the drowned men, who lay among the muddy debris, some with their clothing still intact. In all, many thousands of objects, each relating to a particular aspect of life aboard a seventeenth-century warship, were recovered. When they have been recorded, treated for preservation, and reassembled, the *Wasa* will live again, not only for maritime historians but for all who visit this dramatic survival of a past age.

There are many other old ships known to have sunk in the fresh waters of the Baltic and the Great Lakes, and some may well be in a fine state of preservation. But the most interesting wrecks from a historical standpoint tend to be in the teredo-infested salt

After Sweden's *Wasa* was raised in the early 1960s, the hull was towed to a dry dock where, supported by beams and scaffolding (top left), it underwent repair and restoration work. The vessel was then moved to a more convenient spot and an aluminum framework was built around her. Thus protected, she was given further treatment for permanent conservation (above), including constant hosing to prevent the wood from drying out too soon.

Top: Any effort to raise the Dutch East Indiaman *Amsterdam*, which sank in the English Channel in 1749, will involve removing a vast weight of sand from her hull. So far she has yielded such artifacts as the above wine bottle and cannon bearing a crest of the Dutch East India company. Right: Survey and exploration work suggest it may be even harder to rescue Henry VIII's warship *Mary Rose*, which lies in swift-flowing muddy water.

oceans, where the hull is worth salvaging only if completely buried in sand or mud – and the technical problems and costs of raising such vessels are very much greater than for the *Wasa*. That is why work is progressing rather slowly on two historic ships in British waters that are known to be well preserved. One of them is the *Amsterdam*, a Dutch East Indiaman that ran aground on an English Channel beach near Hastings in 1749 and swiftly settled in quicksand. Her top hull timbers are exposed during very low tides, and in 1969 an enterprising contractor dug into the wreck with a mechanical grab and extracted quantities of still intact cargo before his destructive – though illuminating – exercise was halted.

Given the necessary financing and technical expertise, the whole ship could perhaps be raised, *Wasa*-like, with its remaining cargo. But it would not be an easy job. Because the *Amsterdam* lies between high

and low tide, conventional lifting or working vessels could not be floated around her; and the weight of the hull, full of sand as it is, would make it impossible to move the ship without either strengthening the hull or emptying it, or doing both. Various schemes have been suggested for coping with the difficulties, and the city authorities in Amsterdam, the vessel's home port, are taking a direct interest in salvage proposals. So it is possible that work will go ahead one of these days in spite of the high costs.

In the deep mud of the Solent – the narrow channel between England and the Isle of Wight – lies a great warship that is older and potentially even more exciting than either the *Wasa* or the *Amsterdam*: the *Mary Rose*, a battleship of Henry VIII's navy built in 1509, which capsized and sank during a battle with the French in 1545. She was discovered during wreck-clearing operations in the 1830s, when several of her guns were

raised; and she was relocated in 1967 by a determined British historian and writer, Alexander McKee, who organized a search that took three years of echo-sounding and diving by team after team of amateur divers – one of the great success stories of undersea detective work. In the early 1970s, McKee and an archaeologist colleague, Margaret Rule, carried out survey operations that indicate that a considerable part of the hull survives, and that the *Mary Rose's* contents are exceptionally well preserved. The ship, however, is much more deeply embedded in the mud than was the *Wasa*. Moreover, since she was a fully operational fighting ship with some 500 men on board, she probably contains a much greater number of priceless artifacts, which would pose vast problems of recovery and conservation. It will be some years, therefore, before the *Mary Rose* takes the 20-mile trip back to Portsmouth, where she was built.

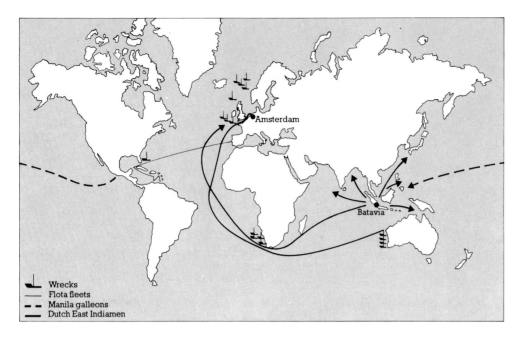

Wrecks
Flota fleets
Manila galleons
Dutch East Indiamen

America's Kip Wagner (the white-haired man) is an outstandingly successful treasure hunter. Here he helps haul a basket of coins and other material from Florida's coastal waters, where Spanish treasure ships were often wrecked. Unlike many other amateur salvors, Wagner voluntarily cooperates with archaeologists.

Treasure Galore

The voyage of Columbus to America in 1492 and the opening of trade routes to the Far East via the Cape of Good Hope caused an explosion of maritime activity around the globe. Between the late fifteenth century and the end of the seventeenth, a rather small number of men motivated chiefly by self-interest, though sometimes by patriotism and religion, laid the foundations of European empires that were to dominate the world until modern times. They achieved this feat by means of guns and sailing ships. Foremost among them, of course, were the Spanish conquistadors of Central and South America; but while the Spanish were carving out their empire in the Western Hemisphere, a lucrative spice trade with the Far East was also being carried on as a monopoly – first (for a short time) by Portuguese seafarers, then by the more aggressive Dutch. The New World and the ancient Orient were, in a way, linked by what one historian has called "a river of silver" flowing from west to east.

The source of this gleaming "river" was the rich mines of America, from which the conquistadors took vast quantities of silver, as well as lesser amounts of gold. They transported their treasure across the Atlantic to Europe, and much of it was then borne off to the East to purchase spices and other exotic goods. There was also a westward flow of the New World's treasure – from Mexico across the Pacific to the Philippines – in the latter years of the sixteenth century; the so-called Manila Galleon sailed once a year with a cargo of silver for the East, and it brought spices back to Mexico. This annual trade was restricted by the fact that the Spaniards wanted most of their silver to go to the homeland. Even so, the various ships that made the run over the years were reputed to be loaded with fabulous riches, and

Throughout the sixteenth and seventeenth centuries, the trade routes that linked Europe with her colonies and customers were a watery graveyard for frail sailing vessels. If the sea could be drained all along the routes drawn on this map, they could be viewed almost literally as being paved with gold and silver.

several of them are known to have been wrecked off the western coasts of the Americas. But it is the enormous losses of the America-to-Europe traffic that have tempted the treasure hunters of later centuries to act upon their dreams of finding a fortune under the waves.

Treasure hunters, as has been pointed out, are not archaeologists. Their activities often run counter to the interests of marine historians. Sometimes, though, the adventurous amateurs do lead the way to valuable archaeological discoveries. It is at least partly owing to their imaginative and single-minded daring that we now know as much as we do about the ships and men that created the "river of silver."

What a treacherous river it was! Oceanic sailing before the nineteenth century was such a hazardous business that losses of 10 per cent on each voyage of a Spanish *flota* (the armed convoy that sailed regularly from Caribbean ports) were accepted as commonplace. Even as late as 1743, a director of the Dutch East India Company coldly observed that the loss of five or six ships a year on the spice-trade route could easily be written off against profits. It is no wonder that the well-defined trade routes are still littered with the remains of richly laden ships. Hurricanes were, of course, a major menace in and around the Gulf of Mexico and off the coast of Florida (as they are today), and individual *flota* wrecks have been found all over the area. Though, sadly, archaeologists have not always been permitted to study and record the findings, there is increasing cooperation these days

between treasure hunters (who are too often entitled by law to keep their booty) and government authorities whose primary interest is documentary.

To be sure, it is hard for the most sober person not to get excited at the prospect of dredging up some of the conquistadors' long-lost riches. Consider, for example, the case of Kip Wagner, a middle-aged building contractor who lives near Cape Canaveral in Florida. Twenty years ago, he became a dedicated beachcomber when he discovered that old Spanish coins kept turning up on the beaches whenever heavy storms had roughed the sands about. By 1957, convinced that the coins were coming from wrecks that lay substantially intact offshore, he imparted his enthusiasm to a number of energetic romantics like himself, and they started in on a long period of very hard and unromantic diving – which resulted, after several years, in the recovery of the greatest amount of Spanish treasure ever found! Wagner and his associates located the wreckage of several ships and salvaged not only thousands of gold coins and silver wedges, but all sorts of items – clothing, buckles, porcelain, jewelry, ropes, muskets, daggers, grenades, naval guns – that tell us something about life aboard the ships. Under Florida's laws, the salvors could profit from selling most of the treasure, and they did. But the state posted an archaeologist to work with Wagner's team, and Wagner himself did everything he could to ensure that the findings were properly surveyed as archaeological relics.

Treasure ships are not the only vessels

that have been found in the area. Small trading craft, and sometimes pirates, thronged the Caribbean islands and seaways during the first centuries of European conquest, and many were lost. What is believed to be the wreck of a small pirate vessel dating back to the mid-sixteenth century has been excavated in the Bahamas under the guidance of Mendel Peterson, curator of coins at the U.S. Government's Smithsonian Institution. Peterson's study of the remains would be quite helpful to any modern designer of pirate sailing craft: the vessel was narrow-beamed and pointed at bow and stern for extra speed, with a formidable armament of quick-firing breechloaders.

Apart from a large warship that sank in the Indian Ocean close to the African coast in 1696, few wrecks dating from Portugal's great trading days have been found. But the Dutch East Indiamen that made the two-year-long round trip from Holland to Indonesia left many of their representatives in the waters off the European, African, and Australian coasts, and divers have located several of these. Some of the wrecks, whose dates span the Dutch East India Company's two centuries of existence, have given archaeologists much new information about the day-to-day workings of early commerce and industry. Aside from their intrinsic value, coin hoards provide economic his-

On display here are priceless objects found in the Straits of Florida by Kip Wagner in the 1960s. A Spanish fleet sailing from Havana in 1715 was battered by a hurricane, and Wagner's recovery of coins, jewelry, pewter, and ceramics from the wrecked ships adds up to an incomparably rich find of submerged treasure.

torians with samples of trading currency frozen, as it were, in transit. An examination of salvaged everyday goods, whether for trade or use on board, can be equally informative. Recovered pieces of evidence concerning the ships and their crews range from building materials carried as ballast to tools, domestic hardware, scientific instruments, and even knickknacks or toys.

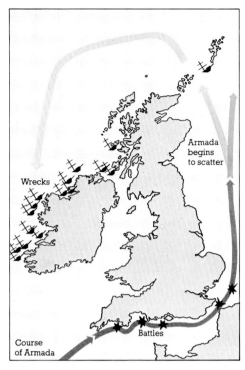

Was the Armada a Bluff?

Thirty wrecks of the Spanish Armada, the fleet dispatched by Philip II to invade England in 1588, lie under the stormy waters that surge against the western and northern shores of the British Isles. As everyone knows, the invasion failed, but contemporary records are so infused with propaganda and special pleading that the reasons for the failure have never been entirely clear. Now, thanks to divers who have found four of the wrecks, some missing pieces of the puzzle are being put in place.

The four wrecks, in order of discovery, are the *Girona*, a Neapolitan-built galleas found off the north coast of Ireland by a famous Belgian diver, Robert Stenuit, in 1967; the *Santa Maria de la Rosa*, a converted armed merchant ship of 945 tons, found in 1968 by British divers Sydney Wignall and Colin Martin in deep waters off the southwestern Irish coast; the *Gran Grifon*, the 650-ton flagship of a squadron of cargo hulks, which was wrecked near Fair Isle far to the north of the northernmost tip of Scotland, found by Colin Martin in 1970; and finally, *La Trinidad Valencera*, a 1,100-ton merchantman converted to carry heavy guns, discovered in 1971 by members of the City of Derry Diving Club in one of the many bays of northwest Ireland. Partial excavation of the last of these under the direction of Colin Martin suggests that it is the most important find so far.

The Armada invasion plan was, for its day, as ambitious an operation as the 1944 D-Day landings of World War II. Warships were pressed into service from the fleets of all Spain's allies. Large merchantmen were transformed into warships by building them up with superstructures and fitting them with heavy guns. There were rowed galleys and Mediterranean galleases, and there were heavy cargo hulks from the Baltic ports, equipped with light guns to protect them while they transported all necessary stores. But of the 130-vessel-strong total fleet, only 20 were large galleons built specifically for fighting in heavy seas. It was extremely hard to find enough guns to arm such a fleet. Europe was scoured for suitable cannon, while new ones were hurriedly made and extra iron shot of all calibers was cast at an unprecedented rate. When the motley fleet sailed from Lisbon in May, 1588, it was superficially a mighty force. The 130 ships, bearing a total of 8,000 seamen and nearly 20,000 soldiers, were fitted with 1,500 bronze guns and more than 900 iron guns, and their mission was not only to defeat Elizabeth's Royal Navy but to pick up a huge Spanish army at Calais for a full-scale invasion of southern England.

As it turned out, however, nothing went right. Bad weather caused long delays, and the enemy were waiting when the Armada finally reached the English Channel late in July. Then, as the Spaniards were trying to embark their army at Calais, the English sent in fire ships, and the Spanish fleet scattered and was blown into the North Sea. Retreat through the Channel was now blocked by the wily English navy, and so there was no way back to Spain but northward – a route of nearly 3,000 kilometers around Scotland and western Ireland.

The invasion had been a failure, but the return was a disaster. Forty ships were lost in the savage storms of the northern seas, and although the other ships picked up more and more of the survivors, only about a third of the total of almost 30,000 men lived to see the following spring. The wreck of the *Girona* provides a thumbnail sketch of this part of the story. Although she was designed to carry only 600 men, a total of 1,300 were drowned when she sank.

It is possible that the Armada was a gigantic bluff – a bluff that failed. We know, from Philip II's instructions to the man who was to have commanded the invading army, that the king was aware of the difficulties of crossing the Channel in the face of the English fleet, and that he hoped his monstrous show of force might frighten the enemy into making political concessions. Marine archaeology has brought up evidence that supports the contention of some historians that the Armada's strength lay more in *apparent* than in *real* power. Much of that evidence comes from *La Trinidad Valencera*. The *Trinidad* ran aground and broke up in heavy surf in mid-September, 1588. Her remains lie in only 10 meters of water, which means that diving archaeologists can work for an unlimited time without risking decompression sickness (the bends). It has thus been possible to carry out the excavation very slowly, with extreme care.

The original metal-detector survey was made on a precise grid laid out in three-meter squares over a rectangular area measuring roughly 30 by 60 meters. Then divers of the Irish club that had found the wreck, supervised by Colin Martin and other professionals, mapped numerous finds, among which were two handsome bronze cannons weighing 2.45 tons each, with lifting handles in the form of beautifully molded dolphins. The guns had been cast in Flanders by Remigy de Halut, a celebrated gun founder. Yet, oddly, their weight is not much more than half the weight of equivalent English naval guns, and close study has shown that they were siege guns, made for mounting on horse-drawn carriages. Nearby, on the sea floor, lay massive spoked wooden wheels rimmed with iron, which seem to have been part of such carriages. (Seawater had so weakened the wood that the wheels had to be encased in special protective boxes to keep them from breaking as they were hauled up to the divers' support ship.) We shall see why this appears to support the "bluff" theory, in spite of the beauty of the cannons.

Thousands of artifacts have been raised from all four Armada wrecks, of course, and they tell us much about the nature of life on board and about the last tragic days of the ships. But the most fascinating detective work relates to the guns and the shot, because these shed light on the Spaniards' true preparation for their spectacular mission. The guns that have been raised represent all combinations of range and shot weight, from 50-pounders to $1\frac{1}{4}$-pounders. And several important facts can be deduced from these findings. Among those facts:

1. The guns were often of poor quality. For example, the *Gran Grifon* carried at least two iron guns of which the barrels were made of heavy iron strips hammer-welded together around a cylindrical former. This was an outmoded technique by 1588. On the same wreck, one of the bronze-cast cannons was made so badly that the hollow bore is off-center and the wall of the breech would have been twice as thick on one side as on the other. Thus, the barrel would have been liable to explode.

2. On *La Trinidad Valencera*, the fact that large siege guns intended for land battles had to double for service on the ships is not surprising, since it would be a waste of firepower to carry them purely as cargo. What is surprising, though, is that the guns

A contemporary artist's impression of the Spanish Armada before it was dispersed and 40 of the 130 ships wrecked in Scottish and Irish waters. The vessel in the center foreground, a Neapolitan galleas, is similar to the *Girona*, on which 1,300 lives were lost when she sank off the northern coast of Ireland.

were mounted in the low mid-section of the ship on their large-wheeled land carriages, which took up an enormous amount of space. Moreover, this type of mount raised the center of gravity of the guns needlessly high above the water line, prevented them from being projected fully beyond the side of the ship, made it impossible to swivel them sideways, and must have resulted in excessive recoil. When you consider that these land guns were no doubt manned by soldiers who weren't used to naval warfare, you can see that they probably contributed little to the Armada's fighting power.

3. Recovered iron cannonballs of all sizes show further signs of hasty manufacture. In many specimens, the iron was evidently quenched in water for speedy cooling, and the metal has cracked into concentric rings like the skin of an onion. This correlates with historical evidence that although the Spanish ships fired broadside after broadside at the English, they did almost no damage. The poor-quality iron shot simply shattered on impact against heavy oak hulls.

On paper, the balance of forces suggests that the Armada should have inflicted heavy casualties on the English, that the powerful invasion threat should have wrested political concessions, and that, even if not entirely successful, the Armada should have returned to Spain with dignity and honor. Marine archaeology has revealed some reasons why all ended in disaster.

Cannon recovered from the Armada wrecks reveal some weaknesses of the apparently invincible fleet. Far left: Since the bore of this bronze cannon is off-center, it would have fired off-target and might have exploded. Much more impressive is a cannon found on *La Trinidad de Valencera* bearing the coat of arms of Spain's Philip II. But such guns, meant for land battles, were unlikely to be effective naval weapons. The huge wheel photographed here on the sea bed is from the gun carriage for one such non-nautical cannon. Note the archaeologist's scale rod in the foreground.

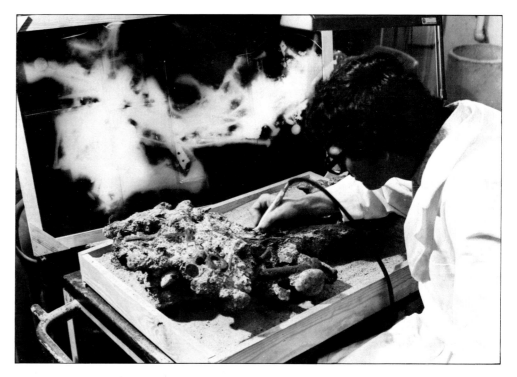

A mosaic of X-ray photographs of a concretion recovered from the sea guides the archaeologist in the delicate process of chipping off the overlay with an electric vibrating chisel (right). Below: Halfway through the cleansing job, coins, nails, and wire begin to appear, and fragments of metal protrude from the hard, shell-like covering. The fully released objects below – including trapped stones – are positioned exactly as they were in the underwater concretion.

Preserving the Finds

Conservation of anything raised from salt water must take place in two stages. First the object must be prevented from drying out or being attacked by atmospheric oxygen or microbes. This is a matter of "first aid." Secondly, if permanent preservation is justified, the material must be treated so as to reverse, as far as possible, the process of decay and preclude future decay.

For the first-aid stage, baths of fresh water containing protective chemicals are normally kept at the site of excavation, either aboard the support ship or ashore. The chemicals involved are so simple that any amateur can carry out most kinds of temporary conservation. (It seems hardly necessary to emphasize that conscientious amateurs never raise antiquities from the sea floor without first consulting a museum or archaeological authority about the best way to handle them once out of the ocean.) While in their protective bath, finds can be labeled and catalogued; if necessary, they

can also be lifted out for a quick photograph. And at this stage the decision as to whether or not an item is worth full conservation must be made.

The techniques used for permanent preservation of organic and inorganic materials are generally very different. Organic remains – for instance, wood, rope, leather, cloth, baskets, bones, or food stores – are likely to be found in salt water only if they are buried under sand or mud, where microorganisms have not been able to attack them. Fortunately for marine archaeologists, the cargo of a large ship quickly accumulates sand and mud in the crevices, and many organic objects are therefore likely to be protected from total decay underneath the main cargo, or trapped deep in it. But they may have become so fragile that they cannot be lifted without careful support, and so they are usually placed in special boxes or frames padded with sand before being raised. As soon as they reach the surface, they are gently washed clean, then placed in the "first-aid"

tank for temporary storage. If allowed to dry out, organic remains shrink, crack, or simply disintegrate, since their natural oils and waxes have been leached out of them. For permanent conservation, though, they must be made capable of withstanding the atmosphere. No matter what the material, the aim is the same: to diffuse into it a stable oily or waxy substance that will replace the water with which it is saturated.

The substance most commonly used for this purpose is polyethylene glycol. The salvaged item is soaked, to begin with, in an extremely weak solution of polyethylene glycol, and then, month after month, the concentration is slowly strengthened until all the water in the material has been forced out and it is impregnated with a waxy preservative. When a ship is being raised timber by timber, storage tanks near the diving site can be built cheaply of plastic or concrete; thus, the initial stages of permanent conservation can be started at once. Where the whole ship is exposed to the air, as with the *Wasa*, it must be subjected to long years

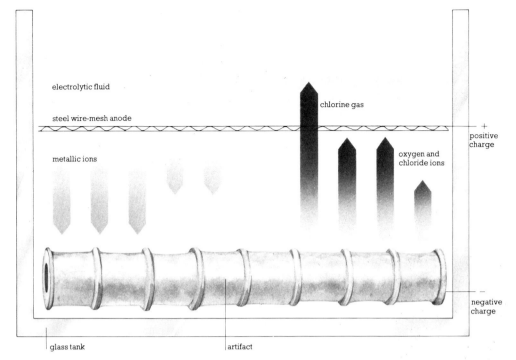

electrolytic fluid

steel wire-mesh anode

metallic ions

chlorine gas

oxygen and chloride ions

+ positive charge

− negative charge

glass tank

artifact

of spraying with the polyethylene-glycol-and-water mixture.

Inorganic remains – iron, bronze, silver, gold, lead, copper, pottery, glass, jewels, such stone as marble or granite – are apt to have been corroded by the salts in seawater, attacked by chemicals secreted by plants and animals, and embedded in concreting organisms. Metal objects are usually the most difficult to conserve, whereas pottery, glass, and precious stones are comparatively easy, for they are likely to be well preserved, though coated in encrusting concretions, and all that is needed is washing in fresh water and careful removal of the crust. Ordinary stone presents no problems, either. Submerged limestone and marble may have been attacked by certain kinds of mollusk that bore into rock and can destroy the surface of a sculpture; but they cannot injure stone that lies under sand or mud, and all decay stops as soon as the stone is raised and washed in fresh water.

There are two kinds of metal, actually, that are even more immune to injury in salt water than stone: gold and lead. Neither of these is corroded by clean seawater, and lead is so toxic to marine organisms that it never gets covered by concretions. That's why many hundreds of ancient lead anchor stocks have been found in almost mint condition on the floor of the Mediterranean. Gold, however, even though not corroded, may need careful cleaning of a kind that involves some of the processes required for conserving such other metals as iron, bronze, silver, and copper. All of these must first be stored, concretions and all, in a weak solution of sodium sesquicarbonate or potassium dichromate. This achieves temporary preservation during an initial

period of cataloguing and photography. If concretion is heavy, archaeologists usually X-ray the mass before breaking it open, since among the objects inside may be a delicate coin, or a pistol or watch, and the X-ray picture can serve as a warning guide.

After the first-aid stage, the various metals require different treatments for permanent conservation. To get an idea of the general procedure, let's concentrate on what's done to iron, which is both the most common metal and the most difficult to preserve. Iron can rust so quickly that when a concretion is cut open, nothing remains but a black sludge. If that happens, the concretion itself is often used as a mold for casting a model of the missing object. But if any chunk of rusted iron survives, the aim is to reverse the rusting process and to extract from the metal all the destructive chloride ions that have diffused into it from the salt in seawater. One cheap – but very slow – way to do this is to leave the rusty iron in its initial bath of sodium sesquicarbonate and change the liquid at monthly intervals for a year or more. This leaches out most of the chloride, but it doesn't reverse the surface rusting. A costlier but more effective method is to reverse the natural processes of both rusting and corrosion by means of electrolysis. A small electric current is run through a special bath in such a way as to prevent the rust layer from being lifted off the metal; as the rust is gradually converted back to metal, the current is increased, forcing out the chloride ions. This process might take several months for something like a large cannon, but it works well.

The fastest method of treating iron, however, is to heat it to over 1000°C in a special furnace. By this highly sophisticated method

Corroded metal objects can be cleaned and protected from further damage by electrolysis. In this process – as shown in the diagram at the top of the page – an electric current passes between a wire-mesh anode and the artifact through a current-conducting bath of electrolyte. As the wire mesh slowly corrodes, metallic ions are transferred to the artifact, replacing the rust produced by seawater and gradually restoring the corroded surface. Top right: Cannon barrels and an iron bar in their electrolyte bath; note the wire-mesh anode at the foot of the tank. Above: A preservationist adds electrolyte to a tank containing salvaged Spanish coins. The anode (which will be lowered into the liquid before the current is switched on) is wide enough to provide even-current density to all parts of all the coins.

the rust is converted back to metal and all chloride is removed in a few hours. Though the technique uses very expensive equipment, it is ideal for treating a great batch of, say, hundreds of similar metal objects, all of which are worth preserving, in one fell swoop. All that remains to be done is to rinse each item in alcohol, dry it, and seal it with a suitable oil, wax, or plastic.

222

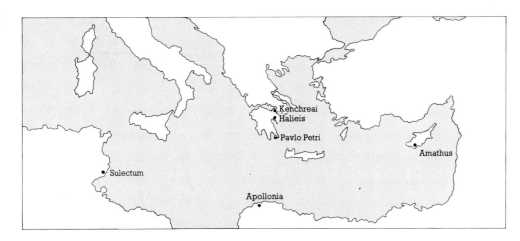

Cities in the Sea

Many early Stone Age caves and Neolithic villages exist on the continental shelf at depths of several tens of meters as a result of the rise of sea level at the end of the Ice Age. But the disappearance of man-made habitats under the sea since then has been largely due to earth movements of various kinds. From the point of view of the undersea historian, the *cause* of the submergence of a ruin or city is less significant than its *rate*. Here's why:

If a city sank slowly over hundreds of years, it is obvious that the economic and military power of the community must also have decayed slowly. The more ambitious citizens had time to move away, and public buildings and harbor installations lapsed gradually into disrepair. This slow decay is the common lot of most archaeological sites on land; it frustrates the archaeologist's natural wish to find evidence of the way life was lived in a given place when it was at its liveliest, for, even though the buildings and their foundations relate to the principal period of civilization, many smaller artifacts and luxury goods will have been removed. Nevertheless, the fact that the city was not destroyed as the result of a sudden catastrophe does mean that the buildings are structurally more or less intact, and the stratigraphy – that is, the layers of accumulated building material, artifacts, and rubbish from different periods – is preserved.

In contrast, if a coastal city was sub-

Submerged walls may be remains of houses or of breakwaters that were once only partly under water; divers try to relate them to an earlier sea level and the original city plan in order to deduce their function. Left: These narrow walls, now on the sea bed near Sfax in Tunisia, belonged to houses in the ancient town of Sulectum. Below: The broad harbor wall of Amathus, a Phoenician port of the eighth century BC, has large stone blocks stacked across its width as an extra measure of wave resistance.

merged dramatically in an earthquake or landslide, perhaps accompanied by a tidal wave, the drowned ruins are, from the archaeological standpoint, like a shipwreck in that all buildings and artifacts, together possibly with human skeletons, are preserved from one moment of time. This is a large gain – somewhat offset, however, by the fact that the ruins of a suddenly stricken city are likely to be shattered, overturned, and hurled from their original positions. And where pottery, furniture, jewelry, and bones are mixed chaotically with fragments of fallen walls and layers of mud or sand, stratigraphy is very poor.

Most underwater ruins are in the "gradual" category. These include Stone Age sites found by divers off the coast of Gibraltar, Italy, Israel, Florida, and California, as well as most of the hundred or more Mediterranean cities that are now either entirely or partly submerged. In the "catastrophic" category are underwater cities destroyed by volcanic eruptions as well as by violent earthquakes, as we shall see in the next section. (Earth movements can be upward as well as downward, of course, and there are a few examples of ancient harbors that have been raised out of the water. But this book is about the undersea.)

Divers have documented more than a hundred Mediterranean cities that lie under from one to five meters of water. Of these, four in the "gradual" category are particularly notable because of their size and the completeness with which they have been surveyed or excavated: the Bronze Age site of Pavlo Petri in the extreme south of Greece; Apollonia, near Benghazi in Libya; Kenchreai, near Corinth; and Halieis in the western Peloponnesus. The four are of different dates, with different arrangements of harbor basin and town plan, but if you could have a diver's-eye view, you would find that these drowned cities have many things in common.

Imagine you're diving with an aqualung in water only three or four meters deep over one of these cities. The sunlight produces some highlights of color – yellow, green, and silver – but there's a bluish tinge over everything except the brown of mosslike and encrusting algae. The sea floor is a mass of irregular stones, patches of sand, and clumps of weed. Where is the city? Well, if you stare straight downward, you'll see a line of square-cut stones flush with the surrounding rubble, stretching across the bottom. Swim along the line and you notice curved shards of pottery jammed between some of the stones on either side. After ten meters or so, the line – a wall – disappears under the rubble. Now, wherever you turn, you find more such lines, and sometimes corners or even a whole rectangle. Occasionally you see two or three courses of block protruding from the bottom or out-

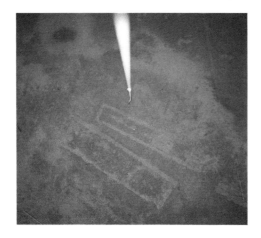

Sacred sites in three submerged cities of ancient Greece: temple foundations at Halieis seen from a balloon (above, left); stone-lined "cist," or "box," graves of Pavlo Petri (above); and temple walls at Kenchreai (left). All were found in shallow water, but the water in the Kenchreai temple was pumped out during excavation. A notch on the wall shows the original water level. The glass mosaics being labeled here lay in a pile against the wall, undamaged by the water, for many centuries.

crops of the solid bedrock quarried into rectangular masses. But you're so close to the ruins, and visibility is so limited, that it's hard to perceive the shape of a single room or house, let alone the whole city. To understand the broad pattern of the site you need to see it from much higher up.

Aerial photography is now a standard technique in the early phases of surveying underwater cities; it was used in all our four examples. At Apollonia in 1958–59, Nicholas Flemming had to rely on high-level military-survey photographs and color pictures taken from a helicopter – none of which showed all the details required, since they were not taken for archaeological purposes. At the other sites, cameras suspended from tethered balloons took special photographs. This may sound like a primitive method, but it actually gives the archaeologist complete control, and the shots can be made when the light is right and the wind is not ruffling the water. (It would cost a fortune to use aircraft for that kind of photography.) With the help of aerial pictures, the marine historian can rough out a map of the site; and on the basis of his map he decides upon the control points for the survey and the baselines from which he will make all measurements.

At Apollonia, which was a busy seaport under Roman rule in the first century after Christ, a baseline was laid out for a kilometer along the now nearly barren beach, and optical surveying instruments were used for mapping both the underwater ruins and the ruins of a later period on shore. Snorkel swimmers began by sketching the rough shapes of buildings, walls, and rock cuttings and numbering corners and intersections. Then a second team placed survey poles on key underwater points in numerical order so that a surveyor on land could plot them without getting confused. Gradually, the general outline of the town – most of it submerged, some of it not – was revealed. Meanwhile, divers were working on buildings, slipways, and docks, measuring stone blocks and wall thicknesses, examining how blocks were keyed together, and so on. The detailed survey also turned up some unbroken pottery, an early Greek stone anchor, a lead ship's-ballast weight, and a marble statue of a faun that lay half-buried in the sand on the floor of a large stone tank. The Romans used such tanks, which were often decorated with colonnades and sculpture, for fish farming; young fish caught at sea were fed and fattened in them. Unlike the buildings of

224 Apollonia, the Bronze Age structures at Pavlo Petri did not have straight walls, and so the technique of plotting corners and joining them up would not work.

The Cambridge University team that surveyed this ancient city in 1968 placed a grid of numbered plastic markers on the sea floor at 20-meter intervals. Plastic tapes could then be laid along the sides of any one square to delineate a compact area that could be measured in detail. The divers discovered no specific harbor structures, but the houses lay in an arc that must surely have enclosed a sheltered basin. It is easy to deduce that the town was a landfall for Mycenaean ships trading with Crete. And from dated pottery found in the ruins we can gather that it must have been built about the same time as the Cycladic wreck discovered by Peter Throckmorton in 1975. After a long period of gradual submergence, it was probably completely abandoned by 1400 BC – just about the time of the Bronze Age shipwreck excavated by George Bass off Cape Gelidonya.

At the Peloponnesian cities of Halieis and Kenchreai, teams of diving archaeologists from several American universities recently carried on extensive excavations over the course of a number of seasons. Because of their methodical and exhaustive efforts, we now have a clearer-than-ever view of gradually submerged ancient cities. As a direct result of all such expeditions, we can speak with some certainty about how harbors were built, how civilizations launched and recovered their ships, how they loaded and unloaded cargoes, and how they protected military fleets. Moreover, there is a worthwhile by-product of this branch of archaeology: the mass of statistics being accumulated on submerged cities – and on uplifted cities, too – tells us much about which parts of the coast are stable and which are not. Such information can prove valuable to engineers who are considering the feasibility of building factories or nuclear power stations in coastal areas.

Catastrophes

There are two ways in which a coastal city can be suddenly precipitated into the sea. If it is built on accumulated sand or gravel along a steep coast, an earthquake may start a landslide or series of landslides, so that the ground almost dissolves under the buildings and they collapse into the depths. Alternatively, cities near a large volcano may be submerged as the earth's crust readjusts to the release of huge quantities of molten rock ejected during an eruption. The fallen ruins are swiftly covered by sand and mud in the first case, and by ash and possibly lava in the second. In either event, the recovery of artifacts and information is bound to pose problems for the underwater historian.

Divers have searched for four cities lost in famous catastrophes but have found only two (one of which is actually six cities, as we shall see). Of the two that are still undiscovered, one does not even have a name. We know only that a huge cataclysm in about 1400 BC wiped out the advanced Minoan civilization of the island of Santorini (now known as Thira) in the Greek Cyclades. Some people believe that this could be the seed from which sprang the ancient legend of the lost city of Atlantis, and so various expeditions have searched for submerged ruins in the region, but so far to no avail. The second failure might seem more surprising, since the catastrophe is well documented: several reliable ancient historians and geographers describe the inundation of Helike, a city on the shore of the Gulf of Corinth, which was destroyed by an earthquake-landslide followed by a tsunami in 373 BC. Numerous search teams with diving equipment, echo sounders, and mud-penetrating sonar have tried to find Helike, but the coast in the area is so steep that the drowned city's buildings and streets are probably 20 meters or more below the surface and buried in perhaps 10 meters of mud – a truly daunting situation.

The other two catastrophes, however,

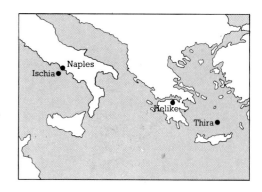

have given divers plenty of real work. The first of these involves a site on the northern shore of the Bay of Naples; the second is in the Caribbean – Port Royal in Jamaica. Vesuvius, to begin with, isn't the only volcano near Naples, for there are several small volcanoes in the region called Solfatara, and a large one on the beautiful island of Ischia, just outside the bay. Ischia has not erupted for centuries, but at Solfatara you can see craters of hot mud bubbling and hissing continuously. Major eruptions of Vesuvius or Solfatara coincide with rapid vertical movements of the land nearby, and local fishermen have long known that there are houses and ruins under the water in many places between Naples and Miseno (the ancient city of Misenum), 20 kilometers away. They do not know the date – or dates – of submergence, though, and we don't either. All we can be certain of is that the several cities submerged on that shore were still active in AD 410, in the last days of the Roman Empire, and that they must have gone under well before 1538, since some of the ruins actually rose out of the water at that date. We'd know much more if we could make detailed surveys, but underwater work there can progress only very slowly.

Earthquakes in volcanic regions can result in swift submergence, or in equally swift uplift, of a city. In 1538 these temple columns at Pozzuoli near Naples rose from the water into which they had once suddenly sunk.

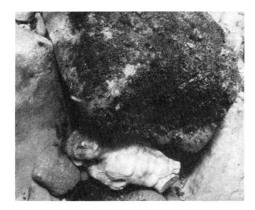

Roman fish-preservation tanks – *piscinae* – were decorated with columns and statues. This first-century-AD faun was found in the sand on the floor of one such tank at Apollonia in Libya.

In spite of the uplift of 1538, the net effect of all the movements in the Bay of Naples has been to leave what appear to be six separate cities submerged in as much as seven meters of water. That may not seem deep, but the sewage and industrial waste from modern Naples make underwater visibility in the bay less than two meters in nearby waters and no more than five at Miseno. Aerial photography is useless in such conditions, and a casual diver could totally fail to realize that he was swimming around fallen buildings and drowned streets. The clues are there, of course. You can swim through the gloomy rooms of the submerged ground floor of a three-story-high Roman house even without diving equipment. At the outer edge of the bay, at Miseno, there is a vaulted cave with an underwater tunnel leading into it, and on the floor of the cave is a perfectly preserved fish tank, smaller than the one at Apollonia but complete with channels for controlling the flow of water and slots for sluice gates. In the walls of the cave are cut small niches, which must once have been occupied by statues or other objects. But individual finds of this sort are not enough. Because of the poor visibility, the muddy bottom, and the sheer size of the task, we still lack an over-all view of any one of the drowned communities.

We have had much more success in our search for the lost city of Port Royal, a Jamaican pirate town of 8,000 people, which was struck by an earthquake in 1692 and disappeared under the waves in a matter of minutes. A key figure of this archaeological story is an American, Ed Link, who began to probe for the ruins in 1959 by means of echo sounders, precision navigation equipment, and other devices. There was plenty of documentary evidence for the catastrophe recorded in maps and eye-witness accounts, and Link and his team knew where to look. Here too, though, underwater visibility was poor, and echo-sounding revealed an apparently smooth muddy bottom, and so the only possible procedure was to dig straight into the mud at the likely places. Link anchored his boat securely and went to work with an airlift – a powerful suction pump – to remove mud and sand from the spot where he hoped to find the ruins.

This may seem to have been a reckless thing to do, since there was no accurate preparatory mapping, and material would, if found, come up the airlift pipe in a chaotic fashion. But the sudden collapse of Port Royal meant that there would be little order in the ruins anyhow, and so the "rules" of archaeology need not apply as strictly as they should in most cases. At any rate, Link lowered his airlift pipe at a position where, according to contemporary maps, he hoped to find the remains of a warehouse. Divers guided the suction head to cut a broad pit in the sea floor, but after several days of

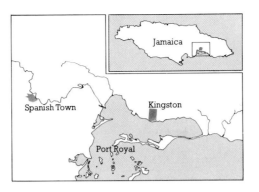

A broadside giving the people of England their first news of the catastrophe that had struck Port Royal, Jamaica, on June 7, 1692. This is, of course, only an imaginative reconstruction of the scene, but there is no doubt that nearly 70% of the town slid into the Caribbean and 2,000 people died as the result of an earthquake and ensuing tsunami or storm wave.

probing they had pumped up only mud, gravel, and some modern bottles and broken china. They again drew a blank at a second location, but their third try paid off. After bringing up great quantities of wine and rum bottles, clay pipes, broken dishes, bones, and cooking utensils, they at last found leaning sections of the town's brick walls. No doubt about it: this was Port Royal. And as if to corroborate all accounts of its destruction, one of the finds was a telltale brass watch. Missing were its glass and hands, but when the calcareous growth covering the face was X-rayed, the position the hands had been in could be seen – 17 minutes to 12 noon, exactly the time of the 1692 earthquake.

Ed Link proved that Port Royal was worth excavating, but he also proved that it would take many years of work. In 1963 the Jamaican Government appointed another American, Robert Marx, to continue the project on a systematic basis. Marx divided the area into manageable squares and got teams of divers working carefully with airlifts, plotting the finds in each square. After three years, they had raised a staggering number of artifacts – but from only a tenth of the 140,000 square meters of the submerged city! A few statistics indicate both the richness and the sheer repetitiveness of the finds: 20,000 iron objects, 2,000 glass bottles, 6,500 clay pipes, 500 objects made of tin and silver, 3 tons of animal and human bones. . . . Further excavation was halted some years ago, ostensibly to permit a breathing space for cataloguing, preserving, and interpreting the finds. Will excavation start again soon? It seems unlikely. Such operations are expensive, and interest in them tends to wane after the initial drama of discovery.

A Few Harbors in One

If an archaeologist of the year AD 4000 excavated the remains of a 2,000-year-old racing car, he would find it hard to understand how it was used. Surely this couldn't be the only means of transport for the primitive people of 1977? If he later unearthed a family car, a small bus, and a motorcycle, he would begin to get a general idea of the modes of transportation in our era. Similarly, every ancient harbor is unique, and we today can understand how it functioned only when we compare its layout and installations with those of others. By now, diving archaeologists have swum over hundreds of submerged harbors, and we are starting to build up a general picture of what they were like in the days of ancient Greece and Rome. Instead of glancing too briefly at a few specific harbors, let's try to bring all the elements of the best-preserved ports into one imaginary place and describe it rather thoroughly. We'll call it Portus Romanus.

Aerial photographs and preliminary plans made by a team of amateur divers show that the original Greek harbor was built by narrowing the mouth of a bay that was already almost enclosed by clawlike headlands. In the first century AD the Romans abandoned this small bay with its narrow slipways and constructed two big harbors on either side of it – one for commercial vessels, one for the military fleet. After three seasons of mapping and excavation by teams of modern experts, we now know enough about Portus Romanus, which was gradually submerged in about three meters of water, to take what amounts to a guided underwater tour of the city.

An area of streets and building foundations fringes the present shore. Swimming across these – divers have cleared weed and sand away from many of them – we head for the outer breakwater of the commercial harbor. This huge structure has been somewhat flattened and scattered by winter storms, and the irregular masses of stone that extend outward for more than 500 meters seem almost like a natural reef; but divers have discovered a regular occurrence of squared blocks, and geologists have identified them as coming from nearby quarries.

The original structure was far from simple. Its bottom layers were composed of small stone blocks only about half a meter long. These, piled together, were designed to give the breakwater a broad foundation below the depth of vigorous wave action. They were probably dropped from barges that brought them from the quarries, and the outer end of the breakwater was originally in 10-meter-deep water. The upper sections of the mound were built of increasingly larger stones, until finally, near the ancient waterline, enormous blocks weighing from 100 to 200 tons formed the outer flank, which had to withstand the largest waves. At intervals all along this strong wall there used to be square platforms for ships to moor against. One such platform survives. If you swim down and examine the blocks of which it is built, you will see that some of them have slots cut into the corners so that they could be secured together by keys of iron and lead. At the outermost end of the breakwater is a pile of colossal blocks, some of them over five meters long; they are probably the remains of a tower or lighthouse.

Swimming back toward the shore across the mouth of the commercial harbor, we notice a massive outcropping that looks like natural rock, but that is actually a 15-meter-square block of Roman concrete of the kind that sets well underwater. There are 12 of these positioned at intervals in a straight line toward the shore, each of them rising to within a meter of the present surface. This was a clever device to keep water circulating through the harbor and prevent the accumulation of sand and silt: the blocks are close enough together to disperse waves that might approach the shore obliquely, but far enough apart to allow the wind and tide to drive currents through the harbor. A catwalk of timber probably ran from block to block in Roman times.

Inside the basin are the remains of a network of tanks and channels – a group of fish tanks that were operated as a single system. The tanks have all been excavated and studied in a successful effort to determine the original sea level in this area, and we're beginning to understand how the channels and sluice gates were opened and closed so

Portus Romanus

as to get maximum water circulation out of the small tidal range and variable wind directions. It may seem surprising that tanks for fish rearing were put in a harbor, where they were undoubtedly exposed to sewage from the city. In practice, though, the dilute sewage probably promoted planktonic growth, which helped feed the fish (and the risk to human health was something that the Romans were not aware of).

Finally, toward the back of the commercial basin we swim over numerous quays and buildings that formed the trading heart of the city. Perforated stones that project from some of the quays are the mooring rings to which grain ships used to be secured. And in the maze of foundations and streets bordering the quays in the shallow water, we can identify warehouses, stores, and workshops, as well as dozens of small buildings that were probably offices for the merchants, money changers, and dealers in credit without whom no commercial harbor could possibly function (in Roman times as well as ours).

The military harbor, which is reached by swimming across the small harbor dating from early Greek times, has an entirely different type of breakwater from the other one. An unsung Portus Romanus engineer evidently decided that this harbor should be protected by narrow vertical walls of exceptional strength. The wall is almost intact to this day, though the top is partly broken. Its central core is six meters thick and consists of well-cut rectangular blocks, each keyed by dovetails to all its neighbors. The iron keys have rusted away, but thousands of slots bear witness to their former presence. As if this were not enough, the whole wall was covered in concrete a meter thick, and it is only where heavy breakers have stripped this away along the top that we can see the inner structure.

Outside the mouth of the military harbor, a group of concrete structures look at first glance like a natural island. Divers have cut away the weeds, and so we can examine this final example of Roman ingenuity: the submerged structures are, in fact, several ships loaded with concrete. They were sunk next to each other to form a barrier, so that any ship entering the harbor had to make a dog-leg turn – a maneuver that would give the harbor's defenders time to ascertain the identity of the approaching vessel.

There is, as we've said, no actual Portus Romanus. But everything we have described can be found somewhere among the harbors of antiquity.

Some Vital Questions

The success story of undersea archaeology in the last two decades has led to an increase in the number of intended surveys and a determination to achieve increasingly high standards of excavation. But the proliferation of activity brings in its wake a string of questions that are begging for answers. Among the hundreds of wrecks and drowned settlements that are discovered every year, how do we select the best targets for excavation? How do we preserve sites not yet excavated and protect them from looting? What about legal problems – specifically, who owns offshore archaeological remains? Then, too, with the increasing complexity of techniques, the increased duration of undersea excavations, and consequent soaring costs, where will the money come from? How can enough trained people be found to do all the work both under and at the surface? And how can we better synthesize our fragmentary bits of information so as to create a history of seafaring rather than just an inventory of finds? In short, where do we go from here? We cannot fully answer such questions, but we can at least *begin* to answer them.

First, the targets. Undersea archaeology has flourished all over the world in recent years, and the priorities obviously differ from country to country, but there are several outstanding target areas for diver-historians, wherever they live. For instance, it is a bizarre fact that we have not yet found a Mediterranean wreck that retains structural elements defining it incontrovertibly as an ancient warship, and most undersea historians would agree on the desirability of intensifying our search in such areas of known sea battles as Syracuse or Pylos. There are still only two known wrecks of any kind dating back beyond 1000 BC, and only one of these has been fully excavated; this is another enormous gap. Outside the Mediterranean, we have much to learn about the European colonizations of India and the Far East, as well as about local shipping in Asian waters. Spanish and Portuguese wrecks in the Caribbean have been well studied, but we still know far too little about life on the sea route around Cape Horn a few centuries ago. As for submerged settlements, harbors such as Apollonia and Kenchreai have been known for years, but we need more detailed work on individual dockside structures if we are to re-create the daily routine of operations in a great harbor of the past. One probable target area for investigation of harbors that existed in the early days of systematic seafaring is to be found in the river mouths of several Middle East countries. And going even farther back, we must admit that we have barely begun to look into the archaeology of the continental shelf throughout the Stone Age. All these are important targets for future exploration.

The legal problems of ownership of antiquities and the protection of sites from looting differ in every country. Where there is strong national feeling about the matter, a simple law tends to prevail: all antiquities onshore and off belong to the state, and you must have a permit before you can excavate or even survey them. The law is usually toughest in countries that have most to lose, like Greece and Turkey; and it is these areas, of course, that attract the greatest numbers of undersea lawbreakers. It is almost impossible to keep an eye on divers in remote bays, and there is always someone prepared to handle undersea loot on the black market. Britain has recently passed laws designed to prevent diving on wrecks of probable archaeological interest except by teams approved by a committee of respected archaeologists. U.S. law varies from state to state, but more and more of the coastal states are taking an active interest in conserving their marine heritage and now have state-funded archaeologists em-

powered to supervise offshore activities. Congress, too, has passed a law that obligates offshore contractors to take into account the archaeological effects of their proposed work.

The best advice for anyone who finds antiquities under the sea is as follows: photograph or map the area, but don't lift anything; instead, report your discovery to a museum or official department of antiquities, preferably one with an interest in marine finds – and to nobody else. If you want to be associated with further study of the site, find out what laws apply to it. It's possible that you can persuade the authorities to let you assist them; in some places, in fact, you may be able to acquire legal ownership of the site so that you can excavate it with official approval.

Our improving standards for accuracy in undersea excavations and the associated complexity of equipment have naturally meant rising costs, and the current unwillingness to waste money may produce the impression that marine activity is slowing down. Understandably, universities and other sponsors hesitate to commit themselves to expensive projects that can last for several years. Certainly there is some reassessment going on as to the value obtained from such an investment of money

and time. Still, with an intelligent selection of targets there is no doubt that enthusiasm will continue. And modern developments don't really support the notion that the great days of undersea archaeology were back in the late 1950s and early '60s and that it's all over now. Whether you look at Armada wrecks off Ireland, submerged Neolithic villages off Israel, or Dutch wrecks near Australia, the future looks full of excitement.

Technical developments from now on may well be limited by shortage of money. Ideally, improved accuracy in surveying by means of photogrammetry, sonar, and underwater television would be a sure thing in the near future, and the introduction of advanced diving equipment would enormously broaden the archaeologist's horizons. But to be realistic, it is hard to envisage *any* noncommercial target that would persuade a sponsor in these last years of the grim 1970s to part with thousands of dollars per day to maintain submersibles and complex diving systems on a deep ocean site.

As for finding people to carry on the work, there need be little concern about this aspect of undersea archaeology. The very fruitful cooperation between serious amateurs and professional archaeologists of the recent past is almost certain to continue. The

medical and technical problems of diving with aqualungs have not deterred amateurs, and university student bodies provide a steady flow of eager underwater assistants for professional diving teams. Although women divers are almost unheard of in the commercial world, women have shown great competence in marine archaeology; indeed, several expeditions have been led by women. Future excavations will no doubt demand an increased degree of training even among amateurs, but that should be no problem: many colleges and diving schools are already offering short courses in underwater archaeology and its methods.

The last question is perhaps the hardest: do their findings give undersea archaeologists a real right to call themselves archaeologists? As we've noted, professional archaeological institutions used to treat undersea work with some reserve, but attitudes are changing as we build up an ever-clearer picture of the interaction of past civilizations with the sea. It is up to the few professional marine archaeologists and the many dedicated amateurs who are all but professional to make sure that the interpretation and synthesis of their finds can stand increasing comparison with work on land. It seems to the authors of the foregoing pages that this can be, and is being, done.

Facing page: This pewter plate recovered from the *Hollandia*, a Dutch ship sunk in British waters in 1743, will go to the highest bidder in one of London's noted sales rooms – but the competitive buyers may well be more intent on probable cost-and-profit figures than on the piece itself. As yet there are no strict laws in Britain against selling salvaged treasures for gain, but marine archaeologists are making headway in their campaign for archaeological supervision of offshore salvage operations on wrecks of historical interest. For instance, unauthorized diving is now legally banned within several hundred meters of this buoy, which marks the site of *Mary Rose*, an English warship that sank during a battle with the French in 1545.

8
Diving and Divers

H. G. Delauze
X. Fructus
C. Lemaire
A. Tocco

Most people think of the diver as a man in a baggy suit with a ponderous-looking helmet and massive lead boots, or else as a slender, nearly naked fellow wearing fins and a mask. These images, though valid as far as they go, are a vast oversimplification. Today's professionals do so many kinds of essential underwater work, and at such unprecedented depths, that scores of specialists are engaged in a worldwide effort to develop sophisticated equipment for them. The great difficulty – apart, of course, from the obvious stresses and dangers of living and working in a cold, dark, and often unpredictable medium – stems from the fact that the deep-sea diver is exposed to surrounding water pressure. The pressure increases progressively with increasing depth, and so does the problem of how to provide the diver with a safe supply of air or some other gas to breathe at a given depth for a given length of time.

In considering modern techniques, tools, and types of equipment, it is important to differentiate among the various kinds of diving, each of which may be best adapted to one or another of the possible underwater tasks (which include such widely differing jobs as working on engineering projects, carrying out scientific research, or attacking an enemy in wartime). How today's scientists and technologists solve the problem of keeping men not merely alive but comfortable, alert, and able to work for economically feasible periods of time depends in part on their answers to questions like these: Is the diver required to move freely, or will he be connected to surface support equipment? Will he have to swim in mid-water or walk on the bottom? At what depth must he work, and for how long? Is he working alone or with a team?

Though free diving is popular among sportsmen and is useful for some scientific and naval projects, it has little importance for industry. Since a free diver, for instance, must carry not only his own air supply but everything else he needs, he gains complete mobility in three dimensions only at the sacrifice of dive duration. In addition, he cannot move far without losing track of his whereabouts; navy frogmen and scientific researchers often have additional equipment to help them navigate – but the more they carry, the more limited their maneuverability. Most nonrecreational diving, therefore, is done with the diver connected to a surface supply of gas, electric power, and expert supervision.

Broadly speaking, there are two techniques for supplying a breathing gas to the diver. It can flow through a hose from the surface, either to a valve on his back or into his helmet and suit, out of which the gas that he exhales bubbles through a relief valve; or it can be compressed into a diving bell, within which divers are lowered into the

Though he is free to focus his camera on any nearby object, this diver's movements are restricted by his dependence on air fed from the surface through a flexible hose. His photographic gear includes a floodlight, powered by a pack of batteries carried on his back, and an exposure meter around his neck.

sea. Modern diving bells are known as SDCs (submersible decompression chambers), and they are fitted with a continuous gas supply to balance increasing water pressures. When a diver leaves the chamber underwater, he remains attached to it by a hose, which is appropriately named the umbilical. A diver working from an SDC can either swim with fins or wear boots and walk, whereas the man whose air flows directly into his suit from the surface is more restricted in what he can do. The reason for this is that the air-filled suit is so buoyant that the diver who wears it must have heavy lead boots and carry about 50 kilograms of lead on a harness attached to his chest in order to keep down.

In the following pages we shall see how modern technology is refining and improving on these long-established methods of furnishing underwater workers with life-sustaining gases. But note our avoidance of the word "air." Only divers in depths shallower than 60 meters (roughly 195 feet) are generally able to breathe ordinary air, which is composed of about 78% nitrogen and 21% oxygen, with traces of other gases, including carbon dioxide. Oxygen is, of

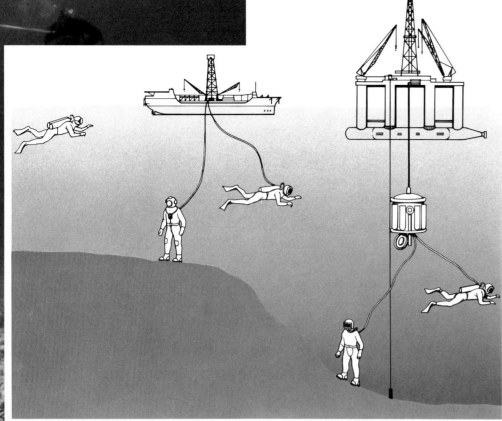

Three basic life-support systems. Cylinders of air on a molded backplate (left) keep the free diver alive. Or (center) air is pumped from the surface, and a regulating valve on the mouthpiece of the hose provides breathing gas on "demand"—i.e., each time the diver inhales. Or (right) the diver is lowered in a chamber containing gas maintained at the correct pressure; when he emerges from the chamber (often called the "bell"), he is fed this breathing mixture through a hose (technically termed the "umbilical"). In all three systems, exhaled gas bubbles out into the water through an exit valve.

Right: The standard uniform for a "hard-hat" diver, who wears heavy boots and walks on the sea floor, is a canvas-and-rubber suit with a metal helmet bolted to the collar. The nose clip helps him clear his ears, and a lead weight worn on his chest offsets the buoyancy of air-filled clothing, so that he can stay down.

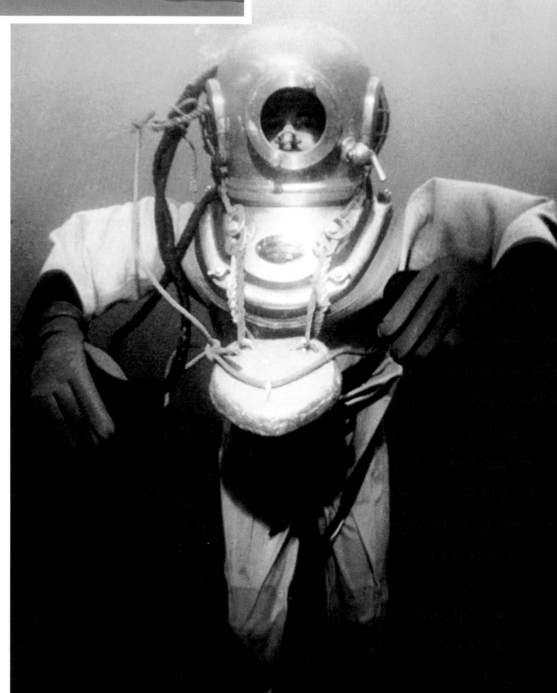

course, essential to life, but the nitrogen that comprises such a large portion of our atmosphere is an "inert" or "neutral" gas: that is, it plays absolutely no part in our body metabolism. Under high pressure, however, it dissolves in the body and can make deep-sea divers dangerously ill. It is the physical presence of the nitrogen, not its chemical properties, that causes the trouble, and so the term "inert" still remains valid.

Because of this fact, and for other reasons that we shall look into, ordinary air is out of the question at depths of much more than 60 meters. Even the amount of oxygen that can be safely tolerated varies according to depth. Yet there must be some oxygen, and it must be mixed with some sort of inert gas as a substitute for the nitrogen in the air we breathe on land. With divers reaching great depths and staying down for long periods – as they have been doing in recent years – the problems of selecting and training suitable men, as well as keeping them alive and healthy, are urgent. To the credit of modern technology, astonishing progress has been made at the cost of very few serious accidents.

Pushing Down the Barriers

The human body, a frail, soft, porous structure, is easily damaged by changes in pressure. Gas spaces such as the sinuses and lungs can collapse or explode; in liquids such as the blood, narcotic gases can be dissolved and transported through the body; and such delicate mechanisms as the brain, nerves, and individual cells can be swiftly injured by changes of the chemical balance in the fluids that surround them. What we are naturally equipped for, of course, is to withstand the normal pressure of air on land, and we use a unit of measurement that reflects this physical fact when we say that the human frame is subjected to a pressure of one "atmosphere" at sea level: which means specifically about 14.7 pounds on every square inch of the body.

In metric terms, an atmosphere equals a kilogram of pressure per square centimeter, and the body must withstand an additional atmosphere for every 10 meters of water. Thus, a diver working beneath the sea at a depth of 450 meters (nearly 1,500 feet) would be subjected to a pressure of 46 atmospheres – 45 of water, plus 1 for the air above it. Such a pressure is equivalent to more than a third of a ton on every square inch of the body!

Only recently have we been able to push the diving frontier as far down as several hundred meters. In fact, until some thirty years ago it was not possible to make an over-60-meter dive that was truly useful – in other words, that permitted a long enough stay for the diver to accomplish a job worth going down for. Progress was restricted – and still is – by the constant need for a better understanding of the physiology of man under abnormal pressures, and also, as it has become medically feasible to keep men alive in deeper water, the need to invent engineering devices that make efficient work possible within the defined medical limits.

By the beginning of the twentieth century we had not advanced much farther along the road toward solving such problems than had the remarkable engineers who, in the seventeenth century, managed to salvage 53 of the 64 cannon in a Swedish warship, the *Wasa*, which had sunk in 33 meters of water. The divers whom they lowered in a diving bell were supplied with additional oxygen by means of weighted barrels of air that were "tipped" into the bell, and the diving bells of the early 1900s were only slightly more sophisticated than that. Similarly, the diving suit and helmet that had been perfected in the nineteenth century continued in use almost unchanged until our own time. Free-diving techniques improved gradually after the invention of a self-contained breathing apparatus (for escaping from damaged submarines) during World War I, and then again with the adoption of swim fins in the 1930s. And World War II brought the invention of the compressed-air aqualung, which, although strictly limited in its ability to function for long periods of time or at great depths, is now used by millions of sports divers. But the really big breakthrough in deep diving has come since 1960.

The chief reason can be stated in three words: offshore oil wells. As oil wells go deeper and deeper, so must the divers who are essential for the processes of drilling and maintenance. And necessity has been the mother of research and experiment as well as invention. In the years between 1930 and 1960, experimentation with various breathing mixtures made it possible for occasional divers to reach depths of from 100 to 160 meters. But they were seldom able to

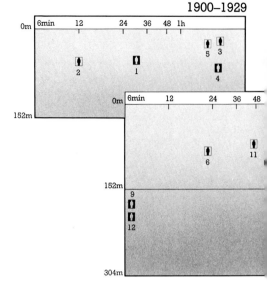

1900–1929

stay down for useful working periods; and the problem of decompression as the deep diver is brought back to the surface proved a particularly hard nut to crack, since the human body can suffer lethal damage during the transition from extremely high to low pressures. When, in 1956, a British naval diver named George Wookey managed to stay on the bottom of a Norwegian fjord for a few minutes at a depth of 180 meters, decompression took so long that his dive was regarded as proof that diving beyond 100 meters was a pointless exercise.

Five years later, this "proof" was definitely disproved by a young Swiss mathematician, Hannes Keller, who made a series of dives, using secret breathing mixtures, that culminated in a U.S. Navy-sponsored dive to a depth of 300 meters. The date of this last dive – December 3, 1962 – stands as something of a milestone. Keller's achievement gained wide publicity for two reasons. First, it pointed up the heroism of the men who risk their lives undersea, for Keller's companion in his diving bell, the British diver Peter Small, died during the ascent, and another man was killed in the recovery operation. Secondly, and more affirmatively, it served as an incentive for physiologists and engineers (as well as the financial interests that back them) to redouble their efforts to break the depth barrier.

Since the early 1960s, national governments and private industry, particularly in France and America, have been investing heavily in research and experimental equipment. And the enormous expenditures are beginning to pay off so handsomely that, for example, the pioneering French organization *Compagnie Maritime d'Expertises* (known throughout the diving world as Comex) has proclaimed its belief that divers will soon be provided with sufficient control and power to work for hours on end at depths of 600 meters.

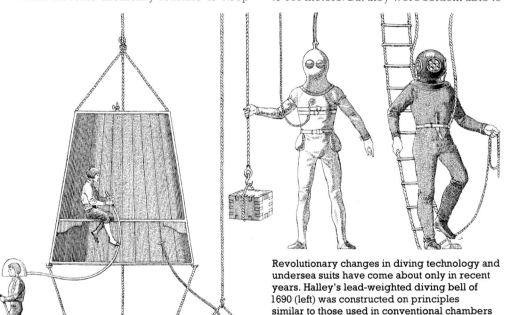

Revolutionary changes in diving technology and undersea suits have come about only in recent years. Halley's lead-weighted diving bell of 1690 (left) was constructed on principles similar to those used in conventional chambers today. And modern hard-hat divers resemble the two pictured above – though the suit on the left dates back to 1811 and the other to 1855.

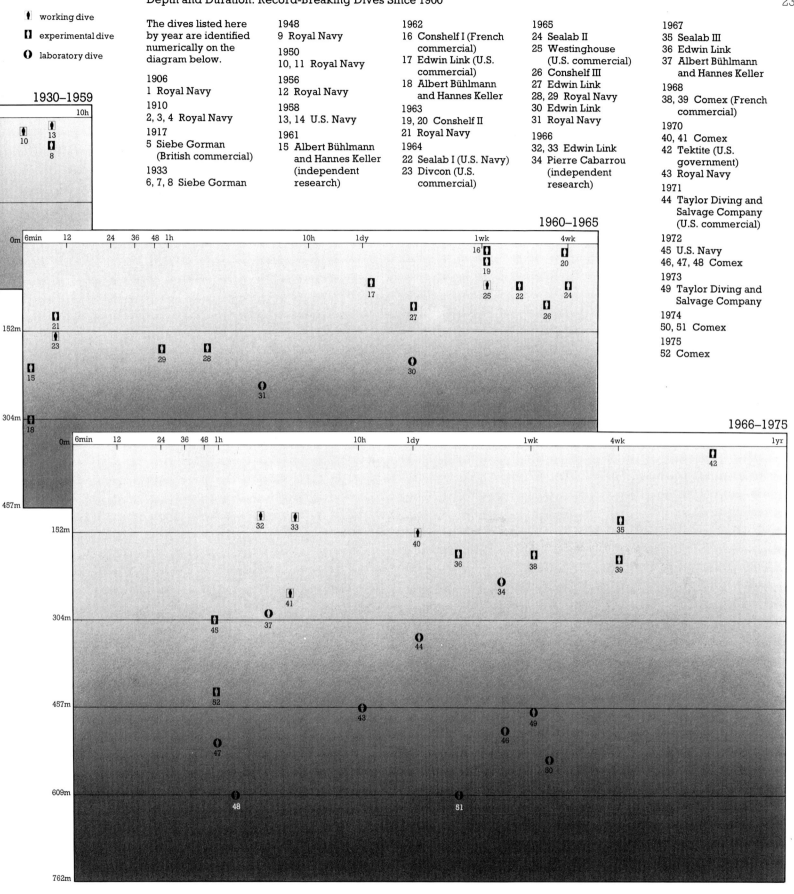

working dive

experimental dive

laboratory dive

The dives listed here by year are identified numerically on the diagram below.

1906
1 Royal Navy
1910
2, 3, 4 Royal Navy
1917
5 Siebe Gorman (British commercial)
1933
6, 7, 8 Siebe Gorman

1948
9 Royal Navy
1950
10, 11 Royal Navy
1956
12 Royal Navy
1958
13, 14 U.S. Navy
1961
15 Albert Bühlmann and Hannes Keller (independent research)

1962
16 Conshelf I (French commercial)
17 Edwin Link (U.S. commercial)
18 Albert Bühlmann and Hannes Keller
1963
19, 20 Conshelf II
21 Royal Navy
1964
22 Sealab I (U.S. Navy)
23 Divcon (U.S. commercial)

1965
24 Sealab II
25 Westinghouse (U.S. commercial)
26 Conshelf III
27 Edwin Link
28, 29 Royal Navy
30 Edwin Link
31 Royal Navy
1966
32, 33 Edwin Link
34 Pierre Cabarrou (independent research)

1967
35 Sealab III
36 Edwin Link
37 Albert Bühlmann and Hannes Keller
1968
38, 39 Comex (French commercial)
1970
40, 41 Comex
42 Tektite (U.S. government)
43 Royal Navy
1971
44 Taylor Diving and Salvage Company (U.S. commercial)
1972
45 U.S. Navy
46, 47, 48 Comex
1973
49 Taylor Diving and Salvage Company
1974
50, 51 Comex
1975
52 Comex

As a result of recent advances in medical science and diving technology, divers have been able to go deeper and deeper and to stay down longer and longer – even if more often in controlled laboratory tests than in the open sea. Here we see how depth and duration records were made and shattered in the 45-year period from 1930 to 1975. Symbols keyed in the diagram indicate whether a given dive was an actual working one or was made experimentally either in the sea or in a laboratory; and divers are identified by either their own names or the names of sponsoring organizations. Note the tremendous acceleration of the barrier-smashing process from the early 1960s on. It is, of course, because of the high medical and financial risks of sending men to great depths that most of the dives deeper than 300 meters have been simulated in laboratories.

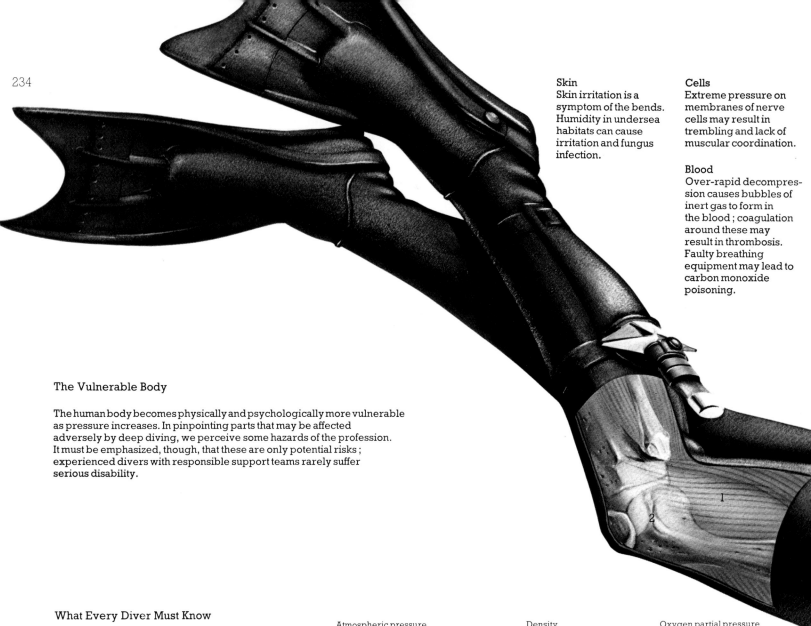

234

Skin
Skin irritation is a symptom of the bends. Humidity in undersea habitats can cause irritation and fungus infection.

Cells
Extreme pressure on membranes of nerve cells may result in trembling and lack of muscular coordination.

Blood
Over-rapid decompression causes bubbles of inert gas to form in the blood ; coagulation around these may result in thrombosis. Faulty breathing equipment may lead to carbon monoxide poisoning.

The Vulnerable Body

The human body becomes physically and psychologically more vulnerable as pressure increases. In pinpointing parts that may be affected adversely by deep diving, we perceive some hazards of the profession. It must be emphasized, though, that these are only potential risks ; experienced divers with responsible support teams rarely suffer serious disability.

What Every Diver Must Know

Shown here are some of the effects on the body of increasing pressures in a dive down to 30 meters. Column one: As depth increases, so too does the surrounding ("ambient") pressure exerted on the entire body surface. At sea level – 1 atmosphere – ambient pressure amounts to 1 kilogram per square centimeter. As the diver descends, it increases by 1 atmosphere for every 10 meters. Thus, at a depth of 30 meters he is subjected to 4 kilograms on every square centimeter of his body. One consequence of this increase in pressure is that the gas the diver breathes is also compressed within his lungs. With each doubling of ambient pressure, it compresses by half its volume (as shown in the middle column), until by 30 meters the same quantity of breathing gas takes up only one quarter of the space that it does at sea level; the lungs are therefore filled with 4 times as much air as they would be at the surface. Each gas in a breathing mixture exerts an independent pressure on the lungs (called "partial pressure"), as indicated in the third column by black dots for oxygen and red dots for nitrogen (the inert gas); total pressure of the mixture is, of course, the sum of the partial pressures. Because partial pressures of both oxygen and inert gas are nearly doubled with every doubling of depth measurement, each creates a separate physiological danger that limits the depth to which a diver may go on a given breathing mixture.

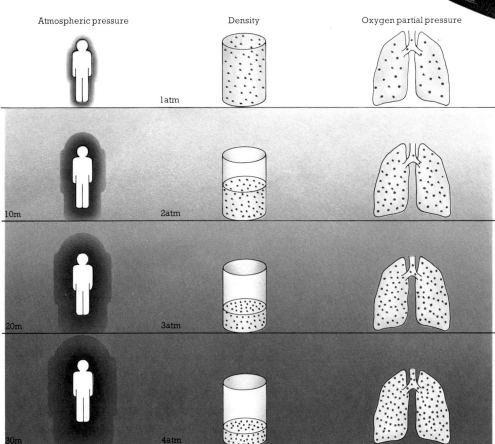

Atmospheric pressure   Density   Oxygen partial pressure

1atm

10m   2atm

20m   3atm

30m   4atm

1
Because the muscle and fat tissues absorb gases more slowly than does blood, they determine decompression rates. Extreme cold can cause cramp by slowing breakdown of the muscle stimulant acetylcholine.

2
Repeated dives with inadequate decompression probably cause aseptic bone necrosis.

3
Extreme cold results in diminished dexterity; and trembling hands are a major symptom of high-pressure nervous syndrome.

4
Upset stomach from overeating or seasickness may cause vomiting; a resultant blockage in the breathing system can suffocate the diver.

5
Rapid pressure changes at depths below 300 meters can bring on abnormal heart rhythms.

6
Holding the breath during rapid ascent can expand lungs so drastically that the pleural sac ruptures. If a diver has a cold, the blocked part of the lung may expand under pressure and damage the alveoli. Severe heat loss from lungs can be fatal.

7
Because pressure increases the density of breathed gases, respiration through the trachea can become dangerously inefficient.

8
Helium in the larynx distorts sound, making the speech of helium-breathing divers unintelligible.

9
The neck and head are sensitive to heat loss because there are many surface veins and little insulation from fat. Severe heat loss may reduce brain efficiency.

10
If a Eustachian tube is blocked through infection or inflammation, the eardrum may burst under pressure. Ear infection is a common ailment of divers who spend long periods in damp undersea habitats.

11
Excess oxygen in the breathing mixture can cause brain convulsions. Over-rapid compression at depths below 100 meters disturbs nervous conduction, with consequent lack of coordination. Other factors can cause extreme mental stress.

12
If mucus blockage or inflammation closes the sinuses during ascent or descent, a resultant imbalance of pressure may lead to internal bleeding and pain.

13
Narcosis, low temperature, and fatigue may affect the sight, causing a diver to experience "tunnel vision."

Air Diving

It is possible for almost any diver to breathe ordinary air down to a depth of 60 meters, and some air-breathing men have even managed to work for brief periods at depths of up to 100 meters. But the safe limit for air diving has now been established at about 70 meters. To understand why this is so, we must consider what happens to the gases that we breathe as we inhale them under increasing pressures. (Until the late nineteenth century almost nothing was known about why divers fell ill and sometimes died. We owe much of our present understanding of diving physiology to two great physiologists, France's Paul Bert, 1833–86, and Scotland's John Scott Haldane, 1860–1936, whose pioneer efforts to discover what happens to the body under pressure often involved taking extreme risks with their own health.)

The human body exposed to water pressure needs an adequate supply of oxygen, just as every living animal does, but the big medical problem stems from the difficulty of defining the word "adequate" in an underwater context. To reach a meaningful definition, we need to calculate what is called the "partial pressure" of the oxygen in any mixture of gases at a given depth. At sea level, for instance, the partial pressure of oxygen is, obviously, 21% of one atmosphere, and the partial pressure of nitrogen is about 78% of an atmosphere. At a depth of 10 meters, therefore, the partial pressure of the oxygen in the air breathed by a diver becomes 42% of one atmosphere, since an additional atmosphere of total pressure is added for every 10 meters of water. At 20 meters the partial pressure increases to 63% – and so on. Similarly, the partial pressure of nitrogen at 10 meters is close to 1.57, and at 20 meters roughly 2.35, atmospheres.

This concept of partial pressure as the pressure exerted by any one gas in a mixture is very important in diving, because we use it to calculate the effects of each gas separately, and hence the safe limits within which the gas can be used. We have learned that the *minimum* safe limit for the partial pressure of oxygen is about 17% of one atmosphere. If the oxygen partial pressure falls below 17%, unconsciousness results – and with no forewarning. Any breathing gas must therefore contain enough oxygen so that, at a given pressure, there is at least 17% – preferably 20% – of one atmosphere partial pressure.

There is a *maximum* safe limit, too. Actually, it is perfectly safe to breathe 100% pure oxygen at sea level for a short time; and a diver can breathe it comfortably for an hour or so if he does not go below 7 to 10 meters. Deeper than that, pure oxygen produces violent convulsions that can lead to death. And this sort of oxygen poisoning happens rapidly if the oxygen partial pressure in a gas mixture exceeds 2 atmospheres. For prolonged dives, in fact, the partial pressure is best kept below about 50% of one atmosphere. Thus, the safest concentration of oxygen in any gas mixture is as much as will exert a partial pressure of no less than about 20%, and no more than about 50%, of one atmosphere.

The nitrogen in air presents a different problem. Although it is an inert gas and has no effect on the body at normal atmospheric pressure, it dissolves in the blood under higher pressures and is carried around to the tissues, muscles, and nerves, where it can upset nerve function so as to produce a narcotic anesthesia – which can obviously spell catastrophe for the diver. Nitrogen narcosis may begin to slow down a diver's reactions at a depth of only 40 meters. How soon and how profoundly he reacts depends, of course, on the individual. The same person may react differently, too, at different times. But for almost everyone who breathes air below 60 meters there is a high risk of perilous lethargy, hallucinations, and loss of control.

Thus, the depth limit for air diving is dictated by the physiological effects under pressure of the two major components of air. It is significant that the proportions of oxygen and nitrogen are such that the diver faces problems from both gases simultaneously when he descends about 60 to 80 meters: the partial pressure of the oxygen that he is breathing approaches the limit of 2 atmospheres and could irretrievably damage his lungs; and the partial pressure of nitrogen approaches 7 atmospheres and could bring on lethal narcosis. Moreover, nitrogen is a heavy gas, with a molecular weight of 28, and increasing pressure makes it denser. The resultant strain on a diver's lungs causes shortness of breath and muscular inefficiency. At some depth greater than 60 meters, therefore, air diving becomes impossible for everybody, and a substitute mixture of gases is necessary.

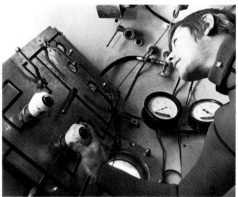

Left: With technical precision, an engineer on a surface-support barge blends an oxygen-helium breathing mixture from cylinders containing pure oxygen and pure helium. Helium, used as the inert gas for dives beyond 60 or 70 meters, averts the problem of nitrogen narcosis – but at the risk of "helium tremors," which are brought on by cell imbalance when helium permeates some cells faster than others. Above: During compression and decompression in a diving chamber, the bell man blends the breathing mixture in response to the increase or decrease of pressure. It is his responsibility to determine correct proportions of oxygen and inert gas.

The Use of Helium

A diver going much below 60 or 70 meters needs to breathe a mixture in which the percentage of oxygen is reduced and nitrogen is replaced by a less troublesome inert gas: preferably one that does not, under pressure, produce narcosis, is light enough to be comfortably inhaled and exhaled, and does not readily dissolve in the blood and body tissues. So far, alas, we have found no such paragon. The inert gas that we *have* settled for is helium, which is not narcotic and is far less dense (molecular weight 4) than nitrogen. Its one drawback is that it is even more soluble in the blood and body tissues than is nitrogen, and so it creates some very special problems in the processes of compression and decompression, as we shall see. First, though, let us consider the oxygen concentrations that may be safely included in an oxygen-helium mixture as we go to 100 meters and beyond.

Even for very short dives, the oxygen partial pressure must be kept below 2 atmospheres; for long ones, it may not safely exceed 50 to 60% of 1 atmosphere without risk of lung injury. Thus, as the surrounding pressure is increased from 1 atmosphere at the surface to 10 atmospheres at 90 meters and to 31 atmospheres at 300 meters, the amount of oxygen in the mixture must be continuously reduced from an initial 20% to less than 2%. And the safe *minimum* at 300 meters is around 7%. Less than that will inevitably bring on loss of consciousness from anoxia (lack of oxygen). The narrow band of safety between the high and low limits keeps getting narrower as depth increases. This means, of course, that enormous technical control and accuracy are required of the engineer who directs any deep-diving operation.

Getting the proportion of oxygen and helium right is only one of many engineering tasks. The pressure in the diving chamber within which divers are lowered must be raised to the approximate pressure of the depth where they are to work, and this must obviously be done by degrees. The process of gradual compression presents no great difficulties down to 100 meters, for the diver can quite safely be subjected to increasing pressures at a rate of 30 meters per minute. But when engineering companies tried to lower diving bells below 100 meters in the early 1960s, they found that severe fits of trembling seized the work teams as they approached the working depth unless the descent was immensely slowed down. In a typical project carried out by a diving subsidiary of the American Westinghouse company in 1965, the men had to be compressed slowly for several hours before they could repair a dam at a depth of 150 meters. This, of course, seemed an expensive waste of time.

Since then, laboratory studies carried out in several countries have shown that the 30-meters-a-minute compression rate simply will not work once the 100-meter mark has been passed. Among the ill effects of such rapid compression are not only trembling of the limbs but pains in the joints, dizziness, and loss of balance, with some divers being more severely affected by these ailments than others. It is not the pressure that causes such so-called helium tremors, but the over-swift rate of compression, for the highly soluble helium in an oxygen-helium mixture dissolves at different rates in different body tissues.

Thus, if a deep diver is compressed slowly, each tissue has time enough to take up the helium at its own rate, and so the tissues remain in balance. But with fast compression the helium dissolves rapidly in the "fast" tissues, while the "slower" tissues lag behind. This imbalance adversely affects both the synovial fluid, which bathes the joints, and the endolymph fluid of the inner ear. The greater the overall pressure, the more severe the effects of the imbalance.

As a result, the basic procedure for descent in the 100-to-300-meter range involves holding the diving-chamber pressure steady for periods of several hours at various stages of compression, so as to regain equilibrium in the tissues. Naturally, this is a costly business in terms of lost working time, and efforts to solve the problem are constantly under way. In 1974, Dr. Peter Bennett, an Englishman working at Duke University in America, experimented with adding a trace of nitrogen to the conventional helium-oxygen mixture. Since helium tremors take the form of super-excitability, he argued, the addition of a small amount of a narcosis-inducing gas ought to counteract this effect. Bennett's mixture was tried out on mice with apparent success, and human divers breathing it have been compressed to 300 meters in only 33 minutes.

The technique is still far from proven. One difficulty is that the addition of nitrogen makes the gas heavier, and therefore harder to breathe. But Bennett has opened a promising door to the future.

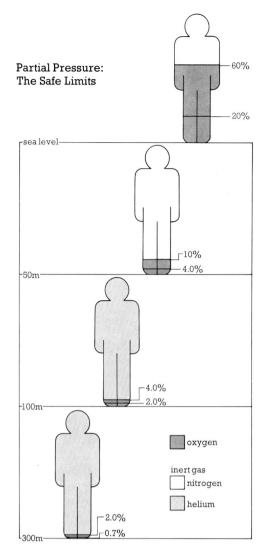

Partial Pressure: The Safe Limits

oxygen

inert gas
nitrogen
helium

Too much or too little oxygen can damage the lungs or brain. The figures above suggest approximate maximum and minimum safe percentages at various depths. Note the extreme narrowing of the safety band with increasing oxygen partial pressure.

Experiments at Duke University suggest that a trace of nitrogen in an oxy-helium mixture counteracts some of the bad effects of helium. These profiles of one such dive compared with a U.S. Navy dive using a conventional oxy-helium mixture show that, in comparable dives to 300 meters, the Duke dive required only 33 minutes for compression and 4 days for decompression, as against Navy times of 1 day and 11 days.

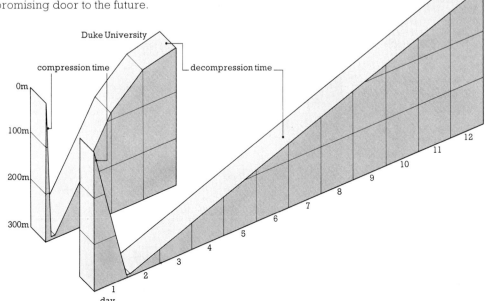

The Helium Barrier

By 1975 undersea men could work profitably for several days at a time at depths of 300 meters. But although pressures equivalent to 610 meters had been reached in laboratory experiments, the limited endurance of the divers who took part appeared to indicate that the frontier could not be pushed farther down. Just as the possibilities of breathing pure air are exhausted below 60 or 70 meters, so the helium-oxygen mixture runs into trouble soon after the 300-meter mark is passed.

There are virtually no insoluble engineering problems in putting men far down in the ocean and bringing them up again. The medical problems, however, remain baffling; and many hard-working experts are trying to solve them in laboratory pressure chambers, where the "divers" are safely in the hands of doctors, physiologists, and psychologists. It was in 1968, when the first tests of dives beyond 300 meters were being made, that the subjects began to show signs of what the supervising physicians in a French laboratory described as the "high-pressure nervous syndrome." This illness affected the movements, balance, and dexterity of the divers, who were breathing an oxygen-helium mixture. Moreover, the rhythm of their normal electric brain waves was disturbed. These symptoms were so severe that for a year or two researchers suspected the existence of an insurmountable "helium barrier" at a depth of roughly 360 meters.

The barrier is there – but it is not, we now know, insurmountable. Once again, the rate of compression is at least partly to blame. In March, 1970, two British divers made a successful experimental dive down to 457 meters at the Royal Naval Physiological Laboratory near Portsmouth, England, at an unprecedentedly slow rate of compression.

Extreme care was taken in the control of the gas composition, removal of carbon dioxide, and general comfort of the men, who reached 457 meters at the end of the fourth day. After remaining there for 10 hours, they made a gradual ascent to normal atmospheric pressure in 11 days. Since then, over a dozen men have supported pressures of from 500 to 610 meters during experiments in France and the United States.

So the 300-meter helium barrier has been overcome – not only, by now, in the laboratory but in practice. However, though various kinds of "barrier" that obstructed the development of deep diving in the past

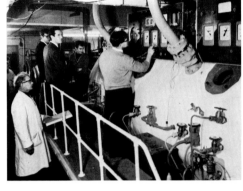

have been broken through, it does look as if we are approaching a group of extremely knotty problems at the 600-meter mark.

We have noted that the increasing density of gas under pressure makes it harder to breathe. A liter of air, which weighs 1.3 grams at sea level, weighs ten times as much at a depth of 90 meters. And a liter of helium-oxygen, which is much lighter than air at the surface, weighs 10.5 grams at 590 meters, where it is under 60 atmospheres of pressure. Clearly, this mixture is as hard to breathe at 600 meters as it is to breathe air at 70 meters. The dense gas is bound, therefore, to result in a reduced

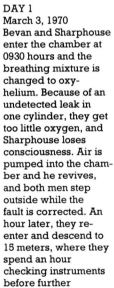

DAY 1
March 3, 1970
Bevan and Sharphouse enter the chamber at 0930 hours and the breathing mixture is changed to oxy-helium. Because of an undetected leak in one cylinder, they get too little oxygen, and Sharphouse loses consciousness. Air is pumped into the chamber and he revives, and both men step outside while the fault is corrected. An hour later, they re-enter and descend to 15 meters, where they spend an hour checking instruments before further pressurization to 152 meters.

DAY 2
March 4
The divers descend from 152 meters to 304 meters (i.e., pressure is increased to simulate water pressures at these depths). They remain here for the next 24 hours, during which time they keep busy carrying out mental arithmetic and hand-eye coordination tests. The results of these earlier tests provide detailed information of the divers' body functions for subsequent comparison with any changes that occur at deeper phases of the dive.

DAY 3
March 5
The most important day of the dive. Slowly they drop to what has been thought to be the helium barrier – about 360 meters. They pause at 366 meters for further tests and acclimatization. The temperature rises as pressure increases, and they sweat heavily. Bevan begins to move jerkily, suffering from helium tremors as the descent continues to 396 meters. They have broken the helium barrier and must undergo an hour of tests before proceeding.

Before 1970, scientists suspected that the danger of high-pressure nervous syndrome made the use of oxygen-helium mixtures impracticable at depths below about 360 meters. In 1970, however, two British divers breathing oxy-helium descended to a simulated depth of 457 meters in an experimental "dive" at the Royal Naval Physiological Laboratory. Their experience opened the way to further proofs that oxy-helium dives are feasible all the way down to 600 meters if compression and decompression periods are long enough to give the body time to adjust.

An hour and a half before their dive began, John Bevan and Peter Sharphouse reported to the lab, where several spots on their heads and chests were shaved, so that silver electrodes could be attached to the skin. The electrodes were to measure and record brain functions and respiratory and heart activities during the dive. The results of all such tests were highly encouraging; and although the divers had unpleasant moments, particularly during the long decompression period, they suffered no lasting ill effects from their 16 days in the pressurized chamber.

This graph shows the slow course of their "descent" and "ascent." Above it are abridged extracts from John Bevan's journal of the dive, along with a random selection of photographs taken by laboratory observers and the divers themselves. The final photo shows Bevan (left) and Sharphouse jauntily emerging from their ordeal.

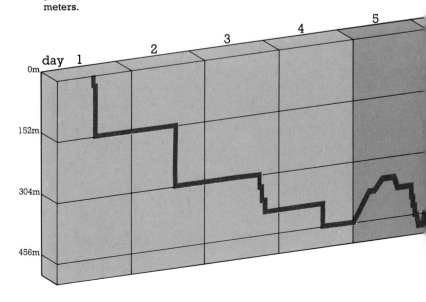

supply of oxygen through the blood to the muscles. So the diver's output of work would suffer accordingly.

What is less obvious is the effect that a dense gas mixture has on sleep. During a diver's sleeping time under high pressure, when his breathing is shallow, physiological changes can take place without his reacting to them. After a while, though, inefficient breathing causes a buildup of carbon dioxide, and the sleeping man wakes up gasping. Even if inefficient breathing does not directly aggravate the high-pressure nervous syndrome, disturbed nights enfeeble the man, so that he may actually tremble more and have slower brain waves in the morning than at the end of a working day.

There are obvious advantages in keeping workers in compressed diving chambers for long periods; but the sleep problem means that even when it becomes feasible to work at 600 meters, divers will have to ascend a hundred meters or more for their hours of sleep. In practice, such partial ascents and descents, though time-consuming, should not be hard to manage. We have perfected a technique (known as "excursion diving") that permits divers in shallower waters to work routinely at a greater pressure than that of the chamber in which they rest between making dives.

Still, the high-pressure nervous syndrome may itself constitute an absolute barrier to going beyond 600 meters. The onset of the illness can be postponed, and its severity reduced, by slow compression, but some symptoms will occur in any diver who stays a while at the 600-meter mark. The trouble here seems to come from the pressure, not the rate of compression. Nor could it apparently be avoided by using some other gas than helium. What happens, quite possibly, is that the diver's cell membranes are compressed at great depths, and this has a bad effect upon the nerves.

239

  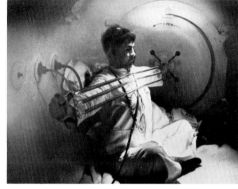 

DAY 4
March 6
The dive continues down to 457 meters. Neither man has much trouble breathing, but even small changes of temperature cause either significant overheating or coldness in their bodies. The most comfortable temperature is 30°C. Bevan still has helium tremors, but both men can work well, and neither has convulsions (as scientists feared they might).

DAY 5
March 7
After 10 hours of mental and physical tests at 457 meters, the ascent begins. Sharphouse soon feels dizzy, vomits, and loses his sense of balance because the ascent is too rapid. To ease the symptoms, the divers have to go back down some 11 meters beyond the 457-meter mark. Bevan becomes hot and drowsy, but perks up when ascent is resumed.

DAYS 6 and 7
March 8 and 9
Sharphouse remains in a constant state of dizziness and is unable to eat solid food. Only after the first few days of the 11-day ascent does his sense of balance return and his appetite improve.

DAYS 8 through 11
March 10 through 13
Daily physiological and psychological tests go on; but the intensity of experimental activity has lessened, and both Bevan and Sharphouse catch up on lost sleep. Sharphouse's dizziness decreases, and he resumes a normal diet and is able to move about comfortably.

DAYS 12 through 15
March 14 through 17
Bevan's right knee begins to hurt because of mild decompression sickness. To ease the pain, they go back down from 9 meters to 18. Pressure is low enough by now so that they can breathe pure oxygen (a classic therapy for the bends).

DAY 16
March 18
At 1600 hours the divers surface to an enthusiastic reception from everyone involved in the experiment. Both men feel fine except for the slight pain in Bevan's knee. Intensive medical examination shows them to be in good shape.

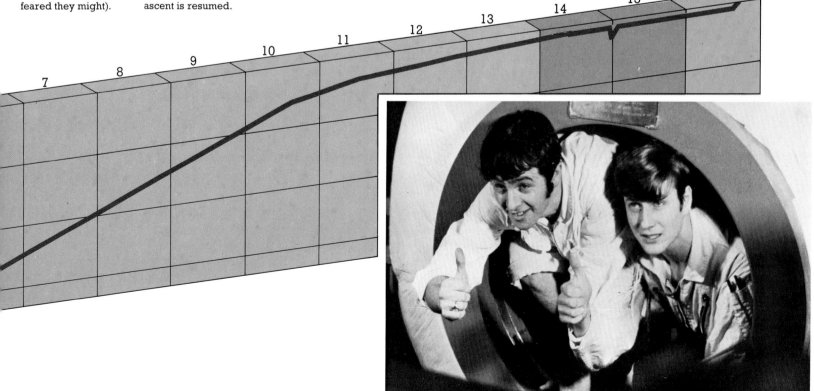

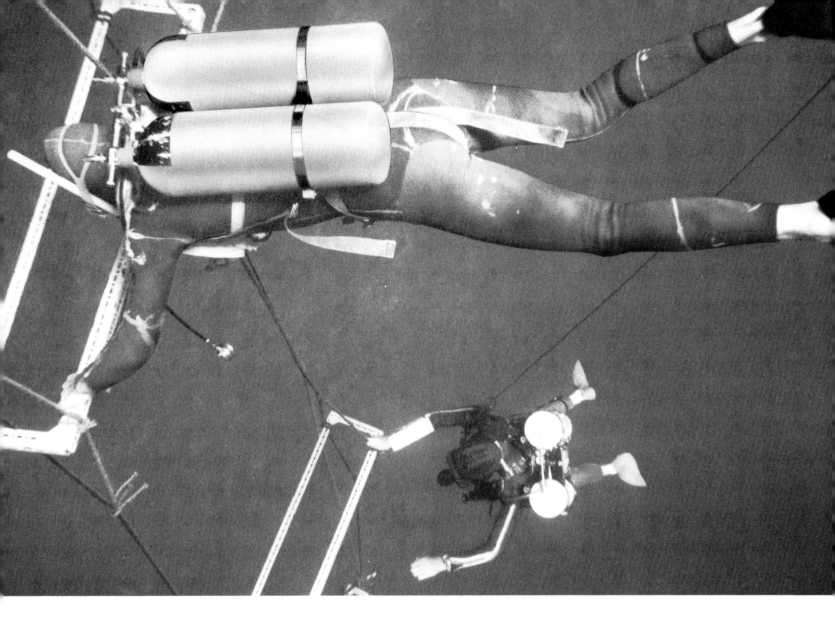

Bringing Them Back

Decompression is an even slower and more delicate process than compression. Under pressure, the inert gas in a breathing mixture has dissolved in the blood and tissues; if the pressure is reduced too quickly, the nitrogen or helium will form bubbles, and these can cause serious illness by choking veins and arteries. This illness, the plague of divers for many centuries, is variously called caisson disease, decompression sickness, or "the bends." Among its exceedingly painful symptoms are aching joints, skin rash, respiratory difficulties, nausea, and paralysis. Severe affliction can mean death. And even slightly inadequate decompression, if it happens often enough, may be the cause of bone deterioration – a disease known as aseptic bone necrosis.

The only way for a diver to avoid decompression sickness is to come up so slowly that the bubbles never form, or at least never get big enough to do damage. This prolongs the dive to a remarkable degree. For instance, if a man stays at a depth of 180 meters for only an hour, his decompression time will be 38 hours. And a brief dive to 300 meters requires many days for the safe re-

turn to normal atmospheric pressure. From the economic standpoint, naturally, such an enormous loss of man-hours is almost too costly to contemplate. So there has been a good deal of relatively recent research directed toward devising ways of increasing the length of underwater working time in proportion to the time required for descending and ascending.

The initial problem, of course, has been to discover the various rates at which nitrogen (for shallower dives) and helium (for greater depths) dissolve within the various tissues. This is a complicated matter, since the answers depend on the nature of the gas, the immensely differing tissues, and the many possible pressures and durations of dives. The time that it takes for a given tissue to become saturated with an inert gas or to give up the dissolved gas varies enormously from one part of the body to another, and no diver is either safely compressed or decompressed until the moment when total bodily equilibrium is reached at the required pressure.

As a result of extensive work by researchers in many countries, we now have both compression and decompression tables for a very wide range of possible

dives. So we know, for example (using figures from the French tables, which differ slightly from those worked out in Britain and America), that an air-breathing diver can remain underwater indefinitely at a depth of 10 meters without needing to pause for the sake of his tissues when ascending. But if he spends an hour and a half at 25 meters, his ascent must take at least 54 minutes, for he needs to stop on the way up and give his body time to adjust to the lower pressure. And after an hour-long dive at 100 meters, where he must breathe a helium-oxygen mixture, the decompression process will require 8 hours.

Air divers can improve their work-time/decompression-time ratio by reducing the amount of nitrogen and increasing that of oxygen in the breathing mixture – though this is done, of course, only at depths below 50 meters, where there is no risk of oxygen poisoning. Decompression time is minimized this way because the partial pressure of the smaller quantity of nitrogen at the required depth is equivalent to the partial pressure of the nitrogen that would be contained in normal air at a shallower depth. Navy frogmen use such oxygen-enriched mixtures when they engage in lengthy and delicate tasks like mine removal. And the fact that the technique permits not only long dives and relatively short decompression periods but also smaller, and therefore less bulky, gas cylinders is proving more and more useful for the sort of industrial operations that can be done by free divers in under-50-meter waters.

For deeper dives, whether on air or helium-oxygen mixtures, we have been perfecting and refining another technique for improving the economics and efficiency of underwater work. This method, known as "saturation diving," exploits the physiological fact that the body becomes saturated with an inert gas after a certain length of time at any pressure. Once a diver has been compressed at a given depth for about 10 hours, his tissues absorb no more inert gas, and he can stay there indefinitely without needing extra decompression time at the end of the dive. It makes no difference at a depth of 200 or 300 meters whether you stay down for 12 hours or 12 days; your decompression time for each depth remains unchanged.

Thus, for commercial work requiring many hours on a deep worksite, we have found a way to keep teams going for days without returning to atmospheric pressure. We do this by transferring them back and forth from a compressed diving bell (the SDC) to a surface chamber (the "deck decompression chamber" or DDC) where they can sleep, eat, and relax while under the approximate pressure of the working area.

Saturation diving represents a great advance over earlier methods. Even so, div-

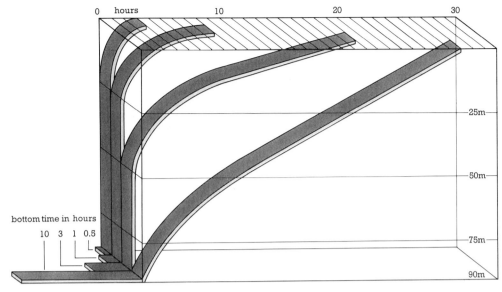

ing remains time consuming and expensive. Obviously, a team must not be cooped up too long. Once a diver has been decompressed, it is risky for him to go under again within a day or so, since some of the inert gas remains in the body after the ascent has been completed. There is no danger if a diver stays on the surface until the gas has escaped through his blood and lungs. But if he descends again within a few hours, he starts the second dive with an excess amount of gas still in the tissues, and he needs a longer than normal decompression period at the end of that dive. "Repeat-dive" tables that give safe rates of ascent for second and third dives have been worked out, but only down to 60 meters. For *any* repeat dives, there are physical and psychological strains on the diver.

If any symptoms of the bends occur during ascent or after completing decom-

Depth and amount of time spent at the bottom together determine the length of the decompression period. Here we see how much longer it takes to bring a diver back to atmospheric pressure after a long dive than after a short one. If, for example, a man works at 90 meters for half an hour, he can ascend comfortably in 3 hours; but if he stays for 10 hours, he needs 27 hours of gradual decompression to prevent dangerous bubbles of inert gas from forming in his blood. If the depth measurement on the graph were extended down to, say, 300 meters, the decompression-time curve would run approximately parallel to those shown here, but it would ribbon off the page.

pression, an increase in pressure is the prescribed treatment. If the symptoms persist, the diver may have to be recompressed even deeper than the depth of his actual dive.

Two ways to decompress safely. Facing page: After diving to 60 meters, these free divers pause at intervals on hanging metal frames, in accordance with decompression schedules printed on small boards attached to their wrists; by the time they reach the surface they're fully decompressed. Below right: SDC divers, who need a much longer decompression time, may spend several days in the decompression chamber. During this period, there's little to do but read and eat. After the cook passes a tray into the supply lock (below), the divers must wait until the outside hatch has been sealed and pressure has been equalized in the lock before they open the inner door and collect their food.

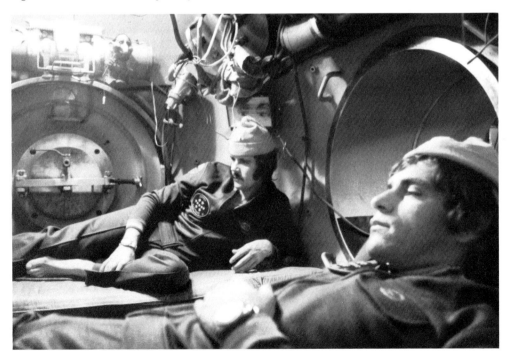

The SDC in which divers are transported to the bottom has a spherical or cylindrical hull and is equipped with portholes that allow the men to observe their worksite before entering the water. Suspended by a support cable, the bell is guided during descent-ascent maneuvers by two cables held in constant tension by automatic winches. In addition to the diving team (which often includes a stand-by man in case of need), there is a nondiving "bell man" inside the chamber; he is in communication with the surface as well as with the divers, and is responsible for much of the technical control of the operation.

For a single dive of short duration – no more than an hour or so – compression time is unlikely to be very long. So, although the divers are equipped and ready, the chamber remains at normal atmospheric pressure until the men have examined the working area through the observation windows. If, as sometimes happens, they decide not to enter the water, pressurization is unnecessary. Otherwise, compression now begins, with the gas being fed into the chamber from cylinders affixed to its external walls or – much less frequently – from a surface control station. When the correct pressure has been established, hydraulic controls open the escape hatch and the divers go out immediately, for the dive has started, and each minute spent under pressure lengthens the time of decompression. The bell man pays out communication cables as well as the umbilical cords that supply the divers with breathing gas.

The divers work in the beams of searchlights from the SDC, and they keep the bell man informed of their progress. As soon as their task is completed, they rapidly return to the chamber and decompression begins, with the bell man himself controlling the first part of the process during the ascent. The lift through the wave zone can be lively, not to say hair raising, in a choppy sea, and the men are generally not at all unhappy when they feel the diving bell being lifted aboard, say, a drilling ship or surface control station by the two powerful hydraulic jacks that swing it up to a position just above the deck decompression chamber. The two chambers, in which the pressure is equalized, are now locked together, and the diving team moves into the DDC.

The DDC is much roomier than the SDC. All its fittings are arranged so as to provide as much comfort as possible for several men in a limited space. Since they may need to spend many hours there while the pressure is gradually lowered – an hour of work at 180 meters, remember, requires 38 hours of decompression time – the DDC has washing and toilet facilities as well as beds. But the

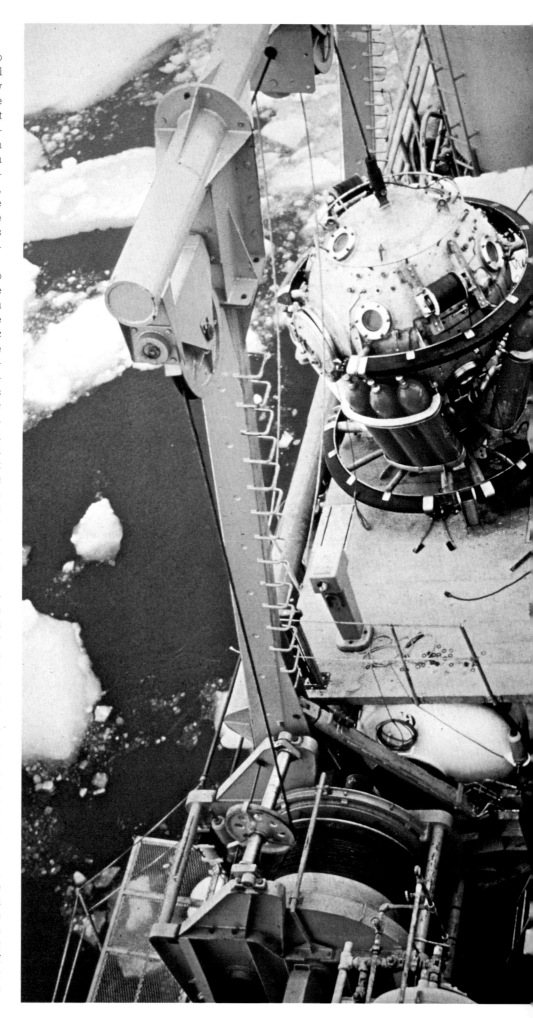

fact that it must be occupied for a long time in comparison with the time spent in the SDC creates a problem; while it is tied up, no other team of men can be sent to the bottom unless there is a spare DDC in which to house *them* when they surface. The use of two DDCs in tandem is becoming fairly common, but this is sometimes not economically feasible.

Saturation diving solves this problem at least to some extent by keeping the men in the DDC at or near working pressure. Thus they are always ready to descend again without slow compression, and they can be transferred back and forth from DDC to SDC during the course of several days, with only one slow compression phase at the beginning and one slow decompression at the end. If the DDC is large enough, two teams of three men each can perform such dives in shifts, with, say, two 3-hour shifts per team in every 24-hour period. Obviously, there is a limit to human endurance under such conditions, but a 15-day-long saturation dive seems to be quite feasible, as long as the divers get extensive rest periods between their 15-day stints.

The final decompression period for saturation diving can be considerably shortened through a procedure perfected by an experimental group working for the French organization Comex, which can take credit for many remarkable advances in modern diving techniques. This procedure – "excursion diving" – involves an initial compression of the divers in the DDC to a somewhat lower pressure than that of the working depth. Thereafter, at the bottom of each trip in the diving bell, pressure is raised to surrounding water pressure; and it is dropped back to DDC pressure during the ascent. Because the difference in pressures is only as much as permits ascent from the work depth without stopping, the transfer to the lower pressure in the DDC does no harm to the divers. Yet a good amount of decompression time is lopped off the total required at the end.

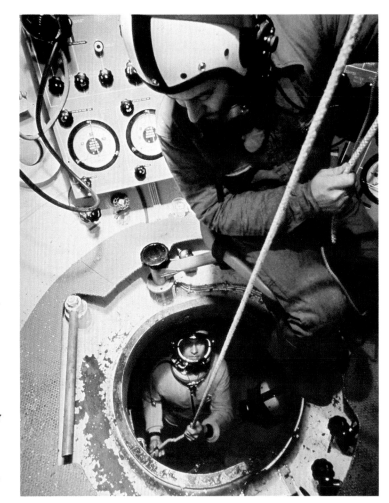

Key moments in any SDC operation are the beginning of the descent, the start of the working period, and the final breakthrough to the surface. Facing page: Suspended on a support cable and stabilized by guide wires, a spherical chamber is about to plunge into the ice-clogged waters of Hudson Bay; the winch in the foreground controls its rate of descent. Right: At the worksite, while the watchful bell man pays out communication lines and a life-preserving umbilical, the diver drops through a hatch into the water. Below: The SDC shown here is being hauled out of rough North Sea waters by its heavy support cable. The guide wires in the foreground prevent swinging.

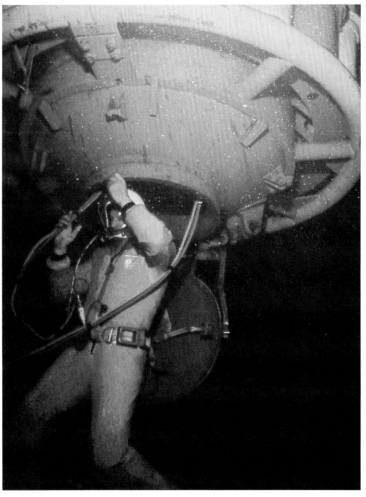

Normally, except in far-northern waters, sea-water temperatures hover around 5° to 15°C (41° to 59°F), which means that the water is chilly, not freezing. It is, nevertheless, cold enough to kill a man unless he is well protected. The sort of cold that all swimmers know is at worst only a slight discomfort, but even this causes loss of efficiency. The loss is far greater for a man working underwater. If he permits himself to get progressively colder, his efficiency drops, his body temperature sinks, and his heart stops beating. This would happen, slowly but inevitably, to any unprotected diver. Yet modern men have worked comfortably underneath the sea ice of the Arctic.

Canvas-and-rubber diving suits that kept the diver warm enough to be fairly comfortable were developed during the nineteenth century. Today, though, we know that a fair amount of comfort is not enough. Even minor losses of heat can cause a 20% drop in mental or manual efficiency; and this, combined with slight narcosis, reduced vision in dark water, and other stresses, can cost not only money but lives. So we try to use every technological trick in the book to keep out the cold.

A modern equivalent of the old-fashioned rubber suit is made of foam neoprene, a kind of synthetic rubber, and this type of suit is quite adequate for shallow-water spearfishing and sports diving. But it has little value for working dives, since it does not insulate well enough to provide warmth for hours on end. Furthermore, the foam neoprene compresses and thins out under pressure, and its insulation actually decreases when the diver needs it most. More practical for diving at depths of 10 to 60 meters is "constant-volume" waterproof clothing. This is a sealed "dry" suit – "dry" because, unlike the "wet" neoprene suit, no water penetrates to the skin – which has an air supply that inflates the suit to counteract increasing pressures. The diver, surrounded by a constant volume of insulating air, may wear a supple, close-fitting rubber hood; but this part of the outfit is gradually being replaced by a light, molded-plastic helmet, which can be equipped with a telephone, a gas-flow regulator, and an anti-mist arrangement for the face mask.

One or another variant of the constant-volume suit is worn today by virtually all air-breathing divers other than sportsmen. The traditional baggy garment and metal helmet of what is known as the "standard" or "hard-hat" diver's uniform have been almost entirely superseded by the inflatable suit and plastic helmet.

The man who breathes a helium-oxygen mixture needs something different. Because helium conducts heat rapidly and lacks the insulating properties of air, the inhaled gas

How can a diver keep warm enough to work well in cold water? To help solve the problem, the Royal Navy, pinpointing the parts of the body that most need insulation, has experimentally determined the optimal amounts of electric heat for each one. The experiments were conducted in about 3 meters of water, at an average temperature of 4°C, with divers wearing electrically heated undersuits. Recommended amounts of electricity are indicated as watts per square meter, the standard measurement for comparing heat needed for different body-surface areas.

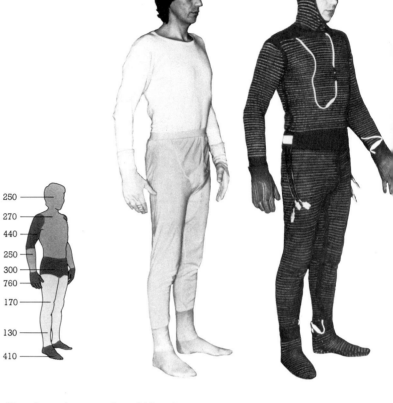

Dressing to keep out the cold is a slow procedure for this diver, whom we see getting into several layers of protective clothing: first, a cotton undergarment; then an electrically heated woolen suit, which has insulating filaments knitted into it; over that a nylon-pile insulating garment; and, finally, the ordinary dry suit. Suitably clad at last, he is ready to be connected to his lifeline, the umbilical, which carries a telephonic communication line and gas hose as well as the electric-power supply for his heated undergarment.

is at the same temperature as the water, and so the deep diver loses heat over his whole body surface as well as from his lungs and respiratory passages. With long-duration helium dives now being required for work on oil rigs in such waters as the North Sea, Hudson Bay, and Labrador, where the water temperature at 180 meters may be as low as –2°C, sophisticated protective equipment is a necessity.

Bodily heat loss increases with increasing depth as a result of the increase in gas density. Thus, whereas the density of a helium-oxygen mixture is about 0.17 grams per liter at the surface, it rises to 6 grams per liter at 300 meters, and so the gas carries away much more heat with each breath than it would on the surface. To counteract such losses, the temperature of the inhaled gas and of the area immediately surrounding the diver's body must be increased in proportion to the depth. There is, in fact, a very narrow thermal zone that is both comfortable and safe. At a depth of 600 meters, for instance, the safe working temperature could not be much lower than 34°C, which is only 2.8° below blood heat.

So far, we have found two possible ways to combat heat loss for helium-breathing

divers: a heated suit, and a heater that warms the breathing gas before it is inhaled. It is fairly easy to provide a diver with heated gas. He simply carries on his back a small cylinder that contains an electric heater, through which the gas fed from the SDC flows. But the heating of clothing presents problems that have not yet been entirely solved. With a dry, constant-volume suit, the diver wears underclothing made of a synthetic material in which are embedded electric elements that can be powered by means of an umbilical from the SDC. Unfortunately, these heating wires tend to be unreliable. Although, as in all underwater equipment, their low voltage eliminates the risk of electrocution, they can go wrong to a point where a given part of the diver's body is occasionally either scorched or chilled. So, though safe enough, this underwater "electric blanket" can be quite uncomfortable.

An alternative possibility is a suit that resembles a hot-water bottle with the diver inside. It is made of rubber, and warm water constantly circulates around its occupant; as the water cools, it flows out through exit tubes near the gloves and boots. In such suits, thermal comfort is excellent, but con-

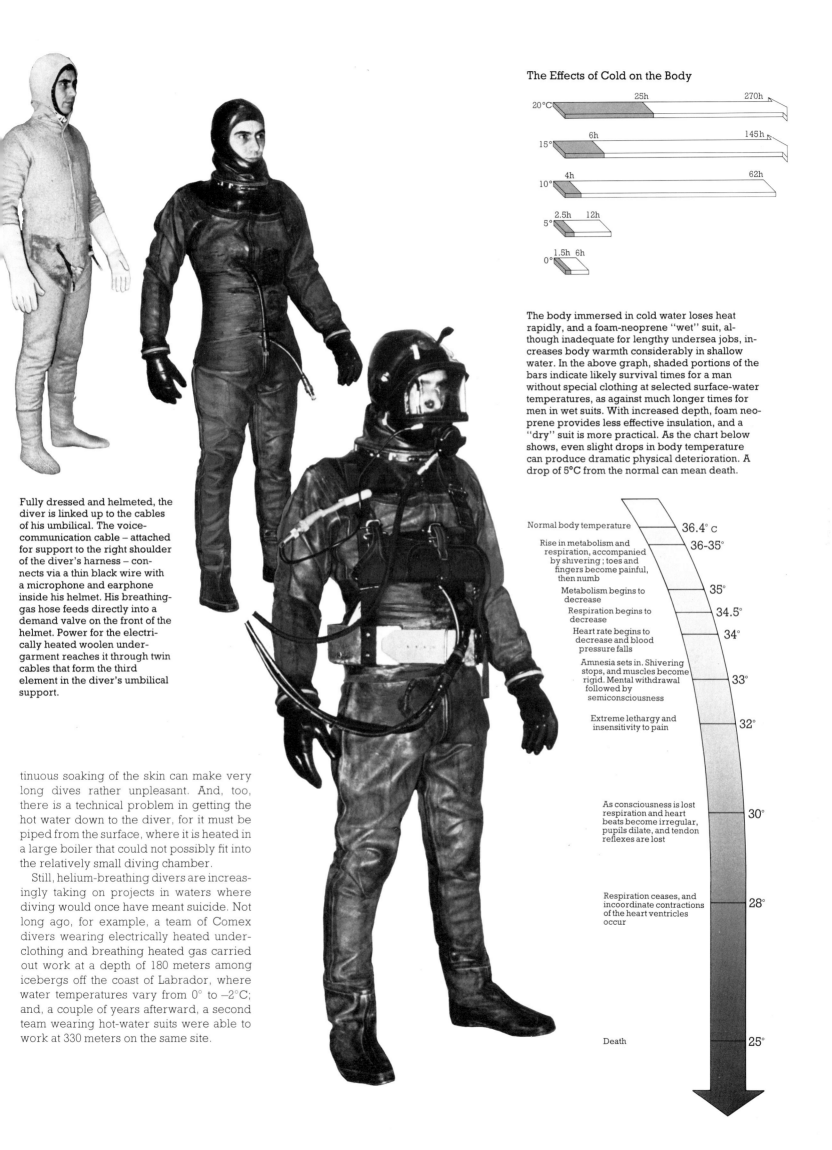

The Effects of Cold on the Body

20°C — 25h — 270h

15° — 6h — 145h

10° — 4h — 62h

5° — 2.5h — 12h

0° — 1.5h — 6h

The body immersed in cold water loses heat rapidly, and a foam-neoprene "wet" suit, although inadequate for lengthy undersea jobs, increases body warmth considerably in shallow water. In the above graph, shaded portions of the bars indicate likely survival times for a man without special clothing at selected surface-water temperatures, as against much longer times for men in wet suits. With increased depth, foam neoprene provides less effective insulation, and a "dry" suit is more practical. As the chart below shows, even slight drops in body temperature can produce dramatic physical deterioration. A drop of 5°C from the normal can mean death.

Normal body temperature	36.4° C
Rise in metabolism and respiration, accompanied by shivering; toes and fingers become painful, then numb	36–35°
Metabolism begins to decrease	35°
Respiration begins to decrease	34.5°
Heart rate begins to decrease and blood pressure falls	34°
Amnesia sets in. Shivering stops, and muscles become rigid. Mental withdrawal followed by semiconsciousness	33°
Extreme lethargy and insensitivity to pain	32°
As consciousness is lost respiration and heart beats become irregular, pupils dilate, and tendon reflexes are lost	30°
Respiration ceases, and incoordinate contractions of the heart ventricles occur	28°
Death	25°

Fully dressed and helmeted, the diver is linked up to the cables of his umbilical. The voice-communication cable – attached for support to the right shoulder of the diver's harness – connects via a thin black wire with a microphone and earphone inside his helmet. His breathing-gas hose feeds directly into a demand valve on the front of the helmet. Power for the electrically heated woolen undergarment reaches it through twin cables that form the third element in the diver's umbilical support.

tinuous soaking of the skin can make very long dives rather unpleasant. And, too, there is a technical problem in getting the hot water down to the diver, for it must be piped from the surface, where it is heated in a large boiler that could not possibly fit into the relatively small diving chamber.

Still, helium-breathing divers are increasingly taking on projects in waters where diving would once have meant suicide. Not long ago, for example, a team of Comex divers wearing electrically heated underclothing and breathing heated gas carried out work at a depth of 180 meters among icebergs off the coast of Labrador, where water temperatures vary from 0° to –2°C; and, a couple of years afterward, a second team wearing hot-water suits were able to work at 330 meters on the same site.

How Divers Breathe

All diving operations, from the simplest to the most advanced, are planned around the need to provide the diver with a supply of the right gas mixture at the right pressure at the right times. He can use no equipment and make no decisions that might result in an interruption of the gas supply.

The free diver, who generally swims at shallower depths than 60 meters, and for no more than an hour at a time, carries light, self-contained equipment – a cylinder that holds a limited amount of air – and he may require a decompression period of up to half an hour. So he must plan ahead to finish his work and get to the surface well before his gas runs out. In most such sets, the diver inhales through a mouthpiece held between the teeth, or through a small mask pressed over his mouth and chin. In either case his nose and eyes are protected by a glass-fronted mask, into which the small mouthpiece may be built. The simplest method of feeding the gas from cylinder to lungs is through what is technically termed an open-circuit demand system – popularly known as an aqualung or scuba (which is an acronym for "self-contained underwater breathing apparatus"). Why "open-circuit"? Because the gas is not re-used; it flows through the diver's lungs, is exhaled, and vents through a nonreturn valve into the sea. Why "demand"? Because a valve in the cylinder supplies gas only when the diver inhales; it shuts off as he exhales.

The standard system of supplying air from the surface to a diver wearing an old-style helmet and canvas-and-rubber suit is also open circuit, but the air flows freely, not just on demand. For such surface-supported divers, the gas passes continuously from the hose to the helmet and suit, and is vented through a relief valve. A compromise between the old-fashioned and the aqualung methods is a surface-demand system in which the breathing gas is fed down a high pressure hose, with a demand valve that supplies gas to the diver only when he breathes in.

Each type of equipment has its own safety device. The scuba diver must have a highly visible watch, a gauge to indicate depth and hence the rate of gas consumption, and a pressure gauge attached to the cylinder to show how much air is left. In addition, many cylinders are fitted with a reserve lever which must be pulled when only 20 atmospheres of pressure are left. The helmeted suit has a built-in margin of safety because of the large amount of air in it. If surface-supplied air fails, the diver can close the relief valve and continue to breathe for a while. And a surface-demand set has a reserve gas cylinder that comes into use automatically if hose pressure drops.

Open-circuit breathing systems are uneconomical in that the diver uses only a fraction of the oxygen in the air, for the rest is wasted when he exhales. Two other systems now in use permit a free diver to rebreathe the gas he exhales – not indefinitely, of course, but several times. Both systems involve the use of soda lime within the equipment that the diver carries. In the more frequently used system – semi-closed-circuit breathing – the exhaled gas mixture is cleansed of its carbon dioxide as it passes through the soda lime and can then be rebreathed, while only a small amount of oxygen-depleted gas bubbles off through the nonreturn valve. In the other method – closed-circuit breathing – no gas at all escapes into the water; the diver breathes pure oxygen (which restricts his activities, obviously, to a depth of less than 10 meters), and since there is no nitrogen in his cylinder and the soda lime constantly absorbs carbon dioxide, the supply of breathing gas lasts a long time. Navy frogmen prefer closed-circuit breathing because no bubbles signal their presence to the enemy.

In theory, all three methods of feeding gas – open circuit, semi-closed circuit, closed circuit – are feasible not only for free divers but also for those who work out of submersible chambers. In practice, though, open-circuit systems consume so much gas that they are uneconomic for deep dives. At 250 meters, for instance, an open-circuit-fed diver would use up 750 liters of oxygen-helium mixture in one minute. So the preferred method is more nearly closed circuit. Oxy-helium fed to the diver through an umbilical is fed back to the bell via a return hose. Within the chamber, the gas is cleaned and, with the waste of very little helium, made ready for further use. Submersible chambers carry gas cylinders on their sides, and the regenerated oxy-helium is pumped back into these. For almost all saturation dives, it is from the SDC's own cylinders, not from the surface, that the divers' breathing gas comes. Only occasionally – and only for very deep, lengthy dives – is a hose from the surface affixed to the chamber.

A free diver swimming in relatively shallow water breathes air from an aqualung (scuba) carried on his back. This is technically termed an open-circuit demand set; air flows from the aqualung only when the diver inhales, and exhaled gas is wasted into the water, where it causes a cloud of bubbles. Below: The orange life jacket around this scuba diver's neck can be inflated with air from a separate cylinder. He also has a snorkel tucked into his mask band; when swimming at the surface he can get the snorkel into his mouth and shut off the supply of gas from his aqualung. The general principle of an open-circuit demand system is illustrated in simplified form below right. Gas, usually pure air, is stored in the cylinder (1) at a pressure of 150–200 atmospheres. When the diver inhales, the demand valve (2) opens and gas flows into his lungs. When he exhales, the valve closes, and the gas he breathes out flows away (3).

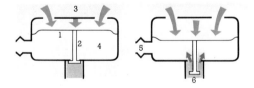

Above: How the demand valve in an open-circuit system does its job. It operates, as illustrated in the two diagrams (very much simplified), by means of a synthetic-rubber diaphragm (1) to which a piston (2) is linked. The diaphragm is exposed to ambient water pressure on the outside (3) and to a gas-filled chamber on the inside (4). When the diver removes air by inhaling through the breathing tube (5), the gas pressure in the chamber drops, and water pressure on the diaphragm forces down the piston, thus permitting gas to flow in from the neck of the cylinder (6). As the cylinder gas enters the chamber, it again builds up internal pressure; and the piston, pushed back to its original position, closes the demand valve.

247

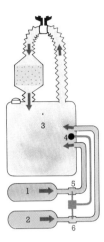

It is apparent that the free diver gains mobility in the water at the expense of duration, whereas surface-supplied and chamber divers can stay down longer, but with much less mobility. This is not just because free divers can swim freely but have a limited amount of gas, and the others have much more gas but are tied to it umbilically; it is also a question of safety. In an emergency, the free diver can swim to the surface relatively fast because the air he carries is not enough to permit him to go down far or to stay long under pressure, and so

even if he gets the bends from too swift an ascent, the effects are not likely to be dangerously severe. So he can swim a kilometer or so away from home base without worrying too much about the need for immediate decompression services in case of a sudden need to ascend. But the chamber diver – especially if he is in saturation – dare not roam far, for he cannot surface rapidly without risking an appalling case of the bends. Thus, if the gas supply from the chamber fails, he must be able to get back aboard at once. This means that he must never swim more than twenty or thirty meters away from the SDC. The brute reality of diving, unlike the popular image of the "conquest of the deep," permits no maneuvers that defy the laws of physics and physiology.

Above: A diver in an underwater habitat helps his mate don a streamlined plastic back pack containing all necessary equipment for a closed-circuit breathing system with electronic gas-mixture controls (a recently perfected innovation). The diagram gives a much-simplified idea of how such electronic systems, which actually mix correct proportions of gases within the breathing set, function. Two gas cylinders are worn, one containing pure oxygen (1), the other an inert gas – either nitrogen or helium (2). Positioned inside the rubber breathing bag (3), where the gases are combined, is an electronic sensor (4), which measures the partial pressure of oxygen. The electronic output from the sensor is arranged so that if partial pressure is too low, the valve on the oxygen cylinder (5) opens; if partial pressure is too high, the valve on the other cylinder (6) permits a flow of inert gas. In this way partial pressure of oxygen is automatically maintained for the diver at optimum levels for all depths.

Far right: A navy frogman wearing a closed-circuit breathing set prepares to dive. The component parts of any such closed-circuit system are illustrated (right). The breathing gas flows from a cylinder (1) through a regulator valve (2) into a rubber bag (3); the diver exhales through a canister of soda lime (4), which absorbs carbon dioxide. For shallow dives he can inhale and exhale through a single tube (5) – called "pendulum" breathing – but at high pressures the residue of carbon dioxide from exhalations into the tube can make inhaling dangerous, and so a "run-around" system (6) has been devised; its separate tubes for breathing in and out are controlled by one-way valves. Closed-circuit systems use pure oxygen, none of which is expelled. Semi-closed-circuit systems, which work on the same basic principles, use a mixture of oxygen and inert gas, and some of the oxygen-depleted mixture escapes through a blow-off valve (7). Below right: How the regulator valve ensures a flow of gas at constant pressure. Gas supplied through the nozzle (1) is controlled by a spring-set diaphragm (2) that shuts off gas whenever pressure rises above the required value. Since pressure remains constant, gas escapes through the hole (3) at a constant rate. The speed of flow is so fast that it remains unaffected by the change of pressure in the breathing bag.

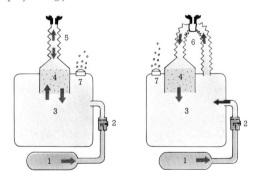

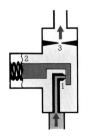

Those who provide diving support for the oil industry sometimes say it's a bit like running a fire service. The divers aren't usually needed until something goes wrong, and then they're expected to be super-fit, in the peak of training, with all equipment in perfect order, ready to descend to a depth of several score meters. The costs of both exploratory drilling and setting up the machinery for eventual production are so astronomical that the oil companies do everything they can to avoid the loss of a single day's operations due to an underwater accident or fault. So they do not balk at the idea of paying up to 2% of total costs in order to have a diving team on permanent stand-by.

The divers are apt to have less free time after drilling of production wells begins. Once a production platform has been fixed over an offshore oil field, several individual wells are drilled from it, and there must be regular underwater checks of the equipment fitted to the top of each well. Divers are also on call for joining together, inspecting, and servicing the pipes through which oil and gas may be carried ashore. Such routine tasks demand a combination of technical precision, physical stamina, and teamwork that can only be found in highly trained diving crews working in conjunction with a smoothly functioning SDC-DDC system.

Many modern drilling rigs have such a system built into the rig structure; those that do not must hire one from a diving contractor. Not even the most gigantic of oil companies can get along without the commercial diving contractor, for the business of training and equipping diving teams is so specialized that it can be managed only by organizations – such as France's Comex, America's Taylor Diving, or Italy's Micoperi – that exist for this purpose alone. The big diving companies maintain bases with personnel and stocks of equipment in all major ports that serve underwater oil fields, from New Orleans to Aberdeen to Jakarta. And, given a few weeks' notice, they are prepared to load a whole SDC-DDC system

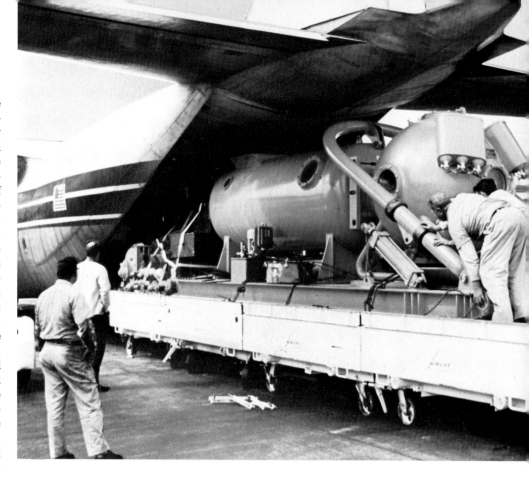

onto an oil-field work boat for shipping to the rig. Depending on the length of the contract and the type of work, the saturation system can be either lifted onto the rig by crane or operated from a pipe-laying or derrick barge. The biggest contractors can provide diving assistance almost immediately in the event of an emergency.

A rig that routinely uses shifts of paired divers in saturation at 150 to 200 meters requires a total team of at least eight or nine men, and there must be a replacement team every two weeks or so, to give the divers a restorative period on land. Each full team is headed by a nondiving supervisor, who assigns work and sees to it that no frictions develop. Diving is a very personal sort of job, since a man's life is constantly in the hands of his mates, and so the supervisor must make sure that his team lives together harmoniously. Of the six working divers,

one or two may have only underwater duties, but the others usually help out with surface support, acting as part-time electricians, mechanics, gas-mixing engineers, etc. To complete the team, there are two saturation-system attendants, who monitor the diving procedure in eight- or twelve-hour shifts.

Boredom is an ever-present menace. Although there are occasions when continuous diving keeps everyone busy for days, at other times – especially in the lengthy exploratory phase – a month may pass without a need for more than one or two short dives. Morale can sag unless the men are given something to do, and so a conscientious supervisor sees to it that they keep fit, that they do not allow chambers and fittings to corrode in the salty atmosphere, and that they maintain pressure seals, valves, and gauges in working order. If no

Attentive students at Italy's Marco Polo school for commercial diving training, near Rome. Marco Polo is an independent school; increasingly, though, the large contracting companies are running their own training schools, which offer aspiring divers the physical training necessary for saturation diving, as well as practical courses in such techniques as cutting, welding, and blasting.

Offshore-oil companies can hire divers and SDC–DDC systems from contracting companies, which generally keep reserves of equipment in major oil ports. Left: An SDC-DDC system, designed so as to be air-transportable, is loaded on to a plane. The skid on which the system is mounted will later be welded to the deck of an offshore oil rig, and the SDC will be swung over the side of the rig by its own arched gantry and hydraulic rams. Below: The crew of a surface-support vessel maneuver an SDC into position for locking on to its DDC, with the "skirt" of gas cylinders fitting snugly on either side of the DDC's extended hatch.

obvious routine exists, he invents one, so that the men devote at least half of every day to useful work.

Another problem concerns communications among men at the worksite, diving chamber, and rig. Few things matter more than this, and there are no difficulties in telephonic intercommunication as long as the divers are breathing air; but trouble begins at 80 meters and below, where a helium-oxygen mixture is used. The words of men breathing helium are distorted (as a result of certain peculiarities of voice formation in helium) to a point where the loss of clarity produces what has been called a "Donald Duck" effect. Until recently, helium breathers could be talked to, but they had to limit their responses to single words like "affirmative" or "negative." Even then they were often impossible to understand. Several companies have developed electronic devices for unscrambling the distorted sounds. Until recently, none of these proved successful, but some type of voice decoder is certain to come into general use soon.

As part of the oil people's contract with the diving company, the rig provides divers with meals and beds, and they soon get to know the oil men, roustabouts, and tool pushers. In a well-managed team, there tends to be a willingness to lend a hand around the rig at painting, chipping rust, welding, and other nondiving jobs. In return, rig crews often help underwater men when the going gets tough. For instance, if the divers are beset by mechanical problems during a stretch of full-scale activity,

Once oil production is underway, full-time diving support is needed – even if only on a stand-by basis – for swift response to emergencies. Divers not required for special tasks fill their days with routine inspections and maintenance work. Right: Undersea men being lowered in a cradle over the side of a floating drilling platform in the North Sea. A diver who has been trained in calm seas may have considerable difficulty when faced with engineering problems in rougher waters. For this reason, such companies as France's Comex, which has its headquarters at Marseille in the Mediterranean, often maintain additional training facilities in northern ports.

the rig's workshops and engineers may help out with repairs, thus averting the need to wait for spare parts to be flown in from the company's land base.

The value of diving services to the offshore-oil industry depends ultimately on the individual diver. Recruitment and training of underwater men used to be a rather haphazard business, but it is no longer so.

The diving contractors who provide their clients with saturation diving systems and teams are more likely than not to run first-rate training schools in association with their experimental physiological centers and laboratory pressure chambers. Divers who have learned their craft at such schools must measure up to extremely high standards of fitness and expertise.

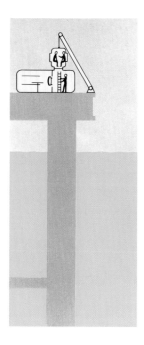

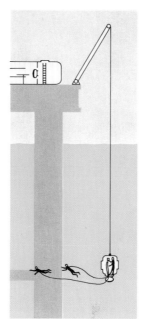

Left: The operation in a nutshell. With divers and bell men on board, the SDC is unlocked from the DDC and swung out over the side. Just below the surface, it is tested for seaworthiness before the bell man signals surface control to lower it to the worksite. At the site, after a careful check through the portholes reveals no unforeseen problems, the bell man pressurizes the chamber to the required pressure, and the divers leave the bell. The bell man's major task now is to feed them breathing gas through their umbilical hoses. The job completed, the SDC returns to the surface and is again locked on to the DDC, where the men can either decompress or remain in saturation.

The SDC-DDC System

Though all SDC-DDC systems are not exactly alike, their basic design and functions are pretty much the same no matter how they differ in details. A system like the one pictured here can be found on many a modern offshore oil rig. It has three main components: a diving bell, technically known as the submersible decompression chamber (SDC); a deck decompression chamber (DDC) – or, as in this case, two such chambers; and a surface control and support area. Here are brief explanatory notes about the various elements of the system, keyed to the numbers.

SDC

1 Rubber fenders help protect the bell during ascent and descent.
2 The main exit hatch, which the divers use when entering and leaving the bell, is at the base. It locks on to the DDC at point 14.
3 The side hatch (not necessarily a feature of all SDCs) can be used for locking the bell into position on a side-mating DDC.
4 Gas-storage cylinders for use during dives are stored on outer walls. (For very deep and long dives, gas would be fed from the surface.)
5 Within the lift cable that supports the SDC there is an internal electric-power conductor.
6 The bell carries about a ton of ballast to weigh it down during a dive. The suspended weights can be jettisoned from inside the SDC for an emergency float back to the surface.
7 The bell is suspended from the "A" frame during dives.
8 Hydraulic rams swing the "A" frame and SDC out over the sea for the dive, and back to the DDC when the dive is completed.
9 The winch for controlling the lowering and raising of the bell is powered by the motor alongside it.
10 A man operates the winch from this control.
11 This emergency buoy can be swiftly secured to the bell by a cable and umbilical if an emergency requires evacuation of the rig – provided of course, there is time enough to move pressurized men from the DDC into the SDC and/or to bring the bell back from a dive. Once in the sea, the bell hangs about 15 meters below the buoy which has flashing lights, radio transmitter, and batteries sufficient to keep the bell going for several days.

DDC

12 The two DDCs in this system permit a degree of operational flexibility: one team can remain in saturation in one chamber while a second team is being decompressed in the adjacent chamber.
13 Porthole windows provide visual communication.
14 The bell is locked on to the DDC at this transfer hatch. Once it is mated with the DDC, pressures are equalized and the divers can move freely throughout all SDC-DDC areas.
15 Divers remove and store their diving gear and wet equipment in this transfer area.
16 Through this hatch the divers can make their way into the second decompression chamber.
17 This work and storage area corresponds to the transfer area (15) of the other DDC.
18 The main living area for divers who are either decompressing or remaining in saturation is furnished with bunks only. Stereo music earphones are provided, and divers can chat with members of the support team over two-way communicators.
19 Toilets and showers are provided for both living chambers.

Surface Controls

20 Cylinders of gas – pure oxygen and helium – are stored here at high pressure.
21 In this mixing bank, oxygen and helium – pumped through hoses from the pure-gas storage cylinders via the gas-control panel at point 23 – are combined into breathing mixtures.
22 The gas-mixing unit. From here an operator controls the combining of pure gases in the mixing bank; he also checks pressure of stored mixtures, and feeds these through a hose to the SDC's storage cylinders.
23 This gas-control panel has gauges that record pressures in the main cylinder banks.
24 Through this high-pressure membrane compressor, gases are pumped to any cylinders whose pressure needs to be restored or whose gas needs replenishing.

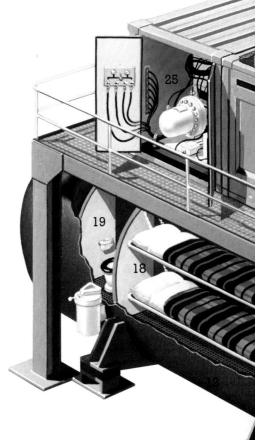

25 This gas-regeneration unit cleans DDC gas, and controls its humidity, by circulating it through chemical filters.
26 This cabin controls all parts and functions of the SDC-DDC complex.
27 Voice-communication facilities keep the control room in touch with all key personnel.
28 Gauge dials on this control console display gas-pressure levels throughout the system.
29 This voice unscrambler is used for communication with helium-breathing divers.
30 These are the main electric-power switch-boxes for the entire system.

It has been found that astronauts in training for space flights can learn a lot by emulating divers and going underwater – a convenient way to simulate weightlessness within the confines of the earth's gravity. Here a Skylab crew member prepares for a mission in a simulation tank at the Marshall Space Flight Center, Huntsville, Alabama. In addition to giving him a chance to test tools and techniques under conditions approximating those he will be subjected to in space, such practice sessions allow the astronaut to repeat each anticipated space-walk procedure over and over again, so that he will be reasonably familiar with the problems of coping with the sensation of weightlessness before experiencing it in outer space.

Salvage and Research

There is a saying among divers that anything you can do on land you can do underwater. This may be true, but underwater jobs take longer. Let's consider what has been involved in a few of the more arduous tasks (apart from "routine" engineering feats) that divers have accomplished in this century. Their most dramatic work, perhaps, has been in the field of salvage, which enables society to recover sunken treasures – and even, occasionally, the ships that carried them. Classic twentieth-century salvage stories are those of the *Laurentic*, *Egypt*, and *Squalus*. Though spread over the period 1917 to 1940, these achievements are still talked about among divers.

The *Laurentic* was a British liner converted to an armed cruiser. At the height of World War I, in 1917, loaded with about $25,000,000 worth of gold bullion, the ship set sail from Liverpool bound for Halifax, Nova Scotia, but struck a German mine off the coast of Ireland and sank in a depth of 40 meters, with a loss of 200 lives. An attempt to recover the gold, which was earmarked for military expenditures, was launched immediately. As soon as a salvage vessel had been anchored at the site of the wreck, divers in standard suits and helmets went down to survey the ship, which was lying on its side, and to blast open its hatches and doorways. Within weeks a diver had reached the ship's strongroom and recovered a box of gold weighing 60 kilograms and worth about $40,000. It seemed that the operation would be over in a month – but the sea is an unpredictable adversary. A succession of storms drove the salvage ship off its moorings and broke up the wreck.

When, after several months, divers were able to return, they found that the gold was buried underneath five decks of crumpled steel, the strongroom had broken apart, and the gold bars were so scattered that they had to be searched for one by one. Thereafter, month after month, year after year, divers worked at cutting, blasting, and driving out drifting sand with powerful hoses as they patiently brought up bits of the treasure. Each dive was limited to 35 minutes on the bottom, with 30 minutes of decompression in the water. By 1924, when work stopped, the British divers had raised 3,186 of the original 3,211 gold bars. Though it was a seven-year job, the salvage cost only 2% of the value of the gold.

The undersea heroes of the *Egypt* story were Italian, not British, but once again the treasure was precious metal – five tons of gold and two of silver, worth more than $5,000,000, that went down when the liner collided with another ship in the Bay of Biscay in 1922. The water where the *Egypt* lay was 130 meters deep, far too deep for working divers at that time. What they could do, however, was to observe and direct the work, which was carried out by the Italian salvage vessel *Artiglio*. In a closed diving bell, where pressure was maintained at one atmosphere, divers kept an eye on the salvage operations through portholes. A heavy crane on the *Artiglio* lowered a claw grab, which could place explosive charges and lift metal plates or cargo in accordance with instructions that the divers telephoned to the crane operator. In 1932, after ten years, the grab at last began to bring up gold and silver, and three-quarters of the treasure was recovered.

Seven years later came the greatest of all recoveries – though not, this time, of gold and silver – when the U.S. submarine *Squalus* sank in 73 meters of water as a result of the failure of an air intake to close during a dive. Within hours, the U.S. Navy brought into action an oxy-helium diving system (its first actual use), with a remarkable rescue chamber that had been designed by Commander Allen McCann. The divers fastened a cable onto a hook near the forward hatch of the submarine, and the chamber was winched down this cable until it locked onto the hull. The hatch could then be opened, and several men at a time transferred into the chamber at one atmosphere pressure (the surface pressure maintained in submarines). The crew of 33 men was rescued in an operation that took only a few hours to complete. Subsequently, divers attached lifting pontoons to the hull and brought up the entire craft.

Salvaging a ship's hull involves enormous forces and skill. In outline, the technique is simple, but in practice it requires extraordinary judgment. Divers must seal all major holes with steel, wood, or cement; then they must pump in air so as to float the hull. To increase the lift and compensate for air lost through leaks, they must sink hollow steel pontoons, lash them to the wreck with steel hawsers, and pump them full of air. To get the hawsers under the hull, the divers have to tunnel through the sand and mud by means of powerful water jets. Modern salvage companies do not usually need an SDC-DDC saturation system, since it would hardly be worth the effort to try to raise a ship in very deep water. But they use every other underwater device, including television. TV communication with the divers allows salvage experts on the surface to control every aspect of the delicate operation.

No salvaging job, of course, is as hazardous as the work of military frogmen – a kind of attack force, incidentally, that was pioneered by the Italian navy during World War II. But any sort of diving, even for pleasure, involves taking risks, and the divers who do peaceful scientific research must often work not merely as hard but under as difficult circumstances as those who salvage treasure,

sow mines, or blow up warships. Though chiefly concerned with observing fish, collecting specimens, photographing and sampling sediments, or with monitoring the performance of instruments, scientist-divers also pursue tasks that permit them to share in the heady dangers of undersea life. When, for instance, academicians such as Glen Egstrom and his team at the University of California in Los Angeles try to find out how narcosis, cold, anxiety, weightlessness, poor vision, and stress affect diving performance, they often do it themselves. Or the scientist-diver might be asked to assist in some such bizarre project as helping to put men into space! This actually happened: in order to accustom them to weightlessness,

American astronauts wearing space suits were trained to maneuver around a submerged skylab rocket. Because a spaceman cannot "swim" in space, the trainees were forbidden to use swimming motions and could move about in the water only by pulling or pushing on the rocket. If they drifted away from it, they had to be maneuvered back by teams of fin-swimming divers acting as thruster rockets.

In spite of the obvious differences between space and the underwater world, the fact of weightlessness is strangely unifying. Any diver who saw films of men drifting in space would, if the films were tinted blue, believe he was watching undersea colleagues.

An important salvage operation in progress off the coast of Australia since 1973 is aimed at bringing up the *Batavia*, a Dutch East Indiaman that sank in 1629 with a cargo of silver. Most of the treasure was recovered in the seventeenth century, but archaeologists of the Western Australian Museum, in Perth, consider the ship's structure and its household artifacts of more lasting value than bullion. During more than three centuries under water, the wreck has been coated with a concretion of coral and mineral matter, which divers employed by the museum must blast away before lifting operations can begin. The two divers (above right) are laying charges for an explosion strong enough to crack some of the stony casing without damaging the ship. Once a segment of the centuries-old concretion has been removed, the ship's heavy side strakes are being cut into short sections (above) for transfer to the museum, where the *Batavia* will be reconstructed.

As in salvage operations, divers are essential personnel for most marine-research projects. The *Gvidon* (below) is a Russian underwater apparatus for studying fish shoals and fish tackle. It can submerge to a depth of 250 meters, is extremely maneuverable, and has an air-regeneration system that can support 2 to 3 divers for 3 days. Below left: Working on biological research at Perry Hydrolab in the Bahamas, an American diver charts current and temperature changes.

Saturation diving, which requires the complex machinery of the SDC-DDC system, has been largely developed for the oil industry. Except in that industry divers still go down only in relatively shallow water and are fed breathing gases from the surface, or are dropped to the worksite in a conventional diving bell. Thus, for ordinary construction and maintenance jobs, as well as for salvage and military operations, where there is seldom a need for going below 50 or 60 meters, the logistics of a dive are fairly uncomplicated. Many divers are self-employed and contract their skills on a short-term basis to commercial diving companies that recruit men for civil-engineering projects. Military divers and marine scientists are, naturally, a breed apart. They work in so many different ways, on such varied (and sometimes secret) missions, that we cannot generalize about what they do. But certain aspects of commercial diving are alike the world over, whether the divers help to build bridges and tunnels, maintain harbors and communications, or whatever.

For instance, take the battle against deterioration. Metal structures corrode and rust in sea water; waves cause the welded tubular legs and joists of platforms to vibrate, so that the joints may crack; and other stresses may develop. Divers must, therefore, constantly inspect structures and make repairs where necessary. Today such jobs are not quite as hard as they once were, for modern divers can test the thickness of metal and detect cracks by means of acoustic ultrasonic devices, underwater TV, and other achievements of this century's technological magic.

An abridged list of the tasks performed by commercial divers might read this way: they pour concrete; inspect permanent anchors and chains for emplacement and freedom from corrosion; take samples of sediments and rocks to check the strength of foundations; drill holes and put down explosive charges in order to make navigation channels; survey the sea floor for pipe routes. For these and many other jobs, the diver of today must often rely on the slow procedures of yesteryear in spite of the latest technical aids, for the work still demands a degree of mental and manual dexterity for which no technology provides a reliable alternative. Undersea surveying, for instance, is especially difficult. In low visibility a compass and tape measure may be the only dependable instruments, even though there are precision tools with telescopic sights and laser rangefinders that emit a water-penetrating frequency of green light.

Many oil-rig jobs are very similar to those just listed – particularly during the exploratory phase and in early stages of production. Special skills become necessary for the drilling itself and for connecting sea-floor pipes, called lines, to the underwater "Christmas tree" (a huge branched stack of valves, connections, and rams at the top of a well). A few years ago, for example, six saturation divers worked in shifts throughout a five-day-long period fitting five lines to such a Christmas tree at 100 meters in the Gulf of Mexico. To do the job, they had to locate the ends of a bunch of pipes that had been dropped during a storm, to clear up a tangle of cables, to attach new cables for positioning the pipes, to separate the pipes and drag each to a specified location, to repair the damaged tree, and finally to connect each pipe with a flange that required the tightening of eight bolts!

Power tools designed for undersea use are obviously invaluable. The tools must be corrosion-resistant, and their controls modified so as to make them easier to manipulate with cold fingers. Then, too, if they're powered by compressed air, there must be some means of releasing the ex-

Most underwater jobs can be done with familiar land tools adapted for use in the sea. Thus, for example, painting (left) is made possible by having the paint pumped through a hose and into the brush; and blow-torch cutting (right) is achieved by protecting the flame of burning gas in a jet of compressed air. But some tasks – welding, for instance – are particularly difficult, because they are impeded by the rapid cooling effect of water. For welding, one solution is to build a dry shelter at the worksite (like the unit pictured below). Inert gas pumped into the unit drives out the water, enabling the diver – who still breathes from his surface-support hose – to weld in dry conditions.

Left: This diver, at work on a pipeline, has been dropped to the shallow worksite in an open bell. Although impeded in his task by relative weightlessness, he can vent air from the surface-support hose or the supplementary cylinder on his back into a "buoyancy bag" (here anchored to the pipe) for lifting heavy pieces of equipment.

haust gas at a distance from the diver as a protection from the noise and turbulence it creates in water. But compressed-air power is not practical below 60 meters because of increasing pressure. So hydraulic oil at very high pressure generally replaces compressed air for power tools farther down. The oil is fed from a surface pump via a strong hose, or it can be transmitted by a short hose to an electric pump mounted either on the diving chamber or in a separate unit at the worksite.

One further problem is that the tools exert a powerful force on the relatively weightless diver. Since he cannot steady himself with his feet, the thrust of a power hammer or saw forces him backward, and the rotation of a power drill tends to whirl him

Gas-powered welding equipment is generally replaced at great depths by electric-arc tools that melt metals by means of very high temperatures produced by a powerful electric current. Above left: A deep-sea diver using this method to repair the underwater structures of an offshore oil rig. Above right: Not electricity but compressed air provides power for the rock drill with which these divers are collecting samples for geological analysis; exhaust gas escapes through the white hose in the foreground.

Hydraulic oil is pumped through a hose at high pressure to drive power tools with rotating or vibrating parts, especially at depths where high pressures make compressed air ineffective. The pump may be either on the surface or on the sea floor near the diver. Above left: A powerful hydraulic scrubber with metal spikes speeds up the tedious job of cleaning a ship's hull. Right: With a small hydraulic rotating saw, this diver is cutting the cable off a fouled propeller.

around. Such handholds as rails or rungs of a ladder at the worksite are not the best of supports because the diver has the use of only one hand while he holds on with the other. To overcome this handicap, undersea power tools are designed to have the smallest possible reaction force; and working divers can, if necessary, be anchored down with clamps and quick-release hooks.

At depths less than 60 meters, the processes of cutting and welding are carried out by oxyhydrogen or oxypropane blowtorches, very much as on land; although water cools metal, the blowtorch maintains a high enough temperature to overcome the cooling effect. Below 60 meters, though, water pressure prevents steady burning of the gases, and so electric-arc welding with

oxygen becomes necessary. In addition, because a weld that cools too rapidly tends to be porous and brittle, underwater welding equipment often includes a device that automatically feeds the welding rod into the weld with an electric current that assures an even flowing together of the molten metals.

Divers must be prepared not only to handle sophisticated tools, but to do so in any sort of water. The hardest jobs are those that combine depth, technical complexity, and bad weather. Consider what it is like, for instance, to dive for an oil-drilling project off the coast of Norway, where operations can be interrupted by sudden storms or approaching icebergs, forcing the drilling ship or rig to stop work and seal off the well until conditions improve. When work

begins again after a long lapse, all the equipment has to be tested and restored to working order. That is one of the many times when the commercial diver, however well paid, more than earns his money.

Many men in diver-training programs have to drop out because they lack the psychological toughness to endure the discomforts and dangers of prolonged underwater work. We should not confuse the abilities and attitudes of the typical scuba-diving sportsman, who dives when he wants to, in conditions of his own choosing, with those of the rugged undersea technician. Every professional diver must be ready and willing to brave the rigors – and even terrors – of daily work in cold water with zero visibility.

Experimental Diving Labs

It is an odd experience to look through the portholes of a laboratory compression chamber at a warmly clad man dozing on his bunk or pedaling a bicycling machine at a pressure equivalent to 500 meters of water. In terms of the time it would take for him to get safely out of the chamber, or for a doctor to get in, he is farther away than a man on the moon. If he seems remote, as if seen through a telescope, it is because he is temporarily in a world of his own. As an experimental diver, he is serving as a guinea pig – but he remains very much a man, who may well resent being stared at as though he *were* a guinea pig.

To learn how to dive deeper and stay longer, we must constantly experiment with possible solutions to medical and engineering problems, and there are always risks in experimentation. That is why compression and decompression schedules on new gas mixtures are tested first on animals. But no animal reacts physiologically exactly like a man, although the behavior of a monkey or goat does give an indication of conditions that might be dangerous. Eventually, therefore, the animal must be superseded by a human being. The diver who volunteers for the job must be extremely fit and emotionally stable, for he has to endure long hours of stress in cramped conditions. And no matter how willing he is, he will not usually be permitted to make more than a few experimental dives; there are too many risks involved in trying out new depths and durations. Sometimes, in fact, when scientists want to study long-term effects, laboratory divers have to forgo all deep diving after an experimental run. That is the only way in which medical observers can detect the possibly long-delayed aftereffects of any exposure to unprecedented strains.

Experimental diving labs are functioning today in many countries – particularly France, Britain, the United States, Russia, and Japan – and are often, as might be expected, part of a nation's military establishment. But several universities – among them Duke and the University of Pennsylvania in America, Newcastle in England, and Zurich in Switzerland – have compression chambers and are doing fine work in diving medicine. And, of course, the commercial diving contractors (the largest single source of employment for deep divers) are especially active in the field. If we glance at some of the activities of France's Comex, we can see not only what goes on in one of the world's largest diving labs, but also what probably happens in such other commercial giants as Oceaneering International and Taylor Diving and Salvage Company, both of the United States.

The various groups of pressure chambers in the laboratory complex that Comex operates in Marseille are arranged so that men can be compressed all the way down to 72 atmospheres (710 meters) with separate chambers for sleeping, working, and storage of equipment. Naturally, a trial "dive" with the men lying on bunks doing nothing would not be a sound test of how given gas mixtures and rates of compression and decompression would affect divers in a cold sea. So when the subjects arrive at maximum pressure, they undergo physiological tests to check their condition. Then they carry out planned exercises to test their reaction to working at different rates as well as their alertness, hand-and-eye coordination, and all-round stability. To test respiration and muscular efficiency, the men work out on rowing and bicycling machines when the dive lasts for several days, they are watched for any changes in diet, temperature, and the daily routine of sleeping and working.

All this testing in what is called a dry chamber, however, tells us little about what will happen to men similarly exposed in the water. To get answers, a laboratory compression chamber is half-filled with water; and when this "wet" chamber is linked at the same pressure to the dry one where the

This exploded view of an idealized experimental laboratory shows all the necessary components for simulating deep dives. Each of the four compression chambers of this particular system – two antechambers, or locks, at either end, one leading into a horizontal living chamber, the other into a vertical diving chamber – can be sealed off by pressure-tight doors. This makes it possible for members of the outside control team to enter either of the locks at surface-level pressure during the dive, for such purposes as repairing, supplying, or removing equipment. During any period when the pressure in the chamber is decreasing for a simulated ascent, the oxy-helium mixture that is pumped out of the chamber is first stored in the recovery bag, then, as an economy measure, separated and returned to the storage bank for further use.

1 Living-chamber lock
2 Living chamber
3 Diving chamber
4 Diving-chamber lock
5 Control console
6 Mixed-gas bank
7 Recovery bank
8 Pump to supply gas to chambers
9 Water-cooling tower
10 Heating/refrigeration unit
11 High-pressure air bank
12 Food/medical locks
13 Television camera
14 Viewing portholes
15 Data and medical control room

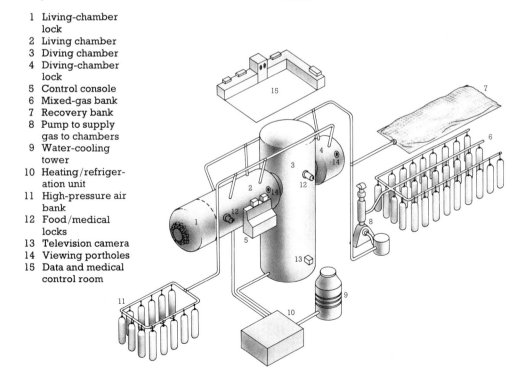

divers rest and put on their diving gear, the experimental subjects can get in and out of the water and test different pieces of equipment as easily as if they were using a surface swimming pool.

The only drawback to most wet chambers is that they are small – 2 to 3 meters in diameter – and so the diver can work only on rather small pieces of equipment. Fortunately, much of the experimental work at the Comex laboratory is partly financed by a government research department known as Cnexo (*Centre National d'Exploitation des Océans*); and Cnexo has built an outsized pressure sphere with a diameter of 5 meters. The bottom half of this sphere can be flooded and the whole thing pressurized to 30 atmospheres. Because its top entry hatch is 2 meters wide, some fairly big pieces of machinery can be lowered inside for the divers to practice on. Thus, welding, cutting, and other techniques for repair of vital oil-industry equipment can be tested under realistic conditions, and divers can be given valuable experience in the sphere while tools and working methods are also being tested. During a saturation dive, up to four men can live on the balcony that projects from the walls of the dry upper half of the sphere.

The diving companies' facilities are rivaled, and sometimes surpassed, by those that the richer nations provide for their armed forces. For example, the Royal Naval Physiological Laboratory near Portsmouth, in England, has a number of cylindrical chambers; one of them, which can be pressurized to 700 meters, was the scene of the notable 457-meter dive by John Bevan and Peter Sharphouse in 1970. Not far away stands the Navy Deep Trials Unit, which has two large chambers, each about 5 meters long and 2.5 meters in diameter. One cylinder is installed vertically, with the bottom half full of water, and the other chamber is horizontally mated to the dry top half of its wet twin – an arrangement that allows teams of up to 10 divers to be pressurized in the dry chamber for experimental work in the water. But the world's largest diving lab is the U.S. Navy's Ocean Simulation Facility, which began operations in the early 1970s at Panama City, Florida. Its biggest chamber is 15 meters long and 4.5 meters wide, and it connects with five smaller ones, all of which can be pressurized to 660 meters. This flexible system can be used both for testing divers and for trying out such sophisticated oceanographic equipment as small submarines or data recorders.

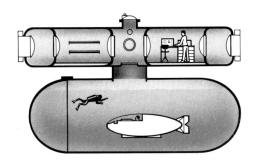

The hard work of those who pioneer diving technology is largely carried out not in the sea, but in the laboratories of the world's navies, universities, and commercial companies. These French (below left) and Japanese (below) "dry" labs are typical of complex pressure-chamber systems where medical and engineering problems of high-pressure diving are tested and retested, with round-the-clock support teams to watch, monitor, and record. A dry chamber, however, is a limited research weapon for those who aim at testing the effectiveness of a man, a technique, or a machine in the water; much laboratory work, therefore, is done in "wet" labs, where part of a compression chamber is flooded. Above: Within the huge pressure complex of the U.S. Navy's Ocean Simulation Facility in Florida, several dry chambers connect with a wet chamber vast enough to test man and machine together in a simulated ocean environment that can be pressurized down to 660 meters. Bottom right: A British navy diver at the Deep Trials Unit near Portsmouth tries out a semi-closed-circuit breathing apparatus under pressure in a wet laboratory.

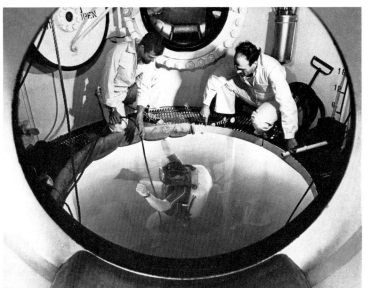

A House Under the Sea?

Considering the difficulties of compression and decompression and the complexities of SDC-DDC systems, wouldn't it be sensible to build underwater houses for divers, where they could live in saturation for long periods? Couldn't any such house have an internal pressure the same as the water outside, so that divers could go in and out freely through a hatch in the floor? Wouldn't all problems be greatly simplified if, after each working dive, they could get back to their undersea hotel in minutes, take off their diving gear, and be at ease in the warm and dry? The answer to those questions is yes, and the practicability of sea-floor houses – or habitats, to use the technical name – has been quite thoroughly tested by now. What we have learned is that the idea is more attractive than the thing itself.

That habitats were basically feasible was proved by Jacques-Yves Cousteau in 1962, when he anchored a barrel-shaped house called "Diogenes" at 10.5 meters depth in the Bay of Marseille, and two divers, Albert Falco and Claude Wesly, lived there for a week. Diogenes was a simple steel cylinder, buoyant because of the gas in it, but secured to the sea bed by four anchors. Inside were two bunks and the equipment necessary to maintain the breathing-gas supply. The two aquanauts, whose food was brought down to them by other divers, showed some signs of anxiety during the first day, but then adapted rapidly; by the fifth day they seemed totally oriented to the underwater world.

After this exciting start in shallow water, experiments with habitats proliferated in France, America, and eight other countries, including Russia, Germany, and Japan. They continued throughout the 1960s, culminating in what were known as Cousteau's "Conshelf" and the U.S. Navy's "Sealab" programs. In 1965 Cousteau's Conshelf III housed divers for 3 weeks at 128 meters, and Sealab III was on the verge of trying to beat the Cousteau record for depth and duration when, in 1969, the U.S. Navy called a halt to its program. One reason for this was that while engineers were installing Sealab III at a working depth of 180 meters off the California coast, one of the aquanauts sent

America's Tektite, in shallow water off the Virgin Islands, was one of the few habitats still in use during the 1970s – pressure and communication problems being not insurmountable at a depth of only 15 meters. Its "bridge" (top of page), where environmental control and communication panels are located, is the habitat's nerve center, spacious enough to double as a sort of dry laboratory for underwater scientists. The narrow passage in the background is a tunnel that connects Tektite's two upright cylinders. These steel "houses" stand on a rectangular weighted base (above), and each of them has an upper and lower floor: bridge upstairs (or rather, up the ladder) and crew quarters down in one; equipment room and observation cupola up and "wet" room down in the other. Divers swim to an entrance in the base.

down to inspect the house was killed by a combination of cold, exhaustion, and carbon dioxide poisoning. The accident dramatized the risks, but the authorities were already doubtful about habitats and felt that the time had come for a reappraisal of underwater living, its goals, and its costs. By 1970 the enthusiasm for habitat diving had waned practically everywhere.

It seems then that we can look back at the 1960s as the era of the habitat. Why? What went wrong? How is it that all the splendid dreams of placing men on the sea floor for as long as they want to stay have failed to come true? Well, first, there is the expense. Because a habitat, being full of gas, is inherently very light underwater, it must be ballasted down with hundreds of tons of iron. This means that moving and positioning the house are tasks requiring large support vessels equipped with heavy lifting cranes. Furthermore, to free the habitat's occupants from needless stresses, no effort must be spared to provide a comfortable and relaxing interior for them. Yet no amount of money seems to solve some problems in the deeper habitats. For instance, the helium atmosphere makes speech difficult, food loses its flavor, and humidity is inescapable, since the atmosphere of the habitat is in contact with the sea at the entrance hatch. The humidity tends to cause skin and ear infections. And the hull of the house needs to be heavily insulated to keep the helium atmosphere from dropping to water temperature and freezing the aquanauts.

A principal goal of all the experiments was to show that men could live and work more efficiently if housed at the worksite. It was hoped that they could be made independent of the surface, thus saving the cost of support ships and crews. But costs rose instead of falling, partly because as more and more functions of diving support, power supply, gas storage, gas mixing, and so on are built into a habitat, the habitat gets bigger, heavier, and harder to install on the sea floor or move to another site. Even more important, though, it eventually became obvious – as perhaps it should have been from the start – that surface support for divers cannot be dispensed with. To sail a ship away and abandon men on the sea floor requires a calculated risk and implies extreme reliability of the safety equipment in the habitat. But how much power reserve should the habitat have in order to be entirely safe? How much reserve breathing gas? What happens if bad weather sets in for a month and the surface-support ship cannot return at the expected end of the working period? After all, a habitat lacks what submarines have in abundance: the ability to bring men to the surface under safe sealevel pressure whenever necessary.

Moreover, the efficiency of habitat dwellers is restricted by their limited mobility

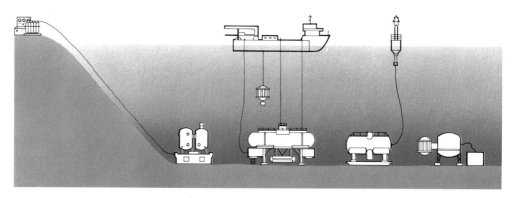

Above: This Japanese deep-sea habitat was still being used for biological studies during the 1970s (though for only a few months a year). It rests on ballast trays, and the flotation tanks that are fixed along the top give it a steadying buoyancy when at the surface. The several gas cylinders stored on its sides are for possible emergency use; and, for safety's sake, an SDC for carrying saturated divers to and from the surface always stands by. Gas and power supplies from the surface flow down through the umbilicals visible in the background. Left: *Chernomor 2* (meaning "Black Sea"), supported from the shore by cables, was used by Russian divers for undersea research in the late 1960s.

when away from home. Just as it is dangerous for a diver to move very far from the SDC when he is saturated, so it is for an aquanaut. Since the weight of the habitat precludes frequent movement from site to site, the total area that the aquanauts can cover is very limited. Expensive experiments with electrically propelled tugs and underwater "chariots" have merely confirmed the suspicion that, although the idea of divers being able to range for miles around a habitat is seductive, the risks of taking saturated men far away from base are too great. If mobility is really required, a solution has been found in a technical development that combines a submarine with a pressure chamber: what is known as a "lock-out submersible." The lock-out submersible, although expensive to build and run, has recently begun to prove its worth for carrying out special diving tasks. It has two compartments, which are securely locked away from each other and communicate only by telephone. One is a pressurized chamber for the transport and release of divers; in the other are the pilot and navigator at atmospheric pressure. So

The above diagram illustrates four ways to maintain a habitat on the sea bed: (1) It can be landbased, linked to the shore by an umbilical; (2) it can be ship-supported – that is, linked to a vessel anchored above it; (3) it can be attached to a buoy that carries a diesel engine, gas supplies, and a radio aerial; or (4) it might be self-contained, with an independent power pack and gas supply. The fourth method, however, has not yet been actually tried.

mobility for divers is gained in this fashion, without subjecting the rest of the crew to the risk of narcosis, bends, or helium tremors. At the end of a stint, the vehicle is lifted out of the water by a mother ship and locked onto a DDC, where the divers decompress in comfort while the other men leave their compartment by another exit.

The habitat era has proved that divers *can* reside underwater – but at a price that is still too high to justify the routine use of habitats, except in quite shallow waters. However, as the scale of offshore-oil activities increases, better days may follow for the undersea house, perhaps in combination with lock-out submersibles to provide more mobility and safety for the aquanauts.

A Look Ahead

What will divers be doing in the year 2000? Will they have overcome the present limits of depth, duration, and mobility? How might their functions change if and when oil-field wellheads are to be found as much as 1,000 meters or more below the waves?

We cannot predict future developments with certainty. We know that it is now possible for a man to descend to 600 meters for a few days or to live underwater for several weeks at a depth of 100 meters, and so we can anticipate that workers will be operating routinely under such conditions within the next few years. But the only other precise forecast that we can make is this: as long as there is economic or military justification for technical progress, some progress is sure to occur. Since oil companies are already studying the problems of producing oil from wells at 1,000-meter depths and laying pipes at that depth – some companies have even taken out concessions for exploring in 3,000 meters of water – it seems certain that there will be a pressing need for much deeper diving than now seems feasible. Naturally, in view of all the difficulties, the oil companies are trying to design their forthcoming machinery in such a way as to preclude the need for SDC-DDC support; a couple of systems now being tested, for example, are designed to lower men in chambers at sea-level pressure (one atmosphere) and to transfer them into underwater houses, also at sea-level pressure, constructed integrally with the oil wellhead. Nevertheless, something is bound sooner or later to go wrong with any underwater equipment, and divers will somehow have to be available in order to cope with such eventualities. So a continual further pushing down of depth barriers seems virtually inevitable.

But let us begin by looking at some major areas of potential progress in other aspects of the diving scene. Constantly evolving marginal changes in the engineering of SDC-DDC equipment, the umbilical, and divers' clothing are to be expected, of course, with consequent improvement in the undersea men's comfort and efficiency. There is still plenty of room, too, for refinements in the design of power tools and portable power sources. For instance, hydraulic power loses much of its efficiency underwater because of the friction that develops in long hydraulic pipes, but high-power electric tools are too dangerous for working divers to handle. A solution already being adopted is the transmission of electric power from the surface or from the SDC to a motor that drives a hydraulic pump, which then powers the tools through a short pipe. Developed further, this method will eventually provide the diver with more powerful drills, cutters, saws,

The one-man submersible known as "Jim" being hoisted over the side for trials on the sea floor. Two meters tall, weighing well over 400 kilograms, Jim has amazing underwater mobility. Its occupant can pick up objects from the bottom, climb a ladder, negotiate fairly steep gradients, and see clearly through large plexiglass (perspex) windows in the domed "hatch." The drawing below shows some interesting features of this metallic monster, within which sea-level pressure is maintained even for dives as deep as 400 meters. The diver exhales through his mask into twin soda-lime canisters (one only visible in the diagram) that act as absorbers of carbon dioxide. The breathing mixture is boosted by a small but steady flow of oxygen from the storage cylinder on Jim's back. Front and rear drop-weight release catches enable the diver to jettison ballast weights (nearly 70 kilos are required for work on the bottom) if he needs to ascend in an emergency.

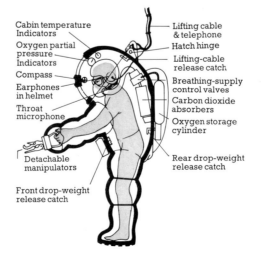

Cabin temperature Indicators
Oxygen partial pressure Indicators
Compass
Earphones in helmet
Throat microphone
Detachable manipulators
Front drop-weight release catch
Lifting cable & telephone
Hatch hinge
Lifting-cable release catch
Breathing-supply control valves
Carbon dioxide absorbers
Oxygen storage cylinder
Rear drop-weight release catch

and nut tighteners. Further improvements in micro-circuit electronics should also bring better control of the atmosphere in the SDC, in addition to better unscrambling of the "Donald Duck" effect produced by breathing helium.

The difficulties of trying to achieve mobility along with depth are already being eased by the development of the lock-out submersible described in the preceding pages. The only big stumbling block in the path of using such vehicles more lavishly is their cost: from half a million to a million dollars for a single lock-out submersible in the 10- to 20-ton weight range that is capable of releasing divers at 200 to 600 meters, not to mention the high operational running costs of both the submersible and its surface-support ship. Still, a number of divers were doing commercial work out of these compartmentalized vehicles by 1975. With a range of 10 kilometers or more, they can actually save money in the long run by transporting divers to a number of different undersea worksites within short periods of time. So we have every reason to believe that the lock-out submersibles will be increasingly in evidence in the years to come, especially since underwater pipelines are growing longer all the time and huge oil-drilling structures are being built farther and farther out on the continental shelf.

Most people have heard or read about spectacular experiments aimed at finding new ways to make men capable of competing with fish as undersea creatures. Professor William Paton of Oxford, for instance, has discovered that if land animals are anesthetized to unconsciousness at the surface, they recover consciousness when compressed to pressures equivalent to 1,000 meters of water; and this indication that narcotics and pressure work in opposition may seem to open new avenues of research into untried techniques for ultra-deep diving. Even more startlingly, the Dutch scientist Johannes Kylstra has flooded the lungs of small animals with oxygen-saturated fluid and found that they can be subjected to high water pressures and brought back without needing to be decompressed (because there is no inert gas in the "breathing mixture"). To understand the significance of his experiments, it must be fully understood that the whole purpose of breathing is to supply oxygen to, and remove carbon dioxide from, the body. The inert gas in any breathing mixture serves merely to dilute the oxygen, thus preventing the occurrence of high-pressure oxygen poisoning. But, as we have seen, it is the inert gas that causes the decompression problem by the way it dissolves in the tissues. What Kylstra has shown is that warm-blooded, air-breathing animals can be attached to a machine that supplies them with a warm liquid containing no inert gas but full of oxygen, and this fluid can be breathed in and out of the lungs like air. Even a human being has had one lung flooded in this manner with no ill effects.

Neither Paton's nor Kylstra's fascinating experiments can be seriously seen as a preview of the future, however. They involve far too much tampering with the human body and psyche. Indeed, although such extraordinary measures might succeed in putting a live diver down below the present 600-meter maximum, he would probably be too groggy to work after undergoing the prescribed treatment.

If most of the proposed science-fiction methods of pushing back the frontiers are too dangerous to contemplate, what other paths are open? Well, there is one device that has been around for a long time without getting much attention: a steel suit that actually protects its air-breathing occupant

from water pressures. The steel suit is, in fact, a sort of form-fitting one-man submarine, with the added advantage that it has flexible steel joints where human arms and legs are jointed, so that the diver can walk and bend at the hips, and the ends of the arms are fitted with claws that are manipulated from inside the suit. These articulated suits were first developed in 1853, but in the early models the joints tended to leak and to stiffen under pressure, so that the diver could not move. Careful engineering of the joints produced steady improvement until the 1930s, but thereafter the whole idea seems to have been abandoned. The general feeling was that they were impractical because, to withstand intense pressures, the steel had to weigh half a ton or more; as a result, divers could walk and keep their balance against the drag of a current only with great difficulty.

A few years ago, however, the idea was resurrected by an English engineer, who interested the Royal Navy in helping out with sea tests of an articulated suit built of light magnesium alloy instead of ponderous steel. Code-named "Jim" (for it does look like a monstrous metallic human being), the suit underwent extensive sea trials for several years. Then, in 1974, it proved itself manfully in a commercial operation to salvage some heavy anchors that had been lost off the Canary Islands in 300 meters of water. Because the diver breathed air at atmospheric pressure from self-contained cylinders carried on Jim's back, there was no need for compression or decompression; and a three-man team – one diver backed by only two surface engineers working from a locally chartered boat – managed to attach lifting lines to the anchors in a few short days. Because of its bulk, Jim requires a fair amount of engineering support, but far less than an SDC-DDC system or lock-out submersible. The future of such jointed metal suits probably lies in just such jobs as the Canary Islands one, where what is needed is a series of brief dives to great depths, because it is in these circumstances that most is gained by avoiding decompression.

In the near future – by 1985 or 1990, say – an undersea oil field may consist of hundreds of square kilometers of continental slope at a distance of 200 kilometers from land, in a depth of 1,000 meters. Over this area, hundreds of wellheads are likely to be producing oil without ever being connected to a platform or surface structure. Machinery to separate oil, gas, and water will work on the sea floor, and pipelines will convey the petroleum products to collection centers, from which they can either be collected by tanker or pumped ashore through large-diameter pipes. A huge array of submarines, underwater one-atmosphere work chambers, robots, and remote-controlled tools will doubtless have to service, maintain, and repair this vast assemblage of underwater machinery. If depth divers are still needed – and they almost certainly will be – it should not be beyond the skill of scientists, of engineers, and of the divers themselves to make sure that they get there as well.

One of the latest advances in undersea technology is the lock-out submersible, which facilitates the transport of divers to and from worksites. Capable of descending to 365 meters, this SDC-submarine combination is divided into three sections. The front end, pressurized at 1 atmosphere, is the spherical control module where the pilot sits; the center section, accommodating two divers, is pressurized to the working depth, because it is from here that the divers "lock out" into the water; and the rear section is the power module.

9
Submarine Craft

John P. Craven

In almost everyone's mind, the word "submarine" evokes fantasies of danger and daring; the imagination conjures up the sound of pinging sonar and the whine of a torpedo slithering through the water, followed by the muffled drumbeat of successive depth charges. It is hardly surprising that this is, to some degree, all that the general public "knows" about underwater craft. Most people, never having seen an actual submarine, have gained their romantic notions through novels, films, and news accounts of submarine exploits and tragedies. Nearly everybody can recall the name and nature of some undersea disaster – *Scorpion*, perhaps, or *Squalus*, or *Affray* – and it is the occasional disaster rather than the steady run of prosaic successes that sticks in our memories.

Why is it that whereas we quickly forget

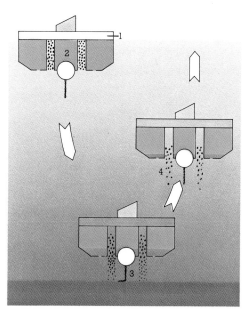

How a Bathyscaphe Dives
Four essentials for depth control: (1) An air tank, used as a float, is flooded for descent; (2) gasoline gives undersea buoyancy; (3) a guide chain keeps the sphere off the bottom; (4) iron-shot ballast is jettisoned for ascent.

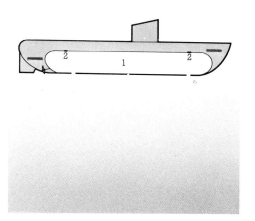

How a Submarine Dives
At the surface, buoyancy tanks (1) on both sides of the pressure hull are filled with air, and the valves (2) are shut.

aircraft accidents, undersea mishaps command world headlines for months after the event? Each year, a number of airplanes crash, with a loss of life far exceeding the loss in the worst of submarine disasters. Every *day*, hundreds are killed on the world's highways, with an annual toll equal to the loss of many, many submarines. Objective analysis, in fact, indicates that a submarine in peacetime is one of the safest means of transport. Yet the mystique of terror persists – and it is so powerful that it has not only infected the general public but has had a measure of influence on the individuals who man the craft.

By its nature, a military submarine must operate either alone or in the company of only one or two others, and it remains detached from normal surface contacts for long periods. This manner of living sets submariners apart and fosters the growth of a tightly knit community of men at sea, psychologically supported by anxious families ashore. Pride of accomplishment and discipline in action are the hallmarks of these men. They make demands that extend their submarines to the limits of operational capability; and because they demand so much, they develop habits of safety and require standards of equipment performance that are extraordinarily severe.

"So much the better!" you might say. Perhaps so – but one unfortunate result of the stress on safety is that the small group of men who design military submarines are also deeply affected by this atmosphere of tension. The combined pressures of public concern and the demands of the submariners have forced designers to be not only meticulous but unadventurous. No change in design is made without rigorous laboratory tests and analyses, and only the most stringent need for new capabilities is seen as justifying a departure from accepted practice. Thus, the notion that submarines are peculiarly vulnerable has delayed, and still delays, advances in design that technology could have made possible long ago.

If aircraft designers were as conservative

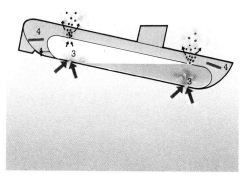

To submerge, valves are opened, permitting water to flood tanks through openings in the bottom (3). Winglike planes (4) control the angle of dive.

as those who build submarines, we'd be flying now at altitudes no higher than 3,000 meters, at speeds no greater than 300 kilometers an hour, with materials no more sophisticated than aluminum alloy – and, no doubt, with a magnificent record of flying safety. If the same philosophy had been applied to automotive design, we'd be driving cars at speeds no greater than about 70 kilometers an hour, employing fuels no more inflammable than kerosene, physically encircled by an envelope of bumpers. That's the brand of conservatism that has dominated the submarines' story. For this reason, I believe that the story of what the military submarine *is* should be told in the context of what it *could be* if daring replaced caution. The first thing we should understand, though, is that military submarines are only one kind of undersea craft.

There are, actually, three main types: the submarine, the bathyscaphe, and the submersible. What all three have in common is the need for a strong hull, a means of controlling buoyancy and depth, a dependable power source, and systems for controlling maneuverability, supporting life within the hull, and communicating with the outside world. The three types, however, are designed to do quite different jobs, at different depths, speeds, and durations, and so their construction and design are also different. It is possible, in fact, to say that the fundamental distinction, which effectively defines each type of craft, is in the design of the *pressure hull* (the part of the vessel that must withstand water pressure, as distinguished from its surrounding structure or other external fixtures).

The pressure hull of a submarine consists of a steel skin strengthened by steel ribs and is more or less cylindrical in shape. This traditional design has long been considered virtually essential for a large vessel that needs to dive deep and travel fast without collapsing under high pressures. Thus, the ribbed hull of a nuclear submarine is identical in principle with the pioneer craft of the American Civil War and the German U-boats of World War I. The hull of a bathy-

The 16 ballistic missiles carried on this nuclear submarine have more total destructive power than all the explosives used by all combatants in World War II, and their range is some 4,600 kilometers. Nuclear power gives an underwater endurance of several months to such lethal vessels – but despite their advanced technology, even the most modern subs continue to achieve depth control according to the unchanging principles illustrated below.

263

scaphe or submersible is very different. Since it is either spherical or egg-shaped, it needs no ribs to strengthen it; its strength lies largely in its shape. It has the advantage, therefore, of being able to go deeper than a comparable ribbed hull can. The bathyscaphe, designed to descend to the greatest possible depths, has to have a very thick steel pressure hull, whereas the hull of a submersible can be built of thinner steel or a lighter material. Thus, the bathyscaphe needs to carry an especially large float of buoyant material to keep it from sinking to the bottom, but the submersible's hull, lighter than the water it displaces, needs very little buoyancy support. In general, the hulls of all submarine craft are among the most finely calculated engineering structures ever made.

There are two aspects of buoyancy control, of course, and every underwater boat

has to have a way of coping with both. It must have *fixed* buoyancy (i.e., it must be able to remain afloat), and it must be able to control its *variable* buoyancy (its power of descent and ascent). In submarines, variable buoyancy is achieved by pumping air into, or venting it out of, external, open-bottomed buoyancy tanks. When under way, a submarine can make minor changes in depth, too, by means of small wings (called "planes") that are tilted to raise or depress the boat. Submersibles also use air for variable buoyancy, but bathyscaphes – again because they go so deep – do not.

As we shall see, the bathyscaphe is mainly of interest as a precursor of the submersible. To sum it all up, military submarines represent the evolutionary end point of a long history of undersea craft, whereas the design and materials of submersibles mark the start of a new era.

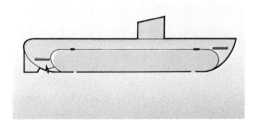

When valves are closed, the flooding process stops and the submarine levels off, with its planes in a neutral position.

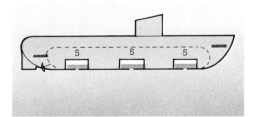

Balance is maintained at the required depth by pumping water (or air, as necessary) into trim tanks (5) positioned inside the pressure hull.

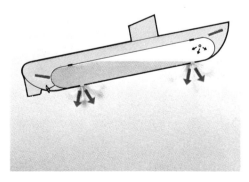

To ascend, a compressor forces air into the main buoyancy tanks, replacing the water. The planes help to angle the ship correctly as it surfaces.

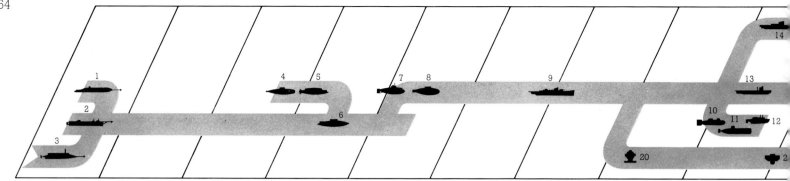

A Family Tree of Submarine Craft
Experiments with submarine craft have been taking place since the eighteenth century. Early boats were propelled by oars or a hand-cranked propeller and could remain submerged only for very brief periods. The "family tree" above traces lines of development from the 1860s on. As the tree shows, some have led nowhere; others have borne fruit either directly or in related "branches." The following list of landmarks in design and/or performance is keyed to the numbers in the diagram. Note that although German U-boats were extremely active during World War I (1914–18), they represented no basic advance over earlier vessels. Even the torpedo, which the Germans developed into a powerful weapon, had been fitted to French subs as far back as 1888.

1 1863 *Plongeur*, France: first non-hand-driven sub, powered by compressed air
2 1863 *David* class, U.S.: steam-powered
3 1864 *Hunley* class, U.S.: hand-cranked, but the first to sink a warship
4 1886 *Peral*, Spain: first electric-battery-powered sub
5 1888 *Gymnote*, France: first sub fitted with self-propelled torpedoes
6 1895 *Plunger*, U.S.: first to use steam on surface; electric when submerged
7 1901 *Holland* class, U.S.-designed: first to be commercially produced in quantity
8 1904 *Aigrette*, France: first diesel/electric submarine
9 1917 *K* class, Britain: oil-fired steam turbines gave fast surface speed
10 1940 *Maiale* class, Italy: known as "Pigs,"

these two-man chariots had detachable warheads
11 1940 *Kairyu* class, Japan: two-man midget subs
12 1942 *X-Craft* class, Britain: four-man midgets
13 1943 *U-264*, Germany: first operational U-boat fitted with a snorkel
14 1944 *U-791*, Germany: experimented with hydrogen peroxide as an ingredient in the fuel
15 1944 Type XXI class U-boats, Germany: increased speed achieved by streamlined hull and more batteries
16 1955 *Nautilus*, U.S.: first nuclear sub
17 1956 *Explorer*, Britain: hydrogen peroxide tested as an alternative to nuclear power
18 1958 *Skipjack*, U.S.: this "pear-drop-shaped" sub set a style for improved maneuverability

War and Peace

Submersibles and submarines are alike in having certain essential characteristics: a sturdy watertight hull, the ability to propel themselves underwater while maintaining sea-level air pressure within the pressure hull, and a variety of sensory devices that help them to operate purposefully in the black night of the deep sea. They are not at all alike, however, in the way they function.

The submarine needs to be speedy enough to track down (or escape from) an enemy, whereas speed is less important than maneuverability for the submersible, since it must do delicate jobs in all sorts of nooks and crannies. Moreover, the big military craft has to be able to remain submerged for very long periods, to range about underwater for thousands of miles, and to be as nearly self-sufficient as possible. The submersible, on the other hand, can generally do its work without an extreme need for secrecy, and it seldom stays down for more than eight hours at a time. So submarines do not ordinarily require surface support; when they want servicing, they head for the nearest military base, but commercial and research submersibles are usually accompanied by surface-support vessels, with which they keep in contact.

Added to the other essentials for a military craft is the weapon requirement. As everyone knows, the basic weapon of World Wars I and II was the torpedo, which is just a small, unmanned underwater boat with an explosive warhead. The torpedo in

its modern form was developed by the Germans for use in the first World War, and the Allies' version merely copied the sophisticated original. That is why the regulation size for the diameter of every nation's torpedo tubes was established at the strange dimension of 21 inches – where it remains. Modern submarine weapons moved a step forward when, after World War II, America and Russia both introduced the deck-mounted cruise missile – a pilotless aircraft that flew to its target on stubby wings. It was soon followed by an even more fearsome instrument, the nuclear ballistic missile, which can be launched underwater from vertical tubes, with a range of up to 6,000 kilometers. And by now there is an advanced cruise missile capable of delivering nuclear weapons in situations where ballistic missiles would be intercepted. So it goes on. Just as the Allies copied the German torpedo, the Soviets and the British have copied America's submarine missile system; the French, characteristically, have developed one of their own.

In America's new *Trident*, which can remain submerged for several months at a depth of more than 1,000 meters, we may be seeing the end of the line of the conventional large-scale submarine. This nuclear-powered leviathan of the deep is equipped with the highest-powered submersible propulsion yet devised. It can carry more missiles and other firepower than have ever been carried in a single military machine. Its accommodations and life-support services are phenomenal. But, like the end of

every evolutionary line in instruments of war, its building and operating costs are also unprecedented. So a new species of sub *has* to come soon. Although the submersible is very different in background and aim, it seems probable that the military craft of the future will evolve from that line.

Progenitor of the submersibles was the bathysphere built in 1930 by William Beebe, an American naturalist who wanted to take a look at the deep ocean. The bathysphere was a spherical, unribbed steel shell, which was lowered from the surface on a cable; when, a few years later, propulsion power was added to this type of craft by the Swiss scientist Auguste Piccard, he called it a "bathyscaphe" (from the Greek words for "deep" and "small boat"). Piccard was determined to take his steel shape down to unprecedented depths, where the use of air for variable buoyancy would be impractical because of its extreme compressibility; and so he used iron shot as a ballast for depth control, and gasoline – which is much less compressible than air and is lighter than water – to achieve fixed buoyancy. It took 70 tons of gasoline to support one sphere just big enough to hold two men! In 1960, Piccard's *Trieste*, piloted by his son, Jacques, and an American Navy man, Don Walsh, descended 11,000 meters – more than 7 miles – into the depths of the Marianas Trench in the Pacific. No other type of undersea craft can approach such a record. Yet the bathyscaphe is too bulky, clumsy, slow, and costly to be of much practical use. The French and the U.S. Navy have made

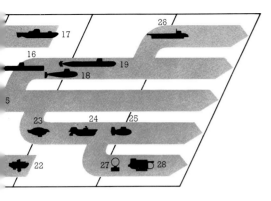

19 1959 *George Washington* class, U.S.: first nuclear-powered missile-firing sub
20 1930 *Bathysphere*, U.S.: a steel sphere lowered to 220 meters on a cable
21 1948 *FNRS 2*, Belgium: the first bathyscaphe
22 1960 *Trieste*, Italy/U.S.: made a record 11,000-meter dive
23 1959 *Soucoupe*, France: the first true submersible – free-flooded machinery
24 1964 *Aluminaut*, U.S.: had a cylindrical aluminum pressure hull
25 1967 *Deep Quest*, U.S.: first submersible with fairing built around a double sphere
26 1969 *NR-1*, U.S.: first nuclear submersible
27 1970 *Nemo*, U.S.: first acrylic pressure sphere
28 1972 *Deep View*, U.S.: first glass hemisphere incorporated as part of pressure hull

improvements in its design, but it seems to have little military or commercial value. Its prime function today is as a stand-by search-and-salvage vehicle in the event of a nuclear sub disappearing in the depths.

The only way for a small craft to dive deep without the expense and drag of a huge gasoline float is to have its strong pressure hull built of such a light material that it needs only a minimum of fixed buoyancy. This was first achieved in 1959, when Jacques-Yves Cousteau built his *Soucoupe*. *Soucoupe*, which could dive to 300 meters, was the first true submersible; and the story of submersible design ever since has been largely a story of successive attempts to find lighter and lighter materials – alloys, heat-treated steels, aluminum, titanium, glass, etc. – with which to make stronger and stronger hulls. As progress has been made on this front, the basic configuration of the submersible has become increasingly clear. It tends to be a spherical pressure hull mounted in a tubular frame that carries at least some of the vessel's batteries, motors, gas cylinders, and their connecting pipes and wires. The hull and its frame are further encased in a streamlined metal or plastic fairing. In the late 1960s, in an effort to enlarge this type of working craft, designs with multiple spheres began to appear. Typical of the bigger modern submersibles is the U.S. Navy's Deep Submergence Rescue Vehicle, which is composed of three interconnected spheres. Since it would be extremely hard to cast or machine a very large ribless sphere – and since such a

sphere would have poor streamlining, and hence high drag – the multiple-sphere design permits an increase in size without sacrificing the strength that a spherical shape provides.

I have now differentiated the three types of undersea craft on the basis of hull and (in very general terms) buoyancy characteristics. I need add nothing further about the bathyscaphe, since only a few such boats have ever been built. Let's concentrate now on submarines and submersibles.

Not all military undersea craft are designed for direct warfare as is the British patrol submarine HMS *Onslaught*, seen above in process of loading a 21-inch torpedo. The U.S. Navy's Deep Submergence Rescue Vehicle, below, is being built for humanitarian purposes. Like most large submersibles, this DSRV has a modular construction of interconnecting pressure spheres. A bell-shaped hatch under its central 2-meter sphere can be mated to a stranded submarine at depths down to nearly 1,100 meters, and 24 endangered crewmen at a time can be taken aboard for transfer to safety.

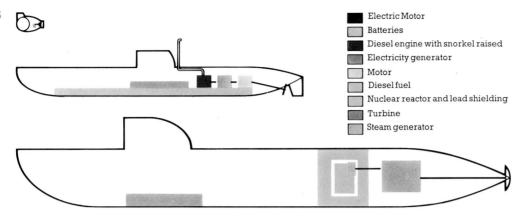

Electric Motor
Batteries
Diesel engine with snorkel raised
Electricity generator
Motor
Diesel fuel
Nuclear reactor and lead shielding
Turbine
Steam generator

The tiny object that seems to be floating over the bow of the smaller of these two boats is a battery-operated working submersible drawn to scale against highly simplified profiles of typical diesel-electric-powered and nuclear-powered submarines. The power and propulsion units of all three vessels are color-coded to indicate their normal placement within each type of undersea craft.

Power and Propulsion

In the primitive submarine of American Civil War days, there were two separate sources of energy: oil lamps provided light; manpower moved the vessel forward by means of an internal handcrank that turned a propeller. Today we try (in our fashion) to be more economical by installing in at least some of our underwater craft a single source of all power. Ideally, one type of energy should be able to provide power enough to run the life-support system, lighting and heating, food preparation, and all the electronics associated with communications and navigation, in addition to driving the propeller or propellers. This, however, is not an easily attainable ideal, for the familiar power-producing fuels cannot burn without oxygen – a commodity that is very scarce underwater.

By 1885, the French were using electric storage batteries in one type of submarine, but the life of batteries on a single charge was terribly limited. Engineers there and in Britain, the United States, Germany, and Italy soon realized the necessity for a combination of power sources in spite of the extra weight; what evolved by the time of World War I was an internal-combustion engine for surface propulsion and surface charging of batteries, while the batteries themselves stored the energy for driving the electric motor that propelled the craft underwater. In the experimental phase, gasoline engines were used, but, for safety's sake, the designers shifted to diesel; and the diesel-battery combination remained unchanged for submarines until the 1950s, when America put her first nuclear boats into service. Through much of World War II, the endurance of a submerged submarine cruising slowly was only eight hours, after which it had to come to the surface, open its hatches to allow air in, and run the diesels to recharge the batteries.

Apart from their other drawbacks, batteries were dangerous, ever prone to liberate deadly chlorine gas or explosive hydrogen. The Germans tried hydrogen peroxide as a compact way of storing oxygen for running the diesels underwater, but the peroxide was found to be unstable – liable to release its oxygen suddenly, with a consequent risk of fire.

It was the Germans again, though, who came up with a truly practical device for solving the fuel-combustion problem when, at the height of World War II, they began to use the now-famous snorkel – a long tube that draws in air by penetrating the surface, but that has a valve which shuts automatically when sea water covers the mouth. The snorkel, which is hinged down on the deck when not in use, makes it possible to use a conventional diesel engine underwater; as a result, World War II submarines with snorkels could stay out of sight for much longer than before. The snorkel's obvious flaw is that it is usable only when the boat is no more than about 10 meters down. Then, too, whenever a wave closes the valve, the diesels suck air from inside the hull and the pressure suddenly drops, to rise again just as suddenly when the valve opens. This strains eardrums and can make the crew very uncomfortable.

The dramatic development of nuclear power was a technical triumph of the years just after World War II. It produced a submarine that can make under-ice or around-the-world voyages and extensive submerged patrols, its endurance limited only by that of the crew. Its drawbacks are its cost, its weight, its technical sophistication, and the hazards of nuclear energy. Only four nations – the United States, the Soviet Union, Britain, and France – have so far had the resources and skill to build and operate such marvels of the deep. Despite the difficulties and expense, however, nuclear power has become the primary means of submarine propulsion in the American and Russian navies.

Nuclear power plants require massive shielding with lead to restrict and limit deadly radiation. Even if the power were scaled down for a small craft of, say, 30 tons, the weight of the lead would be enormous. So working submersibles must still rely on batteries. One model, the German *Tours 300*, has emulated the military design with a combined diesel-electric system, but all other manufacturers reckon that there's no room in small undersea craft for two power sources. Generally, then, submersibles function on the principle that each job can be completed within the life of a single charge of the batteries, which can then be recharged from generators on the mother ship. To conserve power, everything in submersibles is designed to use as little

Built in 1898 by the American naval architect Simon Lake, this bottom-crawling submarine was christened *Argonaut*. Seen here under construction in dry dock at Baltimore, Maryland, it was powered by a conventional internal-combustion engine, which could be geared to either the large side wheels or the propeller; and the same power source also provided energy for driving the lighting generator and sucking air down from the surface. The *Argonaut* actually worked for several years as an underwater salvage vessel.

power as possible, and all equipment is switched off when not in use.

Batteries can be recharged only by feeding electric energy into them. A fairly recent advance is what's known as the fuel cell, a device that produces electricity directly from a supply of chemical fuel. Fuel cells that run on hydrogen and oxygen would, in theory, be much more efficient than batteries, and prototype models have been built. Although the principle on which such cells work has been understood for many years, systematic study of their practicability began only a couple of decades ago. As yet, there are none in commercial use, but they may soon become the power source that frees submersibles from their support ships for days at a time.

In all undersea craft, of course, the major use of power is for propulsion. A typical modern submarine has only one large propeller at the stern; it can cruise underwater at speeds of up to 40 knots, with direction controlled by a rudder and horizontal planes. A submersible, on the other hand, cruises very slowly – 2 to 3 knots – and so a rudder or plane wouldn't bite on the water. For this reason, submersibles usually have several propellers, which can be turned in different directions or switched on and off to make small changes in direction and depth.

Speed is vital for a modern military submarine, but for a submersible such as the U.S. Navy's research vessel *Alvin* (above) maneuverability is more important. Although mainly driven and steered by a reversible stern propeller, *Alvin* also has a pair of lifting motors (one is visible just aft of the superstructure) for slight changes of position and course. The diagram below indicates possible placements for propulsion units; most submersibles are equipped with more than one aid to maneuverability in addition to the main propeller.

The first time you take the controls of a submersible, it almost inevitably bounces about awkwardly, and you get lost. A manned undersea craft is only a mechanical extension of the men who occupy it; like the men, it must breathe and sense its surroundings if it is to function successfully.

To "breathe" means to get enough oxygen not merely to run the engine if the ship is diesel-powered, but to support human life. Since the crew live at a constant pressure of one atmosphere, their consumption of breathing gas does not increase with depth, as is the case with divers. The problem, therefore, is to provide enough oxygen for a given number of men for a given period underwater. The requirements range from the amount needed by 2 to 3 men for 12 hours in a commercial or research submersible, through enough for 75 men for 2 weeks in a conventional sub-

Chemical candles containing sodium chlorate and iron filings are widely used in modern submarines for providing breathable air by means of the oxygen-generating device sketched below. The 36-centimeter-long candle (1) is inserted into a burning chamber (2) through a loading door at (3). A plunger (4) fires a cartridge (5) that ignites the candle. The emitted oxygen passes through a double filter – first slag wool (6), then charcoal (7) – that extracts traces of salt, and is then diffused into the sub's atmosphere through perforated holes in an outflow chamber (8). A submarine may have three oxygen generators; because a candle's life is only about 90 minutes, a supply of hundreds must be carried.

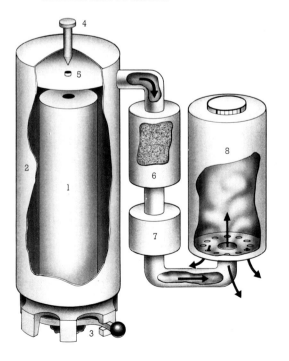

marine, up to the needs of 150 men for 3 months in a nuclear submarine.

Provision of oxygen for a small submersible is simple: the oxygen is bled from a cylinder at a constant rate, and a fan circulates the air through a carbon dioxide absorbent. Since no gas escapes from the sphere, the nitrogen content of the atmosphere is rebreathed again and again. Instruments measure internal pressure, and the pilot can monitor partial pressures and adjust the flow of oxygen accordingly. In military submarines, however, it is not possible to store sufficient quantities of oxygen in high-pressure cylinders. Though some cylinders can be recharged on the surface, the main supply for the full crew in a craft that remains submerged for a long time must be stored chemically as a component of potassium dioxide or chlorate, both of which give off oxygen when heated. (There *is* a way of breaking water down into hydrogen and oxygen by means of electric power, but few non-nuclear craft have enough reserve power to do it.)

If the craft remains submerged for only a few hours, it doesn't matter much if the atmosphere becomes damp, or smells of sweat and food. But during long cruises in nuclear submarines, foul air can impair comfort, morale, and even health. Dangerous contaminants include oil, grease, solvents, chlorine, plastics, fibers, abrasive particles, and many of the chemicals associated with modern materials. And so submarines have mechanical and chemical filters for cleansing the air. As the saying goes, a quick look at a used filter gives you a good idea of where the hole in a pair of socks disappears to.

In the early days, the only way for a submarine to replenish its store of oxygen was to surface, so that the diesel engines could recharge the batteries while high-pressure cylinders were being filled with air. With the increased efficiency of air patrols and radar during World War II, night-time surfacing became almost as risky as doing it by daylight, and that was why the German-invented snorkel came into universal use. Modern sonar and radar, however, can find even a small snorkel, and today's air patrols can detect the heat of a submarine through the water unless it lies much more than 10 meters below the surface. The only safe place for a post-World War II submarine is down deep, and only nuclear submarines can stay there for months at a time.

But how can a submarine have the eyes and ears it requires in order to function purposefully in the depths? The most obvious way to see, of course, is either through portholes or by means of a periscope (which, like the snorkel, is usable at a depth of no more than a few meters). With rare exceptions, submarines have periscopes but no portholes, whereas commercial and

research submersibles have portholes and no periscopes. The reasons for the difference are self-evident. A military craft travels at 30 to 40 knots submerged; since, therefore, it covers the 10 to 20 meters of visible range in about half a second, a window would be useless as a way to see an object in time to avoid collision. But the submarine must approach the surface in order to aim a torpedo at an enemy ship, and here its best eye is a periscope, through which the crew can also take navigational sightings of stars. In contrast, a submersible is designed for working close to things near the sea floor, perhaps, or at the legs of an oil rig, or near the masts of a wreck; its maximum speed is 1 to 3 knots, and so the visible range is traversed in about 10 seconds – a useful period for general observation, or even for avoiding collision or entanglement.

Light and radio waves do not travel far underwater. Even under the best natural-lighting conditions, visual range is never more than 30 meters. Floodlights in deep water, though useful, are ineffectual beyond about 10 meters. Television with image intensifiers provides keener sight than the human eye, but not enough to make a significant difference. Thus, piloting a submersible with your face close to the porthole is like wandering in a strange city in a pea-soup fog, and visual perception of space, distance, and position on this scale becomes quite useless for navigation. How, then, can the submariner get the information he needs?

The answer is "sound." High-frequency sound waves in water can carry everything that radio waves carry in air: radar, television, radio, navigation, range-finding, signal coding and instrument readings, target detection, homing beacons, and echoes to warn of collisions. A man-made submarine craft, like the great whales of nature, progresses through the water enveloped in a cocoon of transmitted and reflected sound, tuned to decipher every ping or bleep. In practice, of course, military submarines try to be as quiet as possible, just listening. Submersibles, which normally have no need of secrecy, use sonar openly for navigation and scanning the sea floor. Most such vehicles in the 10- to 20-ton class carry a depth-to-bottom sonar, a forward-scanning sonar to detect targets in front such as an oil rig or wreck, and a voice-communications set. Such larger nonmilitary craft as America's *Aluminaut*, *Deep Quest*, and *Ben Franklin* and Japan's *Shinkai* also have side-scan sonar, which builds up a map of the terrain on either side.

For navigation, a military submarine needs to know its position on the globe, so that it can launch rockets, avoid the enemy and the sea floor, and get back to port. To do this without detection requires a system that

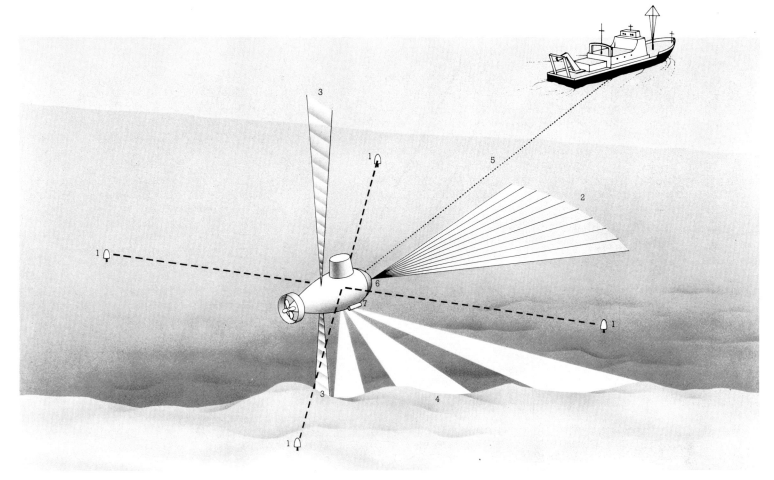

can operate continuously without contact with the outside world – an astonishing feat that is achieved in the same way that space rockets manage it: by inertial navigation. An arrangement of gyroscopes measures every acceleration or change of direction and feeds the data into a computer that continuously plots the sub's course. In theory, such a system would also be ideal for submersibles, but in practice it is too cumbersome and expensive. The first true submersible, Jacques Cousteau's *Soucoupe* of 1959, had only a compass to keep it on course. Since ocean currents often carry a submersible sideways, it's not enough merely to know which way you're heading. Many a submersible of under 10 tons still has a light line and a float tied to it; the surface support ship tracks this and dictates approximate positions to the pilot. Although clumsy, this was the most common method of navigating until about 1973. A few years

earlier, acoustic navigation had been pioneered by the submersible *Alvin*, operated by the Woods Hole (Massachusetts) Oceanographic Institution, and it is now becoming the standard system.

Because sound is the principal sensory channel in the undersea world, military and civilian scientists are continually studying new ways to utilize it, and technical advances have come thick and fast during recent years. In the near future, for instance, listening posts for the detection of military submarines will become ever more sensitive, and engineers will be forced to find new ways of making submarines quieter. On the civilian side, improved microelectronics will mean that small submersibles can carry accurate acoustic equipment, as well as computers to sort out the signals. Deep water remains dark, but modern magic has made the darkness increasingly penetrable.

The Eyes and Ears of a Submersible

1 Transponder
2 Forward-scan sonar
3 Depth-sounding sonar
4 Side-scan sonar
5 Voice communication with mother ship
6 Viewing port
7 Video TV camera

The human eye is a poor tool undersea – at best the visual range is no more than 30 meters – and so the men in a submersible depend largely on an array of electronic sensors to guide them through the water. Very few small craft would need to carry all the devices identified in the above diagram, but all are currently available for use if required. Apart from equipment on the vessel itself, a common navigational aid consists of "bleeping" transponders positioned on the sea floor around the working area; by "interrogating" these the pilot can establish his precise position.

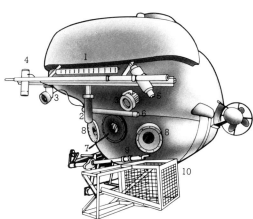

Left: The picture and numbered diagram show how the bow of the submersible *Pisces II* is equipped with the navigational and other aids it requires for doing undersea work, including the collecting of rock and soil samples. Attached to the instrument bars are (1) a forward-scan sonar; (2) a navigation transceiver for "interrogating" transponders; (3) lights; (4) a still camera; (5) its bright strobe flash; and (6) the latter's power supply. Below these are (7) a video TV camera; (8) two viewing ports for the crew; and, finally, (9) a mechanical manipulator and (10) sample cage.

Nuclear reactor
Turbine and steering gear
Missiles
Missile control
Torpedoes
Control and navigation
Crew and storage
Batteries

Above: Layout diagrams of the two major
types of nuclear submarine: the ballistic-
missile sub and the hunter-killer. The function of a ballistic-missile boat is to serve as
an underwater launching platform for land-
directed missiles; the hunter-killer's job is
to track down and destroy enemy vessels,
whether surface or undersea. The top figure
shows the design of an American ballistic-
missile sub of the *George Washington* class.
Vertical missile tubes, with their electronic
controls and launching equipment, are housed
amidships, and there are four torpedo tubes in
the bow. Quarters for the crew of about 140
men are rather cramped, though spacious in
comparison with accommodation in conventional subs. The other layout sketched here is
the U.S. Navy's hunter-killer *Nautilus*, the
world's first nuclear-powered submarine.
Because hunter-killers do not carry ballistic
missiles, they have more living space for their
100-man crews. Their armament consists mainly
of homing torpedoes, with either conventional
or nuclear warheads, which are fired
from the bows. Some – though not *Nautilus* – also
carry SLCMs (sea-launched cruise missiles)
for attacking not only ships and hydrofoils
but also ground targets. The accompanying
chart shows the worldwide line-up of military
submarine forces in early 1976,
identified as conventional (black), nuclear-
powered hunter-killer (blue), nuclear-powered
ballistic-missile (red), and diesel-powered
ballistic-missile (green). Although in sheer
numbers the Soviet Union looks unbeatable,
MIRV (multiple independently targeted re-entry
vehicle) warheads probably give U.S. boats
more true firepower. Many conventional subs
owned by nations other than America, Russia,
Britain, and France are hand-me-downs from the
"big four."

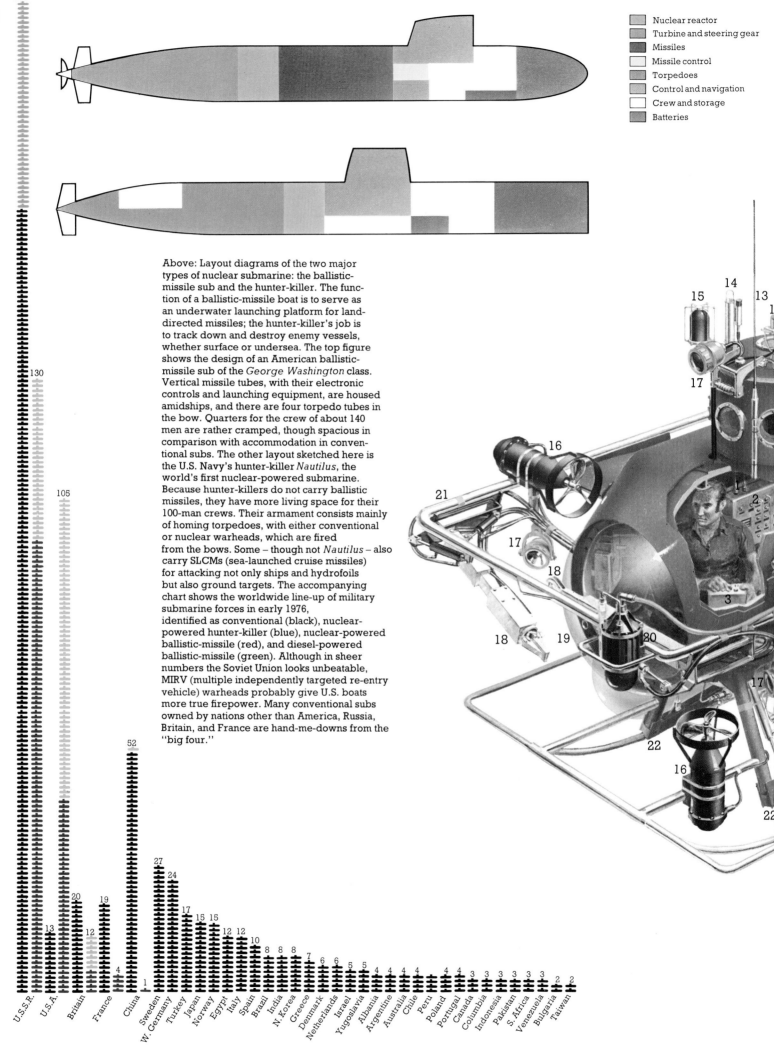

U.S.S.R. 210 / 130 / 105
U.S.A. 105 / 13
Britain 20 / 12
France 19 / 4
China 52 / 1
Sweden 27
W. Germany 24
Turkey 17
Japan 15
Norway 15
Egypt 12
Italy 12
Spain 10
Brazil 8
India 8
N. Korea 8
Greece 7
Denmark 6
Netherlands 6
Israel 5
Yugoslavia 5
Albania 4
Argentine 4
Australia 4
Chile 4
Peru 4
Poland 4
Portugal 3
Canada 3
Columbia 3
Indonesia 3
Pakistan 3
S. Africa 3
Venezuela 2
Bulgaria 2
Taiwan 2

A cylindrical pressure hull is not so inherently pressure resistant as a sphere but is easier and less expensive to construct and is ideal for work at moderate depths. The diver-lockout submersible (built by Perry Oceanographics Inc.) below is designed to operate down to about 360 meters. The forward cylinder of this two-cylindered vessel remains at surface pressure; the rear cylinder, which carries the divers, is pressurized to suit the depth at which they must work. Gas tanks, battery pods, motors, and all other bulky equipment are located outside the pressure hull and covered by fiber glass. Most of the free-flooding machinery can be jettisoned to increase the craft's buoyancy in an emergency.

1 Forward control cylinder
2 Pilot's position (for a fuller view of the background, the pilot himself is not shown)
3 Mechanical-arm control
4 Diving supervisor
5 Diver lock-out cylinder
6 Diver exit hatch
7 Main electric motor
8 Main propeller
9 Compass dome
10 Surface recovery
11 Conning tower and hatch
12 Underwater communication transducer
13 Surface communication aerial
14 Emergency flash and flare
15 Navigation transducer
16 Auxiliary motors
17 Lights
18 Mechanical arms
19 Shielded Plexi-glass (perspex) nose dome
20 Forward-scan sonar
21 Crash bar
22 Extendable legs
23 Jettisonable battery pod
24 Hydraulic release mechanism
25 Batteries
26 Gas cylinders

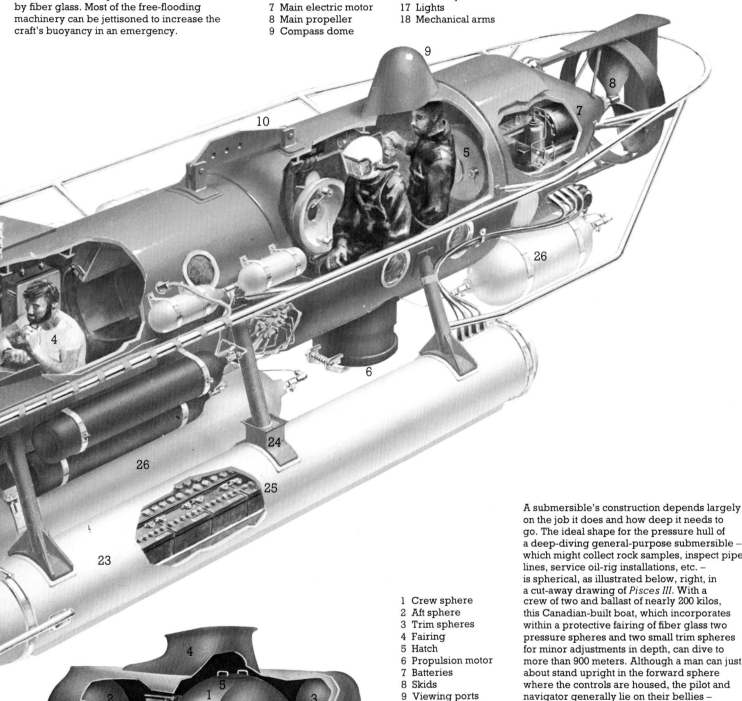

1 Crew sphere
2 Aft sphere
3 Trim spheres
4 Fairing
5 Hatch
6 Propulsion motor
7 Batteries
8 Skids
9 Viewing ports
10 Mechanical arm
11 Auxiliary arm attachment unit
12 TV cameras
13 Still camera
14 Sonar
15 Light
16 Instrument panel
17 Air-purification unit

A submersible's construction depends largely on the job it does and how deep it needs to go. The ideal shape for the pressure hull of a deep-diving general-purpose submersible – which might collect rock samples, inspect pipelines, service oil-rig installations, etc. – is spherical, as illustrated below, right, in a cut-away drawing of *Pisces III*. With a crew of two and ballast of nearly 200 kilos, this Canadian-built boat, which incorporates within a protective fairing of fiber glass two pressure spheres and two small trim spheres for minor adjustments in depth, can dive to more than 900 meters. Although a man can just about stand upright in the forward sphere where the controls are housed, the pilot and navigator generally lie on their bellies – the best position for using the low viewing ports and guiding external manipulators. The smaller aft pressure sphere contains such essentials as oil for use in the trim tanks, an oil pump, and a water-detection device that sets off an alarm if any small leakage develops. Wherever there is an unmanned pressure hull of any kind, some such automatic device is essential for protection of the craft and crew.

Submersibles at Work

Nonmilitary underwater vehicles are – or soon will be – designed for a variety of different jobs. Most of them are built to do specific tasks for the offshore-oil industry, with the rest earmarked for research. From 1960 to 1970, about 50 types of submersible were manufactured in 14 countries, and there was a vast amount of experimentation with materials and methods; but many of the early machines came to nothing and are now laid up, rusting on quaysides. Since 1970, manufacturers seem to have settled down to the task of designing their craft for particular kinds of work, with emphasis on practicability rather than novelty.

Like the French *Soucoupe* and the American *Asherah*, the earliest submersibles (dating back to the end of the 1950s) had very limited power, duration, and depth – and were, therefore, of little practical use. Since then, we have overcome some of the limitations and have been able to build such undersea workers as tracked bottom-crawling vehicles and cable-controlled manned and unmanned machines. Given adequate surface support, the modern submersible is

not a crude exploratory device but a sophisticated working tool, which *has* to be efficient to justify its daily operating cost of many thousands of dollars.

Submersibles are seldom operated by their manufacturers. Operation and support at sea require skills and facilities that can be provided only by specialist companies, all of which buy their craft from the same rather limited range of builders. The manufacturers simply name and sell their various

The working submersibles on this page illustrate two extremes of functional design. The Japanese Komatsu bulldozer (top) has to exert a huge mechanical force without drifting off the bottom. A generator on the mother ship provides the necessary power via an umbilical, and caterpillar tracks help to grip the sea floor. The Komatsu vehicle is electronically controlled, either by a diver with a portable control box or by a surface operator who observes every movement of the submersible on a working model that exactly mirrors its actions (above). Maneuverability, not power, is the key to the design of America's *Johnson-Sea-Link* (left). This manned commercial research vessel is too light to do heavy sea-floor work, but it can manipulate small pieces of equipment with great precision. The diver here is retrieving a package of instruments (attached to a buoy) that *Sea-Link* has brought up from the bottom. The diver takes over just below the surface to prevent possible damage to the valuable package as the submersible is lifted from the water.

models, much as auto firms like Ford or Renault do theirs. The operating companies offer a complete service of machine-plus-support to oil companies, laboratories, and other users. Even the armed forces may take advantage of the service for such jobs as recovering lost weapons and working on naval testing ranges. And although few academic institutions can afford to operate submersibles regularly, they are occasionally used for pure research.

Today's working submersible usually weighs from 10 to 20 tons and is supported by a vessel of 1,000 to 4,000 tons, depending on sea conditions in the work area. The maximum cruising speed of the undersea craft

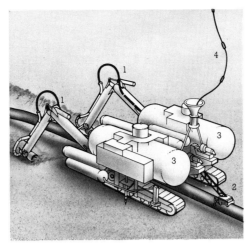

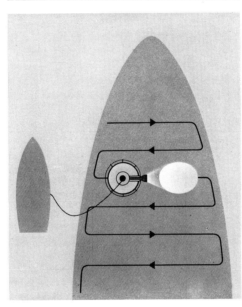

Unmanned submersibles can range in size from such giants as the Italian trenching machine seen here in dry dock (top) to the tiny British-made *Scan* (right). The Italian company built its enormous machine to bury pipes and cables on the sea bed. Straddling the pipeline and controlled electronically from the surface, it digs a trench under the line with two huge arms (1 in the accompanying diagram). Sensors on the gantry (2) guide the machine forward, and electric power for the engines in the flotation tanks (3) is supplied from a mother ship through an umbilical (4). Such bottom crawlers are an advance over the simpler type of cable-laying machine pictured above. This vehicle is towed behind its mother ship and, like a plow, scoops out a shallow trench for a telegraph cable to lie in. *Scan* does its job not on the ocean floor, but on the bottom of a ship, where it sweeps to and fro while its TV cameras record areas of corrosion and other kinds of damage. Once positioned by a diver, it is guided by a cable from the support vessel.

ranges from some 2 to 4 knots, which is just enough to keep it from being driven sideways or backward by strong tidal currents, and the total distance covered in a dive is seldom more than 10 to 15 kilometers. Most dives, apart from those that involve cable and pipeline inspections, require intensive work in a small area, but in a depth range from 150 to 1,000 meters. Since two dives of 4 to 8 hours each can be made in a single day, the total working capability of the craft far exceeds that of divers in endurance as well as depth. Moreover, some types of submersible can carry sufficient reserve power to drive tools such as water jets for digging trenches, hydraulic rock

drills, cutting and grinding tools, articulated arms for lifting equipment, sediment-sampling devices, and a whole range of lighting, listening, and recording instruments. Submersibles equipped in this way have buried cables, repaired pipelines and oil-well heads, surveyed foundations for massive oil-rig legs, placed explosive charges, and assisted salvage operations.

An important development (already noted in the preceding chapter) has been the designing of the so-called lock-out submersible, which can release divers from a pressurized compartment in order to carry out particularly delicate work tasks. Diver lock-out is not a new concept; it became

possible in World War II to put attack divers into the sea near enemy ships and beaches. But the submarines for such tasks weighed over 50 tons and were not designed for repeated exits and re-entries of divers into a pressurized section of the craft. Modern diver lock-out is an extension of the principle of the submersible decompression chamber, which allows divers to leave the hull and come back as often as necessary.

More prosaic, but commercially very valuable, are a wide selection of tracked and wheeled vehicles, some manned, some cable-controlled, that can dig trenches, bury pipelines, dredge sand, and bulldoze the sea floor. All such bottom crawlers are designed to exert a huge mechanical force on the floor, upon which they must sit firmly – a feat that cannot be duplicated by free-swimming submersibles, for the true submersible tends to drift backward in reaction to any force it exerts. The bottom crawler overcomes this problem by friction, but it requires a lot of power to do its work, and this can be provided only by supplying it down an armored electric cable. Since the crawler moves very slowly, if at all, the drag of the cable presents no problem. Machines of this type have been built in Japan, Italy, and Germany, as well as in America.

Some manned submersibles are also attached to cables, either for towing or power supply. This is one way to break out of the restrictions imposed on small undersea craft by the limited power of their batteries. A towed submersible such as the Russian *Atlanta* or Scotland's Aberdeen Fisheries Laboratory machine can usefully examine trawl nets at speeds of 6 knots – a speed beyond the capability of today's self-powered submersibles. The Japanese *Kuroshio II* and the American *Guppy* use power cables to achieve greater maneuverability in strong currents, but the cable itself is something of a drag, and dependence upon it limits the depth of such boats to a maximum of around 300 meters.

Finally, there are now cable-controlled unmanned submersibles that can be equipped with manipulators and TV cameras to do jobs that might be too risky for manned vessels. One such craft is the U.S. Navy's Cable-controlled Underwater Recovery Vehicle (CURV), which was designed to salvage experimental torpedoes after trials. In 1966, one model of this machine achieved worldwide fame when it helped recover an H-bomb lost underwater off the Spanish coast. In 1973, CURV again hit the headlines when it assisted in rescuing the Canadian *Pisces* submersible, which had sunk in the Irish Sea with its two-man crew. Similar unmanned cable-controlled vehicles have been built by the French, British, and Norwegians. Their less spectacular tasks are mainly the inspection and photography of cables, pipelines, and oil-field structures.

A Day in the Depths

Imagine that you're chief geologist for a company called Global Titanic Mining Corporation on a project to drill a rock core from the bedrock 1,500 meters down at a precise location in the North Atlantic. A submersible and mother ship have been chartered; and after being briefed with preliminary surveys made by means of sonar, you're flown by helicopter to join a 4,000-ton surface ship named, say, *Sublaunch*, which is already heading for the work area. During the next two days, you go over the plan of work with the captain of *Sublaunch* and the team leader of the submersible crews. Then, when the ship has completed a series of runs across the work area, using precision depth sounding and side-scan sonar to confirm the location of the rocky outcrops, you ask the captain to position the navigational transponders for the submersible. These acoustic transponders are contained in alloy cylinders, each about a meter long, weighted down with an iron sinker, and held about 10 meters off the bottom by a plastic float. Three transponders are lobbed over the rail of *Sublaunch* in a triangle about 2 kilometers apart, and the submersible will measure its position on the sea floor by means of them.

For the last few hours, engineers have been making sure that the submersible's main power batteries are charged, that all cylinders for buoyancy and breathing gas are at full pressure, and that electric and hydraulic circuits are in order. In particular, they've been checking the hydraulic rock drill mounted on an articulated arm in front of the manned sphere. The submersible is serviced in a hangarlike space aft and is hauled on skids to the stern of the ship for launching.

You're ready to dive.

Equipped with a portable tape recorder, a flask of coffee, sandwiches, notebooks, and possibly seasickness tablets, you step up the aluminum ladder, haul yourself over the rim of the conning tower, and climb

Above: Conditions are satisfactory, and preparations for launching a submersible are under way. The deck crew have attached a cable to the craft – via the hinged "A" frame above it – from the yellow winch in the foreground. Once the pilot, navigator, and scientist-observer are aboard, the boat will be winched off the deck, swung out over the side, and lowered into the sea. Back on the mother ship, surface control (right) keeps in constant touch, checking equipment, giving navigational instructions, and advising the submersible's crew of changing weather conditions. The pilot is kept busy during the dive; but once in position at the working place he has time to chat with the scientist as they watch the external hydraulic drill in action. Even in the glare of strong floodlights, the view through the portholes is limited and television may give the men their best view of what's going on outside.

down into the crowded sphere: crowded because every inch of the wall space is packed with dials, switches, pipes, taps, and boxes of clicking and whirring electronic devices. Low down, there are three portholes through which you can see the legs of people walking about on the deck of the mother ship. You crouch in a corner while the pilot and navigator climb in and the hatch is closed. A powerful winch raises the submersible until she's hanging on a thick cable from a hinged frame, which is rotated out over the stern until you are swinging wildly above the waves. Suddenly you drop down, and waves wash over the portholes as the submersible rocks and wallows. Divers unhook the lifting cable and make a final inspection of the craft as she pitches and rolls in the water. The pilot checks through apparently endless banks of switches and talks to the ship by radio. At last he vents the ballast tanks, and you begin to sink.

Immediately the sickening motion diminishes. Upward the sea appears silvery bright, and below it looks deep blue, almost purple, until, after about 10 minutes, it turns completely black; but you can see dusty white blobs of plankton floating past the portholes, streaking upward like snow in the headlights of a car. To save power, only a small light burns in the cabin. The motors are not being used, and the only noise comes from the gyroscope and air-cleaning fans. The pilot notes the distance from the surface and the distance to the bottom; after half an hour, he reduces the rate of descent and switches on the floodlights. The dark shapes of rock now come into focus, rising from an expanse of pale sand, slightly rippled by the deep-ocean currents. The first rocks you see are small, and might be loose boulders, not parts of the bedrock. So the pilot begins to cruise on a prearranged grid pattern; the navigator, having picked up the echoes of the transponders and switched the tracking onto automatic record, activates the side-scan sonar so that you can see a shadowy map of the terrain for 50 meters to either side.

As the pilot steers the submersible on the planned track, you peer out of the porthole. Ghostly fish hover on the fringe of the globe of light, but seldom come close to the submersible while you're moving. The spectacle is so soothing that you could sit for hours without doing anything useful. But at a cost of $1,000 an hour, you can't be comfortably lazy. After three legs of the grid, you've begun to map out a ridge of rock that rises to the north. On the next leg, you find yourself running straight into a black overhanging cliff. The pilot reverses the motors, and you skid to a halt in a curling cloud of sand. Then, as he adjusts the buoyancy, you climb slowly up the cliff-face to its top, 15 meters above.

You motor along the crest of the ridge until it starts to descend; then you backtrack to the highest point. The pilot ballasts the submersible down and crouches close to his porthole with the manipulator-control box in his hand on a remote-control lead. The hydraulic pump whines as the heavy external arm swings out from its stowed position. At the end of the arm is a diamond-toothed drill, which begins to do its work as soon as unnecessary lights have once again been doused – for power must always be conserved. It takes half an hour to make a 25-centimeter cut in the dense rock. The core is broken off, extracted, and slotted into an exterior storage canister. While the pilot has been controlling the drill, you've been examining track plots and side-scan maps with the navigator and have decided to take two more cores. But *Sublaunch* warns you through the voice-communication system that the weather is deteriorating and that you'll have to be recovered within an hour. That gives you only sufficient time to take one more core.

As soon as the first core is stowed, the pilot steers the submersible to a position that you've mapped out with the navigator. While the second core is being drilled, you relax with your coffee and sandwiches and listen to music from a cassette, hardly even thinking about the fact that you're 1,500 meters down on the sea floor in a sphere with a very limited supply of electricity and oxygen.

Sublaunch calls you back with a warning of rough seas above, and so, once the second core is safely aboard, the pilot lightens the ballast, and you start to ascend. Total darkness slowly gives way to a pale dawn tinge, then a yellowish green light, then a brilliant silvery blue. At last you're bobbing crazily in the waves, with the portholes now deep in foaming water, now twisting skyward. The struggling black figures of divers scramble to attach lifting and towing lines. It takes ages, and you wish you hadn't eaten those sandwiches. The pilot discreetly gets a plastic screw-top jar from under the floorboards and hands it to you. Luckily, though, you don't need it. A sharp jerk means that the lifting cable is being winched in, and you rise slowly upward. A few minutes later, you're out on deck, breathing the sweet fresh air.

Weather and sea permitting, you'll be making another ten dives – similar to, but always different from, this one – during the next week or so.

A submersible's mechanical arm is manipulated from within the boat; seen through the porthole (above), it looks almost like a living monster. Right: In the recovery process pictured here, a diver has attached a hauling line to a surfacing submersible, and he is hitching a ride as she is winched in toward the mother ship.

The Case for Radical Design

If submarine designers had been permitted the boldness of aircraft designers, today's submarines would be made of glass or acrylic instead of steel, which is as inappropriate for a submarine hull as for the fuselage of a high-flying aircraft, since, because of its weight, a submarine made of steel has severely limited diving depths; and, too, in order to preserve fixed buoyancy, the pressure hull has to be kept partly empty, which wastes space. If the designers were as uninhibited as aircraft designers, they would have recognized that batteries are as inappropriate for military submarines as piston engines are for high-flying aircraft, since battery-operated vessels have limited endurance at low speed and no endurance at high speed; and today's big submarines would be propelled by buoyancy gliding rather than by expensive nuclear power. Finally, if submarine designers had been given the leeway of aircraft – or even submersible – designers, they would have found a practical way to place most of the machinery outside the pressure hull. Obviously, the problem of maintaining and servicing external parts during long submerged cruises would not be easy to solve, but imaginative engineering could surely solve it. Internal machinery is as inappropriate – and as dangerous – as would be engines located inside the fuselage of an airplane.

Materials, propulsion, machinery – these three hold the key to the submarine that could have been built yesterday, and *can* be built tomorrow. I'd like to say a word or two now about each of these basic aspects of underwater-craft design.

What are the characteristics of glass and acrylic that make them so appropriate for undersea use? To begin with the more common material, glass is practically indestructible under high pressure. Surprising as it may seem, when engineers tested spherical glass shells by fitting them with trip hammers set to shatter the spheres at various depths, the spheres shattered easily at or near the surface, but in deep water the hammers couldn't break the glass even when they were explosively driven against it! The reason for this resistance is that glass breaks when it is exposed to tension or pulling stresses, but it will withstand almost limitless compression or pushing stresses. Thus, hammer blows near the surface produce a complex pattern of stress – partly compression, partly tension – and the tension forces win; but the uniformly applied pressure of the depths produces compression stresses throughout a glass sphere, and the tension stresses of even a heavy blow cannot overcome them. A glass submarine could be strong enough to carry more than 100 per cent of its weight at a depth of 8,000

meters. Its cost would be a pittance compared with the cost of a steel hull. Moreover, its transparency would provide full visibility of the deep-ocean environment – a useful, if not essential factor. Only inertia has prevented the development of just such a submarine.

The value of transparency is already being demonstrated by a few shallow submersibles made of acrylic, which is actually better than glass for depths of 300 meters or less. Acrylic cannot withstand a great amount of pressure, but it is light in weight, can be made very thick, and does not shatter under tension. Several two-man submersibles capable of descending to depths of 300 meters have been successfully built of acrylic. Large submarines with steel ribs but acrylic shells would be easy to build, cheap, fully transparent, and shock-resistant.

The tragedy of submarine design lies not merely in the failure to use new materials but also in the failure to understand that, even with steel, we could build far more buoyant submarines than we now have. There is, for example, a new material called syntactic foam, which consists of a mixture of resin and tiny glass balls and is incredibly strong and light. A hull made of low-strength, low-cost steel could be very thick and heavy, yet remarkably buoyant, if a float of enough syntactic foam to overcome the weight of the steel were added to it. Glass, acrylic, and syntactic foam are the natural materials for all submarine craft. It may be years before they come into common use, but the end of what we now think of as the inevitable look of a submarine hull is as certain as was the end of the oared galley, the galleon, or the battleship.

As for propulsion, buoyancy propulsion – a simpler, cheaper, and more powerful method than any other known system – was demonstrated as long ago as 1957. In this method, the boat is equipped with wings; negatively buoyant at the beginning of a dive, it glides at an angle down to its maximum depth, where ballast is blown from a set of tanks and the craft becomes positively buoyant, so that it glides at an angle up to the surface many miles away from its starting point. This cycle is repeated as long as the supply of buoyancy gas lasts. The only fuel required is buoyancy gas, and the only machinery consists of ballast tanks, control surfaces, and a buoyancy generator. If a submarine of conventional military size were to be built in accordance with the principles of this system, it could attain an underwater speed of 50 knots, and if it were designed of glass or otherwise had a depth capability of 7,000 meters, it could traverse more than 100 kilometers in a single cycle. Just consider how useful such a boat could be for long-range transport!

Yet, despite its obvious feasibility, the

Some Engineering Truisms

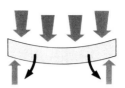

If you build a bridge with a flat span, it has limited strength because an evenly distributed load tends to rotate the two halves about the end supports and bend the middle downward.

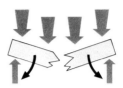

Any flat surface under pressure will therefore have a maximum stress on it at the midpoint, and even a moderate load will force the center in and tear the material apart.

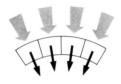

A bowed structure – one built in the shape of an arc with its ends securely anchored – can support heavy loads because downward pressures actually force the sections more firmly together.

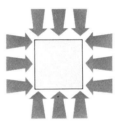

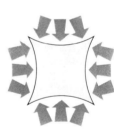

Under equally strong external pressures, all sides of a square structure will eventually give way and cave inward. Thus, a cubic shape could not possibly withstand deep-water pressure and would not be suitable for any type of undersea vessel.

glass-and-buoyancy submarine remains in the realm of the future, while the history of the advanced small submersible has been written by designers less constrained by tradition. They have experimented with such materials as aluminum (in America's famous *Aluminaut*) and titanium (used to replace the steel hull of the oceanographic submersible *Alvin* when, after being lost for a year in an unfortunate accident, she was recovered). As a result, some small craft have been able to operate at depths of as much as 4,000 meters. And another American research submersible, *Ben Franklin*, which is the brainchild of Auguste Piccard's son, Jacques, has actually – even if only for a month-long demonstration period – used the currents of the ocean as the main source of its propulsive power, thus devoting its entire battery load to such life-support requirements as light and heat.

One of the most significant milestones in submersible design is represented by

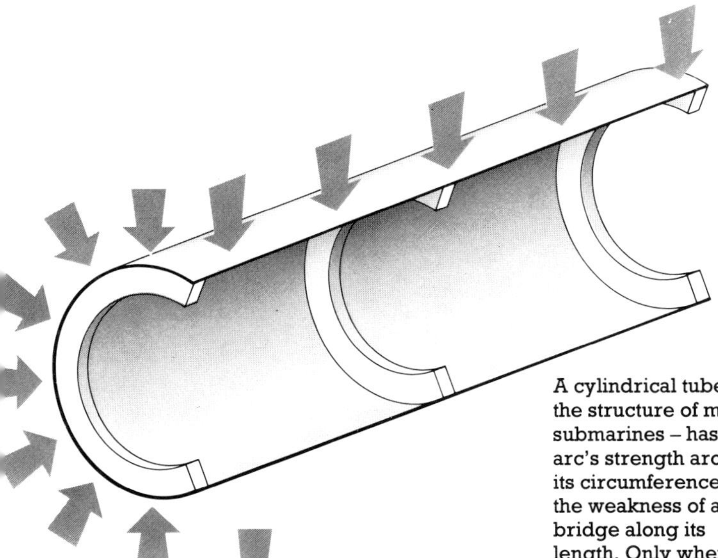

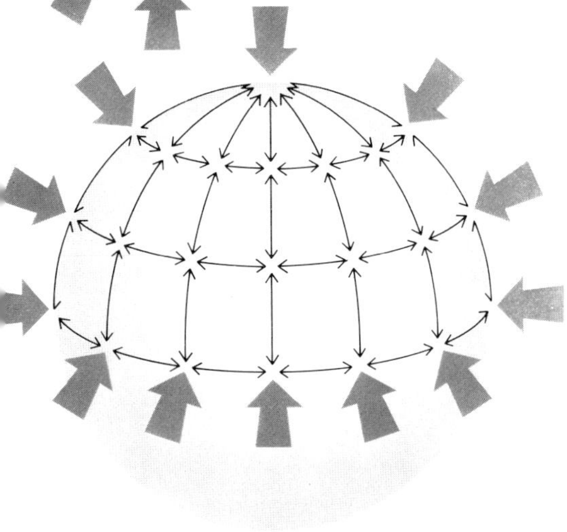

A cylindrical tube – the structure of most submarines – has the arc's strength around its circumference but the weakness of a flat bridge along its length. Only when reinforced by a series of circular ribs can it withstand considerable pressure. The ideal structure for resisting water pressures – and therefore the ideal design for the hull of a deep-diving undersea craft – is the sphere, because the arc principle illustrated on the facing page applies in all directions. Under compression, a sphere may shrink a little; but since no point on the surface is weaker than any other, the equalized pressures inhibit collapse. In practice, bathyscaphes and submersibles, which generally need to go very deep, have spherical pressure hulls, whereas submarines are usually cylindrical. Submarines could dive as deep as submersibles if they too had spherical pressure hulls. But the cost of a large sphere has so far been considered prohibitive – and the width of a large sphere produces drag.

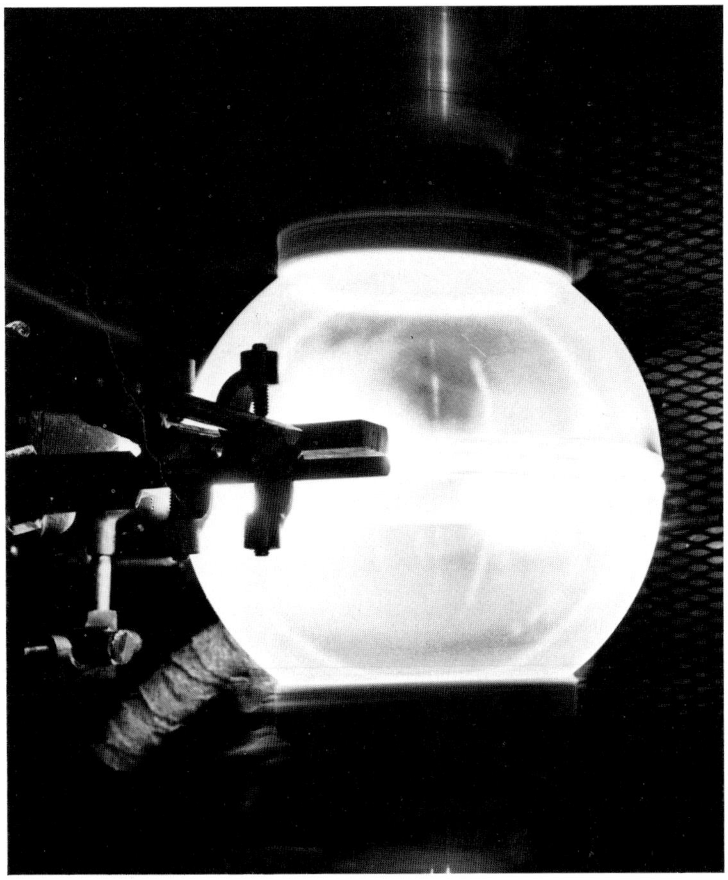

Above: Small hemispheres are flame-sealed to make a sphere for testing the properties of glass underwater. Such tests suggest that a submarine with a spherical glass hull, and with free-flooding machinery, would be almost indestructible at great depths. But although one or two submersibles now incorporate some glass in the hull, it is as yet impossible to mold a large sphere without a flaw, and any defect means a weak point that could collapse under pressure.

Maximum depths that spherical hulls can dive before collapsing depend largely on what the hulls are made of. The graph below shows theoretical limits of spheres built of various materials. The vertical axis gives sea-floor depths; the horizontal shows the percentage of floor accessible to craft built of specified materials. Thus, while a low-grade-steel sphere can explore only 15% of the sea bed, a perfectly machined glass and ceramic hull could go everywhere.

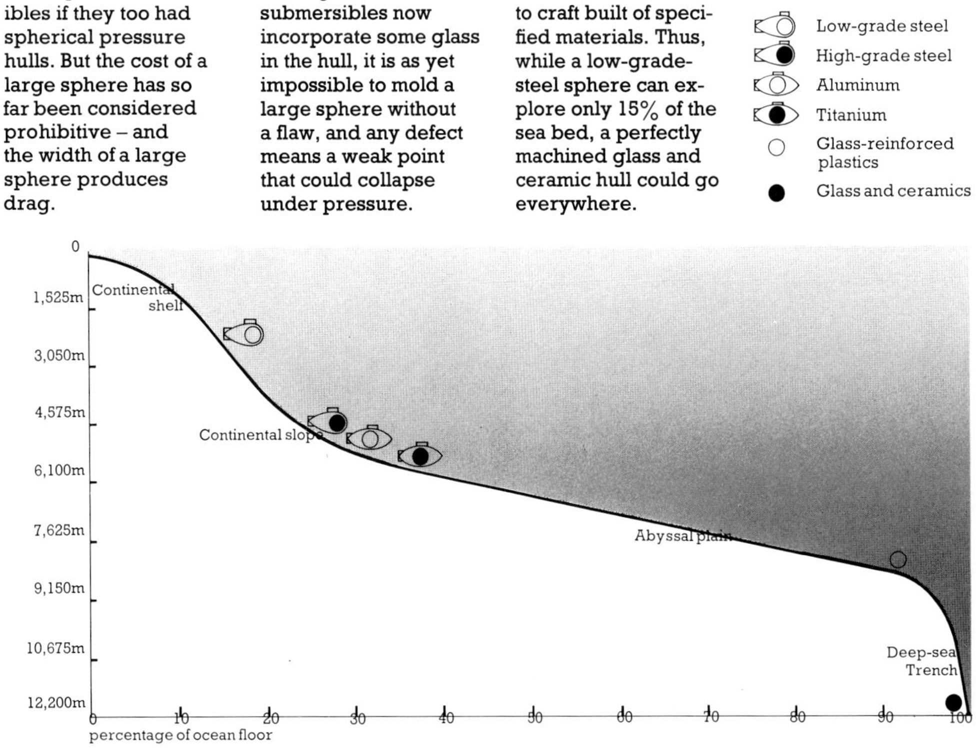

the U.S. Navy's *NR–1*, a 400-ton nuclear-powered oceanographic vessel built in 1970 at least partly to test the feasibility of nuclear power for research craft. The *NR–1* contains the world's smallest nuclear reactor. But it is not its nuclear power that is of chief interest to students of undersea-craft design, fascinating as the idea may be of a nuclear submersible with a crew of only five and an ability to sustain undersea operation for many weeks. (The cost of more than $100 million is prohibitive by ordinary oceanographic standards, of course.) What is especially significant about the *NR–1* from the design standpoint is that its propulsion motors, which are electrically driven by power generated by the internal nuclear plant, are placed outside the hull, where the machinery is directly exposed to the pressure of depth. Such free-flooded machinery, which eliminates the need for a big and complicated shaft to penetrate the main hull, is now a feature of most small submersibles, but it is something rather new for undersea craft of over 20 tons.

Among all the difficulties facing the designer of a conventional military submarine, none is more troublesome than the clutter of often dangerous essentials within the hull. Interior batteries can start fires or emit poisonous gases. Penetrations through the hull to carry cables for sonars and sensors are major sources of leaks. If ballast systems are contained within the pressure hull, they can be emptied only by pumping against the pressure of the sea. And, of course, there are the cumbersome, space-consuming propeller shafts and the inefficient shaft seals that seldom entirely succeed in enabling the shafts to pass through the hull and rotate without causing leaks. The solution to all these problems is free flooding. There is little need for equipment to be put inside the pressure hull. Apart from the crew, the crew's life-support systems, their emergency controls and certain delicate pieces of electronic equipment, nearly everything on a submarine would be better off either right in the water or immersed in a container of oil, which in turn would be located in the water.

An electric motor, for example, can be sealed in a tank of oil that has a flexible bellows attached, so that water pressures are communicated to the bath of oil. With the motor outside the hull, no shaft seal is required for the propeller, and only the control wires need to penetrate the pressure hull. It is just such an external motor that sets the *NR–1* apart from most other submarine craft designed specifically for military service. Until now there has been a sharp line of demarcation in design between vessels of under about 50 to 100 tons and heavier boats. Designers assume that the proper use for the extra space inside the pressure hull of an undersea craft of more than 50-odd tons is to fill it with machinery; in other words, they build a conventional submarine. If the all-up weight is significantly less than 50 tons, the hull is too small to contain much in the way of machinery and batteries, and so the designers put the machinery on the outside; that is, they build a submersible.

Outmoded notions of what is and is not appropriate should have been discarded long ago. There is no overriding reason why larger craft should not be built like submersibles, and there are excellent reasons why they should. With new materials and the design concepts I've been discussing, submerged craft of the future could carry enormous payloads of goods, or even many passengers – and in great comfort, too. The *NR–1* is a step in the right direction. It is, in fact, a step toward the future in many ways. For instance, it is the U.S. Navy's test bed for the most advanced types of underwater computer and navigation systems. Using the inertial navigation package, a computer can store the entire record of the track the boat has covered. Thus, if the pilot wants to return to any previous point or to follow a path parallel to an earlier track, he can feed instructions to the computer, and the *NR–1* will automatically carry out the necessary maneuvers. This is obviously very useful indeed in mapping, making geological surveys, or searching for lost objects.

Pictured here are some examples of advanced submersible design, all of them American. The Navy's *NR-1* (facing page top left), ostensibly a research submersible, is none the less powered by a nuclear reactor; but, unlike nuclear subs, her electrically driven motors are free-flooded – i.e., placed outside the pressure hull. So secret are essential features of her design and construction that the Navy has released no more revealing photograph than this somewhat doctored one. *Deep View* (top right) is the first submersible to have incorporated a glass hemisphere as a significant part of its design. *Nemo* (below left), the Navy's manned observation sphere, was the first submersible to use an acrylic pressure hull. Right: The *Johnson-Sea-Link*, a commercial research submersible, has a manned acrylic pressure sphere; all its machinery – motor, ballast tanks, oxygen cylinders, and manipulators – are free-flooded.

The Past and the Future

On a January day in 1966, two American military planes collided over the sea just off the town of Palomares, Spain. Into the water with some of the wreckage fell one of several hydrogen bombs. Those that fell on land could be swiftly dealt with, but the submerged one posed a horrifying threat to Palomares. Whether it could be found and recovered was a matter of international concern, and the U.S. Navy mustered its forces to cope with the situation. The vehicles and techniques involved in the operation were somewhat less efficient than they might be in a similar situation today, but we have made no dramatic advances in undersea craft since then. Let's compare what happened in 1966 with what *could* happen in the future if vehicles of advanced design were to become generally available.

Although local witnesses were able to give a rough estimate of where the bomb had fallen, the waters in which it was lost posed a variety of search problems. At the edge, a sandy beach sloped into a Mediterranean bay, with scant indication of the rugged terrain that lay to seaward. On the left-hand side of the bay facing seaward, the bottom was smooth and gently sloping, reaching depths of about 100 meters before dropping abruptly into deep water. On the right-hand side a ravine extended from the beach as a gentle depression, with barely a hint at first of its sharp drop to depths of over 300 meters, where it divided into branching chasms between narrow canyon walls. This rugged terrain persisted until it joined the flat part of the bottom at depths of more than 1,000 meters. The bomb could have sunk anywhere, at any depth, in this widely varying terrain. There was fine silt, easily stirred, in one place; hard exposed rock in another; tailings from a lead mine all over the shoreline area; assorted anchors, traps, and trappings of generations of fishermen strewn about everywhere; and the debris of the wrecked aircraft, shattered, extended, and dispersed, to provide an additional field of false contacts and misleading clues.

Ironically, although some of the U.S. Navy's finest nuclear submarines lay at anchor in Rota, Spain, they could not help to look for the bomb, because not one had a single underwater window, or the television equipment required for optical search, or the right type of sonar for acoustic search. None of them was even suitable as a command post. A massive guided-missile cruiser was chosen for this job; although equally ill-equipped for search and recovery, it at least *looked* impressive to the watching world.

At the Navy's disposal were only three small submersibles capable of search in the deepest portions of the area: the *Alvin*, the *Aluminaut*, and the *Cubmarine* (a product of the Perry submarine-building company). Of these, only the *Alvin* was mobile enough and could go deep enough to search in narrow ravines, but it was thousands of miles away; and although it could be transported by air, the job of taking it apart and reassembling it was time-consuming. It took a month for it and the *Aluminaut*, which was carried across the Atlantic in a military landing ship, to join the search. Meanwhile, the *Cubmarine*, which was air-transportable without disassembly, had been at work, but it could go no deeper than 200 meters. Navy minesweepers were also pressed into service. They are designed, however, to deal with anti-ship mines in very shallow water, and so their efforts were confined to a narrow band of water near the beach.

Added to the array of manned vehicles were two unmanned systems. One consisted of underwater television cameras and acoustic sensors towed by the surface research ship *Mizar*. These devices were excellent for deep water where the terrain was relatively smooth, but they could not negotiate many of the difficult canyons and crevices in the area. The other system was the Navy's cable-controlled unmanned vehicle, or CURV, which was brought into action toward the end.

In the shallow water, Navy divers pressed diving tables to the limit, but saturation diving facilities weren't available. And the hard-working armada included a vertical-ascent submersible (one designed to go up and down, not to make headway) with the quaint name of *Deep Jeep*, destroyers, tenders, tugs, salvage ships, small boats, and landing craft. Eighty days after the grueling search had begun, the *Alvin* located the dangerous bomb at a depth of 950 meters. Because of the steep slopes and the danger of entanglement, it would have been too perilous for a manned submersible to try to effect a recovery, and so CURV was brought in to attach a line to the bomb, which was then winched up. Nearly three months and millions of dollars had been expended, and a fleet of ships had been immobilized – all for the lack of more imaginative design of underwater equipment.

Let's rewrite the story. Instead of an armada of hastily assembled ships, a single rescue ship is dispatched to the site. It has decompression chambers large enough to serve as living quarters for saturated divers. It also carries several acrylic lock-out submersibles capable of diving to 600 meters, plus a number of glass-hulled buoyancy submersibles and syntactic-foam work barges with a depth capability of 8,000 meters. The surface ship lays out a network of underwater beacons for navigation purposes; the lock-out submersibles carry divers to a given depth for each day's search. In weekly cycles, they search at progressively greater depths down to about 300 meters for the first week and at progressively shallower depths for the second week, so that decompression is completed with the end of the search effort. On the broad plain to the left of the bay, buoyancy submersibles carry out an optical search, gliding from depths of 300 meters to depths of thousands, and returning up the slope propelled by buoyancy. In the ravines, the syntactic-foam barges conduct a slow but

The current disposition of military submarines is, of course, top-secret information, but there are no secrets about where commercial submersibles are at work at any one time, as this 1976 chart shows. Note the great concentrations of undersea vessels in regions where offshore gas and oil are being produced, particularly in the booming North Sea. Yet despite their growing numbers (about 10 submersibles are built annually), no coordinated system has so far been evolved by commercial interests to mount an extensive rescue or salvage operation, if necessary.

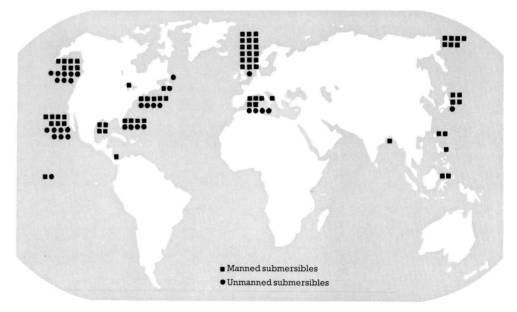

■ Manned submersibles
● Unmanned submersibles

steady search of the crannies and crevices. The effective search is perhaps ten times as efficient as the actual Palomares operation, and it takes a week or two rather than months, at a fraction of the cost. The recovery, in nearly 1,000 meters of water, is effected by the submersible work boats aided by cable-controlled vehicles, themselves controlled from submarine craft. The surface ship has been simply a transporter, and the operation has been unaffected by the weather.

How long will it be before these new undersea craft become commonplace? Who can say? We already know how to build them. All we need now is the will and daring to go ahead.

It took three months, millions of dollars, and even more naval vessels than those pictured here to recover America's lost H-bomb off Spain despite the use of the most advanced equipment available in 1966. If the U.S. Navy had had a custom-built rescue ship equipped with decompression chambers, sonar detection and positioning devices, and its own fleet of diver-lock-out and deep-diving glass and buoyancy submersibles, retrieval would undoubtedly have been achieved more quickly and cheaply. The recovered bomb itself is circled in the photograph at the bottom of the page.

10
Marine Law and Politics

Robin Churchill

Before we can look into the sometimes obscure depths of the international law of the sea, we need a clear understanding of what international law itself is. Many people have rather vague notions about international law; they know the term but are not sure of its precise meaning. So let us begin by stating the simple fact that international law is the legal system that governs relations among different countries (or states, to use the correct legal terminology). The matters that international law is concerned with include the drawing of frontiers between states, diplomatic relations, the framework of international trade, economic and other forms of cooperation, air travel between different states, the settlement of disputes, and the establishment and functioning of such international organizations as the United Nations and its specialized agencies or of such regional organizations as the European Economic Community and the Organization of African Unity.

Some sort of international law is particularly important where the sea is concerned. Because the 30% of the earth's surface that is land is divided into many different states, human activities on land can be regulated by separate legal systems of individual states; but the sea is, as it were, indivisible, and so marine activities cannot be regulated primarily by national laws. While it is true that every state with a stretch of water off its coasts can assert control over that zone for economic or security purposes, the geographical extent of the zone (which, by the way, is conventionally measured in nautical miles) and the nature of the coastal state's rights must still be governed by international law. Beyond these narrow zones, which add up to only a small fraction of the sea's surface area, are what lawyers term the high seas, and no state has a legal right to appropriate the high seas. They are a common area, open to reasonable use by all comers, and, as a common area, they require common rules to govern their use, whether for navigation, fishing, cable laying, or military activities. These rules are provided, whenever possible, by international law. Here are a few examples of the kinds of question that such law seeks to answer:

Would the U.S.S.R. be allowed to place submarine listening devices off the coast of the United States – or vice versa? Was Great Britain justified in bombing the Liberian oil tanker *Torrey Canyon* when, in 1967, it ran aground in an area a few kilometers outside Britain's territorial waters and poured large quantities of oil onto her coasts (the purpose of the bombing was to set fire to the oil still in the ship so that it would burn up before escaping into the sea)? Was Iceland entitled to extend its fishing limits to 50 miles in 1972, and again to 200 miles in 1975, in spite of the protests of other states that fished in these waters (chiefly West Germany and Great Britain)? It is the complexities of such questions that we shall be trying to disentangle in this chapter.

Although international law in many ways resembles a national legal system, there are significant differences. In a national legal system, a central body makes laws, and disputes can be settled by judicial proceedings. In the international legal system, however, no central lawmaking body is empowered to make and enforce the rules. Instead, they are drawn up by states themselves in the form of international agreements known as treaties, or else they evolve as accepted custom. There is no automatic mechanism for settling disputes. Procedures range from informal negotiations between the states concerned, through mediation by a third party, to full-scale judicial proceedings before the International Court of Justice sitting at The Hague – but whatever the chosen method, the consent of the states involved is always required. There is no international policeman to deal with infractions; without consent, the law cannot be enforced – except, ultimately, by force of arms – if any state persists in breaking it. When the United Nations was founded, toward the end of World War II, it was hoped that the world body could take on the policing job, but because of disagreements among the major powers – particularly the United States and the U.S.S.R. – this has not happened.

Nonetheless, international law *is* a system of law, not simply of morality or courtesy. Such spectacular breaches as the Russian

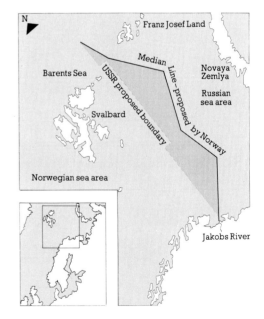

Frontiers at sea often cause friction between states with opposing objectives. Currently in dispute is the area of the Barents Sea off Norway's Svalbard, where Norway, claiming a 200-mile exclusive fishing and sea-bed zone, wants a different boundary line from the one proposed by neighboring Soviet Russia, whose interest in these waters is mainly strategic.

invasion of Czechoslovakia in 1968 or the American intrusion into Cambodia in 1970 should not obscure the fact that international law is every day being quietly observed around the world, helping to make relations among sovereign states run smoothly. But since it governs only the relations among states, it has no direct application to individuals or companies. In order to give such application to any provision of international law, a state must enact legislation that makes the provision part of its own law. Thus, for example, when international rules governing pollution from oil tankers became binding on Japan, they became binding on Japanese-owned tankers only because the national legislative body enacted laws making those rules part of the law of Japan.

As was pointed out earlier, there is no *international* legislative body; the rules of international law are found in custom and treaties. Customary behavior at sea becomes virtually obligatory when there has been a period of fairly consistent practice by most states in regard to a certain matter, and this used to be the source from which most of the law of the sea derived. Nowadays, however, custom is being largely replaced by treaties, for they provide a degree of certainty, speed, and flexibility that custom lacks.

Although treaties are often spoken of as "agreements" or "conventions," this does not mean they are not legally binding. All treaties are composed of sets of rules, and whether they are concluded by two states only (bilateral), or by states of one particular region, or by states on a general, worldwide basis (multilateral), they have real legal force. Normally a treaty is drawn up at an international conference (or, in the case

The International Court of Justice in session at the Peace Palace in The Hague. On the bench sit 15 judges, a registrar, and a deputy registrar. In the current court, the judges, who are elected for a 9-year period by the UN, come from Argentina, Dahomey, France, Great Britain, India, Japan, Nigeria, Poland, Senegal, Spain, Syria, Uruguay, the U.S.A., U.S.S.R., and West Germany. The court processes cases as shown in the diagram, right. Since its inception in 1945, it has given about 15 advisory opinions and has dealt with more than two dozen interstate cases, five of which have concerned marine laws. Most of its decisions have been accepted by the disputants, but it has no power to force compliance.

Types of case submitted:

A. Advisory opinions on points of law requested by the UN.
B. Disputes resulting from special agreements between states.
C. Disputes resulting from general agreements to which states concerned are parties.
D. Disputes not covered by B or C between states that have volunteered to accept the Court's decision as binding.

of a bilateral treaty, at a meeting between representatives of the two governments). The treaty is then signed by the parties involved. Signature alone does not ordinarily bind a state to the terms of the treaty but merely signifies broad agreement with those terms. To become bound usually requires a further act: ratification (i.e., official confirmation by the concerned governments). Most multilateral treaties stipulate a minimum number of ratifications before they assume the force of law, and then, of course, they are binding only on the states that have ratified. In the following pages we shall be examining some of the reasons why marine treaty-making is a slow and difficult process.

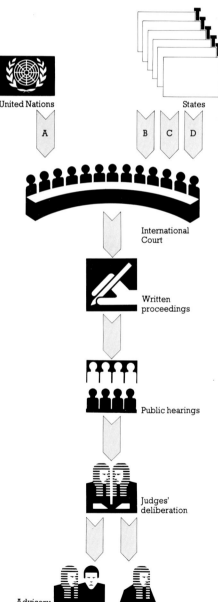

United Nations

States

A

B C D

International Court

Written proceedings

Public hearings

Judges' deliberation

Advisory opinion

Judgment

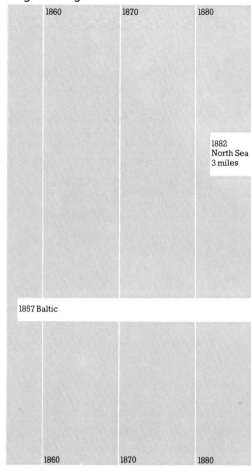

1860   1870   1880

1882
North Sea
3 miles

1857 Baltic

1860   1870   1880

Geneva 1975
New York 1976

New York 1973
Caracas 1974

Geneva 1960

Geneva 1958

Steps Toward a Package Deal

The international law of the sea has evolved over a long time. Its beginnings are shrouded in the mists of the early Middle Ages in Europe. Until the end of World War II it was almost entirely based on custom. The main features of this customary law were acceptance of the existence of narrow "territorial seas" (zones considered as belonging, without strings, to the coastal states they bordered) and the doctrine of "the freedom of the seas" (a virtually unregulated common right to use the high seas). A general acceptance of these two principles was adequate because the uses of the sea were few (chiefly navigation and fishing). There was no real conflict among maritime states in peacetime, no danger of overutilizing the sea's resources.

Soon after World War II, the United Nations decided that it would be useful to codify customary law in the form of one or more treaties, taking into account some postwar technological developments, particularly the possibility of getting oil and gas from the sea bed and the potential danger of overfishing as a result of improved fishing techniques. Accordingly, a United Nations Conference on the Law of the Sea was held in Geneva in 1958. The conference succeeded in drawing up four conventions, one dealing with questions concerning territorial seas and their contiguous zones, one with the high seas, a third with fishing and the conservation of living resources, and a fourth with the continental shelf. Each of these conventions has been in force since the early 1960s and has been ratified by about 50 of the world's nearly 150 states. But the conference (along with another one held two years later) was not completely successful, for the four treaties, although meaningful enough in general terms, left vital problems unresolved – for instance, the breadth of the territorial sea and the question of special fishing rights for coastal states beyond their territorial seas.

By the end of the 1960s, the failure of the two Geneva conferences to agree on those important matters had led to a proliferation of claims to all kinds of zones. Claims to the territorial sea varied in breadth from three miles (mainly the West European states, Japan, and the United States) through 12 miles (some 55 states in different parts of the world) to 200 miles (mainly South American states). In addition to a territorial sea, many states asserted their rights to exclusive fishing zones, ranging from 12 to 200 miles. Some also proclaimed special zones for other purposes, including the prevention of smuggling (for example, Denmark, South Africa, the United States) and the enforcement of pollution control (Canada, Oman, and Great Britain).

There have been two other basic grounds for dissatisfaction. First, many states have felt that the laws as codified by the UN are already out of date or inadequate for dealing with the fruits of modern technology – such new uses of the sea, for instance, as the dredging up of manganese nodules from the deep-sea bed, or fish farming, or the erection of artificial islands, or the large-scale dumping of pollutants. Conventional uses of the sea have also changed in degree; ships have become larger and less maneuverable, traffic in busy waterways has increased, and more dangerous cargoes (oil, chemicals, liquid natural gas) are being carried. Fishing techniques have been refined, too – particularly by the Japanese and Russians – to a point where practically all edible food can be vacuumed out of whole areas of the ocean. And increased and diversifying use of the sea is likely to mean increasing conflict between different uses – for example, offshore drilling rigs can interfere with shipping, or

fishing nets can catch on pipelines. In short, the *laissez-faire* concept that underlay the development of international marine law up to the time of the Geneva conferences of 1958 and 1960 needs to be replaced by planned marine management.

The other major complaint comes from the developing countries. Many of them were not yet independent at the time of the conference and so were not represented there. They have long felt that existing conventions meet neither their needs nor their aspirations.

In December, 1967, the United Nations set up a committee to study the question of the use of the deep-sea bed. By 1970, after this sea-bed committee had been at work for three years, there was a general feeling that many other aspects of marine law ought to be studied as well; and so the General Assembly adopted a resolution calling for a conference to be held in 1973 for the purpose of drafting a treaty dealing with virtually all aspects of the law of the sea. The sea-bed committee's mandate was widened so that it became, in effect, the preparatory committee for the conference. The first session of the Third Law of the Sea Conference was held in New York in December, 1973. This was followed by a meeting in Caracas, Venezuela, in the summer of 1974, and thereafter by annual two-month-long sessions in different parts of the world.

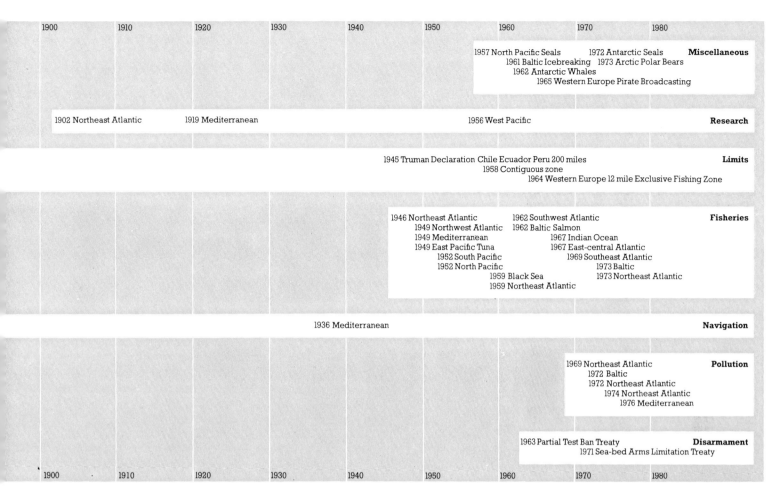

Timeline (1900–1980):

Miscellaneous
1957 North Pacific Seals
1961 Baltic Icebreaking
1962 Antarctic Whales
1965 Western Europe Pirate Broadcasting
1972 Antarctic Seals
1973 Arctic Polar Bears

Research
1902 Northeast Atlantic
1919 Mediterranean
1956 West Pacific

Limits
1945 Truman Declaration Chile Ecuador Peru 200 miles
1958 Contiguous zone
1964 Western Europe 12 mile Exclusive Fishing Zone

Fisheries
1946 Northeast Atlantic
1949 Northwest Atlantic
1949 Mediterranean
1949 East Pacific Tuna
1952 South Pacific
1952 North Pacific
1959 Black Sea
1959 Northeast Atlantic
1962 Southwest Atlantic
1962 Baltic Salmon
1967 Indian Ocean
1967 East-central Atlantic
1969 Southeast Atlantic
1973 Baltic
1973 Northeast Atlantic

Navigation
1936 Mediterranean

Pollution
1969 Northeast Atlantic
1972 Baltic
1972 Northeast Atlantic
1974 Northeast Atlantic
1976 Mediterranean

Disarmament
1963 Partial Test Ban Treaty
1971 Sea-bed Arms Limitation Treaty

To the layman, this must seem a very cumbersome way to make new laws, but the slow progress can be explained by a number of factors. First, consider the vastness of the subject matter: the Law of the Sea Conference is trying to make or revise rules that would regulate every activity, actual or potential, on, in, or under the sea. Secondly, there are about 150 participating states (compared with only 85 at the first two conferences nearly 20 years ago). These states have differing economic, military, and political interests; it is not easy to draw up laws on which they can agree. Thirdly, it has been decided that the final conventions are to be embodied in a single treaty – a giant package deal – and this requires the ability to accept compromise from all participants. Finally, this has been the first standing international lawmaking conference ever. Thus, novel problems of organization and procedure have arisen, and solutions have to be worked out cautiously, without the benefit of earlier experience.

Meanwhile, international law has not stood still. A good deal of bilateral, regional, and specialist lawmaking has been going on apart from the worldwide efforts of the United Nations. On the bilateral level, there have been many agreements on boundaries between the territorial seas or continental shelves of neighboring states. A number of regional agreements on pollution and fishing have been concluded. And the UN's specialized agency for shipping matters, the Intergovernmental Maritime Consultative Organization, has been responsible for drawing up treaties dealing with pollution from ships, the regulation of navigation, and the safety of shipping. It should be emphasized, however, that such legislation, though piecemeal, complements rather than competes with the efforts of the long-standing UN Conference on the Law of the Sea.

Why some sort of marine law "package deal" seems increasingly necessary. Despite the apparently cumbersome process of reaching agreement in international conflicts over use of the sea, more than 30 successful interstate regional agreements have been reached. Such settlements would complement, but they cannot replace, an internationally accepted Law of the Sea.

Secretary General Kurt Waldheim addresses the UN Conference on the Law of the Sea at its 1976 session in New York. By now such delegates as Sri Lanka's H. S. Amerasinghe (center) and Australia's David Hall are fully prepared for long debates without too many surprises.

Conflicting Interests

To function satisfactorily, a law must reflect competing interests within the society in which it operates. An obvious example of this on the national level is labor law, which is often the product of the conflicting views of employers' organizations and trade unions. In this respect, international law is no different; in fact, its need to reconcile opposing interests is even more obvious.

Before 1945, as we have seen, the international law of the sea was based on the doctrine of the freedom of the seas, and this doctrine was largely developed by the military and colonial powers of Western Europe because it best suited their interests: the promotion of seaborne trade between Western Europe and the rest of the world, and the maintenance of maritime communications between the powers and their colonies. Since 1945, this simple situation has become complex. The number of sovereign states has increased dramatically with the emergence into independence of about 80 developing countries (so that there are now nearly 150 states, of which 125 have a coastline). And, at the same time, new and increased uses of the sea have produced numerous different interests. Let us examine some of the chief interest groupings, particularly as they have emerged in preparation for and during the Third Law of the Sea Conference.

To begin with, there are the two traditional groupings to be found in all UN gatherings: East vs. West (i.e., the Soviet Union and its allies vs. Western Europe and North America); and, roughly, North vs. South (i.e., developed vs. developing countries). The first is not particularly significant in the context of the Law of the Sea Conference, but the second becomes important whenever matters arise that concern the distribution of the world's wealth.

Next there are two line-ups based on military interest. Here the superpowers, Russia and the United States, are on the same side of the fence; for strategic reasons, they and their more immediate allies want the greatest possible freedom of movement for surface warships and submarines. This attitude is generally distrusted by the non-aligned states, which would like to see some restrictions on such freedom.

Then there are groupings that stem from economic interests. In questions dealing with fishing, two groups are always potential antagonists. On the one hand, there are many states (the developing countries, Canada, Iceland, and Norway) that do most of their fishing close to their own coasts; on the other hand, such states as Japan, the Soviet Union, the East European states, and some West European states – for instance Belgium, West Germany, and Spain – usually fish off the coasts of other states, far from home. Where sea-bed minerals are concerned, states that have broad continental shelves with exploitable resources (Australia, Norway, Britain, and the U.S., for example) have a special interest. In the case of manganese nodules, the states (mainly developed countries) that are heavy consumers of minerals obtained from the nodules (cobalt, copper, manganese, and nickel) are bound to have differing demands from those of the land-based producers of these minerals (such as Chile, Cuba, Zaire, and Zambia), for the terrestrial producers might be adversely affected by competition from ocean production. There are also conflicting interests in regard to shipping. States such as Liberia, Japan, or Greece, all of which have large merchant fleets, may not worry much about the coastal effects of pollution from passing ships, but this is a very troublesome matter for such states as Denmark, Indonesia, and Malaysia, which border on busy waterways.

As to marine pollution in general, there are notable differences of view and interest. Some states (Japan is an example) take advantage of the sea as a cheap dumping ground, thus cutting the costs of their industry. Some states (Brazil, for instance) simply take no interest in the problems of marine pollution. A few (such as Great Britain, Canada, and Norway) are actually trying to do something to remedy the situation.

Finally, geography creates its own groupings. States with long coastlines fronting the open oceans (South and North American states, Norway, Australia, etc.) stand to gain from laws that extend the rights of coastal states seaward. States with short coastlines in proportion to their land area (the classic example is Zaire) or those that border such enclosed or semi-enclosed seas as the Baltic, Mediterranean, and North Sea are in a different category. And the landlocked states (of which there are about 25) naturally resent missing out altogether on the wealth of the sea.

A very simple example (based on, but not identical to, recent proceedings at the UN conference) may show how some of these interests interact. Let us imagine that at an international lawmaking conference, Oceania, a state with a long coastline fronting the open sea, proposes that every state should have the right to claim a special 100-mile resources zone off its coast. This proposal

Land producers of minerals naturally fear the competition once sea-bed mining by industrial nations gets rolling. Below: An over-all view of the principal exporters of substances found in undersea manganese nodules. Would a glutted market send prices tumbling?

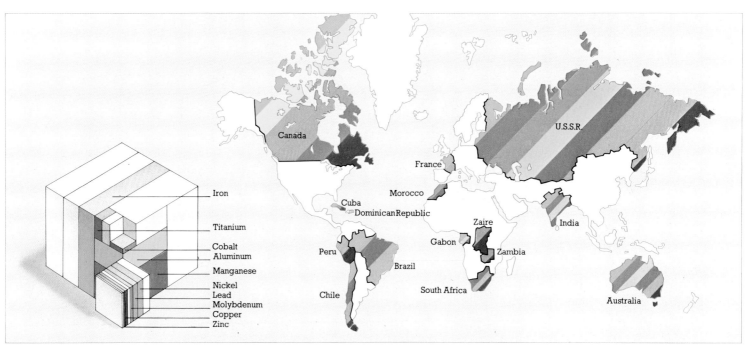

is welcomed by Industria, a distant-water fishing state with a broad continental shelf, which suggests, however, that the resources zone be limited to sea-bed resources. Pescia, a developing country with a growing coastal fishing industry, rejects Industria's suggestion, arguing that the zone should be for both sea-bed and fishery resources, and Industria agrees – provided that in cases where a coastal state does not take all the harvestable fish catch off its coast, distant-water fishing states shall be able to take the remainder. At this point, the delegate of Montania, a landlocked country, leaps to his feet and says that landlocked countries should be given an opportunity to exploit the sea's resources. The conference must now try to accommodate his views, along with alternative suggestions from states that are worried about how the basic proposal might affect the free passage of shipping.

This example is flawed in that it may imply that states have just a single interest to promote. In fact, most have several different – sometimes, even, conflicting – interests. West Coast American fishermen, for instance, do distant-water fishing, whereas East Coast fishing is largely coastal; so the United States has the interests of both a coastal and a distant-water fishing state. Similarly, Britain has one of the world's busiest waterways, the English Channel, and so it has interest in both noninterference with, and the orderly regulation of, navigation. Such situations only add to the complexity of international negotiations.

The diagram below shows line-ups of NATO and Warsaw Pact naval forces, with the official 1976 figures for numbers of operational craft. Obviously, both America and Russia have a vested interest in maintaining the greatest possible freedom of movement for warships.

Vigilance is an overriding concern of the superpowers, and international laws of the sea cannot govern their need to use any and all waters for constant surveillance of each other's marine activities. Above, left: This 15-ton U.S. Navy listening device for tracking the movements of other nations' submarines lay on the Caribbean Sea floor, nearly 700 meters below the surface, throughout the late 1960s and early '70s. Above, right: Two Russian "trawlers" shadow the commando carrier HMS *Hermes*, which is taking part in a NATO exercise – a familiar occurrence during Western naval maneuvers. The need to maintain fleets of warships in distant waters is one good reason why states with such requirements continue to resist demands for further restrictions on the freedom of the seas. Because of the vulnerability of coastal naval bases – and their scarcity in this post-colonial age – servicing and refueling are often done at sea. Below: A cruiser takes on fuel from a 30,000-ton tanker. The new breed of support ships and tankers would be prime targets in the event of war. And servicing at sea – especially fuel transference – can become a pollution hazard in rough weather.

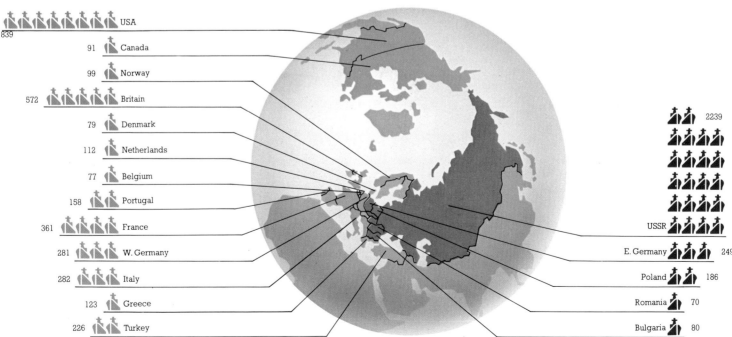

839 USA	
91 Canada	
99 Norway	
572 Britain	
79 Denmark	
112 Netherlands	
77 Belgium	
158 Portugal	
361 France	
281 W. Germany	
282 Italy	
123 Greece	
226 Turkey	

USSR 2239
E. Germany 249
Poland 186
Romania 70
Bulgaria 80

How Wide Are Coastal Waters?

Most marine activities take place in coastal waters, and most of the sea's economic resources are found there, too. So it is not surprising that states place their greatest emphasis on legal regulations that apply to such areas. Over the past fifty years there has been a steady trend toward coastal states' claiming exclusive rights to increasing widths of coastal zone for an increasing variety of functions. Originally, however, there was only one kind of zone: the territorial sea, which was considered part of the state's territory, and within which foreign states had no other right than that of innocent passage for their ships. By the eighteenth century the generally accepted breadth of the territorial sea was three nautical miles – then the maximum range of a cannon fired from a shore battery. As this cannon-shot rule indicates, the prime purpose of the territorial sea was to provide military security.

It was only as states became aware of the economic potential afforded by exclusive ownership of the territorial sea that they began to claim breadths of more than three miles. In the forefront of this movement, which gained real momentum after World War II, was Peru, which set the fashion for other South American states in 1947 by claiming 200 nautical miles. She chose the figure of 200 miles because this took in the Humboldt Current, the home of the anchovy, the main species caught by Peruvian fishermen. Since then, the most popular figure, claimed by about 55 states in different parts of the world, has become 12 miles. But a dozen or so African states claim varying breadths from 12 to 200 miles, and the old three-mile rule is still adhered to by the states of Western Europe, Japan, the U.S., and a few Commonwealth countries, which favor a narrow coastal zone for all states in order to protect their own distant-water fishing and navigation interests. Meanwhile, efforts to overcome some of the drawbacks of a narrow territorial sea have resulted in the establishment, farther out, of various functional zones, which are not regarded as part of a state's territory and in which the state asserts only limited rights, usually of an economic or security nature. The most important such zones are four: an exclusive fishing zone, the continental shelf, the contiguous zone, and antipollution zones.

An exclusive fishing zone is an area beyond the territorial sea where the coastal state proclaims that it, and it alone, may regulate fishing. This does not necessarily mean that it claims the sole right to fish there. Other countries that have traditionally fished the area are often allowed to continue to do so in a manner normally regulated by bilateral agreements. Virtually

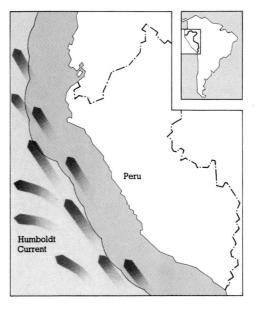

Because of her narrow continental shelf, Peru was the first country to claim (in 1947) an exclusive 200-mile fishing zone. Her main objective was to protect her anchovy fishing in the Humboldt Current, which, as indicated here, sweeps out westward into the Pacific.

all states that retain only three-mile-broad territorial seas are now claiming exclusive fishing zones. In most cases, the claim has been for 12 miles (i.e., 9 miles beyond the outer limit of the territorial sea); the exceptions in mid-1976 were Iceland, which had claimed an exclusive fishing zone of 200 miles since 1975, and the U.S., which was expected to adopt the 200-mile limit in early 1977. Exclusive fishing zones have also been claimed by some of the states that claim 12-mile territorial seas – for example, Oman and Pakistan (both 50 miles), Morocco

A helicopter hovers over Rockall Island during servicing operations on the navigation beacon installed in 1972, several years after Britain had annexed this remote, uninhabitable rocky tip of a sunken mini-continent. As shown in the map, Britain claims full rights to the sea bed for a 200-mile radius, but Iceland, Ireland, and Denmark dispute the claim. Rockall's commercial significance lies in the surrounding waters, which are rich in fish and possibly oil.

(70 miles), Senegal (110 miles), and Costa Rica (200 miles).

The second type of functional zone is the entire continental shelf. Claims to the natural resources of the shelf began in 1945 with a Proclamation made by America's President Truman. This was followed by many similar claims, all of which received international recognition with the adoption of the Convention on the Continental Shelf at the original UN Conference on the Law of the Sea. And that same conference legalized a third kind of functional zone by adopting a Convention on the Territorial Sea and the Contiguous Zone, which gives a state the right to enforce its fiscal, sanitary, immigration, and customs laws in an area – the "contiguous zone" – extending up to 12 miles from its coast. (This provision, obviously, has practical application only for states that claim territorial seas of less than 12 miles.)

A more recent concern has been the question of pollution: should states be authorized to take action in nonterritorial waters against foreign ships that pollute or threaten to pollute their coasts? International agreements of 1969 and 1973 give states certain rights of intervention even on the high seas in such cases, but some countries feel that the agreed-on rights are too limited and have passed stronger laws of their own. Thus, Canada has claimed very extensive rights to deal in Arctic waters with foreign vessels that might pose pollution risks; she even asserts the right to specify vessel design. And Oman and Britain have also taken unilateral – though less drastic – action.

How far, then, can all the diverse claims be justified or regulated by international law? As we have seen, those relating to the continental shelf, the contiguous zone, and pollution control are at least based on existent conventions; and in some cases, notably the 12-mile exclusive fishing zone, there has been enough consistency of state practice to establish a rule of customary law. In other cases, though, there is no uniformity of practice, and so no definite rule of custom has

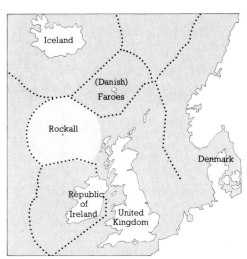

The extent of territorial seas and other zones is conventionally determined by measuring out a given distance from a coastal baseline. Such baselines are generally drawn at the low-water mark. But where the shoreline is deeply indented with bays or estuaries, or where it is fringed with islands or harbor works, the conventions apply as illustrated here.

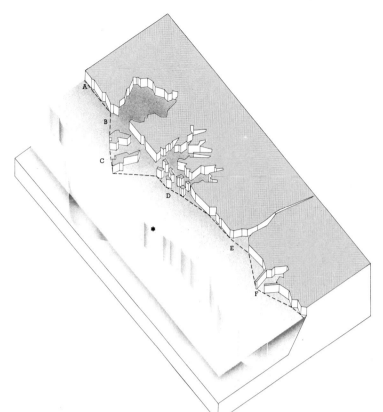

*Outer limit of the territorial sea.
A. The low-water mark is the normal baseline for measuring a territorial sea.
B. A baseline may be drawn across the mouth of a bay when (1) the area of the bay is greater than that of a semicircle drawn on the line across the mouth, and (2) this latter line is less than 24 miles long.
C. An island generates its own territorial sea.
D. When a stretch of coast is deeply indented and fringed with islands, baselines are drawn by connecting the outermost points of the configuration.
E. The baseline is drawn across the mouth of a river that flows directly into the sea.
F. Harbor works are considered part of the coast for baseline purposes.

emerged. Some of the more extreme claims, such as Canada's Arctic-pollution legislation, are clearly contrary to international law because they encroach on other states' freedom of the seas.

By 1976 it had become obvious that the time was ripe for an end to haggling, and that the Law of the Sea Conference must at least devise an acceptable compromise among divergent views about coastal zones. The likeliest such compromise involved declaration of a 12-mile territorial sea for all coastal states, beyond which there would be an exclusive economic zone stretching up to 200 miles from the coast. Within this zone (which has come to be known as the EEZ) the coastal state would control all resources, fish as well as sea-bed minerals, while the ships of foreign states would continue to enjoy freedom of navigation. And noncoastal states would be allowed to fish in the EEZs of their neighbors. This EEZ concept should be an important means of regulating marine activities. Two-hundred-mile EEZs established on a worldwide basis would account for something like 40% of the total surface area of the sea, would include most of the world's sea-bed hydrocarbons and commercial fisheries, and would cover many of the principal shipping routes.

Although the idea of exclusive economic zones originated with the developing countries (and largely, no doubt, in a spirit of self-interest), few of them would be among the immediate beneficiaries. The states that would stand to gain most are those that front the great oceans; and as a glance at an atlas will show, many such states are developed, not developing, countries. The ten leading potential beneficiaries are, in order of magnitude of area gained: the United States, Australia, Indonesia, New Zealand, Canada, the Soviet Union, Japan, Brazil, Mexico, and Chile. At present, moreover, few of the developing countries have the technological or economic capacity to exploit their EEZs, nor could they police them adequately. Still, the idea promises an end to dissension over the extent of coastal zones, and this in itself is an end worth pursuing.

It is very difficult to draw up internationally acceptable territorial zones for states such as those of the Bahamian archipelago, whose collective land masses are much smaller than the surrounding area of sea. The big problems: to what extent should each state enjoy exclusive rights to the waters that separate it from the others; and how should baselines be drawn to ensure equitable distribution?

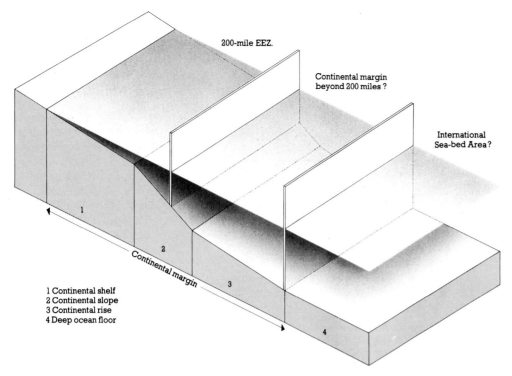

200-mile EEZ.

Continental margin
beyond 200 miles ?

International
Sea-bed Area?

Continental margin

1 Continental shelf
2 Continental slope
3 Continental rise
4 Deep ocean floor

Norway

Norwegian
zone

United Kingdom
zone

Danish
zone

Denmark

German
zone

Dutch
zone

Germany

England

The Netherlands

The numerical key to this diagram identifies
geographical sea-bed divisions. But geography
and economics are not synonymous; legal
boundaries as set by UN conferences may well
give coastal states control over 200-mile
economic zones and contiguous marginal zones.

Most continental-shelf boundaries in the North
Sea are, roughly, equidistant between states. But
when West Germany objected to the median-line
division (dotted lines on the map), the
International Court gave her a larger share (as in
the shaded area).

The Continental Shelf

The first drilling for oil on the bed of the
open sea began in America shortly before
World War II. Although this was only in shallow water close to shore, its potential led
President Truman to issue a Proclamation in
1945 of America's exclusive ownership of
the hydrocarbon resources of all continental shelves adjoining the American coast.
Many other coastal states soon made similar
claims, justifying them on the grounds not
only of the close relationship of the continental shelf to the land, but also of requirements for military security. In granting
international recognition to such claims, the
UN's Convention on the Continental Shelf,
adopted in 1958, set forth a fairly comprehensive set of rules governing the nature
and extent of states' rights. For a number of
reasons, some of those rules already need
to be revised.

Take, for instance, the most basic of issues:
a truly workable definition of the continental shelf itself. Since it varies in
different parts of the world in both depth
and width – for example, the shelf is almost
nonexistent off the Pacific coast of South
America, whereas it extends for several
hundred miles off Australia and Argentina –
a general definition based on the continental
shelf's natural features would hardly seem
fair to "deprived" states. So, after much discussion, the 1958 conference reached a
compromise that defines the continental
shelf for legal purposes as "the sea bed and
subsoil of the submarine area adjacent to the
coast but outside the area of the territorial
sea, to a depth of 200 meters or, beyond that
limit, to where the depth of the super-ad-
jacent (i.e., overlying) waters admits of the
exploitation of the natural resources of the
said area." In 1958 that seemed fair enough
to the nonindustrial states, for it was not

technically possible to drill in more than
200 meters of water. Since then, however,
drilling has taken place in much greater
depths, and offshore exploitation of oil and
gas has increased enormously. As the developing countries see the situation, the
seemingly open-ended definition spelled
out nearly two decades ago could permit
technologically advanced countries to extend their continental shelves seaward until
they met in the middle of the oceans. To
keep this from happening, the current Law
of the Sea Conference must try to establish
an unambiguous outer limit to the shelf.

When the proposal for a 200-mile exclusive economic zone (as discussed in
the preceding pages) appeared to become
generally acceptable, it seemed that the
EEZ concept ought to solve the problem.
But the continental margins of a large and
important group of states – among them
Argentina, Australia, Canada, Great Britain,
India, Norway, and the United States –
stretch beyond 200 miles; and because the
continental margin includes not only the
continental shelf but also the probably
hydrocarbon-rich slope and rise that connect the shelf to the deep-sea bed, these
states have insisted on having control of the
outer part of the margin. It seems likely that
this will eventually be agreed to – but only
on condition that a share of the revenue derived from any production of resources
from the continental slope will be given to
the developing countries.

A further uncertainty about the extent of a
state's continental shelf arises where it ad-
joins the shelves of neighboring states. The
1958 convention provides that in such cases
the boundary is to be the line of equidistance between the coasts of the states
concerned. There are often special circumstances, though, such as islands or sea-bed
depressions that make the line of equidistance either difficult to ascertain or unfair.
In such cases, the neighboring states have
to do their best to agree on a different line.
The problems involved in drawing such
boundaries tend to be greatest in shallow
enclosed or semi-enclosed seas, which are
often bordered by several states. Still,
boundaries have been satisfactorily drawn
for virtually the whole of the North Sea, for
much of the Persian Gulf, and for some parts
of the Mediterranean and Baltic. Altogether,
in different parts of the world, about 35 sea-
bed boundaries have so far been agreed
on, and the great majority are based on the
equidistance principle.

So much for the *extent* of the continental
shelf. Now what about the nature of a state's
rights in the area? The natural resources
over which a coastal state has exclusive
rights, according to the Continental Shelf
Convention, include not only hydrocarbons
and such other minerals as tin, sand, and
gravel, but also of living resources that are

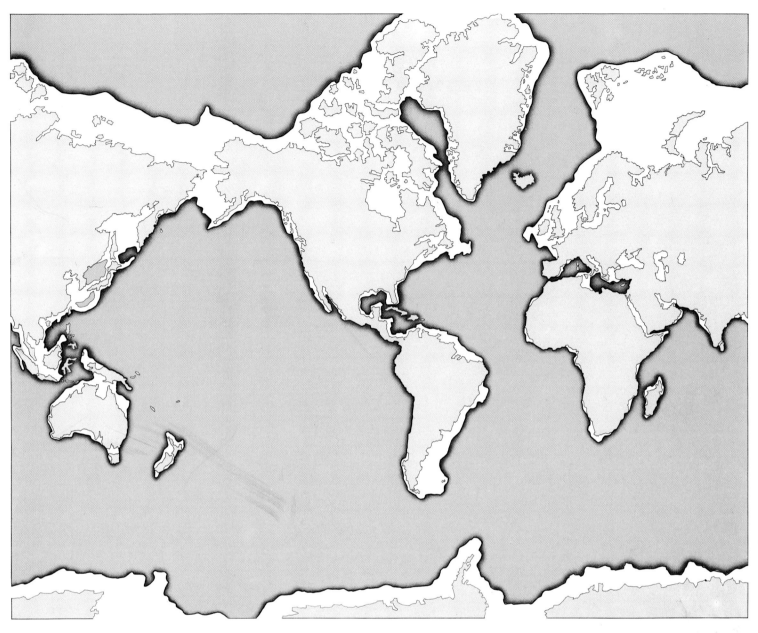

attached to or live on the sea bed, such as crabs, mussels, and seaweed. The state's rights on the shelf are strictly limited to the sea bed and its subsoil, however; they do not affect the legal status of the waters above, which therefore remain open to, for example, other states' fishermen (as was pointed out in an earlier chapter of this book). The possibilities for conflict in such a situation are obvious.

Many of the 1958 regulations, though, are holding up well. Take, for example, the legal problems that can easily arise within a small international community living on a large drilling rig far out on the continental shelf (and there are increasing numbers of such communities every year). What would happen if, say, an American murdered a Frenchman on a Norwegian rig in the middle of the North Sea, or if an Englishman on the rig suffered an injury in the course of his work because of the Norwegian owner's

negligence? Fortunately, the 1958 convention provides a simple solution to such problems by stipulating that the coastal state's full jurisdiction extends to rigs and similar structures on its continental shelf. In other words, the rig is treated as part of the state's land territory. Thus, in our first example, a helicopter would fly a Norwegian policeman out to the rig; he would be empowered to arrest the American and take him to the mainland, where he would be tried by Norwegian courts. In our second example, the Englishman could claim compensation under Norway's industrial-injuries legislation.

Another problem raised by the existence of rigs on the continental shelf is their possible interference with shipping. Here again the convention has provided for most of the foreseeable contingencies. Rigs, to begin with, must not "be established where interference may be caused to the use of rec-

As the white areas of this map show, continental shelves vary enormously in width, and it is the developed nations that are most generously endowed. The task of agreeing on international legislation is made quite a bit more difficult by the fears of less wealthy states that they may therefore get less than a fair share of profits from the shelves' potential riches.

ognized sea lanes essential to international navigation." Wherever rigs are set up, "due notice" has to be given (the convention does not, unfortunately, specify what constitutes "due notice"), and such warning devices as lights and foghorns must be installed. Moreover, to minimize the risk of collision between a ship and a rig, all rigs are surrounded by 500-meter safety zones inside which no shipping other than supply or similar vessels may enter. Clearly, no state can resent rules of this sort, even though they slightly limit the state's *total* control of its continental shelf.

Who Owns the Depths?

As we saw in the chapter entitled "The Ocean's Resources," many people's imaginations were fired in the 1960s by reports of great mineral riches lying in the deep-sea bed. It was mainly the developing states' fear that these riches would be grabbed by technologically advanced states under either the pretext of extending their continental shelves or the doctrine of the freedom of the seas that led to the establishment of the United Nations Sea-bed Committee in 1967. A "Declaration of Principles" eventually produced by the committee and approved by the General Assembly in 1970 declares that beyond the limits of national jurisdiction (which are not defined) lies an area of the oceans that no state may legally appropriate. This area – "the common heritage of mankind" – is to be subject to regulations drawn up by an international body, and its resources are to be exploited for the benefit of all mankind.

The elaboration of this "Declaration of Principles" is one of the major tasks before the Law of the Sea Conference. The first requirement, of course, is to fix the limits of the international sea-bed area, and this can be done only after the limits of national jurisdiction have been set. Since it seems likely that the agreed-on outer limit of coastal states' jurisdiction is to be 200 miles or the edge of the continental margin, whichever is larger, and since all known hydrocarbon deposits are in the continental margin, it appears probable that the only substantial sea-bed resource of the international area is manganese nodules. And the question of mining rights for the nodules has been a particularly troublesome one – even though, as explained earlier, the difficulty of exploiting them means that the deep-sea-bed area is unlikely to become the El Dorado it was once thought to be.

Opinion as to the nature and functions of any international authority that may be given the job of administering the area has been rather sharply divided between developing and developed states. The former originally proposed that the mining of manganese nodules should be carried out directly by such an authority; they felt that this would be the only effective way for them to acquire the revenue and the technological know-how that they wanted, and they distrusted the large American, Japanese, and West European consortia that were working out a technology for nodule mining. The developed countries, on the other hand, thought that the all-powerful authority proposed by the developing countries – certainly the most ambitious international organization ever – was impractical and would effectively result in the nodules' not being exploited at all. Instead, they wanted an authority empowered merely to grant mining licences to states or companies – which would mean, initially at any rate, that the existing consortia would go on doing all sea-bed mining.

Trying to find a compromise between these extreme views has, not unnaturally, been very difficult. By 1976, though, such a compromise did seem to be emerging. This would permit nodule mining by states and companies as well as by the international sea-bed authority, but with over-all regulatory powers in the hands of the authority. In its initial phases, the international body would be unlikely to undertake any mining on its own, but it would work closely with the consortia through such means as service contracts or joint ventures.

How an international sea-bed authority would be organized if current proposals were to be accepted by the UN Law of the Sea Conference. The authority would theoretically exploit the resources of the deep-ocean floor for the benefit of all mankind; but no such body seems likely to materialize – at any rate, not in this cumbersome form.

The problem of reaching agreement is compounded, however, by the fact that nodule mining, when it eventually shifts into high gear, may severely cut into the profits of the land-based producers of the principal minerals found in manganese nodules. It is in the developing countries that most such minerals are mined. But the states now engaging in pioneer dredging operations argue that sea-bed production will always be more expensive than land-based production, and so the current producers have nothing to fear from future competition. It is far from easy to resolve this argument, since nodule-mining technology is still in its early stages and no one can predict the ultimate economic impact of sea-bed production.

Meanwhile, the experimental dredging being done by the various consortia (all of which are privately financed and run) can be legally justified as permissible research under the freedom-of-the-high-seas doctrine. It could be argued too that, as matters now stand, the actual commercial mining of nodules would be justifiable under the same doctrine. Thus, if money-making mining were to begin before the Law of the Sea Conference finished its work – which is unlikely – or if the conference failed to agree on an international sea-bed regime, the developing countries would have little hope of getting hold of a share of the profits.

When the idea of an international sea-bed authority was first put forward, its proponents claimed that it would make a major contribution to the redistribution of the world's wealth. This now seems unlikely. What has gone wrong – apart from the unfortunate fact that there are likely to be far fewer sea-bed resources than was originally thought and that, for the foreseeable future, manganese nodules seem to be the only commercially exploitable resource? The main trouble is that because coastal-state demands have pushed the limits of international waters farther and farther out, those waters will not cover the hydrocarbon resources of the outer continental margin, as once seemed probable. An evident contradiction is worth noting in this context: it is largely the developing states that have clamored for extensive coastal rights, while the same states have been the keenest advocates of a strong international regime; in short, they appear to have fallen victim to their own rhetoric about economic independence. In more concrete terms, the Latin Americans, the staunchest supporters of extensive coastal states' rights, have to support Asian and African demands for a strong international deep-sea-bed authority in return for the others' support of coastal rights. (Out of this bargain the Latin Americans have come off best.)

A further disappointment for those who anticipated a major redistribution of the world's wealth is that the unexpectedly high

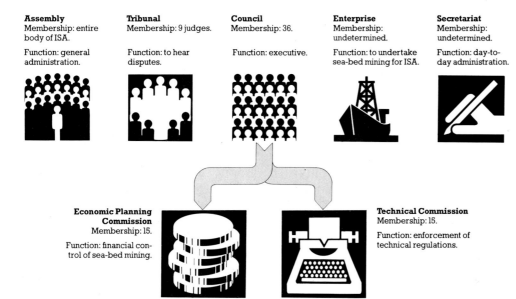

Assembly
Membership: entire body of ISA.

Function: general administration.

Tribunal
Membership: 9 judges.

Function: to hear disputes.

Council
Membership: 36.

Function: executive.

Enterprise
Membership: undetermined.

Function: to undertake sea-bed mining for ISA.

Secretariat
Membership: undetermined.

Function: day-to-day administration.

Economic Planning Commission
Membership: 15.

Function: financial control of sea-bed mining.

Technical Commission
Membership: 15.

Function: enforcement of technical regulations.

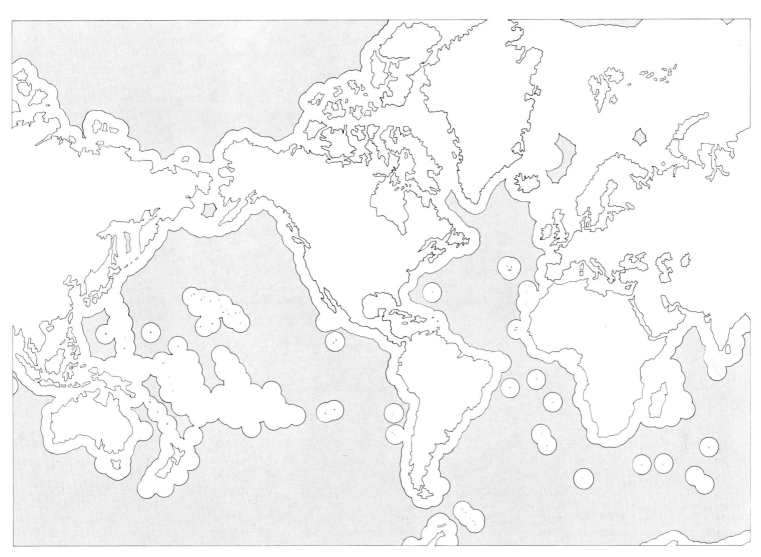

When an international sea-bed authority of some sort is finally set up, it will be expected to administer all areas of the world's oceans not included in the 200-mile-wide exclusive economic zones that will, as indicated in the above map, encircle virtually every land mass. Moreover, with a 12-mile territorial sea, the international authorities would also have to guarantee rights of passage though over 100 straits that have traditionally remained uncontrolled in peacetime, with the accompanying possibility of occasional political fireworks. Right: Some of the more important such straits are singled out here.

costs and technological difficulties of sea-bed mining will mean much smaller profits to be shared with the poorer countries. What money they might get would hardly be enough – even together with shared revenues from the hydrocarbon mining of the continental margin beyond 200 miles – to make any great impact on their economies. They will no doubt profit from picking up some useful technological knowledge through membership in an international sea-bed authority. But even that body will probably be less internationalist than its fervent supporters hoped in the late 1960s when they coined the ringing phrase "the common heritage of mankind."

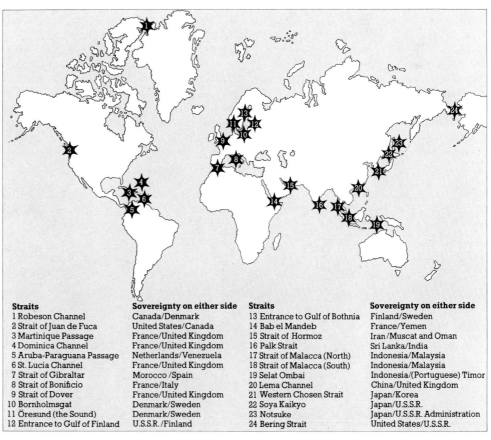

Straits	Sovereignty on either side	Straits	Sovereignty on either side
1 Robeson Channel	Canada/Denmark	13 Entrance to Gulf of Bothnia	Finland/Sweden
2 Strait of Juan de Fuca	United States/Canada	14 Bab el Mandeb	France/Yemen
3 Martinique Passage	France/United Kingdom	15 Strait of Hormoz	Iran/Muscat and Oman
4 Dominica Channel	France/United Kingdom	16 Palk Strait	Sri Lanka/India
5 Aruba-Paraguana Passage	Netherlands/Venezuela	17 Strait of Malacca (North)	Indonesia/Malaysia
6 St. Lucia Channel	France/United Kingdom	18 Strait of Malacca (South)	Indonesia/Malaysia
7 Strait of Gibraltar	Morocco/Spain	19 Selat Ombai	Indonesia/(Portuguese) Timor
8 Strait of Bonifacio	France/Italy	20 Lema Channel	China/United Kingdom
9 Strait of Dover	France/United Kingdom	21 Western Chosen Strait	Japan/Korea
10 Bornholmsgat	Denmark/Sweden	22 Soya Kaikyo	Japan/U.S.S.R.
11 Öresund (the Sound)	Denmark/Sweden	23 Notsuke	Japan/U.S.S.R. Administration
12 Entrance to Gulf of Finland	U.S.S.R./Finland	24 Bering Strait	United States/U.S.S.R.

Several "cod wars" between Britain and Iceland and periodic arrests by Ecuadorian authorities of U.S. tuna vessels fishing in Ecuador's claimed 200-mile territorial sea are dramatic symptoms of the major problem facing fishing today: how to regulate allocation of a resource threatened by over-exploitation. The general background of the problem is explained in chapter 5 of this book. Our chief aim here is to see what is being done – and what, unfortunately, is not being done – to solve it. Let us begin, though, by stressing the fact that when we speak of "overexploitation," what we mean is that the world's fishermen are catching more of the popular fish species than can be replaced by breeding, and that if the international community allows this situation to persist, it will result in the severe depletion of stocks – or even the extinction of species, as has all but happened with the Atlanto-Scandian herring and the California sardine.

Such marine fishing regulations as exist have mostly been developed since World War II and can be designated under two broad headings, national and international. National rules affect the international picture, of course, since they primarily involve either extensions of the territorial sea or the establishment of exclusive fishing zones. By these means, coastal states have sought to protect local fishermen from the competition of distant-water fishing fleets; where the claims have been sufficiently extensive, they have brought many stocks of fish within the claimants' exclusive regulation. International rules, on the other hand, are less likely to deal with states' claims than with their responsibilities. The main sources of international regulation are international fishery commissions – there are now about 25 of these – which deal either with a given area, such as the Baltic or Mediterranean, or with a particular species (of whales, seals, tuna, etc.). Each such com-

mission is established by agreement among the states engaged in the fishery in question.

The traditional regulatory techniques of a fishery commission consist of prescribing minimum mesh and fish sizes (which are designed to protect immature fish) and closed seasons and areas (to protect spawning fish). Such regulations make good sense, for overexploitation of certain stocks has led to the catching of immature fish, many of which haven't even spawned once; in the case of the Icelandic cod, for example, over 40% of those caught in 1950 were more than 10 years old, as against less than 2% today! But overfishing goes on in spite of the regulations. Not only is it difficult to apply measures regulating mesh sizes and closed seasons (which must be mainly enforced by the national authorities of the fishing vessels concerned), but these measures do not really come to grips with the central problem – how to control the *rate* of exploitation. To achieve such control, some fishery commissions (notably the two concerned with the North Atlantic, where the danger of overexploitation is particularly acute) have developed a quota system under which the total allowable catch for a given fish stock is decided on and divided among the participating states.

This method is by no means trouble-free, however. To illustrate some of the general problems of fisheries regulation, let's look at recent attempts to impose a quota system on the fishing of the northeast Arctic cod. Before 1974 this fishery was open to all-comers, except within the Russian 12-mile territorial sea and the Norwegian 12-mile exclusive fishing zone – areas that covered only a minute fraction of the area of the stock. In March, 1974, by which time the fishery was becoming noticeably over-exploited, the countries responsible for an overwhelming proportion of the catch – Norway, the U.S.S.R., and Great Britain – signed a quota agreement that established an annual total allowable catch (TAC) of 500,000 tons, a figure considerably lower

than the catch of previous years. The big loophole was this: if catches by states *not* party to the agreement ever exceeded 10% of the TAC, any of the signatories could withdraw. Only a few months later, in August of 1974, catches of third states had already exceeded 10%; and so the U.S.S.R., availing itself of the loophole, withdrew from the quota agreement.

Why did the catches of third states (which had always before taken well under 10% of the total) suddenly leap upward? There seem to be two main reasons. First, many fishermen had been driven to the northeast Arctic because most other grounds in the North Atlantic were already being regulated – through quotas established since 1972 by the International Commission for the Northwest Atlantic Fisheries, through a quota agreement for the Faeroe Islands signed by all the states that fish in this area (Belgium, Britain, Denmark, France, West Germany, Norway, and Poland), and through Iceland's 1972 extension of its fishing limits to 50 miles. Secondly, these fishermen were keenly aware of the newly arranged Norwegian-Russian-British quota system, and so they hastened to fish the northeast Atlantic in order to lay claims to be included in any future agreement, even though most of them were doing hopelessly uneconomic fishing in those already busy waters.

In mid-1974, the Northeast Atlantic Fisheries Commission, which has responsibility for the northeast Arctic, was empowered by its member states to establish codfish quotas, and it therefore took over the job of setting quotas for the northeast Arctic cod. Because so many states had by now become engaged in the fishery, the TAC had to be raised from 500,000 tons to 810,000 for each of 1975 and 1976 – a much higher level than the one recommended by fishery scientists, who had suggested a *maximum* total allowable catch of 550,000 tons. Even with this dangerously high TAC, many countries were soon exceeding their quotas (the chief

Cod older than ten years in the Icelandic catch

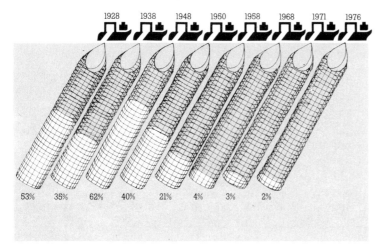

1928   1938   1948   1950   1958   1968   1971   1976

53%   35%   62%   40%   21%   4%   3%   2%

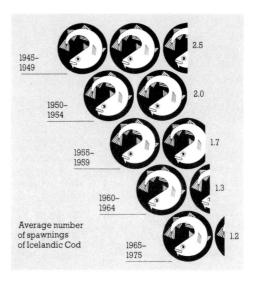

2.5

2.0

1.7

1.3

1.2

1945–1949

1950–1954

1955–1959

1960–1964

1965–1975

Average number of spawnings of Icelandic Cod

Two diagrams that reveal the results of overfishing as related to only one of many threatened species. Far left: How the catches of mature Icelandic cod have varied over the past half-century. Note the drastic decline in recent years. And, as illustrated in the other diagram, current catches of immature fish brought about by the need to maintain increasing catch levels create a vicious circle by sharply reducing spawning.

culprits are thought to include East Germany, Portugal, and Spain), and the total annual catch reached well over a million tons (which must be nearly as great as the catch would have been if no quotas had been established).

The trouble is that quotas can be controlled only through national authorities' accurately and honestly reporting the catch figures of their fishermen. Even where there is a system of international inspection, as there is for mesh sizes, it cannot always be effective. Russian fishermen have become notorious for "losing" their nets overboard when approached by international inspectors. The over-all situation, moreover, is so fluid that yesterday's regulations are outdated by tomorrow morning. For example, where northeast Arctic cod are concerned, the 1976 quota system is almost certain to undergo fundamental changes before the end of the decade. Norway and the U.S.S.R. will probably have established 200-mile exclusive economic zones off their coasts, thus embracing most of the fishery, and other states will have been forced to negotiate the quantity of cod that they may take from the EEZs of those two countries. A degree of international regulation will still be required, though, since some cod will be swimming in open Arctic waters.

This is likely to be the future pattern in most of the world: a broad zone of coastal-state regulation for most stocks, with international regulation of such oceanic species as whales and such highly migratory species as tuna. Distant-water fishing will probably decline drastically and be limited to either fishing under licence in other states' EEZs or entering into joint ventures with developing countries (as Spain and Japan, for example, have already done). The costs of distant-water fishing have increased so astronomically since 1973 as a result of higher-priced oil that some such fishing pattern should make better economic sense. It should make better food sense, too, for the developing countries will be able to get more protein from the sea, provided they're given the necessary technology (and here the UN's Food and Agriculture Organization has made a useful beginning in coordinating international technical cooperation). A more rational system of fisheries management may therefore be in the offing.

But we should not let optimism blind us to the possible pitfalls of increased coastal-state management and a consequently decreasing need for international regulatory bodies. Not long ago, after a survey of the world's fisheries, the Food and Agriculture Organization made public its conclusion that national regulation has been no more effective than international regulation in conserving fish stocks.

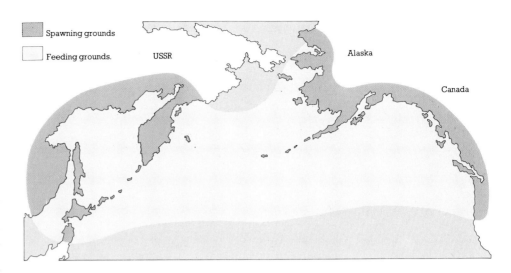

International disputes about overfishing are not confined to the Atlantic. Above: The spawning and feeding grounds of North Pacific salmon, stocks of which are being reduced, say Canadian authorities, because the Japanese catch immature fish before they can spawn.

Below: Even attempts to conserve stocks by applying quotas do not necessarily work. The figures for northeast Arctic cod show agreed-on quotas for 1974 and '75 as against actual catches throughout the '70s. A three-party agreement has been vitiated by too many loopholes.

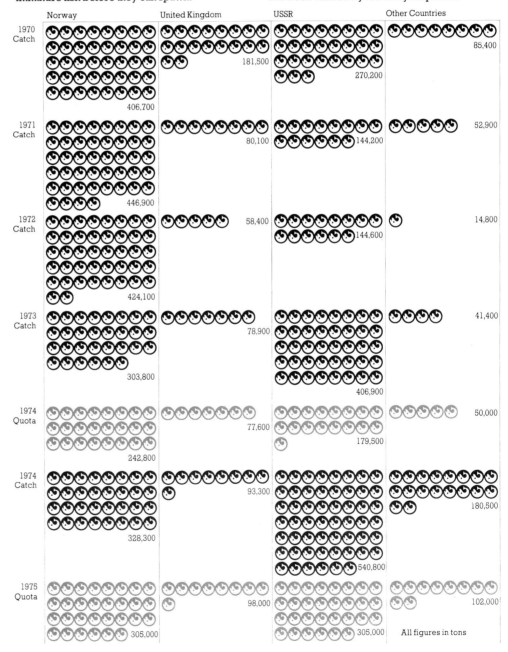

	Norway	United Kingdom	USSR	Other Countries
1970 Catch	406,700	181,500	270,200	85,400
1971 Catch	446,900	80,100	144,200	52,900
1972 Catch	424,100	58,400	144,600	14,800
1973 Catch	303,800	78,900	406,900	41,400
1974 Quota	242,800	77,600	179,500	50,000
1974 Catch	328,300	93,300	540,800	180,500
1975 Quota	305,000	98,000	305,000	102,000

All figures in tons

Traffic Rules

Along with fishing, navigation is the oldest use of the sea, and it remains one of the most important. Ninety-five per cent of all international trade is seaborne; on the military side, the uneasy balance of terror between the superpowers is largely sustained by nuclear submarines.

In territorial seas, as we saw earlier, foreign ships enjoy the right of innocent passage. In other words, they are free to sail across territorial waters as long as they threaten neither the peace, good order, nor security of the coastal state. This is true even of warships, with the proviso that submarines must navigate on the surface. The Third U.N. Law of the Sea Conference has not substantially altered this traditional understanding. For the most part, it has simply tried to spell out more precisely than before the powers of coastal states to prescribe compulsory shipping lanes and to deal with ships that abuse the right of innocent passage within their waters. One aspect of the situation, though, has caused major controversy: the question of straits.

If the territorial sea is extended to 12 miles on a worldwide basis, as now seems probable, some 120 straits will become either largely or fully state-owned. Many of them – for instance, Dover, Gibraltar, Bab el Mandeb, Hormuz, Malacca – have great strategic and commercial significance. The superpowers have been unwilling to see navigation through these straits subjected to the right of innocent passage, for this would mean that their submarines would be required to navigate on the surface, and their aircraft would be denied the right of overflight (since aircraft do not enjoy such a right in territorial seas). Furthermore, they have feared that coastal states might abuse the right of innocent passage so as to restrict the movements of warships, oil tankers, and other potentially dangerous vessels. So both the United States and the U.S.S.R. have pressed for entirely free transit through all straits. Despite strenuous opposition, this concept (somewhat watered down) has been grudgingly accepted by most of the participants in the Conference.

On the high seas, the doctrine of the freedom of the seas has always meant free and open navigation for all. However, as with most other uses of the sea, this unrestricted freedom now needs to give way to a more orderly system of regulation. One reason is that there has been a tremendous increase in the number of ships – total tonnage of world shipping trebled between 1955 and 1975! – creating serious traffic problems in our busiest waterways; some 500 ships a day now make use of the Strait of Dover, for instance. Ships have also increased in size. In the mid-1960s 100,000-ton tankers were the largest vessels afloat. By now 200,000-

tonners are common, and the biggest of tankers is nearly half a million tons. With size has come a lessening of maneuverability; a supertanker going at full speed takes several miles to stop. Ships are carrying more dangerous cargoes – oil, liquid natural gas, toxic chemicals – so any accident may become an international catastrophe.

A further complicating factor has been the rapid spread since World War II of flags of convenience: i.e., registration of ships in low-tax states with which the shipowners have no other connection. Popular flags of convenience include Liberia (which has the largest merchant shipping fleet in the world), Panama, and Honduras. The disadvantage to the international community of such arrangements is that the country of registration has no real control over its ships and may accept undesirably low standards of qualification for crews (it was chiefly a helmsman's failure to understand the workings of the automatic steering mechanism that caused the disastrous stranding of the Liberian *Torrey Canyon* off the British coast in 1967). Current studies of accident statistics show conclusively that ships flying flags of convenience suffer more than their fair share of losses at sea.

In recent years, a special United Nations agency, the Intergovernmental Maritime Consultative Organization (IMCO) has been trying hard to bring some order into the dangerous potential for chaos of traffic conditions. One of its most important accomplishments has been the gradual establishment of more than 70 traffic-control schemes that separate shipping into one-way-only lanes in territorial waters as well as the high sea. Observance of one-way-traffic rules is mandatory for states that have ratified the 1972 Convention on the International Regulations for Preventing Collisions at Sea – and by 1976 such states accounted for over 70% of the world's shipping. As in other matters affecting ships on the high seas, however, only the national authorities of a given ship can take action to see that it observes the rules. And some idea of the difficulties of enforcement can be gained by looking at the experience of the Strait of Dover, where there has been a well-publicized traffic-separation scheme since 1967. Now, about 10 years later, 5 to 10% of all ships still go the wrong way along one-way lanes. Since 1971, the British and French authorities have had radar screens monitoring shipping in the strait; when a rule-breaker is spotted, a helicopter or fast launch goes out and tries to identify it. Often, though, fog or darkness makes identification impossible, and even when ships are identified and national authorities informed, there is still no guarantee of corrective action against the transgressors.

Traffic-separation schemes are likely to be supplemented in the future by more

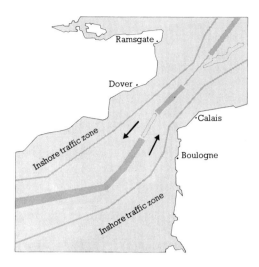

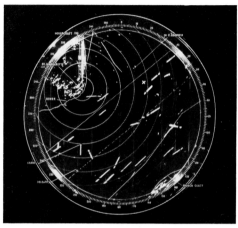

The traffic-routing scheme mapped out above has been in effect since 1967 for the 500-odd ships that use the Strait of Dover on an average day. Sandbanks in the middle of the Channel form a natural "traffic island." Whether voluntary or compulsory, such schemes are hard to enforce, and some 5 to 10% of vessels still appear to ignore this one. Several radar screens for monitoring traffic were established by British and French authorities in 1971. The view on this screen is a 20-minute time exposure; dots indicate position of ships when the exposure began, and continuous lines show their subsequent course. The vessel marked with an X is disregarding the scheme.

comprehensive forms of traffic control in very busy waterways or in areas with a high proportion of navigational hazards. For instance, the IMCO and the Baltic states have been working out a plan for a compulsory deep-water transit route for larger vessels through the Danish straits. The plan would also require virtually all ships in the Baltic to radio their positions at frequent intervals.

International efforts to improve the general safety of shipping are by no means a new development. Ever since the *Titanic* disaster in 1912, there have been various conventions on the safety of life at sea, and their regulations are constantly being updated or replaced in the light of technical advances. Among the things with which the conventions deal are standards for ship construction, radar, life-saving and fire-fighting

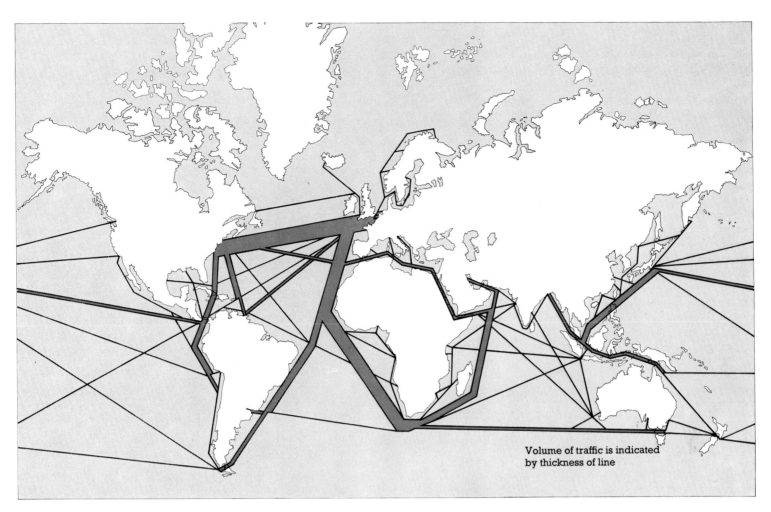

Volume of traffic is indicated by thickness of line

Above: The world's major shipping lanes are shown above. Note the great number that lie primarily in narrow coastal areas – an important reason why many maritime nations fear proposed extensions of the power of coastal states to control traffic within their exclusive economic zones.

equipment, and cargo loading. Most ship-owning states have accepted convention regulations, which the IMCO now supervises.

Where international supervision can achieve practically nothing, though, is in the military use of the high seas. Here, the principle of the freedom of the seas is nearly unbridled. Warships and nuclear submarines are free to roam where they will. Nor is there anything to prevent the deployment of underwater listening and other anti-submarine devices. The only measure so far taken to demilitarize the oceans is the Sea-Bed Arms Limitation Treaty of 1971, which prohibits the emplacement of nuclear weapons or other weapons of mass destruction on the sea bed at any point beyond 12 miles from the coast. Over 50 states have ratified the treaty, including the United States, the U.S.S.R., and Britain, though not – possibly significantly – two other nuclear powers, China and France. For obvious reasons, however, it is almost impossible to ensure observance of any such treaty, and the world has little, if any, precise information about undersea military traffic.

How Flags of Convenience Can Muddy the Waters *(A Hypothetical Case)*

Assume that a group of Japanese, Australians, and Colombians form two companies, Jacco and Cajco.

Jacco (incorporated in the Bahamas) buys the *Maverick*, a tanker registered in Liberia. Its master is Egyptian.

Cajco (incorporated in Bermuda) buys the *Houdini*, a bulk carrier registered in Panama. Its master is Greek.

Maverick is wrecked off the coast of Spain; oil from the ship ruins Costa Brava beaches. Spanish authorities sue Jacco, whose only ship was *Maverick*. Insurance on *Maverick* is insufficient to cover the damage. If *Houdini*, which happens to be in a Spanish port at the time, belonged to Jacco, it could be seized as compensation, but it belongs to Cajco. Nor can the authorities take action to deal with the Egyptian captain, whose negligence was responsible for the accident. Disciplinary and penal action can be taken only by Liberia or Egypt.

Readers of this book are surely aware by now that there is one aspect of the sea that every marine scientist, regardless of his special interests, feels impelled to emphasize: vast as the oceans are, they are neither deep nor wide enough to withstand indefinitely the many kinds of pollutant being dumped into the water by twentieth-century technological man. In earlier chapters we have seen how pollution comes not only from spills of oil and other substances from ships, but also from sea-bed drilling operations and directly from land – for example, through rivers containing industrial waste and domestic sewage that flow into the sea, or from the precipitation into the sea of the chemical contents of smoke from factories. In fact, most of the pollutants in the water come from land-based sources, though their effluents are seldom as startlingly apparent as those that result from the wreck of an oil tanker. Such substances as plastics, radioactive waste, heavy metals, and chemicals like DDT, which are widely used in agriculture, may pose an even greater long-term threat to the creatures that live in the sea – and ultimately to the general well-being of our planet – than does oil.

Still, oil remains the most familiar (and generally feared) marine pollutant, probably because tanker mishaps invariably make the headlines. And the first efforts of international law to deal with the growing menace of marine pollution were concerned with the deliberate discharge of oil from tankers. A convention on the subject was drawn up in 1954 at a conference in London attended by representatives of 32 states. This convention, as amended in 1962, again in 1969, and once more in 1971 (in each case, to make its provisions stricter), is now binding on about 95% of the total world fleet of tankers. The original purpose of the convention was to restrict the practice of washing out tanks at sea after unloading cargoes of oil, which meant that great quantities of oily waste water containing lumps of crude oil were continually being pumped overboard into the world's oceans. This is no longer the source of pollution that it used to be. Thanks to the development of what's known as the "load on top" system, such wastes are now retained on board; when the tanker reaches port, they are discharged into a special receptacle, whose contents are then disposed of at refineries. Today, the convention concerns itself chiefly with such matters as ensuring that tankers operate the load-on-top system and that they keep proper records of all operations involving their cargoes. It also deals with the size of individual oil tanks aboard the vessels, for when a tank is reasonably small, less oil is spilled if it is pierced as a result of collision or stranding.

So much for oil. Rules to curb the deliberate discharge at sea of other substances – especially chemicals, garbage, and sewage – are contained in a convention adopted at a 1973 conference sponsored by the IMCO (the UN's Intergovernmental Maritime Consultative Organization). This convention has not yet come into force. When it does, all deliberate discharges from ships will be regulated by international law. But the very broadness of the agreement's coverage has worked against swift implementation. So far, indeed, the world's states have shown little enthusiasm for ratifying the convention; by early 1976 only two nations, Jordan and Kenya, had ratified. A related problem, however – the practice, prevalent since the late 1960s, of getting rid of industrial waste by loading it onto ships and dumping it at sea – has been dealt with more successfully in a convention that was adopted in 1972 and ratified by enough states to come into force in 1975. It lays down strict guidelines as to how and where such wastes may be dumped – if, in fact, they may be dumped at all, for the agreement prohibits dumping of certain toxic substances. There are also supplementary regional conventions dealing with the dumping of industrial wastes in the northeast Atlantic and North Sea (1972), the Baltic (1974), and the Mediterranean (1976).

Thus, there are plenty of international rules governing discharge standards. But how are they enforced? Traditionally, as we've seen, only the state whose flag a ship flies has jurisdiction over that ship on the high seas. This is not entirely satisfactory. For one thing, the flag state is often only a flag-of-convenience state, with inadequate control over its ships. For another, if a ship is far from home, the flag state may have no means of knowing whether it is observing antipollution regulations. There is, therefore, a growing feeling among participants at the UN Law of the Sea Conference that a legal right to deal with ships that break the pollution laws should also be given to states near whose coasts such infringements take place or in whose ports offending vessels arrive, even when the offences have not occurred within their territorial waters.

Another difficult question concerns the remedies open to victims of pollution. What action, for example, can be taken by the proprietor of a hotel in Jamaica if his beach, one of the hotel's chief attractions, is ruined by oil discharged by a passing tanker en route for New York from Venezuela? Let's imagine that the tanker is actually owned by American citizens who, for tax purposes,

Some pollution from industrial wastes may seem a merely national problem, as symbolized by this pathetic victim of mercury poisoning at Japan's Minamata Bay. But can we be sure that apparently local damage to marine food chains won't have international "domino" effects?

The load-on-top system of washing out tankers is drying up one source of pollution. Instead of being drained directly into the sea, oily washings from all the cargo tanks are pumped into a single tank (the slop tank) where the mixture of oil and water settles and separates. The water is drained off into the sea, while the oil remains on board and some of the next cargo is loaded on top of it.

operate as a company incorporated in the Bahamas, with the tanker registered in Panama. Before 1975, the hotel proprietor would have had to sue the shipowner in either a Bahamas or Panamanian court and to prove negligence or fault on the owner's part. This might have been very hard to do, and the process would certainly have been costly and time-consuming. Now, thanks to an IMCO-sponsored convention of 1969 (which came into force in 1975), the aggrieved man can bring suit in his local Jamaican court. Furthermore, he need not prove fault or negligence; presented with clear evidence that a given ship was the offender, the court will assume the ship-owner's liability and grant compensation to the hotel proprietor. At present, unfortunately, these simplified procedures apply only to damage caused by oil, but the IMCO is considering the possibilities of extending them to cover other kinds of marine pollution.

The situation as it involves pollution from offshore drilling and from land-based sources is – not surprisingly – far less satisfactory. As yet there is little international marine law in this area. Where sea-bed operations are concerned, there are at present no plans at all for a worldwide agreement governing the question. Eventually there may be some regional agreements (regulations for the North Sea, Baltic, and Mediterranean areas are already being considered), but until real action is taken, pollution from offshore drilling operations will continue to be almost wholly governed by each state's own laws. And international law has barely begun to tackle the formidable administrative difficulties that would be involved in regulating pollution from land. The only thing we have so far is three regional agreements – again for the North Sea and northeast Atlantic, the Baltic, and the Mediterranean – that are really just *frameworks* for agreement. In other words, they establish commissions with the task of drawing up rules governing the discharge of pollutants from land into the sea. Since by 1976 these commissions had hardly begun to get down to work, it is still too soon to predict what effect, if any, they will have on marine pollution from land-based sources.

To minimize effects of spillage from tankers during loading, harbor authorities at Gävle in Sweden surround berthed ships with a "bubble barrier" – an underwater hose which, by releasing compressed air through a series of outlets, prevents spilled oil from spreading.

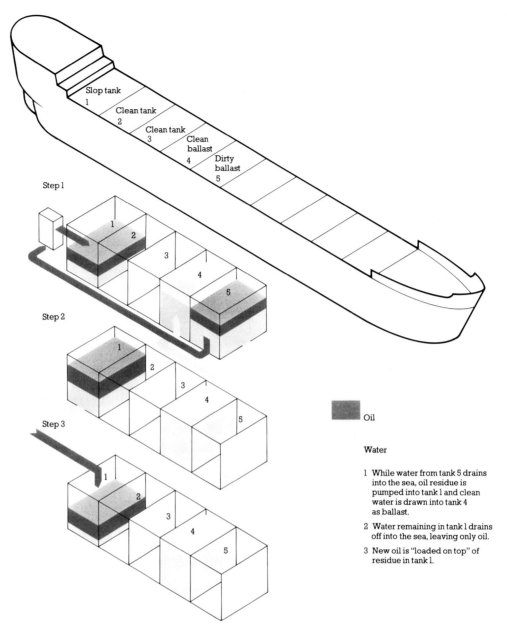

Step 1

Step 2

Step 3

Oil

Water

1 While water from tank 5 drains into the sea, oil residue is pumped into tank 1 and clean water is drawn into tank 4 as ballast.

2 Water remaining in tank 1 drains off into the sea, leaving only oil.

3 New oil is "loaded on top" of residue in tank 1.

Order or Anarchy?

Current negotiations aimed at establishing an internationally acceptable corpus of marine law are one aspect of what is perhaps the major challenge facing mankind in the last quarter of this century: the devising of a framework of international cooperation that might make it possible for an ever-increasing population to live in reasonable harmony on a planet endowed with only limited resources. As far as the sea is concerned, the vital question of whether or not we can meet this challenge depends largely on the results of the Third United Nations Conference on the Law of the Sea.

As these words are being written (in mid-1976), it is impossible to predict the end result of the conference, which has been meeting once a year for the past few years without reaching final agreement on several basic points. The only satisfactory result, of course, would be the adoption of a convention whose general acceptability would

be demonstrated by its swift ratification by most, if not all, states. Such a convention would give a certainty to the law that has been badly needed for more than a generation, and it would put an end to such modern bones of contention as extravagant national claims to offshore zones and international disputes over the ownership of undersea natural resources.

The alternatives – the adoption of a convention that only a small number of states would subsequently ratify, or the failure to adopt any convention at all – are both bleak. Either of these depressing results would lead to increasing unilateral action (particularly claims to extensive coastal zones); to an unedifying scramble for resources beyond coastal zones that would parallel the nineteenth-century scramble for Africa by the European colonial powers; and to more interference with foreign shipping in coastal states' waters, including perhaps the levying of tolls in the world's more important straits. Above all, failure to agree upon

a widely acceptable set of rules would mean a severe and possibly irreparable jolt to international cooperation – especially to the very important current negotiations between developed and developing countries over the establishment of what is called "the new international economic order."

However, even if the conference succeeds in producing a satisfactory convention, there will be little cause for rejoicing among internationally minded observers. At best, the convention will be much less internationalist than many once hoped it would be or than it perhaps ought to be in this age of global interdependence. It will not provide for any extensive redistribution of the world's wealth, and the states that are commonly described as "geographically disadvantaged" are likely to gain very little from it. This relative lack of a truly internationalist approach has mainly resulted from the new nationalism of coastal states, as well as from the fact that the international community is still not sufficiently

community-minded to accept the necessary degree of international organization that a broad and liberal agreement would involve.

There is, nevertheless, one notably hopeful indication that a more internationally conscious world community is in the making: the widespread support for the proposal that coastal states should contribute part of all revenue from their sea-bed mining operations to developing countries. If finally agreed (as now seems probable), this regulation would mark the first beginnings of a system of global taxation. And if we extend this type of share-the-wealth thinking into other areas, we might someday arrive at a situation where all sorts of economic aid for developing countries came from the revenues of an international system of taxation and the poorer states were not, as they are now, dependent on the fickle charity of the rich.

Whatever the results of the Law of the Sea Conference, we can be sure that the use of the oceans for their natural resources and as a means of communication will continue and almost certainly increase. This means that there will be an increasing need for more effective, less damaging management of marine resources and the marine environment. If no acceptable convention comes out of the UN conference, the need can to some extent be met by regional agreements; and even if a new and workable convention does emerge, regional agreements will continue to be useful because, by its very nature, a worldwide agreement could not possibly provide the necessary detailed regulations required for, say, local fisheries management. In such areas as the Baltic, the North Sea, the Persian Gulf, or the Gulf of Mexico and the Caribbean, where exceptionally heavy use of the sea's resources could bring on potential or actual conflict between different uses and users, we may well see the ultimate establishment of regional marine authorities with full responsibility for over-all management of the waters. To make such international authorities truly authoritative, of course, much more attention will have to be given to ways of controlling and policing users.

Even a superficial examination of the situation described in the foregoing pages should clearly indicate why we are coming to the end of an era in the international law of the sea. The era now ending has been one of tremendous change and upheaval, reflecting the change in international society during the thirty-odd years since the end of World War II – a period of enormous technological developments that have revolutionized man's use of the sea. What the international law of the sea will be like in the years ahead depends very largely on the outcome of the UN conference. It is perhaps rash at this stage to predict, but it seems probable that we will witness either the development of an orderly management of the seas or a worldwide upheaval that is quite likely to follow on the heels of anarchy.

Bibliography

The books listed here are only a selected fraction of the many available volumes, both scholarly and popular, about the sea. Included at the end of the list for each chapter are the names of major journals that print articles on the pertinent subject matter. All material recommended in the lists is in English, but many of the books have been translated into other languages.

General

Carson, Rachel **The Sea Around Us,** Oxford University Press, New York and London, 1951
A brilliantly written classic, readable and informative even though somewhat dated factually.

Fairbridge, Rhodes W. (ed.) **Encyclopedia of Oceanography,** Rheinhold Book Corporation, New York, 1966
A factual A–Z encyclopedia designed for the first-year university student. A useful reference work, it includes valuable lists of suggestions for further reading.

Hill, M. N. (ed.) **The Sea** (5 vols.), Interscience Publishers, John Wiley and Sons, London and New York, 1962–74
This is the basic storehouse of information about the physics, chemistry, and geology of the sea, with a long series of highly specialist papers on every topic.

Varley, Allen (ed.) **Ocean Research Index,** Francis Hodgson, Cambridge, England
A country-by-country listing of the world's marine and freshwater research institutes and university departments. Every laboratory in all of 125 countries is listed, along with its address and topics of research.

1 The Ocean Floor

Bird, J. M., and Isacks, B. (eds.) **Plate Tectonics,** American Geophysical Union, Washington, D.C., 1972
Selected articles dating from 1967 to 1972 that discuss the achievements of the researchers who pioneered the plate-tectonics theory.

Press, Frank, and Siever, Raymond (eds.) **Planet Earth,** W. H. Freeman and Co., San Francisco and Reading, England
A variety of articles from **Scientific American,** all very clearly written and well illustrated, on the structure of the earth's interior and crust, as well as some discussion of the earth's origin and of continental drift.

Shepard, Francis P. **Submarine Geology,** Harper and Row, New York, 3rd ed. 1973
The classic on the subject by a man who has been called the father of submarine geology. Superbly informative for the student – and good value, too, for general readers.

Tarling, D. H., and Runcorn, S. K. (eds.) **The Implications of Continental Drift to the Earth Sciences** (2 vols.), Academic Press, New York and London, 1973
An exhaustive survey by numerous scientists of the impact of the theory of plate tectonics on many aspects of geology, oceanography, and biology.

Journals
Marine Geology, published monthly (in English) by Elsevier, Amsterdam

Tectonophysics, published quarterly (in English) by Elsevier, Amsterdam

2 The Water Itself

Haines, G. **Sound Underwater,** David and Charles, London, 1974
A simple, clear, nonmathematical outline of the properties and uses of sound in the sea, with some helpful illustrations and diagrams.

King, C. A. M. **Oceanography for Geographers,** Edward Arnold, London, 1962
This is one of the very few easy-to-read general introductions to the oceanography of waves, tides, and currents. Even readers who dislike mathematics can find their way through its pages.

Neuman, G., and Pierson, W. J. **Principles of Physical Oceanography,** Prentice-Hall Inc., Englewood Cliffs, New Jersey, 1966
One of the great textbooks, this is an immensely thorough treatment of the properties of seawater, ice, waves, tides, instrumentation, currents, and air-sea interaction. It requires a knowledge of mathematics at university level, however.

Journals
Deep Sea Research, published monthly by Pergamon Press, New York and London
The standard international academic journal for deep-water multidisciplinary oceanography.

Estuarine and Coastal Marine Science, published bimonthly by Academic Press, New York and London
A journal specializing in multidisciplinary papers on topics within the area of the continental shelf, coastal waters, and estuaries.

Oceanus, published quarterly by the Woods Hole Institution of Oceanography, Woods Hole, Massachusetts
A popular magazine that brings the latest news of advances in oceanography to the general reader.

3 and 4 Plant Life and Salt Water Animals

Note: Most popular books and textbooks up to university level treat marine plants and animals together, under the general heading of Marine Biology.

Catala, R. L. A. **Carnival Under the Sea,** R. Sicard, Paris, 1964
A remarkable book of photographs of corals and fish of the South Pacific. Some fascinating color effects with ultraviolet light.

Friedrich, H. **Marine Biology,** Sidgwick and Jackson, London, 1970
This book for the student treats animals and plants together in each type of marine environment: surface, mid-water, deep-ocean, sea-floor, and coastal.

Hardy, Alister **The Open Sea: Its Natural History** (2 vols.), Houghton Mifflin, Boston, 1971; Collins, London, 1956 and 1959
Still the most delightful introduction to the world of marine life. Illustrated with many drawings and watercolors by the author, who pioneered much of the research.

Herring, P. J., and Clarke, M. R. (eds.) **Deep Oceans,** Arthur Baker Ltd., London, 1971
A general introduction to the undersea world, but with emphasis on descriptions of plants and animals. Many excellent illustrations.

Raymont, J. E. G. **Plankton and Productivity in the Oceans,** Pergamon Press, New York and Oxford, England, 1963
A comprehensive review of primary productivity and zooplankton growth, aimed at the university student.

Journals
Journal of the United Kingdom Marine Biological Association, published quarterly by Cambridge University Press, New York and London

Marine Biology 16 issues a year (in English), Springer Verlag, Berlin and New York

5 The Ocean's Resources

Barton, Robert **Oceanology Today: Man Exploits the Sea,** Aldus Books, London, 1970
A compact, well-illustrated review of the resources of the sea and the techniques of exploiting them, with special emphasis on the oil and fishing industries.

Culland, J. A. (ed.) **The Fish Resources of the Ocean,** Fishing News Books Ltd., London, 1971
An ocean-by-ocean survey of the quantities and species of fish caught and of fisheries research and fishing methods in each area.

Hull, E. W. Seabrook **The Bountiful Sea,** Prentice-Hall Inc., Englewood Cliffs, New Jersey; Sidgwick and Jackson, London, 1964
A journalist's survey of the resources of the sea, peaceful and military, amply illustrated.

Mero, John L. **The Mineral Resources of the Sea,** Elsevier, London and New York, 1965
An optimistic study of the possibilities of mining manganese nodules and extracting minerals from the continental shelf and chemicals from seawater. Particularly interesting because it has provoked a good deal of controversy.

Woodland, Austin W., and Cole, H. A. (eds.) **Petroleum and the Continental Shelf of North West Europe** (2 vols.), Institute of Petroleum and Applied Science Publishers, Barking, England, 1975
Volume 1 is a review by many experts of the structure of the continental margin in terms of plate tectonics, followed by descriptions and cross-sections of individual oil and gas fields, based on seismic records and commercial drilling. Volume 2 surveys the risks of oil pollution arising from various drilling and production methods, and discusses pollution prevention.

Journals
Fishing News International, published monthly by Arthur J. Heighway Ltd., London
The standard international magazine for reports on fisheries activities, political developments, vessel design and construction.

Ocean Industry, published monthly by Gulf Publishing Co., Houston, Texas
A mine of up-to-the-minute information on the offshore oil industry, with some treatment of related technical subjects such as dredging and specialized ships. Superb color photography.

Offshore Services, published monthly by Spearhead Publications, Kingston-upon-Thames, England
A news-packed magazine concentrating on the North Sea oil scene, with coverage of related technical subjects. Spiced with some nice humor.

6 Using Ocean Space

Goldberg, Edward D. (ed.) **A Guide to Marine Pollution,** Gordon Breach Science Publications, New York and London, 1972
A brief review of the chemical and ecological characteristics of the most important classes of pollutants.

Hood, D. W. (ed.) **Impingement of Man on the Oceans,** Wiley-Interscience, New York and London, 1971
A wide-ranging review of the various things that we dump in the sea, as well as of the effects upon the coast of construction and alterations.

Horsfield, B., and Stone, P. **The Great Ocean Business,** Coward McCann and Geoghegan Inc., New York, 1972; Hodder and Stoughton, London, 1971
A topical account of the developments and personalities in oceanography and marine exploitation.

Ketchum, Bostwick H. (ed.) **The Water's Edge,** Massachusetts Institute of Technology Press, Cambridge, Massachusetts, 1972
A highly readable report of a symposium at which a group of experts discussed the interacting problems of overexploitation of the coastal zone and made some practical recommendations.

Marine Science Affairs: Report of the President to Congress on Marine Resources and Engineering Development,
US Government Printing Office, 1970
A fascinating insight into the way a great nation stimulates marine technology and exploitation. Written in plain, direct language, with lots of illustrations and tables of data.

Wiegel, Robert L. **Oceanographical Engineering,** Prentice-Hall Inc., Englewood Cliffs, New Jersey, 1964
A technical book dealing with the effects of waves and currents on a wide range of engineering and coastal structures as well as on the coast itself.

Journals
Marine Pollution Bulletin (news, views, and book reviews), published monthly by Pergamon Press, New York and London

7 Underwater Archaeologists

Bass, George F. (ed.) **A History of Seafaring,** Walker and Co., New York, Thames and Hudson, London, 1972
This is a collective work by George Bass, Peter Throckmorton, Michael Katzev, and many of the other "greats" who established the modern style of underwater archaeology in the 1960s. Splendidly produced, with beautiful color photographs, this is an excellent introduction to the subject. Includes extensive bibliographies.

Blackman, David J. (ed.) **Marine Archaeology,** Shoe String Press Inc., Hamden, Connecticut; Butterworths, London, 1973
An authoritative collection of 24 papers by different authors, covering subjects like ancient harbor technology, ancient lighthouses, underwater survey methods, and the excavation and salvage of ancient ships. Amply illustrated, and with full bibliography.

Flemming, N. C. **Cities in the Sea,** Doubleday, New York, New English Library, London, 1972
A diver's description of underwater cities in the Mediterranean from the Bronze Age to the Roman Empire. Locations and maps of many submerged ruins, with plenty of photographs.

Marsden, Peter **The Wreck of the Amsterdam,** Stein and Day, New York, 1975; Hutchinson, London, 1974
The story of the wrecking of this Dutch East Indiaman and of the first attempts to salvage it.

Martin, Colin **Full Fathom Five,** Chatto and Windus, London, 1975
A dramatic analysis of the sequence of events that led to the destruction of the Spanish Armada, with much new evidence resulting from the excavation of wrecks off Ireland and Scotland.

Throckmorton, Peter **Shipwrecks and Archaeology,** Atlantic Monthly Press, Boston; Gollancz, London, 1970
A first-hand account of the exploration of ancient wrecks by a man who has a flair for making new discoveries.

Journals
The International Journal of Nautical Archaeology, published quarterly by Academic Press, New York and London
The only journal in the world on the subject.

8 Diving and Divers

Bennett, P. B., and Elliott, D. H. (eds.) **The Physiology and Medicine of Diving and Compressed Air Work,** Baillière Tindall, London, 1975
The definitive book on the subject, with contributions by experts from many countries

BSAC Diving Manual
Official instructional manual published by the British Sub-Aqua Club, London. An essential book for beginners.

Cousteau, Jacques-Yves **The Silent World,** Harper and Row, New York; Hamish Hamilton, London, 1953
The popular classic by the inventor of the self-contained aqualung, which brought diving within the reach of millions of enthusiasts.

Drew, E. A., Lythgoe, J. M., and Woods, J. D. (eds.) **Underwater Research,** Academic Press, New York and London, 1976
A professional review by a group of American and British scientists of the many ways in which diving can be *used* in underwater scientific research.

Dugan, James **Man Explores the Sea,** Hamish Hamilton, London, 1956
A thorough inquiry into the early origins of diving and its development through the nineteenth century to the middle of the twentieth. Immensely readable.

Miles, Stanley **Underwater Medicine,** J. P. Lippincott, Philadelphia; Staples Press, London, 1966
The only book on diving medicine that presents the facts in nontechnical language.

NOAA Diving Manual
The official instructional manual published by the National Oceanic and Atmospheric Agency for diving scientists. In spite of its specialist background, this volume is one of the best general books on the subject. Well illustrated with photographs and diagrams. Can be obtained from the US Government Printing Office, Washington D.C.

Journals
Skindiver, a monthly magazine for sports divers, published by Peterson Publishing Co., Los Angeles

Triton, the monthly journal of the British Sub-Aqua Club, London

Undersea Biomedical Research, a quarterly medical journal published by the Undersea Biomedical Society, Bethesda, Maryland; devoted entirely to the study of man – and sometimes other organisms – underwater.

9 Submarine Craft

Cohen, Paul **The Realm** [UK: **Province**] **of the Submarine,** Macmillan, New York; Collier-Macmillan, London, 1969
A popular history of the technical improvements in submarines over the years, written by an insider. Oddly devoid of illustrations.

Piccard, Jacques **The Sun beneath the Sea,** Charles Scribner's Sons, New York, 1971; Robert Hale, London, 1973
A day-by-day account of the building of the mesoscaphe *Auguste Piccard* and of the 2,400-kilometer drift in the depths of the Gulf Stream with a crew of six men.

Sweeney, J. B. **A Pictorial History of Oceanographic Submersibles,** Crown Publishers Inc., New York; Robert Hale, London, 1972
A chatty book, full of unusual pictures of early submarines, but not very well organized.

Journals
There are no generally available journals devoted solely to submarine craft, but articles and pictures concerning them appear frequently in:

Ocean Industry, published monthly by Gulf Publishing Co., Houston, Texas

Offshore Services, published monthly by Spearhead Publications, Kingston-upon-Thames, England

10 Marine Law and Politics

Brown, E. D. **The Legal Regime of Hydrospace,** Stevens and Sons, London, 1971
The subject is changing so fast that only newspaper and magazine reports and professional law journals help one to keep up. Nevertheless, this book is a valuable outline of the basic principles of law at sea, and it provides a good account of the early stages of evolution of the Law of the Sea Conference.

Lay, S. H., Churchill, R., and Nordquist, M. (eds.) **New Directions in the Law of the Sea** (2 vols.), Oceana Publications Inc., Dobbs Ferry, New York; British Institute of International and Comparative Law, London, 1973
A facsimile reproduction of numerous claims, international treaties, conventions, giving exact texts of laws and documents relating to national limits, fisheries, pollution, continental-shelf rights, conservation, nuclear shipping, etc.

Wenk, Edward **The Politics of the Ocean,** University of Washington Press, 1972
A realist's analysis of the political problems of managing the sea, treated from both a national and international standpoint, with suggestions for future developments in marine law.

Journals
Ocean Development and International Law Journal, published quarterly by Crane Russak and Co. Inc., New York

Ocean Management published quarterly (in English) by Elsevier, Amsterdam

Some Useful Addresses

There are a number of organizations that can help you find out more about the sea, or can tell you how to get professional training in marine biology or oceanography or how to join a diving club. Here are addresses of some major oceanographic and marine-science research laboratories (which, incidentally, are likely to have the best libraries on marine subjects):

Australia
Australian Institute of Marine Science, PO Box 1104, Townsville, Queensland

Commonwealth Scientific and Industrial Research Organisation, Division of Fisheries and Oceanography, 202 Nicholson Parade, Cronulla, New South Wales

Canada
Bedford Institute of Oceanography, Dartmouth, Nova Scotia

Institute of Oceanography, Dalhousie University, Nova Scotia

Institute of Ocean Sciences, Patricia Bay, 1230 Government Street, Victoria, British Columbia

France
Centre Océanologique de Bretagne, Brest

Laboratoire Arago, 66650 Banyuls-sur-Mer

Laboratoire d'Arcachon, 36 Boulevard Deganne, Arcachon

Station Marine d'Endoume, 3 Rue de la Batterie des Lions, Marseille

Italy
Stazione Zoologica di Napoli, Acquario, Villa Comunale, Naples

Japan
Tokyo University of Fisheries, 4-5-7 Konan, Minato-ku, Tokyo

University of Tokyo Ocean Research Institute, 15-1, 1-chome, Minamidai, Nakano-ku, Tokyo

Monaco
Musée Océanographique, Avenue Saint-Martin, Monaco-Ville

Netherlands
Delft Hydraulica Laboratory, Rotterdamseweg 185, Delft

Norway
Institute of Marine Research, PO Box 2906, Bergen

Sweden
Oceanographic Institute, Box 4038, 40 Gothenburg

UK
Department of Agriculture and Fishes for Scotland, Marine Laboratory, PO Box 101, Victoria Road, Torry, Aberdeen

Institute of Oceanographic Sciences, Wormley, Godalming, Surrey

Marine Biological Association of the United Kingdom, The Laboratory, Citadel Hill, Plymouth, Devon

Ministry of Agriculture, Fisheries and Food, Fisheries Laboratory, Pakefield Road, Lowestoft, Suffolk

Scottish Marine Biological Association, Dunstaffnage, PO Box 3, Oban, Argyll

USA
Lamont-Doherty Geological Observatory, Palisades, New York

National Oceanic and Atmospheric Administration,
Atlantic Oceanographic and Meteorological Laboratories,
15 Rickenbacker Causeway, Miami, Florida

National Oceanic and Atmospheric Agency (NOAA),
Pacific Marine Environmental Laboratories,
1801 Fairview Avenue East, Seattle, Washington

Rosenstiel School of Marine and Atmospheric Science,
4600 Rickenbacker Causeway, Miami, Florida

Scripps Institution of Oceanography, La Jolla, California

Woods Hole Oceanographic Institution, Woods Hole,
Massachusetts

USSR
Atlantic Scientific Institute for Marine Fisheries
and Oceanography,
Dmitry Donskoi 5, 236000 Kaliningrad oblastnoi

Marine Hydrophysical Institute,
Ul. Lenina 28, 335005 Sebastopol 5

Pacific Oceanological Institute,
Prospekt of 100th Anniversary of Vladivostok 159,
690022 Vladivostok 22

West Germany
Deutsches Hydrographisches Institut,
Bernhard-Nocht-Strasse 78, Hamburg

Institut für Meereskunde, Niemannsweg 11, Kiel

Some of the institutes on the above list are in fact departments of universities, but all are wholly devoted to marine research. There are hundreds of smaller institutions and university departments in which at least some research effort is marine. A complete list of addresses by country appears in the Ocean Research Index listed in the bibliography above.

In several countries there are societies whose aim is to bring together scientists, engineers, students, businessmen, and politicians to further the exploitation of the sea. The addresses of some of these societies are:

France
Association Scientifique pour l'Exploitation des Océans,
3 Rue La Boëtie, Paris

Sweden
Svenska Mekanikers Riksförening, Malmskillnadsgatan 48a,
Stockholm

UK
Society for Underwater Technology, 1 Birdcage Walk,
London SW1

USA
Marine Technology Society, 1730 M Street, NW,
Washington, D.C.

There are as yet no internationally recognized standards for commercial diving training, though schools have been set up in several countries. Standards of training and safety for scuba-aqualung diving are now regulated internationally, and most national sports-diving federations are affiliated with the Confédération Mondiale des Activités Subaquatiques, 34 Rue de Colisée, Paris, France. If you have had proper training from one of the recognized diving schools or clubs in your own country, you can obtain an international certificate of equivalent training from CMAS, and this certificate, recognized in about 60 countries, will guarantee that you can hire equipment almost everywhere. The CMAS also publishes a yearbook containing the addresses of diving organizations in all member countries. Some addresses of diving clubs:

France
Fédération Française d'Etudes et de Sports Sous-marins,
24 Quai de Rive-Neuve, Marseille

Italy
Federazione Italiana Pesca Sportive e Attiva Subacquea,
Viale Tiziano 70, Rome

Norway
Norges Dykkerforbund, Postboks 6514, Hauger Skoleveien 1, Oslo

Sweden
Svenska Sportdykarförbundet, Box 925, Stockholm

UK
British Sub-Aqua Club, 70 Brompton Road, London SW3

USA
National Association of Underwater Instructors,
22809 Barton Road, Colton, California

West Germany
Verband Deutscher Sporttaucher e.V., D-2000 Hamburg 36,
Bleichenbrücke 10

Each of the above organizations has numerous associated clubs or branches. There is a good chance, therefore, that there will be a diving center near any place you choose for a vacation.

Many people assisted in the preparation of this book. The publishers wish to thank them all, particularly:

Richard Gray	Assistant to the art editor
Karen Gunnell	Diagram research and design
Howard Dyke	Diagram design
Judith Allen	Design for chapters 8 and 9
Harold King	Diagram design
Mark Dartford	Picture research
Sheila McKenzie	Production
Dr. David George	British Museum of Natural History
Dr. Dodge	Birkbeck College, London
Dr. David Irvine	Polytechnic of North London
Spike Knole	Fisheries expert, Marine Harvests Ltd.

Acknowledgments

Artwork is acknowledged in **bolder** type

Introduction

Page	
6–7	NASA/Colorific! Photo Library, Britain
8–9	© Dr. Douglas P Wilson, Britain
11	Lockheed Missiles & Space Co. Inc. USA
13	Anver Raban, Israel
15	Brown & Root (UK) Ltd., Britain
16–17	Comex/Dr. John Bevan, Britain
18–19	Christian Pétron/Seaphot, Britain
20	US Navy Photo

1 The Ocean Floor

Page	
22–23	**Diagram Visual Information Ltd.**
24–25	**Diagram Visual Information Ltd.**
26–27	**Diagram Visual Information Ltd.**
27	© *Scientific American:* **Diagram Visual Information Ltd.**
28–29	**Diagram Visual Information Ltd.**
30–31	**Diagram Visual Information Ltd.**
31	Woods Hole Oceanographic Institution, USA
32–33	**Diagram Visual Information Ltd.**
34	NASA/Camera Press Information Ltd.
34–35	**Diagram Visual Information Ltd.**
36	**Diagram Visual Information Ltd.: Len Whiteman/Gordon Cramp Studio**
37	The Geological Museum, London
38–39	**Diagram Visual Information Ltd.**
40	**Diagram Visual Information Ltd.**
41	**Richard Gray**
42	CNEXO, France
43	Luave/Colorific! Photo Library, Britain
44–45	Lamont-Doherty Geological Observatory, USA

2 The Water Itself

Page	
46	**Osborne/Marks**
48	**Osborne/Marks**
49	**Osborne/Marks** US Navy Official Photo/Robert F. Dill, USA
50	**Richard Gray: Graphters** US Navy Official Photo, USA
51	**Richard Gray: Graphters**
52	**Rose Associates** Crown Copyright (Ministry of Defence RN), Britain
53	**Len Whiteman/Gordon Cramp Studio**
54	**Len Whiteman/Gordon Cramp Studio**
55	**Len Whiteman/Gordon Cramp Studio** *The Daily Telegraph* Colour Library, Britain
56	**Rose Associates**
57	left: Professor J. D. Woods, Britain right: Institute of Oceanographic Sciences, Britain
58	**Len Whiteman/Gordon Cramp Studio**
59	top: Michael Newton, Britain bottom: Canadian Government Office of Tourism, London
60	Robert L. Wiegel, USA/Modern Camera Center, Hawaii
60–61	**Richard Gray: Osborne/Marks**
61	**Richard Gray: Len Whiteman/Gordon Cramp Studio**
62	**Richard Gray**
63	top: Institute for Marine Environmental Research, Britain center: Institute of Oceanographic Sciences, Britain bottom: Institute of Oceanographic Sciences, Britain
64	Woods Hole Oceanographic Institution, USA/Robert Munns
65	NOAA, Rockville, Md., USA

3 Plant Life

Page	
66–67	Department of Zoology, University of Oxford, Britain
67	Institute for Marine Environmental Research, Britain **Frank Kennard**
68	John D. Dodge, Britain
69	top left: John D. Dodge, Britain top right and bottom: Dr. Gerald T. Boalch/Seaphot, Britain
70	John Lythgoe/Seaphot, Britain
70	**Howard Dyke: Rose Associates**
71	John Lythgoe/Seaphot, Britain
71	**Howard Dyke: Rose Associates**
72	**Howard Dyke: Osborne/Marks**
73	Tom Smith/Camera Press Ltd., Britain **Howard Dyke: Osborne/Marks**
74	**Howard Dyke: Richard Gray**
74–75	John Lythgoe/Seaphot, Britain
75	**Osborne/Marks**
76	top and bottom left: © Dr. Douglas P. Wilson, Britain top center, left center and bottom center: John D. Dodge, Britain right center: © Dr. Douglas P. Wilson, Britain right: John D. Dodge, Britain
77	Peter Scoones/Seaphot, Britain
78	© Dr. Douglas P. Wilson, Britain
78–79	© Dr. Douglas P. Wilson, Britain
79	Gillian Lythgoe/Seaphot, Britain
80	© Dr. Douglas P. Wilson, Britain
81	**Richard Gray: Rose Associates: Chris Sorrell**
82	© Dr. Douglas P. Wilson, Britain
83	© Dr. Douglas P. Wilson, Britain
84	**Howard Dyke: Rose Associates**
85	top and bottom: Keith Hiscock, Britain center: Geoff Harwood/Seaphot, Britain
86	John D. Dodge, Britain
87	Walt Deas/Seaphot, Britain
88	Crown Copyright (Natural Environment Research Council), Britain
89	Michael Francis Wood, Britain

4 Salt-water Animals

Page	
90–91	**Howard Dyke: Osborne/Marks**
92	**QED**
92–93	Oxford Scientific Films, Britain
93	top right: British Museum (Natural History), Britain **Osborne/Marks:** © Dr. Douglas P. Wilson, Britain
94	**Osborne/Marks** Ken Vaughan/Seaphot, Britain
95	top left: David George, Britain **Osborne/Marks** center left: Heather Angel, Britain center right: Christian Pétron/Seaphot, Britain bottom: David George, Britain
96	top left: Heather Angel, Britain top center: Heather Angel, Britain **left: Chris Sorrell** **bottom: Linden Artists**
96–97	Peter Scoones/Seaphot, Britain
97	**Osborne/Marks** Heather Angel, Britain
98	Peter David/Seaphot, Britain
98–99	Japanese Information Service, Britain **Frank Kennard**
99	top right: David George/Seaphot, Britain center left: Heather Angel, Britain center right: Heather Angel, Britain
100	Dick Clark/Seaphot, Britain

100–101 © Dr. Douglas P. Wilson, Britain
101 top: Peter David/Seaphot, Britain
center: © Dr. Douglas P. Wilson, Britain
Osborne/Marks
102 top: David Mallot
left: Christian Pétron/Seaphot, Britain
right: Heather Angel, Britain
bottom: **Osborne/Marks**
103 top: **Frank Kennard**
center: **Linden Artists**
Peter David/Seaphot, Britain
104 top left: Walt Deas/Seaphot, Britain
center left: Heather Angel, Britain
top right: **David Mallot**
center right: Heather Angel, Britain
bottom right: **Osborne/Marks**
105 top: Dick Clarke/Seaphot, Britain
bottom: Ben Cropp/Seaphot, Britain
106 **Richard Gray: Frank Kennard**
107 top left: Neville Coleman/Seaphot Britain
top right: Heather Angel, Britain
bottom: Neville Coleman/Seaphot, Britain
108 **QED**
109 **Osborne/Marks**
top: Walt Deas/Seaphot, Britain
center: Walt Deas/Seaphot, Britain
bottom: Jeff Foott/Bruce Coleman Ltd., Britain
110 **Richard Gray**
111 **Frank Kennard**
F. Erize/Bruce Coleman Ltd., Britain
112 **Osborne/Marks**
Heather Angel, Britain
113 top left: Heather Angel, Britain
top right: Peter David/Seaphot, Britain
bottom: © Dr. Douglas P. Wilson, Britain
Osborne/Marks
114 top: © Dr. Douglas P. Wilson, Britain
Frank Kennard
bottom: David George/Seaphot, Britain
115 **Frank Kennard**
Neville Coleman/ Seaphot, Britain
116 **Osborne/Marks**
Heather Angel, Britain
117 **Michael Copus**
bottom left: Heather Angel, Britain
bottom right: © Dr. Douglas P. Wilson, Britain
118 top left: Dr. George Warner, Britain
top right: Guy Powell, Alaska, USA
bottom: Heather Angel, Britain
Osborne/Marks
119 **Osborne/Marks**
top right: Peter Vine/Seaphot, Britain
bottom right: Peter David/Seaphot, Britain
120 **Linden Artists**
Professor B. B. Boycott, Britain
121 **Osborne/Marks**
Camera Press Ltd., Britain
122–123 **QED**
123 **QED**
124 **QED**
125 top left: Dr. W. G. Potts, Britain
top right: Heather Angel, Britain
bottom: Michael Newton, Britain

5 The Ocean's Resources

Page
126–127 Fjellanger Widere AS, Norway
127 **Arka Graphics**
128–129 ELAC-Electroacustic Kiel, W. Germany
129 top: **Arka Graphics**
center: Peter David/Seaphot, Britain
bottom: **Rose Associates**
130 left: **Rose Associates**
right: B. L. Henn/Zefa, W. Germany

131 top left: **Rose Associates**
top right: O. Mustad & Søn AS, Norway
bottom: Are Dommasnes, Institute of Marine Research, Norway
132 top: **Richard Gray**
bottom: **Rose Associates**
133 top left, top right: Michael Francis Wood, Britain
bottom: ELAC-Electroacustic Kiel, W. Germany
134 Novosti Press Agency, Britain
135 **Rose Associates**
Novosti Press Agency, Britain
136 Composite photograph
top: Michael Francis Wood, Britain
center: Crown Copyright (Ministry of Agriculture, Fisheries and Food), Britain
bottom: Michael Francis Wood, Britain
137 Jack Thayer/Camera Press Ltd., Britain
138 **Arka Graphics**
138 **Arka Graphics**
139 top: *The Daily Telegraph* Colour Library, Britain
bottom: Victor Kennett, Britain
140 Australian Information Service, Britain
141 Marine Harvest Ltd., Britain
142 Offshore Supply Association Ltd., Britain
143 **Arka Graphics**
British Petroleum Ltd., Britain
144 Shell Photo Service, Britain
145 **Rose Associates**
145 **Rose Associates**
ceenter: Novosti Press Agency, Britain
bottom: Shell Photo Service, Britain
146 **Osborne/Marks**
Shell Photo Service, Britain
147 Shell Photo Service, Britain
148 British Petroleum Ltd., Britain
148–149 British Petroleum Ltd., Britain
149 British Petroleum Ltd., Britain
Rose Associates
150 An Exxon Photo
Osborne/Marks
151 **Osborne/Marks**
Aker, Norway/Harry Nor-Hansen
152 Brown & Root (UK) Ltd., Britain
152–153 top: Shell Photo Service, Britain
center: Brown & Root (UK) Ltd., Britain
153 **Arka Graphics**
Shell Photo Service, Britain
154 M. St. Gill/Transworld Syndication Ltd., Britain
155 top left: Dr. W. G. Potts, Britain
top right, center and bottom: BBN – Geomarine, USA
156 De Beers Consolidated Mines Ltd., Britain
157 Aokam Tin Berhad, Britain
158 Leslie Salt Corporation, USA
159 **Rose Associates**
159 center: UK Atomic Energy Authority, Britain
bottom: Courtesy of the Kaiser Company, USA
160 Bos Kalis-Westminster, Britain
161 Bos Kalis-Westminster, Britain
Osborne/Marks
162 Dr. D. S. Cronan, Imperial College, Britain
QED
163 **Arka Graphics**
164 Deepsea Ventures Inc., USA
165 Deepsea Ventures Inc., USA
Osborne/Marks

6 Using Ocean Space

Page
166–167 NASA, USA

167 Colorific! Photo Library, Mary Fisher
168 Novosti Press Agency, Britain
169 **Osborne/Marks**
General Dynamics Corporation, USA
171 left: Kockums Shipyard, Sweden
right: Marine Explorations Ltd., Britain
172 left: Picturepoint Ltd., London
right: Nancy Moran-NYT Pictures, John Hillelson Agency Ltd.
173 top: Freeman Fox and Partners, Consulting Engineers, London
center: Japanese Information Service, Britain
Osborne/Marks
174 Picturepoint Ltd., London
175 left: Picturepoint Ltd., London
right: Ian Vickery, Irish Republic
Osborne/Marks
176 **Osborne/Marks**
177 Standard Telephones and Cables, Britain
178 Crown Copyright (Central Office of Information), Britain
179 **Osborne/Marks**
180 General Dynamics Corporation, USA
181 top: US Navy Photo
print from J. G. Moore Collection, London, Britain
bottom: Associated Press Ltd., Britain
182 Alain Nogues-Sygma/John Hillelson Agency Ltd.
183 Alfred Eisenstaedt, Life © Time Inc., 1976, from Colorific! Photo Library
184 **Len Whiteman/Gordon Cramp Studio**
185 top: Leroy Gramuis/Camera Press Ltd., Britain
bottom: Associated Press Ltd., Britain
186 US Atomic Energy Research and Development Administration, USA
186–187 US Atomic Energy Research and Development Administration, USA
187 Central Electricity Generating Board, Britain
188 **Howard Dyke: Osborne/Marks**
189 top: Courtesy of Sotramer, France
bottom: Camera Press Ltd., Britain
190 Novosti Press Agency, Britain
191 **Richard Gray: Len Whiteman/Gordon Cramp Studio**
Michael Newton, Britain
192 **Howard Dyke: Osborne/Marks**
S. H. Salter, University of Edinburgh, Scotland
193 **Osborne/Marks**
194 UK Atomic Energy Authority, Britain
195 **Richard Gray: Osborne/Marks**
UK Atomic Energy Authority, Britain
196 Flip Schulke/Seaphot, Britain
197 left: Flip Schulke/Seaphot, Britain
right: Keith Hiscock, Britain
198 **Len Whiteman/Gordon Cramp Studio**
199 top: Patrick Ward/*The Daily Telegraph* Colour Library, Britain
bottom: John Loengard, Life © Time Inc. 1976, from Colorific! Photo Library, Britain
200 **QED**
201 © The British Post Office

7 Underwater Archaeologists

Page
203 **QED**
left: Ernst Wrescher, Israel
right: Photo Czartoryska, Britain
204 left: Whittlesey Foundation/Argolid Exploration Project/Michael H. Jameson, University of Pennsylvannia
right: Argolid Exploration Project/University of Pennsylvania-Indiana University/Michael H. Jameson
Osborne/Marks

205 **Rose Associates**
Peter Marsden, Britain
206 **Osborne/Marks**
top: University of Chicago/J. W. Shaw
bottom: Dr. Peter H. Milne, University of Strathclyde – taken during Shetland Viking Expedition, 1972
207 top and bottom right: Western Australian Museum, Australia/Patrick Baker
bottom left: Robert A. Yorke, Britain, Courtesy of Miss Honor Frost
208 Photo Herb Greer and Peter Throckmorton, Greece
209 Photo Herb Greer, Greece
Rose Associates
210 © National Geographic Society, USA
211 left: Michael L. Katzev, Cyprus
right: Peter Throckmorton, Greece
212 The Viking Ship Museum, Roskilde, Denmark
212–213 The Viking Ship Museum, Roskilde, Denmark
213 Focke-Museum Bremen, W. Germany
Richard Gray: Rose Associates
214 Wasavarvet/Sjöhistoriska Museet, Sweden
Osborne/Marks
215 top left: Wasavarvet/Sjöhistoriska Museet, Sweden
bottom left: Photo Czartoryska, Britain
top right and center: Peter Marsden, Britain
bottom right: Alexander McKee, Britain
216 Lord Kilbracken/Camera Press Ltd., Britain
217 © National Geographic Society, USA
218 **Osborne/Marks**
219 top and center: Colin Martin, Britain
bottom: National Maritime Museum, Britain
220–221 The Texas Antiquities Committee/The Texas Archaeological Research Laboratory, USA
221 **Rose Associates**
222 top: Robert A. Yorke, Britain
bottom: Photo Czartoryska, Britain
223 top left: Whittlesey Foundation/Argolid Exploration Project/Michael H. Jameson, University of Pennsylvania, USA
top right: Photo Czartoryska, Britain
bottom: University of Chicago/J. W. Shaw
224 Photo Czartoryska, Britain
225 Robert Marx, USA
226–227 **Peter Morter**
228 Sotheby Parke Bernet & Co.
229 Alexander McKee, Britain

8 Diving and Divers

Page
230–231 Walt Deas/Seaphot, Britain
231 **Rose Associates**
Barry Gorman/Seaphot, Britain
232 **Robin Jacques**
232–233 **Rose Associates**
234 **Rose Associates**
234–235 **Hargreaves Hands**
236 left: Exxon Production Research Company, USA
right: Comex, Britain
237 **Rose Associates**
238–239 John Bevan/R.N.P.L., Britain
Rose Associates
240 Photo Czartoryska, Britain
241 **Rose Associates**
left: Exxon Production Research Company, USA

right: Comex, Britain
242 Comex, Britain
243 top right: Comex, Britain
bottom left and right: Jean Lattes, France
244–245 John Bevan/R.N.P.L., Britain
245 **Rose Associates**
246 **Rose Associates**
Warren Williams/Seaphot, Britain
247 **Rose Associates**
top: Mike Davies/Seaphot, Britain
bottom: Flip Schulke/Seaphot, Britain
248 Oceaneering: A. D. Stone
248–249 Oceaneering: A. D. Stone
249 Oceans Systems International Inc., USA
251 **Rose Associates**
250–251 **Roy Coombes/Artist Partners Ltd.**
252 Flip Schulke/Seaphot, Britain
253 top left: Western Australian Museum, Australia/Jeremy Green
top right: Western Australian Museum, Australia/Patrick Baker
bottom left: Flip Schulke/Seaphot, Britain
bottom right: Novosti Press Agency, Britain
254 top left: Jean Lattes, France
bottom: Comex, Britain
right: Oceaneering: A. D. Stone
top: Peter Scoones/Seaphot, Britain
255 second from left: John Gifford
second from right and far right: Admiralty Experimental Diving Unit, Britain
256 **Rose Associates**
256–257 Comex, Britain
257 **Rose Associates**
top: Japan Marine Science & Technology Center, Japan
bottom: Admiralty Experimental Diving Unit, Britain
258 Flip Schulke/Seaphot, Britain
259 Mitsubishi Heavy Industries, Japan
Rose Associates
260 D H B Construction Ltd., Britain
Rose Associates
261 Vickers Oceanics Ltd., Britain

9 Submarine Craft

Page
262–263 **Rose Associates**
263 Camera Press Ltd., Britain
264–265 **Rose Associates**
265 top: Crown Copyright (Ministry of Defence), Britain
bottom: Lockheed Missiles & Space Co., USA
266 **Rose Associates**
US Navy Official Photo, USA
267 Photo Czartoryska, Britain
Rose Associates
268 Crown Copyright (HMS *Dolphin*), Britain
Rose Associates
269 **Rose Associates**
Photo Czartoryska, Britain
270 **Rose Associates**
270–271 **Peter Warner**
272 top: Komatsu Ltd., Japan
center: Komatsu Ltd., Japan
bottom: Dick Clarke/Seaphot, Britain
273 top left: Tecnomare SpA, Venice, Italy
bottom left: Flip Schulke/Seaphot, Britain
right: UK Atomic Energy Authority, Britain
Rose Associates
274 top left: Photo Czartoryska, Britain
bottom left: Vickers Oceanics Ltd., Britain

bottom right: Photo Czartoryska, Britain
275 bottom left: Photo Czartoryska, Britain
bottom right: Vickers Oceanics Ltd., Britain
276–277 **Rose Associates**
277 Corning Glass Works, USA
278 top left: General Dynamics Corporation, USA
top right: US Navy Underwater Center, USA
bottom: Flip Schulke/Seaphot, Britain
279 Dick Clarke/Seaphot, Britain
280 **Rose Associates**
281 top: US Navy Official Photo, USA
center row left: US Navy Official Photo, USA
center row right: Associated Press Ltd., Britain
bottom center row left: Associated Press Ltd., Britain
bottom center row right: Associated Press Ltd., Britain
bottom: US Navy Official Photo, USA

10 Marine Law and Politics

Page
282 QED
283 J. T. H. Piek Studio/The International Court, The Netherlands
QED
284 **Peter Marshall**
284–285 **Richard Gray**
285 Popperfoto Ltd., Britain
286 QED
287 top left: Aluminum Company of America, USA
top right: Crown Copyright (Ministry of Defence), Britain
middle: Photo Czartoryska, Britain
QED
288 QED
Photo Czartoryska, Britain
289 QED
NASA, USA
290 QED
291 QED
292 QED
293 QED
294 QED
295 QED
296 QED
Crown Copyright (National Maritime Institute), Britain
297 top: QED
bottom: Peter Marshall
298 John Hillelson Agency Ltd./W. Eugene Smith
299 **Richard Gray: Rose Associates**
By kind permission of Atlas Copco (Great Britain) Ltd., Hemel Hempstead
300 **Peter Marshall**
301 **Peter Marshall**

A

Index

Numbers in **bolder** type refer to illustrations.

Aber-Wrach, 190
Aberdeen Fisheries Laboratory, 273
abyssal plain, 23, **23**, 48
accidents:
  in deep-sea fishing, 132
  in oil industry, 154
  to shipping, 170, **170–1**, 296
acorn barnacle, **99**
acoustic devices, 41; **see also** sonar; sound waves
acrylic, in submersible design, **265, 276, 278, 279**
Adair, "Red", **154**
Aden, Gulf of, 28
Aegean Sea:
  hydrocarbon reserves of, 142
  volcanoes in, 29
aerial photography, in archaeology, 223
aggregate, 160, **160, 161**
aggregate dredger, **161**
agreements (treaties), 283; **see also** law, international
Ahoy (suction dredger), **160**
Aigrette (submarine), **264**
air, breathed by divers, 230, 230–1, 236, 246
albacore, 129, **129**
algae, 68; **see also** phytoplankton
Algiers, **198**
Alpha buoy, **169**
Alps, origin of, 40
Aluminaut (submarine), **265,** 268, 276, 280
aluminum:
  marine reserves of, 157
  in submersible design, 276
Alvin (submersible), **267**, 269, 276, 280
Amathus, **222**
ambient pressure, **234**
Amerasinghe, H.S., **285**
amoeba, 92
amphioxus (lancelet), **104**
Amsterdam, tidal gauge at, 58
Amsterdam (wreck), **205**, 215, **215**
Anatolian Fault, 28
anchovies, 119, 122, 131, 288, **288**
Andes Mountains, 35, 37
Andrea Doria (liner), 170
anemone, **see** sea anemone
animal kingdom, phyla of, 90, **91**
animals, marine, 90–125
  food chains, 79, 82–3, 88, 115, 122–3, 124, 128
  former land animals, 108–9
  intelligence of, 120–1
  physiology of, 90
  predation among, 112, 114–15
  regeneration of, 117
  reproduction of, 116–17, 118–19
  senses of, 112
  social organization of, 118–19
  varieties of, 90–1, **91**
  vertebrates, 104
  see also types of animal
Antarctica, 181
anthozoans, 94
Antibes, 210
anti-pollution zones, 288
antisubmarine warfare, 178
Apollonia, 223, **224**
aqualung (scuba), 232, 246, **246**
Aquapolis, 185, **185**
Arabian Gulf, 174
Arafura Sea, 47
archaeology, 202–29
  cities, ancient, **204**, 222–7
  future of, 228–9
  prehistoric sites, 202, **203,** 206
  science of, 203
  technique of, 205–7, 220–1, 223
  wrecks, **13**, 202, 205, 206, 208–19, **228–9**
Archimède (submersible), **42**
Architeuthis, 103
Arctic Ocean, 46, 47
  hydrocarbon reserves of, 142, 145
  ice of, **50, 51, 143**, 145
Argonaut (submarine), **266**
Armada, Spanish, 205, 218–19, **219**
arms control, 180–1
arms-limitation treaty (1971), **see** Sea-Bed Arms Limitation Treaty
arms race, 178, 180
arrow worms, 115
arsenic, 158
arthropods, 98
Artiglio (salvage vessel), 253
Ascension Island, 108
aseptic bone necrosis, 240
Ashdod, 174
Asherah (submersible), 272
asthenosphere, 24
astronauts, 253
ASW, **see** antisubmarine warfare
Atlanta (submersible), 273
Atlantic Ocean, 46, 47, 53
  conventions on pollution in, 298, 299
  currents in, 54, 56
  fishing in, 126, 136, **138**
  floor of, **22–3**, 23, 34, 35; **see also** Mid-Atlantic Ridge
  layers of, 52–3
  origin of, 34, 40
  power of waves of, 192
  shipping in, 170, 171
  transatlantic cable, 177
Atlantis, 224

atmosphere (unit of pressure), 232
Australia:
  fishing limits declared by, 139
  mineral resources of, 286
Azores, 23

B

Bab el Mandeb, strait of, 296
bacterial decomposition, 80–1
Baffin Bay, 47
baleen whales, 110, 124
ballistic missile, nuclear, 264, **270–1**
ballistic-missile submarine, **270–1**
Baltic Sea, 47
  continental shelf boundaries in, 290
  conventions on pollution in, 298, 299
  eutrophication in, 185
  tides weak in, 59
Baltimore, pollution at, **166**
Bantry Bay, 174, **175**
Barents Sea, 47, **190, 282**
barges (oil industry), 144, **145**
barnacle, 98, 118
  acorn, **99**
  larva of, **82,** 98
barrage, Thames, **see** Thames barrage
basalt, 23, 24
bases, military, 178–9
Bass, George, 208–9, 213
Batavia (wreck), **207, 253**
Bathypterois, 107
bathyscaphe, **262,** 263, 264–5, **265**
bathysphere, 264, **265**
bathythermograph, **52**
batteries, for submersibles, 266–7
beaches, 188–9
Beaufort Sea, 47
bêche-de-mer, **see** sea cucumber
Beebe, William, 264
Belgium, fishing industry of, 286
bell, diving, **see** diving bell
bell man (diving), **236,** 242, **243**
Ben Franklin (submersible), 268, 276
bends, the (decompression sickness), 240, 247
Benioff, Hugo, **29**
Benioff Zone, 29, 35
Bennett, Peter, 237
Benoît, Fernand, 208
Bering Sea, 47
Bermuda Triangle, 11
Bert, Paul, 236
Bevan, John, **238–9**, 257

bioluminescence, 119
bivalves, 102–3, 115
black scabbard, **136**
Black Sea, 47
    breakwater on coast of, 188
    origin of, 35
blood, human, salt content of, 90
bloom (phytoplankton), 68
    food chain dependent on, 78–9
    pollution and, 84, 124
    "red tides", 124
    seasonal, 72, 74, **75, 78, 79, 82, 83**
blow-out preventer stack (BOP), 146
blowtorches, 255
blue whale, 90, 110(2), 115
boats, **see** ships
Bodrum, 209
bone necrosis, aseptic, **see** aseptic bone necrosis
BOP, **see** blow-out preventer stack
boron, 158
Bosporus Bridge, 173, **173**
Bothnia, Gulf of, 189
bottle-nosed whale, 110
bottom-crawling vehicles, **266,** 272, **272,** 273, **273**
bottom-dwelling animals, 81, 96–7, 122
breakwaters, 188
    in ancient harbors, 226
breathing:
    of bivalves, 103
    of divers, 230, 230–1, 232, **234, 236,** 236–9, 244, 246–7
    of marine mammals, 108
    in submersibles, 268
    of whales, 110
breeding, **see** reproduction
Bremen, wrecked cog from, 212, **213**
bridges, 173, **173**
brine shrimp, 98
Britain, **see** Great Britain
brittle stars, 97, 118, **118**
bromine, 126, 158
Bronze Age wrecks, **208,** 208–9, **209**
Brunel, Isambard Kingdom, 176
bucket dredging, **157,** 164, **165**
Bühlmann, Albert, **233**
Bullard, Sir Edward, 37
bulldozer, submarine, **272**
buoyancy:
    of fish, 106
    of marine animals, 100
    of plankton, 77, 100
    in seawater, 90
    of submersibles, 263, 265
buoyancy propulsion, 276
buoys:
    oil industry, 153, **153**
    meteorological, 168, **169**

wave-power, 193
burrowing, by worms, 96, **96**
Byzantine ships, 213

C

Cabarrou, Pierre, **233**
Cable-controlled Underwater Recovery Vehicle (CURV), 273, 280
cables, submarine, **176,** 176–7, **177, 273**
caisson disease, **see** bends, the
California, fishing off, 72
Canada:
    fish farming in, 140
    fishing industry of, 128, 286
    marine parks in, 197
canals, 172–3; **see also individual canals**
Canary Current, 54
cancer, 85, 117
cannon, from Armada wrecks, 218–19, **219**
canyons, on continental slope, 23, **23,** 48
capelin, 131, 136
carbon, 68
carbon dioxide, 68, 74
carbon dioxide poisoning, carbon-14, 69
carcinogenic compounds, 85
Caribbean Sea, 24, 47
    shipping in, 170–1
    tides weak in, 59
    volcanoes in, 29
Carlsberg Ridge, 37
carotenoids, 68
carp, 140
Carpathian Mountains, origin of, 40
Carr, Archie, 108
Cascade Range, 35
Caspian Sea:
    oil industry of, 144, **145,** 149
    origin of, 35
catfish, 140
catspaws, 60
Caucasus, origin of, 40
caves, prehistoric, 202, **203,** 206, 222
cement, 160
Centre National pour l'Exploitation des Océans (CNEXO), 200, 257
cephalopods, 102, 103
cetaceans, 110; **see also individual species**
Challenger (research ship), 66, **67,** 162
Channel, **see** English Channel
Channel Tunnel, 199
chemicals:
    essential to life, 68
    for plant growth, 74–5

    as pollutants, 85, 182, **199,** 298
    from sea creatures, 136
chemosense, 112
Chernomor 2 (habitat), **259**
Chesapeake Bay, 174
Chevron Oil Company fire on rig, 154
Chile:
    earthquakes in, 29
    fishing industry of, 128
    mineral resources of, 286
chitons, 102
chlorine, 158
chlorophyll, 68
Christmas tree (oil rig), **148,** 149, 254
ciliates, 92
cities:
    coastal, redesign of, 185
    sewage control in, 184–5
    shipping and industry of, 174
    submerged, 206, 222–5
Civil War, American, 266
clam, razor, 103
clams, farming of, 84, 89
clingfish, 107
clinker construction, **213**
closed-circuit breathing, 246, **247**
clothing, protective, for divers, 230, 232, 244–5, 246–7
clown fish, **95**
CNEXO, **see** Centre National pour l'Exploitation des Océans
coast, coastline:
    changes in, 48–9, 188–9
    world's total, 166, 198
coastal zone, 166–7, 198–9, 288
Coastal Zone Management (CZM), 166–7, 198–9
cobalt, 127, 157, 163
coccolithophores, 76, **76,** 77
cod, fishing for, 131, 138, 139, **139, 294, 294, 295, 295**
"cod wars", 139, **139,** 294
coelenterates, 94–5, 114; **see also species of coelenterate**
cog (medieval ship), 212, **213**
cold, divers and, 244–5
cold wall, in North Atlantic, 56
color:
    of plants, 68, **70,** 77
    protective, 114–15
Comex, **see** Compagnie Maritime d'Expertises
Committee for East-Central Atlantic Fisheries, 139
communications, telephonic/telegraphic, 176–7
Compagnie Maritime des Expertises (Comex), 232, **233,** 243, 245, 248, 256–7
compensation point, in photosynthesis, 71
compression, of divers, 237

concrete, Roman, 226
Conference on the Law of the Sea, **see** Law of the Sea Conference
conquistadors, 216
conservation, 164, 200–1
    of archaeological finds, 220–1
    of fish resources, 138–9
    marine parks, 197
    of mineral resources, 164
Conshelf (habitat), 233, 258
constant-volume suit, 244
containers, 174–5, **175**
continental drift, 23, 26, 37–8; **see also** plates, plate tectonics
continental margin, 23, **23,** 48
    "active" and "passive", 34, **36**
continental rise, 23, **23**
continental shelf, 23, **23,** 48, 288
    boundaries of, 290, **290, 291**
    definition of, international, 290
    fishing concentrated on, 136
    international law and, 284, 288, 290–1
    UN convention on, 284, 288, 290
continental slope, 23, **23,** 28, 48
continents:
    origins of, 28
    rocks of, 24
contractor, diving, 248, 249, 256–7
Convention on the Continental Shelf, 284, 288, 290
Convention on the International Regulations for Preventing Collisions at Sea, 296
Convention on the Territorial Sea and the Contiguous Zone, 284, 288
conventions (treaties), 283; **see also** law, international
Cook, Captain James, 52, 197
Cook Islands, 197
copepods, **79,** 98, **98,** 100, **101,** 118
copper, 127, 163
copulation, 116, 117
coral, coral reefs, **18–19,** 86, **86,** 94–5, **95**
    and algae, 86, 94–5
    damaged by crown-of-thorns starfish, 97
    protection of, 197
    reproduction of, 117, 118
Coral Sea, 47
Corinth Canal, 172, **172**
Coriolis, Gaspard de, 54
Coriolis effect, 54
Cosmonaut Vladimir Komarov (research ship), **1668**
courtship displays, 117
Cousteau, Jacques-Yves, 211, 258, 265

crab, 98, **98, 99, 116, 119**
  larvae of, 98, **101**, 116
Crepidula fornicata (slipper limpet), 117, **117**
crinoids (sea lilies), 97
crown-of-thorns starfish, 97
cruise missile, **180,** 181, 264
Crumlin-Pedersen, Ole, 212
crust, of the earth, 22, 23, 24, 26, 41
crustaceans, 98
  larvae of, **79,** 98
  see also types of crustacean
Cuba, mineral resources of, 286
Cubmarine (submersible), 280
currents, oceanic, 54–7
  direction of flow of, 54, **54**
  energy from, 190, 276
  fishing affected by, 72
  meanders and eddies in, 56–7
  speeds of, 54, 55
  two types of, 54
  wind's effect on, 54, 55
  see also individual currents
CURV, **see** Cable-controlled Underwater Recovery Vehicle
custom, and international law, 282, 283, 284
cuttlefish, **103,** 114, 117
Cyana (submersible), **42**
CZM, see Coastal Zone Management

D

David class submarine, **264**
DDC, **see** deck decompression chamber
DDT, 84–5, **124,** 125, 283
deck decompression chamber (DDC), 241, 242–3, **250–1**
  in oil industry, 248–9
decompression, of divers, 232, 240–1, 242–3, **250–1**
decompression sickness, **see** bends, the
Deep Jeep (submersible), 280
deep-ocean floor, 48
Deep Quest (submersible), **265,** 268
deep scattering layer (DSL), 101
Deep Sea Drilling Project (DSDP), 44–5
Deepsea Miner (research vessel), **165**
Deepsea Ventures Inc., 164
Deepstone (suction dredger), **160**
Deep Submergence Rescue Vehicle (DSRV), 265, **265**
Deep View (submersible), **265, 279**
deep-water waves, 61

Defense (American Revolutionary brigantine), 205
Denmark:
  fishing limits declared by, 139
  marine reserves of, 197
desalination of seawater, 158, **159**
detergents, oil spills treated with, **125**
deterrent, nuclear, 178
deuterium, 195
developing countries, 284, 286, 289, 290, 292, 301
devilfish, **see** manta
diatoms, **69,** 74, 76, **76, 79, 80**
Dieldrin, 125
Dietz, Robert, 37
Dinaric Mountains, origin of, 40
dinoflagellates, 76, **76,** 77, **78, 88, 101,** 124
Diogenes (habitat), 258
director (fish), **136**
disarmament, 180–1
Divcon (submersible), **233**
divers, diving, **105,** 230–61, **275**
  archaeological research by, 202–29 **passim, 253**
  breathing of, 230, 230–1, 232, **234, 236,** 236–9, 244, 246–7
  clothing of, 230, 232, 244–5, 246–7, **260,** 261
  compression of, 237
  decompression of, 232, 240–1, 242–3, **250–1**
  depths reached by, 232, **233,** 238
  excursion diving, 239, 243
  experimental work in, 256–7
  free diving, 230, 241
  frogmen, 230, 240–1
  future developments in, 260–1
  "hard-hat", **231,** 244
  nineteenth-century, 232, **232**
  and oil industry, 232, 248–9, 254, 255, 260, 261
  record-breaking diver, **233**
  saturation diving, 241, 242–3
  scientific research by, 68–9, 253
  steel suit ("Jim"), **260,** 261
  tools used by, 254–5, 260
  types of diving work, 230
diving bell, diving chamber, 230, **232, 236,** 237, 242; **see also** submersible decompression chamber
dogwhelk, **113**
dolphin, 109, 110(3), **111,** 112, 119, **121**
Domesday Book, 190
"Donald Duck" effect, 249
Dover, cliffs of, 92
Dover, Strait of, 173, 296, **296**
Dover Harbor, 190
downhole safety valve, 154

dredging, for minerals, **157,** 160–1, **161,** 163, 164, **165**
drift-netting, 130, **130**
drilling, for oil:
  exploratory, 144–7
  for production, 149–50
  see also oil
drillships, 144–5, **145**
drumfish, 112
DSDP, **see** Deep Sea Drilling Project
DSL, **see** deep scattering layer
DSRV, **see** Deep Submergence Rescue Vehicle
duck (wave-power device), **see** rocking boom
duck rearing, Long Island, 84
dugong, 109
Duke University, diving research at, 237
Dungeness, 188
Dutch East Indiamen, 215, **215,** 216, 217, **253**

E

earth:
  core of, 24
  crust of, 22, 23, 24, 26, 41
  density of, 24
  heat flow from, 41
  internal structure of, 24, 25
  magnetic field of, 24, 41
  magnetic poles, 28, 33
  mantle of, 24
earthquakes, 28, **28,** 29, 31, 35
  cities destroyed by, 224, 225
  data gleaned from, 24
  "tidal" wave caused by, 58
East African Rift Valley, 28, 35
East Indiamen, **see** Dutch East Indiamen
echinoderms, 96–7, 115; **see also individual species**
echo sounding, 41, 47
Ecuador, U.S. tuna vessels arrested by, 294
eddies, in ocean currents, 56, 56–7
eels, 112, **112,** 115
EEZ, **see** exclusive economic zone
eggs, of marine creatures, 116
Egstrom, Glen, 253
Egypt (liner), 252–3
electric power, **see** power, electric
electric ray, 112, 114
electrolysis, in conservation, **220,** 221
elements, chemical, in seawater, 158
elephant seal, 109
energy:
  loss of, in food chain, 88, 123
  for organic growth, 68

see also power
English Channel:
  cross-channel cable, 177
  resonant tides in, 59
  shipping in, 171; **see also** Dover, Strait of
  Tunnel project, 199
  wave power from, 192
Erie, Lake, 185
erosion:
  of coasts, 188
  of continental rocks, 36–7, 48
euphotic zone, 71, 77, 84
eutrophication, 84, 184–5
evaporation, 53
evaporation ponds/stills, 158, **158**
evolution, 95, 108, **108,** 116
Ewing, Maurice, 37
exclusive economic zone, 289
exclusive fishing zone, 284, 288, 294
excursion diving, 239, 243
experimentation, in diving, 256–7
Explorer (submarine), **264**
explosions, in empty oil tanks, 170, 171
Exxon system (subsea production), **151**
eyes, of cephalopods, 103, 112, **113**

F

factory ships, 134, **134, 135**
Faeroe Islands, fishing quota agreement for, 294
Falco, Albert, 258
fallout, radioactive, 183
FAMOUS, Project, **42,** 42–3
fan worm, 115, **115**
FAO, **see** Food and Agriculture Organization
farming:
  of algae, 88–9
  of clams, 84, 89
  of fish, 140–1
  of oysters, 84, 89, **89, 140,** 141
fault (geological term), 28
  transform faults, 31, 32
fertilizers, artificial, 84, 183, 193
fetch, 60
filamentous green algae, 87
finds, archaeological, 205ff
  conservation of, 220–1
fire, in oil industry, 154, **154, 155**
First World War, **see** World War I
fish, 106–7
  distribution map of, **127**
  jawless, 104
  learning and memory in, 121
  schools of, 119
  senses of, 112, 119
  species of, 90

314

swimming motion of, 104
see also individual species
fish farming, 140–1, 187
Fisheries University, Tokyo,
129, 134
fishing, 67
centers of, 72, 136, **137**
commissions, international,
294
fleets, 134–5
future of, 126
international law and, 286,
288, 294–5
limits, 136, 139, 282, 284,
288, 294
nets and lines, 130–1
overfishing, 88, 107, 124,
138, 138–9, 294, **294, 295**
trawlers, trawling, **129**, 130,
130, 130–1, **132**, 132–3, **133**
world catch, 88, 126, 128
fission, nuclear, **see** nuclear
fission
fixed platforms (oil industry),
148, 149, **149**
flags of convenience, 170, 296
flatfish, 114
flatworm, 117, **117**
fleet system, in fishing industry,
134–5
Flemming, Nicholas, 223
flesh-bone separator, 136
floodlights, submarine, 268
floods, **188**, 189, **189**
Florida, 172, 197
flying fish, 106
FNRS 2 (bathyscaphe), **265**
foam neoprene, 244
food, from the ocean, 66, 67, 89,
122, 123, 125; **see also** fishing
Food and Agriculture
Organization (FAO), 136,
137, 201, 295
food chains/webs, 79, 82–3, 88,
115, 122–3, 124, 128
foraminifers, 92, **93**
Fos-sur-Mer, 174
France:
fishing industry of, 136
nuclear submarines of, 178
oil industry of, **151**
free diving, 230, 232, **240**, 241,
246, **246**, 247
free-fall sampler, 163
free flooding (submersible
design), 278
freedom of the seas, 284, 286
freezer trawler, 132
freezing, in fishing industry,
132
frogmen, 230, 240–1, 246, **247**
fronts (meteorology), 56
fuel cell, as submersible power
source, 267
Fundy, Bay of, 58–9, **59**
fusion, nuclear, **see** nuclear
fusion

G

Galapagos Islands, 23
garbage, pollution from, 298
gas:
breathed by divers, 230;
see also divers, breathing of
natural, 126, 142, **150**
gastropods, 102, 102–3
Gävle, Sweden, **299**
Gelidonya, Cape, **208**, 208–9,
209
General Dynamics Company,
168
Geneva Convention on
Conservation of Biological
Resources of the High Seas,
139
George Washington class
submarine, **265**, 270
GESAMP, **see** Group of Experts
on the Scientific Aspects of
Marine Pollution
Gibraltar, prehistoric caves at,
203
Gibraltar, Strait of, 53, 57, 173,
296
gills, of bivalves, 103
Girona (Armada wreck), 218
glass, in submersible design,
265, 276, **279**
globefish, 107
Glomar Challenger (drillship),
43, 44
Glomar Coral Sea (drillship), **145**
gobies, 107, 116
gold:
from ancient wrecks,
conservation of, 221
marine reserves of, 157, 158
from modern wrecks, 252, 253
goldfish, 121
Gran Grifon (Armada wreck),
218
gravel, 126, 160
gravity, 41–2
seawater reduces effect of, 90
gravity structures (oil
industry), 149–50
gray nurse shark, **105**
Graythorp II (oil rig), **148**
Great Barrier Reef, **86**, 197
Great Britain:
fish farming in, 140
fishing industry of, 128, 136,
139, **139**, 294
mineral resources of, 296
nuclear submarines of, 178,
178
oil industry of, 142
Great Eastern (steamship), 176
Great South Bay, 84
Greece, ancient:
harbors, 226–7
submerged cities, 222–4
wrecks, **210**, 210–11

Greenland Sea, 47
grenadier (fish), 107, 136
grey (gray) mullet, 140
gribble, 98
grid, in archaeological
measurement, 206, **206**
Group of Experts on the
Scientific Aspects of Marine
Pollution (GESAMP), 182
Gulf of Aden, **see** Aden, Gulf of
Gulf of Bothnia, **see** Bothnia,
Gulf of
Gulf of Mexico, **see** Mexico,
Gulf of
Gulf Stream, 54, 56, **56**, 72, 190
Guppy (submersible), 273
gurnard, 106
Gvidon (submersible), **253**
Gymnote (submarine), **264**

H

habitats (underwater
dwellings), 178, **247, 258**,
258–9, **259**
hagfish, 104
Haiti, fishing limits declared
by, 139
hake, 107
Haldane, John Scott, 236
Halieis, **204**, 223, **223**, 224
Hall, David, **285**
harbors and ports, 174–5, 188
in ancient world, 226–7
"hard-hat" diver, **231**, 244
Hardy, Sir Alister, 66, 100
Hastings, shipwreck off, 215
hearing, sense of, in marine
animals, 112
heat, waste, 186–7, 195
Heezen, Bruce, 37
Helike, 224
helium, breathed by divers,
237, 238, 244
helium barrier, 238
helium tremors, 237
hermaphrodites, 117
Hermes, HMS (commando
carrier), **287**
hermit crab, 98, **99**
herring, **116**, 119, 122, 123
fishing for, 124, **127**, 130,
131, 138, 139
Hess, Harry, 37
high-pressure nervous
syndrome, 238
high seas, 282
UN convention on, 284
Himalayas, 37
Holland, **see** Netherlands
Holland class submarine, **264**
Hollandia (wreck), **228**
Honduras, flags of convenience
provided by, 296
horizons (archaeological term),
205

Hormuz, strait of, 296
Hudson Bay, 47
diving in, **242**
Humboldt Current, 288, **288**
Hunley class submarine, **264**
hunter-killer submarine, 270–1
hurricanes, and oil industry, 154
Hydra, wreck off, 209
hydraulic tools, **255**
hydrocarbons, **see** gas, natural;
oil
hydrogen, 193
hydrogen bomb, 194
retrieval of, at Palomares,
280–1, **281**
hydrozoans, 94

I

ice, 50
Arctic, **50**, 51, **143**, 145
extent of, 48
ice ages, 48, 202, **203**
icebergs, 145, **171**
and oil industry, **143**, 145, 155
Iceland, 23
fishing industry of, 131, 136,
139, 282, 286
IMCO, **see** Intergovernmental
Maritime Consultative
Organization
"imprisoned plankton", **see**
zooxanthellae
India, fishing industry of, 136
Indian Ocean, 46, 47
fishing in, 126
Indo-Australian plate, 35
industrial waste, pollutant effect
of, 84, 125, 182–3, 298
industry, coastal sites for, 174–5
inertial navigation, 269
ink, emitted by cephalopods,
114, **114**
insecticides, pollutant effect of,
84–5
instinct, 120, 121
intelligence, 120–1
of cephalopods, 103, 120
(octopus)
of dolphins, 120, **121**
of fish, 121
intercontinental seas, 46
Intergovernmental Maritime
Consultative Organization
(IMCO), 170, 183, 201, 285,
296, 298, 299
Intergovernmental
Oceanographic Commission
(IOC), 201
International Commission for
the Northwest Atlantic
Fisheries, 294
International Court of Justice,
282, **283**
International Decade of
Oceanic Exploration, 201

international law, **see** law, international
International Phase of Ocean Drilling (IPOD), 45
intracontinental seas, 46
IOC, **see** Intergovernmental Oceanographic Commission
IPOD, **see** International Phase of Ocean Drilling
Irish Sea, 47
    wave power from, 192
iron objects, from wrecks, conservation of, 221
island arcs, 24, 37
isostasy, 24, **26**, 48
Israel, fish farming in, 140

J

jack-up barge/rig, 144, **145**, 154
jacket (oil rig), 149
Japan:
    bridges and tunnels in, 173, **173**
    earthquakes in, 29
    fish farming in, 140, 141
    fishing industry of, 128, 130, 131, 134, **135**, 136, **138**, 139, 286
    marine parks in, 197
Japan, Sea of, 47
Japan Ocean Resources Association, 164
jawless fishes, 104
jellyfish, 94, **95**, 112
    lion's mane, 100
"Jim" (diving system), **260**, 261
John Pennekamp Coral Reef State Park, 197, **197**
Johnson-Sea-Link (submersible), **272**, **279**

K

K class submarine, **264**
Kairyu class submarine, **264**
Kapitän, Gerhard, 213
Kapkin, Mustafa, 208–9
Katzev, Michael, 210, 211
Keller, Hannes, 232, **233**
kelp, **77**, 136
Kenchreai, 223, **223**, 224
Kenya, marine parks in, 197
Key Largo, Florida, marine park at, 197
killer whale, **111**
king crab, **119**
Kislaya Guba, **190**
Komatsu (submersible), **272**
krill, 89, 100, 119, 136
Kuroshio Current, 72
Kuroshio II (submersible), 273
Kylstra, Johannes, 260
Kyrenia, 210, **210**

L

Labrador current, **53**
lagoon, artificial, **199**
lake, freshwater, not frozen solid in winter, 50
Lake, Simon, **266**
laminaria kelp, **77**
lamprey, **104**
lancelets, 104, **104**
land bridges, 202, **203**
land hemisphere, 46, **46**
land reclamation, 189
lantern fish, **119**
lanthina (snail), 100
larvae, 100, 116
    of barnacle, **82**
    of crustaceans, **79**, 98, **101**
    of sea squirt, 104
Laurasia, 40
Laurentic (liner), 252
lava, volcanic, 31
law, international, 166, 200–1, 282–301
    archaeological finds, 228–9
    coastal zones, 288–91
    fishing, 136, 294–5
    Law of the Sea Conference, 200–1, 284(2), 286, 289, 292, 296, 300
    mineral resources, 164, 292–3
Law of the Sea Conference, UN, 200–1, 284(2), 286, 289, 292, 296, 300
lay-barge, 152, **152**
lead:
    not corroded by seawater, 221
    in nuclear submarines, 266
    pollutant effect of, 125
Liberia, flags of convenience provided by, 170, 296
limpet, **79**, 102(2), 117, **117**, 120
line-bucket dredging method, 164, **165**
lines, fishing, 130–1
Link, Edwin, 225, **233**
lion's mane jellyfish, 100
lithosphere, 24, 26
"load on top" system, 298, **299**
lobster, 98, **99**, 116
    spiny, 112
lock-out submersible, 259, 260, **261**, **271**, 273
London:
    flood protection at, **188**
    port of, 174
Long Island Sound, 84
long-lining, 130, 131, **131**
longshore drift, 188
Lowestoft Fisheries Laboratory, 129
Lundy island, nature reserve around, **197**
lungs, of marine mammals, 108

M

McCann, Allen, 253
McKee, Alexander, 215
McKenzie, Dan, 37
mackerel, 106, 115, 131
magma, 31
magnesium, 126, 158
magnetic field:
    of the earth, 24, 41
    in nuclear fusion, 195
magnetic poles, of the earth, 28, 33
magnetic stripes, 33, **33**
magnetometer, 205, **205**
Maiale class submarine, **264**
Malacca, Strait of, 171, 296
manatee, 109
manganese, 127
    nodules, 156–7, 162–4, 286, **286**, 292
Manila Galleon, 216
manta (devilfish), 105
mantle, of the earth, 24
mantle, of mollusks, 102
Manuae, 197
Marco Polo diving school, **248**
marginal seas, 24, 46
Marianas Trench, 24, 35, 47, 264
marine parks, **see** parks, marine
Marshall Space Flight Center, **252**
Martin, Colin, 218
Marx, Robert, 225
Mary Rose (wreck), 215, **215**, 229
Masuda, 193
Matthews, Drummond, 37
Maury, Matthew Fontaine, 47, 171
meanders, in ocean currents, 56–7
measurement, in archaeology, 205–6
median rift valley, 23, **23**, 31, **31**, 42–3
medieval ships, 212
Mediterranean Sea, 47
    cities submerged in, 223; **see also individual sites**
    continental shelf boundaries in, 290
    oil prospecting in, 145
    origin of, 35
    pollution in, 85, 172, 298, 299
    salinity of, 53
    shipping in, 171
    tides weak in, 58, 59
medusae, 115, 123
Melbourne, **198**
mercury, 51, 183
mermaids, 109
Mero, John, 156, 157

mesh size, of nets, regulations concerning, 138
metal objects, from wrecks, conservation of, 221
meteorological stations, 168
Mexico, Gulf of, 47
    oil deposits forming in, 142
    oil industry of, 144, 149, **150**, 152, **152**, **153**, 154
    shipping in, 171
Mezenskaya Guba, dam projected at, **190**
Micoperi, 248
Mid-Atlantic Ridge, 23, **23**, 28, 42–3
mid-ocean ridges, 23, **30**, 48
    earthquakes and, 28
    new ocean floor created at, 26, 42
    peridotite found along, 24
    transform faults in, 31
Middle East, oil from, 142
migration:
    of eels, 112
    of salmon, 112, 121
    of zooplankton, 100, 118
Mikhail Tukhachevsky (factory ship), **134**
Milford Haven, 174
military research, 178–9
milkfish, 140
Milos, obsidian from, 202
Minamata Bay, 183
minerals:
    difficulty of mining, 160–1
    extraction of, from salt water, 158
    international relations concerning, 286, 292–3, 301
    ocean resources of, 126–7, **127**, 156–7
    ownership problems concerning, 164, 292–3
    see also individual minerals
minesweepers, 280
mining, **see** minerals
MIRV, **see** multiple independently targeted re-entry vehicle
missiles, nuclear, **see** nuclear missiles
Mizar (research ship), 280
Mohole Project, 44
Mohorovičić, Andrija, 24
Mohorovičić discontinuity (Moho), 24, **25**
mollusks, 102–3; **see also species of mollusk**
monitoring programs, 168–9
Mont-Saint-Michel, **59**
moon-pool, 144
Morro Castle (wreck), 170
Morse code, 176
mother ships, 134
mountain ranges, origin of, 27

mullet, 122
  grey (gray), 140
  yellowtail, 140
multiple independently
  targeted re-entry vehicle
  (MIRV), **270**
multistage flash distillation, **159**

N

Nagasaki, **198**
Naples, Bay of, 224–5
narcosis, in divers, 236
National Oceanic and
  Atmospheric Administration
  (NOAA), 200
NATO, **see** North Atlantic
  Treaty Organization
nature reserves, marine *see*
  parks, marine
Nautilus (nuclear submarine),
  264, **270**
navigation, of submersibles,
  268–9
Navy Deep Trials Unit, 257
  257
Nazca plate, 35, 36
neap tides, 58
nekton, 100
Nemo (submersible), **265**, **278**
Neolithic sites, **see** Stone Age
  sites
nests, of fishes, 116
Netherlands (Holland):
  coastline of, 48–9
    land reclamation in, 189
    marine park in, 197
nets, fishing, 130–1, 138
New Jersey, coast of, 189
New Orleans, port of, 174
Newport News, Virginia, **198**
New York, port of, 174(2)
nickel, 127, 157, 163
Niger Delta, oil industry in, 152
96-hour toxicity limit (TLm96),
  182
nitrate, 51
"nitrifying" bacteria, 80
nitrogen:
  breathed by divers, 230–1,
  236
  as phytoplankton nutrient,
  74–5, 80
NOAA, **see** National Oceanic
  and Atmospheric
  Administration
Nomad buoys, 168
normal polarity, 33
Noroit (research ship), **42**
North Atlantic Current, 54
North Atlantic Drift, **53**
North Atlantic Track
  Agreement, 171
North Atlantic Treaty
  Organization (NATO), **287**
North Equatorial Current, 54

North Korea, fishing industry
  of, 136
North Sea, 47
  continental shelf boundaries
  in, 290, **290**
  conventions on pollution in,
  298, 299
  diving in, **243**
  oil industry in, 142, **142**, 144,
  145, **148**, 149, 150, **151**,
  152, **153**
  research concerning, 126
  resonant tides in, 59
  sinking of, 48
Northeast Atlantic Fisheries
  Commission, 294
Norway:
  fish farming in, 140
  fishing industry of, 128, 131,
  136, 139, 286, 294
  mineral resources of, 286
Norwegian Sea, 47
notochord, 104, **104**
NR–1 (submersible), **265**, 277,
  278, **278**
nuclear fission, 194
nuclear fusion, 194–5
nuclear missiles, 178, **178**, 264,
  270–1
nuclear nonproliferation treaty
  (1968), 181
nuclear power, **see** power,
  nuclear
nuclear power stations, **see**
  power stations, nuclear
nuclear submarines, 263, **263**,
  266, 280
  detection of, 181
  Nautilus, 264, 270
  oxygen requirements of, 268
  Polaris, 178(2), **178**
  Trident, 264
  two types of, 270–1
  Whale, 50
nuclear waste, 183
nuclear weapons, orbiting, 181
nutrients, of phytoplankton, 72,
  74–5, 77, 80

O

obsidian, 202
ocean:
  currents in, 54–7
  death of, 35
  depth of, 47
  evaporation of, 53
  extent of, 22, 46, 47, 166
  floor of, 23, 24, 26, 32, **32**,
  41–5, 48
  food from, 66, 67, 89, 122,
  123, 125; **see also** fishing
  formation of, 34
  layers of, 52–3, **57**
  living matter in, 66; **see also**
  animals, marine; plants, marine

  sediments in, 27, 36–7, 48,
  49, 157
  statistics of world oceans, 47
  temperatures of, 52, 59, 90,
  244
  see also ice; seas; seawater;
  water
Ocean Simulation Facility (U.S.
  Navy), 257, **257**
oceanographic commission,
  see Intergovernmental
  Oceanographic Commission
octopus, 102, 103, 114, 120, **120**
Office of Ocean Economics and
  Technology, 201
oil, petroleum, 126, 142–55
  distribution map, **127**
  divers and oil industry, 232,
  248–9, 254, 255, 260, 261
  drilling for, 144–50
  formation of oil deposits, 35
  pipelines, **152**, 152–3, **153**
  pollution from oil, **see** oil
  spills
  producers, major, **143**
  prospecting for, 45, 92, 144–7
  rigs, **see** rigs, oil
  tankers, 153, 155, **155**, 170,
  171, **171**, **287**
oil spills, 125, **125**, 154, 155, **182**,
  183
  international law and, 298–9
  Mediterranean, 172
  phytoplankton affected by,
  84, 85
Okhotsk, Sea of, 47
oligotrophic surface water, 72
olivine, 24
Onslaught, HMS (submarine),
  265
ooze, of ocean floor, **80**, 92, **93**
open-circuit demand system,
  246, **246**
organic remains, in wrecks, 220
otter, sea, **see** sea otter
otter boards, 130–1
ownership:
  of archaeological finds,
  228–9
  of resources, 164, 292–3
oxygen:
  for divers, 230–1, 236, 237, 238
  "oxygen demand", 80
  and photosynthesis, 68
  for submersibles' crews, 268
oysters:
  farming of, 84, 89, **89**, **140**, 141
  larvae of, 116

P

Pacific Ocean, 46, 47, 53
  fishing in, 131, 136
  floor of, 23, 27
  oil deposits forming in, 142
  volcanoes round, 29

Pacific plate, 35
Palomares, hydrogen bomb
  retrieved at, 280
Panama, flags of convenience
  provided by, 170, 296
Panama Canal, 172, **172**
Pangaea, 27–8, **28**, 37–8, 40, **40**
parasitism, 86–7
Paria, Gulf of, 174
parks, marine, 196–7
partial pressure, **234**, 236, 237
Passamaquoddy Bay, 190
Paton, William, 260
Pavlo Petri, 223, **223**, 224
pea crab, **99**
pearls, 102, **102**
pendulum breathing, 247
Pentland Firth, 59
Peral (submarine), **264**
peridotite, 24
periscope, 268
Persian Gulf, 47
  continental shelf boundaries
  in, 290
  oil industry of, **148**, 149, 152
Peru, fishing industry of, 72,
  128, 131, 136, 139, 288, **288**
pesticides, 183
Peterson, Mendel, 217
petroleum, **see** oil
pharmacology, marine, 126
pheromones, 112
Philippine Islands:
  fish farming in, 140
  fishing industry of, 136
  marine parks in, 197
Phoenician submerged port
  (Amathus), **222**
Phoenician wreck, 210
phosphate, 51
phosphorite nodules, 157
phosphorus, 74–5, 80–1
photography, aerial, **see** aerial
  photography
photosynthesis, 68–72, 84
phyla, of the animal kingdom,
  90, **91**
phytoplankton, algae, **8–9**,
  66–89 **passim**, 122–3
  bloom of, 68, 72, 74, **75**, **78**,
  78–9, **79**, **82**, **83**, 84
  buoyancy of, 77, 100
  cells of, 66–7, 68
  death and decay of, **80**, 80–1
  distribution map, **67**
  eaten by zooplankton, 77–9
  ecosystem, stability of, 82–3
  extinction, avoidance of, 82–3
  farming of, **88**, 89, **89**
  food of, 72–5, 77, 128
  as food source, 88–9
  photosynthesis by, 68–72, 84
  pollution and, 84–5, 124–5, 184
  seasonal growth of, 72, 74,
  82, **83**
  types of, **76**, 76–7
  vertical mixing and, 72, **73**

Piccard, Auguste, 264
Piccard, Jacques, 264, 276
pillow lava, 31, **31**
pipefish, 106, **107, 116**
pipes, pipelines (oil industry), **152**, 152–3, **153**
Piraeus, 174
pirate wreck, 217
Pisces (submersibles), **269, 271,** 273
pistol shrimp, 112
plaice, **114**, 122, 140
planets, oceans lacking on, 22
plankton, 66, 92, **93**, 100, **100, 101,** 116
    early studies of, 66
    ecosystem, stability of, 82–3
    "imprisoned", 86–7
    quantity of, 123, 124
    reproduction of, 118–19
    see also phytoplankton; zooplankton
plankton recorder, 66
plankton trawl net, 66
planning, 166, 175, 198–201
    military, 179
plants, land, 78, 123
plants, marine, 66–89
    quantity of, 66
    see also phytoplankton; seaweed
plasma, 194–5
plates, plate tectonics, 23–45
    cause of movement of plates, 24
    collisions between plates, 35
    earthquakes at plate boundaries, 28
    history of theory of, 37–8
    map of plates, **24**
    speed of movement of plates, 23
    testing the theory, 44–5
Plongeur (submarine), **264**
Plunger (submarine), **264**
Poland, fishing industry of, 134
polar front, North Atlantic, 56
Polaris submarine, 178(2), **178**
pollution, 124–5, 164, 166, 182–7, 201
    from chemicals, 85, 182–3, **199**, 298
    from heat waste, 186–7
    from human waste, 184–5, 298
    international law and, 286, 288(2), 298–9
    from oil, **see** oil spills
    and phytoplankton, 84–5, 124
    radioactive, 183
polychaete worms, 96, 120–1
polyethylene glycol, 220
polyps, 94
    coral, 86
populations, coastal, 166, 184
porpoise, 110(2)
Port Grimaud, **199**

portholes, 268
Port Royal, 224–5
ports, **see** harbors and ports
Portugal Current, 54
Portuguese man-of-war, 94, 100
Portuguese wrecks, 217
Poseidon (nuclear missile), 178
potassium dioxide, in submarine breathing apparatus, 268
power, energy:
    electric, 176
    nuclear, 183, **186**, 186–7, 188, 194–5, **195**
    tidal, 59, 190, **191**
    wave, 192–3
power stations, 186–7
    nuclear, 183, **186**, 186–7, 188, **195**
Pozzuoli, **224**
prawns, farming of, 141
predation, 112, **113**, 114–15
pressure, divers and, 230, 232, **234**, 236–7
propeller, in submarine propulsion, 266, 267
propulsion, of submersibles, **see** submersibles, power source for
Prospector (research vessel), **164, 165**
protein sources, 128; **see also** fishing
protozoans, 91, 92, **92**
Puerto Rican trench, **23**
puffer, 140
purse-seining, **127**, 130, 131, **131**
pycobilin, 68
pyroxene, 24

R

radar, 178
radioactive waste, 125, 183, 195
radioactivity:
    pollution from, 125, 183, 195
    as research tool, 69
radiolarians, 92, **93**
ragworm, **96, 121**
rainbow trout, 140
Rance estuary, 190, **191**
ratification of treaties, 283
ray, 104–5, 122
    electric ray, 112, 114
    stingray, **105**
razor clam, 103
reclamation of land, **see** land reclamation
Red Sea, **34**, 35, 47
    mineral resources of, 157
"red tides", 124
reference points, in archaeological measurement, 206
regeneration, 117, **117**
remora, 107, **107**

repeat-dive tables, **see** tables, repeat-dive
repeaters (amplifier stations), 176, **177**
reproduction, of marine animals, 116–17, 118–19
research, scientific, divers and, 68–9, 253
Resolution (Captain Cook's ship), 52
Resolution (nuclear submarine), **178**
resonant tides, 58–9
resorts, coastal, 184–5
respiration:
    of phytoplankton cell, 70, **71**
    see also breathing
rias, 174
rift valleys, 28, **34**, 35; **see also** median rift valley
right whale, 110, **111**
rigs, oil, 12, **15**, **142**, **145**
    accidents to, 154
    mobile, 144
    semi-submersible, 144, **145, 147**, 154
    and shipping, 291
rock drill, **255**
Rockall Island, **288**
rocking boom (duck), 192–3
rocks, of the earth's crust, 22
    continental, 24, 26
    of the ocean floor, 24, 26
Rocksite, Project, 178
Roman harbors, 226–7
Roman wrecks, 208, 211, **211**
Roskilde Fjord, 212, **212**
Rotterdam, port of, 174(2), **174**
routes, shipping, 170–1
Royal Naval Physiological Laboratory, 238, **238–9**, 257
Royal Navy, record-breaking dives by, **233**
Rule, Margaret, 215
Russia, **see** Soviet Union

S

sablefish, 107
safety:
    in diving, 247
    in oil industry, 154–5
    of shipping, 170–1, 172, 296
    in submersible design, 262–3
Saint-Malo, Gulf of, 59, **59**
salmon:
    farming of, 140, **141**
    fishing for, 130, **295**
    migration of, 112, 121
    spawning grounds of, **295**
salt, extraction of, 126, 158, **158**
salt, salinity, 50–1, 90
    Mediterranean, 53
SALT, **see** Strategic Arms Limitation Talks

salvage work, 252–3
San Andreas Fault, 28
San Clemente, California, **186**
San Francisco Bay, 174(2)
sand, 126, 160
Santa Barbara oil leak, 183
Santa Maria de la Rosa (Armada wreck), 205, **205**, 218
Santorini, **see** Thira
Sargasso Sea, **112**
sargassum fish, 114
satellites, space, 69, 169
saturation diving, 241, 242–3
saw, hydraulic, **255**
SBM, **see** single-buoy mooring
scallop, 112, **113**
Scan (submersible), **273**
Scandinavia, 48
scaphopods ("tooth snails"), 102
schools, of fish, 119
scombrid fish, **119**
Scottish Marine Biological Association, 66
scrubber, hydraulic, **255**
scuba, **see** aqualung
scyphozoans, 94
SDC, **see** submersible decompression chamber
sea anemone, 90, 94, **95**
Sea-Bed Arms Limitation Treaty (1971), 180, 181, 297
Sea-Bed Committee, UN, 284, 292
sea cow, 108, 109(2)
sea cucumber, 96, 97, **97**, 115, **115**, 136
Sea Gem (oil rig), 154
sea gooseberry, 123
sea horse, 106, **107**, 116–17
sea lettuce (*wakame*), **137**
sea lilies (crinoids), 97
sea mouse, **114**
sea otter, 109, **109**
Sea Scape (supertanker), **171**
sea slug, 102, **102**, 114
sea snail, 102(2)
    lanthina, 100
sea snake, 108
sea squirt (tunicate), 104, **104**, 117, 118
sea urchin, 96, 97, **97**, 116
sea wasp, 94
seafaring, early, 202ff
seal, 108, 109
SEAL system, **see** Subsea Equipment Associates Ltd. system
Sealab (habitat), 178, 233, 258
seamounts, 178
seas:
    classification of, 46
    see also ocean **and individual seas**
seawater:
    composition of, 158
    desalination of, 158

318

minerals in, 157
  see also ocean; water
seaweed, 77, **77**, 136, **137**, 141
Second World War, **see** World
  War II
Sedco 135-F (oil rig), **147**
sedimentary rocks, 37
sediments, in ocean, 27, 36–7,
  48, **49**, 157
seismic refraction, in ocean
  research, 41
seismic survey, in oil
  prospecting, 144, **144**
semi-closed-circuit breathing,
  246, **247**
semi-submersible oil rig, 144,
  145, 147, 154
senses, of marine animals, 112
Severn estuary, 190
sewage, 84, 124, 184–5, 298
Sfax, **222**
shallow-water waves, 61
shark, 104–5, **105**, 106, 112
Sharphouse, Peter, 238–9, 257
shellfish, 115, 126; **see also**
  individual species
shells:
  of bivalves, 102–3
  of crustaceans, 98
  of foraminifers, 92
  of mollusks, 102
  of radiolarians, 92
  of shellfish, 126
  of squids, 103
Shetland Islands, oil installation
  in, **153**
Shinkai (submersible), 268
shipping, 170–5
  conflicting interests
    concerning, 286
  congestion, traffic rules, 172–3,
    296–7
  harbors and ports, 174–5
  safety of, accidents to, 170–1,
    172, 296
  see also ships; wrecks
shipping lanes, 296, 297
ships:
  barnacles and, 98
  cable-laying, 177
  prehistoric, 202
  see also wrecks;
  individual ships
shipworm, **see** teredo worm
shipwrecks, **see** wrecks
shrimp, 98
  brine shrimp, 98
  pistol shrimp, 112
Siberia, hydrocarbon reserves
  off coast of, 142
Siebe Gorman, **233**
sight, sense of, in marine
  animals, 112, 119;
  see also eyes
silica, 74, 92
silicate, 51
silicon, 158

silver, 158
single-buoy mooring (SBM),
  153, **153**
skate, 105
Skipjack (submarine), **264**
Skylab rocket, **252**, 253
sleep, in divers, 239
slipper limpet (*Crepidula
  fornicata*), 117, **117**
slug, sea, **see** sea slug
Small, Peter, 232
smell, sense of, in marine
  animals, 112(2)
snail, sea, **see** sea snail
snake, sea, **see** sea snake
snorkel, **246**, 266
social organization, in marine
  animals, 118–19
soda lime, 246
sodium, 158
Solent, wreck in, 215
Solfatara, 224
sonar:
  in archaeology, 205
  in fishing industry, 132–3
  in submarine navigation, 268
Soucoupe (submersible), 265,
  265, 269
sound waves:
  in seismic reflection, 41
  in seismic refraction, 41
  in submarine detection, 178
  in submarine navigation, 268
South Africa:
  fishing industry of, 136, 139
  marine parks in, 197
South China Sea, 47
  hydrocarbon reserves of, 142
Soviet Union (Russia, U.S.S.R.):
  fishing industry of, 128,
    134, **134**, 136, 139, 286, 294
  nuclear submarines of, 178
  Strategic Arms Limitation
    Talks, 181
space flight, **252**, 253
space satellites, **see** satellites,
  space
Spain, fishing industry of, 128,
  136, 286
Spanish ships, wrecks, 216, **217**,
  218–19
species:
  of phytoplankton, 76
  threatened, 196
  of vertebrates, 90
species diversity, 76–7
speech distortion, in divers, 249
sperm whale, 103, 110(2), **114**
spider crab, 98, **98**
spiny lobster, 112
sponges, 90, 94, **94**, 114
spongin, 94
spring tides, 58
spudding in, 146, **146**
Squalus (submarine), 253
squid, 102, 103, **103, 113**, 114, **114**
starfish, 96, 97, **97**, 112, **113, 117**

Stavanger, 174
Steller's sea cow, 109
Stenuit, Robert, 218
stern trawling, 132
stickleback, 116
stingray, **105**
stone, from wrecks,
  conservation of, 221
Stone Age sites, 202, **203**, 206,
  222, 223
storm choke, 154
storms, 60–1
  and oil industry, 154
  photosynthesis impeded by,
    72
  prediction of, 189
  and wave power, 192
straits, 172, 172–3, **293**, 296; **see
  also individual straits**
Strategic Arms Limitation Talks
  (SALT), 181
stratigraphy, 205–6, 222
strontium, 92
submarines, 20–1, 178, 262–81
  passim
  American Civil War, 266
  bottom-crawling, **266**, 272,
    272, 273, 273
  breathing in, 268
  design of, 262–3, 276–8
  detection of, **287**
  diving mechanism of, **262–3**
  navigation of, 268–9
  nuclear, **see** nuclear
    submarines
  recovery of *Squalus*, 253
  U-boats, **264**
  weaponry for, 178–9, 264
  World War II, 57, 101
submersible decompression
  chamber (SDC), 230, 241,
  242, 242–3, **243, 250–1**
  in oil industry, 248–9
submersibles, **11, 42**, 262–81,
  265, 266, 267, 269
  breathing in, 268
  cable-burying by, 176
  for cable-laying, **273**
  design of, 262–3, 276–8
  hull of, 263, 265, **271**
  lock-out, 259, 260, **261, 271,
    273**
  maneuverability, need for,
    264
  for military research, 178
  navigation of, 268–9
  nuclear, **265**, 277
  one-man ("Jim"), **260**, 261
  power source for, 266–7,
    276–7
  for scientific research, **253**
Subsea Equipment Associates
  Ltd. (SEAL) system, **151**
subsea production (oil
  industry), 150
suction dredge, **160**, 163, 164,
  165

Suez Canal, 170, 172(2)
suit, diving, 230, 232, 244, 246
  steel ("Jim"), **260**, 261
Sulectum, **222**
sun, 194
sunlight, 68, 70
surges, 61
Swallow, John, 37
swells, 61
swim bladder, 106
swimming:
  by dolphins, 110
  by fish, 104, 106
  by squid, 103
  by whales, 110, **111**
Sydney harbor, 174
symbiosis, 86–7
syntactic foam, in submersible
  design, 276

T

tables, compression and
  decompression, 240
tables, repeat-dive, 241
TAC, **see** total allowable catch
taconite effluent, **183**
Taiwan:
  fish farming in, 140
  fishing limits declared by, 139
tankers, oil, 153, 155, **155**, 170,
  171, **171, 287**
Taylor Diving and Salvage
  Company, **233**, 248
Tektite (habitat), **233, 258**
telegraphic communication, 176
telephone, telephonic
  communication, **176**, 176–7
television, submarine, 268, 280
temperature:
  atmospheric, 50
  body, 90
  changes in, ecological
    effect of, 187
  oceanic, 52, 59, 90, 244
tentacles:
  of coelenterates, 94
  of cuttlefish, 117
  of jellyfish, 100, 112
  of octopus, 120
teredo worm (shipworm), 214
territorial limits, 139; **see also**
  fishing, limits; territorial
  seas
territorial seas, 284, 288, **289**,
  294
  UN Convention on, 284, 288
test-ban treaty (1963), 181
Tethys Sea, 35, 40
Thailand, fishing industry of,
  136
Thames barrage, **188**
thermocline, 53
thermonuclear fusion, 194–5
Thira (Santorini), 224
Throckmorton, Peter, 208–9, 209

"tidal" wave, **see** tsunami
tides, 58–9
  power from, 59, 190, **191**
  prediction of exceptional,
    189
tin, **157**, 158
Titanic, 170
titanium, 157, 158
TLm96, **see** 96-hour toxicity
  limit
Tomahawk (cruise missile), **180**
Tonga-Kermadec Trench, 35
tools:
  for divers, 254–5, 260
  octopus cannot use, 120
  sea otters' use of, 109, **109**
"tooth snails", **see** scaphopods
toothed whales, 110
torpedo, 264
Torre Sgarrata, **211**
Torrey Canyon, 155, **155**, 183,
  282, 296
total allowable catch (TAC; cod
  quota system), 294
touch, sense of, in marine
  animals, 112
tourist resorts, 184–5
Tours 300 (submersible), 266
toxicity, of chemicals, 182
trace elements, 51
tracked submersibles, **272**, 273
traffic congestion/rules, at sea,
  172–3, 296–7
transform faults, **30**, 31, 32
Transocean 3 (oil rig), 154
transparency, of marine
  animals, 115
Transworld 61 (oil rig), **142**
trawlers, trawling, **129**, 130,
  130, 130–1, **132**, 132–3, **133**
treaties, 282, 283
  disarmament, 180, 181
triangulation, 206
Trident (submarine), 264
Trieste (bathyscaphe), 264, 265
trilateration, 206
Trinidad de Valencera, La
  (Armada wreck), 218, **219**
trout, 140, **187**
Truman, President, 288, 290
trunkfish, 107
tsunami, 38, 58, **60**
tumors, 117
tuna:
  fishing for, 129, 131, 136,
    138, 294
  mercury in, 51
tunicate, **see** sea squirt
tunnels, 173
  Channel, 199
tunny, 106
turbidity currents, 48, **49**
Turkey, plate movements near,
  28
Turkey Point, Florida, **186**
turtle, 108, **109**
TV, **see** television

U

U-boats, **264**
umbilical (diver's hose), 230
underclothing, heated, for
  divers, 244
UNEP, **see** United Nations
  Environmental Program
United Kingdom, **see** Great
  Britain
United Nations Convention on
  the Continental Shelf, 139
United Nations Environmental
  Program (UNEP), 201
United Nations Organization,
  139, 180, 200–1, 282, 284,
  292; **see also** Food and
  Agriculture Organization;
  Law of the Sea Conference
United States of America:
  fish farming in, 140
  fishing industry of, 128,
    131, 136, 139, 292
  marine parks in, 197
  mineral resources of, 286
  nuclear submarines of, 178
  oil industry of, **151**
  plate movements near, 28
  Strategic Arms Limitation
    Talks, 181
upwelling, 72
uranium, 158
U.S. National Marine Fisheries
  Service, 66
U.S. Navy, record-breaking
  dives by, **233**
U.S.S.R., **see** Soviet Union

V

vegetation, **see** plants
vegetative budding, 116, 117,
  118
velella, **95**, **101**
Venice, 189
Venus's girdle, 115
vertebrates, 90, 104–5
  evolution of, **108**
  origins of, 104
  see also fish **and other**
    vertebrates
vertical migration, of
  zooplankton, 100
vertical mixing
  (phytoplankton), 72, **73**
Vikings, Viking wrecks, 212, **212**
Vine, Fred, 37
voice formation, peculiarities
  of, in helium-breathing
  divers, 249
volcanoes, volcanic activity, 22,
  29
  cities destroyed by, 224–5
  in mid-ocean ridge, 31
Vostok class (mother ships), 134

W

Waddenzee, 197
Wagner, Kip, 216, **216, 217**
wakame (sea lettuce), **137**
Waldheim, Kurt, **285**
walrus, 109
Walsh, Don, 264
Warsaw Pact, **287**
Wasa (wreck), 214, **214, 215,**
  232
waste, industrial, **see** industrial
  waste
waste, radioactive, **see**
  radioactive waste
waste heat, **see** heat, waste
water, 50–1; **see also** ice;
  ocean; seawater
water hemisphere, 46, **46**
Wavepower Ltd., 193
waves, 60–1
  internal, 57
  power from, 192–3
weaponry, submarine, 178–9,
  264
weather, 57; **see also** storms
weather forecasting, 168–9
weather ships, 168
weever fish, 114
Wegener, Alfred, 26, 27, 37–8
weightlessness, 253
welding, underwater, **10**, 255,
  255
Wesley, Claude, 258
West Germany, fishing
  industry of, 139, 286
Whale (nuclear submarine), **50**
whales, 110
  air breathed by, 108, 110
  classification of, 110
  communication among, 112
  formerly land mammals, 108
  hunting of, 109, 110, 136
  reproduction of, 109
  social behavior of, 119
  swimming of, 110
  see also blue whale; sperm
    whale
Wicked Witch (oil platform),
  154
Wignall, Sidney, 218
wildcatting, 144, 145
Wilson, J. Tuzo, 37
wind:
  floods due to, 189
  sunlight affected by, 70
  and surface currents, 54
  waves caused by, 60
wind set-up, 61
WMO, **see** World
  Meteorological Organization
Woods Hole (Massachusetts)
  Oceanographical Institution,
  269
Wookey, George, 232
World Marine Park, 197

World Meteorological
  Organization (WMO), 201
World War I, 264, 266
World War II, 57, 101, 112, 264,
  266, 268, 273
worms, 96, **97**, 115, **115**, 117,
  117, 120–1, **121**
wrasses, 116
wreck formation, 202
wrecks, 202–21
  Bronze Age, **208**, 208–9, **209**
  Byzantine, 213
  conservation of, 220–1
  divers and, 202–21 **passim,**
    252, **253**
  Dutch East Indiamen, 215,
    215, 216, 217
  excavation of, 206ff
  Greek, **210**, 210–11
  mapping of, **13**
  measurement of, 205–6
  medieval, 212, **213**
  modern, 170, 252–3
  oil tankers, **see** oil spills
  Phoenician, 210
  pirate, 217
  Portuguese, 217
  Roman, 208, 211, **211**
  searching for, **204**, 205
  Spanish, 216, **217**, 218–19
  Viking, 212, **212**

X

X-Craft class submarine, **264**

Y

Yassi Ada, 213
yellowtail mullet, 140
Yonge, C.M., 86
Young, Rodney, 209

Z

Zaire, mineral resources of, 286
Zambia, mineral resources of,
  286
Zamboglia, Nino, 208
zones, functional, in
  international law, 288
zooplankton, 74, 79, 100–1,
  122–3
  barnacle lava, **82**
  ecosystem, stability of, 82–3
  food of, varies, 77
  grazing by, 78–9, 80, 83
  vertical migration of, 100
  see also plankton
zooxanthellae ("imprisoned
  plankton"), **86**, 86–7